Manfred Drosg, Michael Morten Steurer
Dealing with Electronics
De Gruyter Graduate

Manfred Drosg, Michael Morten Steurer

Dealing with Electronics

DE GRUYTER

Physics and astronomy classification scheme 2010
01, 06, 07, 84, 85

Authors
Prof. Dr. Manfred Drosg
University of Vienna
Faculty of Physics
Boltzmanngasse 5
1090 Vienna
Austria
manfred.drosg.@univie.ac.at

Mag. Michael Morten Steurer
University of Vienna
Faculty of Physics
Boltzmanngasse 5
1090 Vienna
Austria
michael.steurer@univie.ac.at

ISBN 978-3-11-033840-9
e-ISBN 978-3-11-034108-9

Library of Congress Cataloging-in-Publication Data
A CIP catalog record for this book has been applied for at the Library of Congress.

Bibliographic information published by the Deutsche Nationalbibliothek
The Deutsche Nationalbibliothek lists this publication in the Deutsche Nationalbibliografie; detailed bibliographic data are available on the Internet at http://dnb.dnb.de.

Typesetting: le-tex publishing services GmbH, Leipzig
Printing and binding: CPI books GmbH, Leck
♾ Printed on acid-free paper
Printed in Germany

www.degruyter.com

I dedicate my share in writing this book to my dear wife Brigitte
for having gracefully accepted being married to a bigamist
who is enjoying her rival – science – too much.

Manfred Drosg

Preface

The roots of electronics, as of several other natural sciences, lie in physics. This is reflected, e.g., in the fact that at the Los Alamos National Laboratory the kernel of the Electronic Division was previously a group (P-1) of Physics Division. In addition, the knowledge of electronics is mandatory for any experimental physicist to get the most out of his instruments, i.e., to succeed in performing complicated experiments. Therefore, it should not surprise that an experimental physicist has a fresh approach to this field of science. In particular, he would rather look for exact explanations rather than for rules of the thumb. Surprisingly, these exact solutions do not necessarily involve tedious calculations or higher mathematics. Already as a student in the sixties I built a plug-in for the Digital Computing Unit of TMC effectively programming this "computer" by hard-wire to control an electrodynamic Mössbauer drive.

There did not exist a textbook teaching how to proceed. I got first-hand knowledge from manuals, the hard way. However, this had the advantage that my electronics knowledge accumulated not in the conventional way, always emphasizing a physical approach to this subject. The wealth of electronics insights gained as a doctoral student of experimental nuclear physics made it possible to co-author the Textbook of Nuclear Electronics (Lehrbuch der Nuklearelektronik) in the late 1960s. Decades of teaching Electronics for Physicists (both by lecturing and giving a practical course) and several publications in this field prepared me for this attempt to strip electronics of all unnecessary garments and to show its rather simple and beautiful bare structure.

By constantly referring to duality only half of the facts must be dealt with. Besides, this kind of mirroring helps in understanding of circuits. The quantity *two* shows up in many facts on electronics; it is a characteristic both for analog and binary electronics. The basic ingredient is the dualism between current and voltage reflecting the electromagnetic nature of electrical signals. There are the Fourier transform duals time and frequency. Dualism in time and space is also based on Maxwell's equation. Then there is the geometric dualism in arranging electronic elements either in parallel or in series. Also, the existence of positive and negative charge reflects a twin character resulting in complementary semiconductor devices. Further, an electric signal may flow either way. Finally, the finite size of the maximum current or voltage results in two distinct electrical states which are the basis of binary (digital) electronics. In digital electronics there is parallel and serial logic, and truth tables deal either with standard ("positive") or inverse logic operators. Then, harmonic signals are characterized by two properties: amplitude and phase. Even (electronic) measurements have two mirror faces: the measurement itself and the calibration of the instrument. Finally, there is the ambiguous duality between binary circuits and binary logic.

Taking full advantage of these dualisms the amount of knowledge to cover most of the field can be drastically reduced. All variations of known basic circuits can be arrived at by applying the simple laws underlying these dualisms, and the completeness of each circuit family can be proved. So, three-terminal components have just one basic circuit, the other two can be explained by feedback. For systematic reasons the reverse closed-loop gain in feedback circuits is introduced, the only way to describe the output impedance correctly. Such a *science of electronics* can be an inspiring complement to the *arts of electronics* that is very difficult to teach because it can only be acquired by doing.

This book does not only cover the basics of analog and digital electronic circuits, but also gives full attention to important principles like analog-to-digital conversion (the basis for electronic measurements) and positive and negative feedback.

The cooperation with my young colleague Michael Morten Steurer who gained proficient electronic expertise by designing unique research equipment for biological experiments has proven to be very successful. He contributed several subjects to this book and eliminated a few mistakes. Above all, he excelled in composing the more than 300 figures found in this book supporting the process of understanding.

Vienna, March 2014 *Manfred Drosg*

Electronics, in particular binary electronics, is the fundamental technology of our time. For that reason, basics are already taught at school. At high school, the concepts of current, voltage, and Ohm's law were presented. There, I experimented with some basic circuits from binary electronics and I remember having severe problems in understanding the bistable multivibrator. Although the teacher correctly argued with the current flow and the different (possible) states of the involved transistors, I had problems because I tried to analyze the circuit based on voltages, only.

Later, at university, in the basic physics course we had to deal with operation amplifiers having parallel–parallel feedback. Again, I experienced the situation of doing something without understanding the matter. Obviously, the tutors did not understand, either. I instantaneously felt the desire to learn electronics the right way and to attend the lecture offered at the university on this subject.

Some colleagues strongly argued against participating because the professor giving the lecture (the first author of this book) was known for dealing with the matter "not the usual way." But I thought "If the common way of dealing with electronics causes difficulties in understanding, it may be a good idea to approach the issue by a different one." So I gave it a try. I have never regretted this decision. The knowledge I gained there was a perfect basis for the development of electronic instruments for the research at the University.

I am very glad about the opportunity to contribute to this book.

Vienna, March 2014 *Michael Morten Steurer*

Prolog: Ten myths adored by electronics rookies

Myth 1. *A deeper understanding of mathematics is indispensible for the understanding of electrical circuits.*
It is true that proficiency in differentiation and in handling complex numbers is very helpful if not indispensable. However, an interested layman will hardly venture to solve two-port matrices or to perform Laplace transforms which can be performed by computer programs anyway. As long as the relation between voltage and current is linear the most important mathematical operations are multiplication and division.

Myth 2. *Voltage is the primary electric parameter and voltage gain is the ultimate aim of amplification.*
There are two reasons why, in practice, voltage is "more liked" than current. Firstly, voltages occur *across* electronic components and are therefore easily accessible. Secondly, batteries which are readily available, are natural voltage sources. This preference of voltage is reflected in the fact that operational amplifiers are commonly voltage amplifiers. However, as the amplification of signal power is the aim of any amplifier, current gain is as important as voltage gain because power is the product of current and voltage.

Myth 3. *To realize a properly working electronic system all components must be supplied by the same supplier. Besides, they should be as expensive as affordable.*
The signals between electronic devices are not supplier specific but their properties are easily understood so that mixed electronic systems can be at least as good as such delivered by one supplier. Increasing the accuracy beyond the need (e.g., measuring a resistance that must be known to 10% with a 1%-instrument) is, in general, a very expensive practice.

Myth 4. *An electronic network should be grounded at as many places as feasible.*
Correct grounding is an art in itself that takes a lot of experience. One proper grounding is definitely better than any number of arbitrary groundings.

Myth 5. *Amplifiers increase electric power.*
If there were circuits that could amplify electric power no power stations would be needed at all. In any kind of amplifiers power from the power supply is converted into signal power, a process called (signal) amplification.

Myth 6. *Introduction of an electronic component into a loop (e.g., an amplifier in a feedback loop) changes the properties of said component.*
It is obvious that a circuit containing a specific component will have properties different from the component itself. However, an electronic component with properties depending on its use, belongs either to the still exotic class of memristors or, more likely, is irreversibly damaged and should be discarded.

Myth 7. *It is important to understand the difference between a dynamic impedance and a static one.*
Impedance is the amount of voltage change for a given current change. Voltage or/and current do not recognize how this ratio comes about, i.e., whether this impedance is an intrinsic property of an electric component or whether it arises from the action of some active device(s).

Myth 8. *Digital electronics is superior to analog electronics in any respect.*
Respect must be paid to digital electronics as far as cost and flexibility is concerned. In general, the response of digital equipment is slower and it is more difficult to understand compared to analog equipment, partly because the result is less direct. Consequently, starting an electronics course with digital devices is not the best way.

Myth 9. *Since the arrival of circuit simulation programs there is no more need of understanding circuits.*
Simulating tools can be a great help for an expert by sparing lots of time. The amount of knowledge needed to use a simulation program efficiently is certainly greater than any rookie can provide.

Myth 10. *In teaching electronics active two-ports should be used from the very beginning because with (passive) one-ports (resistors etc.) you cannot learn anything of importance.*
For very good students the method of teaching is quite irrelevant. It seems to be good practice to start any building of knowledge with the foundation.

Contents

Preface —— vii

Prolog: Ten myths adored by electronics rookies —— ix

1 **Preparing the ground —— 1**
1.1 Hierarchy of electronic systems —— 1
1.2 Basics of electricity —— 3
1.2.1 Current flow —— 4
1.3 Helpful basic procedures —— 5
1.3.1 Prefixes of units —— 5
1.3.2 Circuit diagrams —— 5
1.3.3 Linearization —— 6
1.3.4 Duality —— 7
1.3.5 Electronic elements with memory effect —— 9
1.3.6 Relativity of attributes in electronics —— 10

2 **Static linear networks —— 12**
2.1 One-ports —— 12
2.1.1 Discrete ideal one-ports —— 12
2.1.1.1 Merging two ideal one-ports —— 13
2.1.2 Real linear sources —— 15
2.1.3 Adding electrical signals (Kirchhoff's theorems) —— 16
2.1.3.1 Pairs of basic one-ports in parallel —— 18
2.1.3.2 Pairs of basic one-ports in series —— 19
2.1.3.3 Several one-ports in series —— 21
2.1.3.4 Several one-ports in parallel —— 22
2.1.3.5 Superposition theorem —— 22
2.1.4 Special linear one-ports (meters) —— 26
2.1.5 Simplifying circuit diagrams by fusing one-ports —— 29
2.1.6 i–v-Characteristics of nonlinear one-ports —— 29
2.1.7 Linearizing one-ports (small-signal behavior) —— 32
2.1.8 Graphical methods for dealing with nonlinear elements —— 34
2.2 Two-port network models —— 36
2.2.1 Active two-ports (dependent signal sources) —— 36
2.2.2 Small-signal models of two-ports —— 38
2.2.2.1 Forward transfer is voltage gain (g-parameters) —— 38
2.2.2.2 Forward transfer is transadmittance (y-parameters) —— 39
2.2.2.3 Forward transfer is transimpedance (z-parameters) —— 40
2.2.2.4 Forward transfer is current gain (h-parameters) —— 40

2.2.2.5 Summary — 41
2.2.3 Transforming a two-port into a one-port — 42
2.2.4 Amplification (power gain) — 42
2.2.4.1 Optimum power transfer (power matching) — 43
2.2.4.2 Amplification over several stages — 45
2.2.5 One-port used as two-port — 46
2.2.6 Three-terminal element used as two-port — 47
2.2.7 Passive two-ports — 52
2.2.7.1 Attenuators — 53
2.2.7.2 Nonlinear passive two-ports (clipping) — 55
2.3 Real active two-ports (amplifiers) — 60
2.3.1 Maximum limits for voltages and currents — 60
2.3.2 Characteristics of two-ports — 62
2.3.2.1 (Electronic) switches — 64
2.3.3 Setting the operating conditions (biasing) — 66
2.3.4 Classification of amplifiers according to the operating point — 66
2.3.4.1 Biased amplifiers (long-tailed pair) — 67
2.3.4.2 Comparators — 69
2.3.5 Amplifiers as two-ports — 70
2.3.5.1 Fully differential amplifier — 71
2.3.5.2 Operational amplifier — 73
2.4 Static feedback — 74
2.4.1 General accomplishments by negative feedback — 76
2.4.1.1 Improving the stability — 78
2.4.1.2 Feedback over two stages — 78
2.4.1.3 Improving the linearity — 79
2.4.1.4 Improving the noise immunity — 79
2.4.2 Static positive feedback — 80
2.4.3 Static feedback in circuits — 81
2.4.3.1 Impedances of two-ports with external feedback — 83
2.4.3.2 Other dynamic impedances — 86
2.4.3.3 Transfer properties of two-ports with external feedback — 90
2.4.3.4 Summary of feedback actions on two-ports with external feedback — 94
2.4.3.5 Feedback in circuits with three-terminal components — 95
2.5 Operation amplifiers — 100
2.5.1 Inverting operation amplifiers — 101
2.5.1.1 Summing amplifier — 104
2.5.1.2 Nonlinear amplifiers — 104
2.5.1.3 Active voltage clipping (voltage limiter) — 105
2.5.2 Noninverting operation amplifiers — 106
2.5.2.1 Voltage follower — 107

2.5.2.2 Linear voltage power supplies (power amplifiers) — 109
2.5.2.3 Precision half-wave rectifier — 110
2.5.3 Difference operation amplifier — 111
2.5.4 Operation amplifiers with positive feedback — 113
2.5.4.1 Negative impedance converter (NIC) — 113
2.5.4.2 Applications of negative impedances — 115
2.5.5 Dissecting the term gain — 115
2.5.6 Current amplifiers — 117
2.5.6.1 Current sources — 118
2.5.6.2 Current mirror — 120

3 Dynamic behavior of networks (signal conditioning) — 123
3.1 Decomposition of signals — 124
3.1.1 Fourier analysis — 127
3.1.1.1 Fourier series — 127
3.1.1.2 Exponential Fourier series — 131
3.1.1.3 Continuous Fourier transform — 132
3.1.1.4 Properties of the Fourier transform — 134
3.1.1.5 Application of the Fourier transform — 135
3.1.2 Laplace transform — 138
3.2 Frequency dependent linear one-ports — 139
3.2.1 Capacitors — 141
3.2.1.1 Stray capacitance — 143
3.2.2 Inductors — 144
3.3 Time domain vs. frequency domain — 146
3.3.1 Voltage step applied to a capacitor — 148
3.3.2 Charged capacitor — 150
3.3.2.1 RC circuits with two different time constants — 151
3.3.2.2 Clamping — 154
3.3.2.3 Baseline restoring — 155
3.3.2.4 Rectifying (DC power supplies) — 157
3.3.2.5 Diode voltage multiplying — 158
3.3.3 Current step applied to an inductor — 159
3.4 Dynamic response of passive two-ports (passive filters) — 160
3.4.1 Basic filter configurations — 161
3.4.1.1 Voltage filters — 161
3.4.1.2 Current filters — 162
3.4.2 Low-pass filters — 163
3.4.3 High-pass filters — 166
3.4.3.1 Pole-zero cancelation — 169
3.4.4 Band-pass filters — 170
3.4.5 (Voltage) band-stop filters — 170

3.4.5.1 Twin-T filter — 170
3.4.6 Resonant filters — 172
3.4.6.1 Series resonant circuit — 172
3.4.6.2 Parallel resonant circuit — 173
3.4.7 Cascading of filter sections — 176
3.4.8 General considerations concerning filters — 176
3.4.8.1 Voltage step response of simple RC filters — 177
3.4.8.2 Current step response of simple RLC filters (shunt compensation) — 179
3.5 Interfacing (cascading) — 180
3.5.1 Interfacing of single stages — 181
3.5.2 Interfacing of subsystems — 187
3.5.3 Interfacing of systems (transmission lines) — 188
3.5.3.1 Signal transmission by coaxial cables — 189
3.5.3.2 Impedance matching of cables — 194
3.6 Dynamic properties of active two-ports — 195
3.6.1 Signal power transmitted by a two-port — 195
3.6.2 Dynamic properties of operational amplifiers — 197
3.6.3 Dynamic feedback in linear amplifiers — 198
3.6.3.1 Frequency-independent negative feedback — 198
3.6.3.2 Frequency instability — 202
3.6.3.3 Frequency-dependent negative feedback (frequency compensation) — 208
3.6.4 Dynamic behavior of operation amplifiers (active filters) — 212
3.6.4.1 Active low-pass filter — 214
3.6.4.2 Active high-pass filter — 216
3.6.5 Dynamic positive feedback (gyrators) — 219

4 Time and frequency (oscillators) — 221
4.1 Degenerated amplifier output at two levels (hysteresis) — 222
4.1.1 Introduction to relaxation oscillators — 224
4.1.2 Gated (asymmetric) relaxation oscillators — 225
4.1.2.1 Schmitt trigger — 225
4.1.2.2 Monostable multivibrator (one-shot) — 227
4.1.3 Symmetric relaxation circuits — 229
4.1.3.1 Astable multivibrator — 229
4.1.3.2 Bistable multivibrator — 231
4.1.4 Relaxation oscillators involving transformers — 232
4.1.4.1 Monostable (triggered) blocking oscillator — 233
4.1.4.2 Astable blocking oscillator — 234
4.1.5 Switched (astable) delay-line oscillator — 235
4.2 Harmonic feedback oscillators — 236

4.2.1 Oscillators using three-terminal devices — 236
4.2.1.1 Oscillators of Wien-bridge type — 237
4.2.1.2 LC oscillators — 244
4.2.1.3 Types of LC oscillators — 251
4.2.1.4 Quartz oscillators — 256
4.2.1.5 Phase-shift oscillators — 259
4.2.1.6 Twin-T oscillator — 263
4.2.1.7 Summary on harmonic positive feedback oscillators — 264
4.3 One-port oscillators — 266
4.3.1 Harmonic one-port oscillators — 267
4.3.2 One-port relaxation oscillators — 268
4.4 Time and frequency as analog variables — 269
4.4.1 Trigger — 270
4.4.1.1 Leading-edge trigger (time jitter) — 270
4.4.1.2 Zero-crossing trigger — 271
4.4.1.3 Constant-fraction trigger — 272
4.4.2 Coincidence circuits (logic gates) — 274
4.4.2.1 Bothe and Rossi circuit — 275
4.4.2.2 CMOS technology for logic gates — 277
4.4.2.3 Anticoincidence circuit — 278
4.4.3 Linear gates — 278
4.4.3.1 Sampling — 279
4.4.3.2 Sample and hold (peak detector) — 280
4.4.4 Analog delays — 282
4.4.5 Signal shortener — 283
4.5 Analog conversion of signal attributes — 283
4.5.1 Time-to-amplitude (-pulse-height) converter (TAC or TPHC) — 284
4.5.2 Amplitude-to-time converter (time-length modulator) — 285
4.5.2.1 Direct voltage-to-time converter — 286
4.5.2.2 Compound voltage-to-time converter — 286
4.5.2.3 Time interval magnification — 287
4.5.3 Amplitude (voltage-)-to-frequency conversion (VFC) — 287
4.5.4 Frequency-to-amplitude (-voltage) conversion — 291

5 Fully digital circuits — 294
5.1 Basic considerations — 294
5.1.1 Logic codes — 297
5.1.2 Parallel vs. serial logic — 298
5.2 Integrated logic technologies (miniaturization) — 299
5.2.1 Logic families — 301
5.2.1.1 TT (Transistor–transistor) logic — 302
5.2.1.2 EC (Emitter-coupled) logic — 303

5.2.1.3 CMOS (Complementary metal-oxide semiconductor) logic — 303
5.2.1.4 BiCMOS (bipolar-CMOS mixed technology) logic — 303
5.3 Basic circuits — 304
5.3.1 Basic gate (combinational) circuits — 305
5.3.1.1 NOT circuit — 305
5.3.1.2 AND and NAND gate — 306
5.3.1.3 OR and NOR gate — 306
5.3.1.4 XOR (EOR, EXOR) gate — 308
5.3.2 Basic bistable (sequential) circuits — 309
5.3.2.1 SR (set/reset) flip–flop — 311
5.3.2.2 D (data, delay) flip–flop — 312
5.3.2.3 T (toggle) flip–flop — 313
5.3.2.4 JK flip–flop — 314
5.4 Registers (involved sequential circuits) — 315
5.4.1 Shift register — 315
5.4.1.1 Deserializer (serial-to-parallel converter) — 316
5.4.1.2 Serializer (parallel-to-serial converter) — 316
5.4.2 Cyclic register — 317
5.4.2.1 Straight (overbeck) ring counter — 317
5.4.2.2 Switch-tail (twisted, Johnson) ring counter — 318
5.4.3 Prescaling and binary counting circuits — 319
5.4.3.1 Asynchronous (ripple) counters — 320
5.4.3.2 Synchronous counters — 322
5.4.3.3 Binary-coded decimal (BCD) counters — 322
5.4.3.4 Up-down counters — 323
5.5 Time response (to binary pulses) — 324
5.5.1 Signal synchronization and restoration — 324
5.5.2 Application of timers — 325
5.5.2.1 Pulse lengthener — 325
5.5.2.2 Pulse shortener — 325
5.5.2.3 Digital delays — 325

6 Manipulations of signals (digitizing) — 326
6.1 General properties of digitizing — 326
6.1.1 Duration of the digitizing process — 329
6.1.2 Performing the digitizing process — 330
6.1.2.1 Selecting analog data within a single channel — 331
6.1.2.2 Nonlinearity — 332
6.1.2.3 The sliding-scale method — 333
6.2 Direct digitizing of time difference — 335
6.2.1 Composite time-to-digital converters (TDCs) — 336
6.2.2 The vernier method — 336

6.3 Direct digitizing of frequency — 338
6.3.1 Multichannel count-rate-to-digital conversion — 339
6.3.1.1 Multiscalers — 340
6.3.1.2 Multichannel pulse-height digitizer — 341
6.3.1.3 Multichannel time-interval digitizer — 341
6.4 Direct digitizing of amplitude — 341
6.4.1 Digital-to-amplitude conversion (DAC) — 341
6.4.1.1 Parallel DAC — 343
6.4.1.2 Serial DAC — 345
6.4.2 Voltage-comparing amplitude-to-digital conversion (ADC) — 346
6.4.2.1 Parallel ADC (flash converter) — 347
6.4.2.2 Serial–parallel (parallel-pipeline) ADC — 348
6.4.2.3 Serial ADC — 349
6.4.2.4 Serial–serial (sectioned-serial, serial-pipeline) ADC — 350
6.4.2.5 Successive-approximation ADC — 351
6.5 Differential ADCs — 353
6.5.1 Tracking or delta-encoded ADC — 353
6.6 Composite ADCs — 354
6.6.1 Wilkinson (single-slope or integrating) ADC — 354
6.6.2 Dual-slope ADC — 356
6.6.3 Multislope ADC — 357
6.6.4 ADCs using frequency modulation — 359
6.6.4.1 Delta–sigma (Δ–Σ) ADC — 359
6.7 Ranking of ADCs — 360
6.8 ADCs in measurement equipment — 361
6.8.1 Digital multimeter (DMM) — 362
6.8.2 Digital (sampling) oscilloscope (DSO) — 363
6.8.3 Spectrum analyzers — 364

Solutions — 365

List of examples — 381

Index — 383

1 Preparing the ground

Electronics deals with the generation, transmission, modification, measurement, and all kinds of applications of electrical signals. Analog electronics concentrates on the shape (amplitude) of the signal whereas digital electronics uses signals of standardized shape to perform "logical" operations.

In both fields, electronic circuits are used. Therefore, it is necessary to "understand" such circuits. This is done by means of models, usually by way of (circuit) diagrams that contain all (and only) the essentials that would allow a knowledgeable person to build such a network.

Basically, there are three scopes when dealing with electronics: understanding, designing and building of circuits (networks). This book concentrates on understanding. For the design it is helpful to use reference circuits. Building of (advanced) electronic circuits is done by the industry and not so much by individuals any more. Such individuals can be compared with artists: their knowledge is partly unconscious, i.e., it is based on long practical experience. Although the construction of circuits is outside the scope of this book, there will be several practical hints on how to overcome frequently encountered problems.

1.1 Hierarchy of electronic systems

Since the middle of the last century, everyday life has become more and more dependent on devices based on electromagnetism, and more recently on optoelectronics. The general science on which these devices depend is called electronics. Electronic systems have become so complicated that a user cannot be expected to be an expert in electronics.

A user usually requires a system that performs the task he wants it to perform. Hardly any driver of a rental car will bother, e.g., what type of clutch the rented car has. All the driver wants is that certain relevant parameters (breaks, steering, acceleration, all lights, etc.) are the way the user needs them. Technical details are of little concern.

An example more to the point would be a mobile phone. There, the system is rather complicated because more than one antenna receives the signal of the hand set. Consequently, a computer will be involved to connect the appropriate receiving station with the appropriate sending antenna. Although it is an exceedingly sophisticated electronic system, all a user has to know about electronics is that the battery of his mobile phone must be charged regularly.

Mobile phones are a ready example of a complex *hybrid* system. In Figure 1.1, such a system is broken down into its three basic subsystems:

- a (source) one-port in which the conversion of sound waves into electrical signals takes place (called microphone, having the function of a *transducer*),

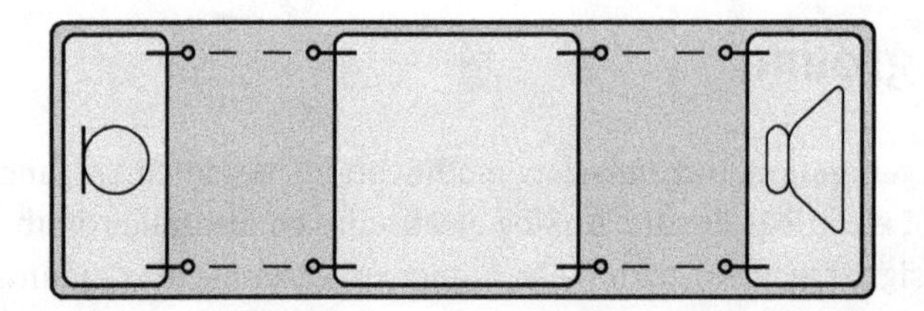

Fig. 1.1. The three basic components of a complex electronic system: source one-port, transfer two-port, and load one-port.

- an electronic two-port that conditions the electric signals and transports them to
- the load one-port that converts electric signals back into sound waves (called loudspeaker, having the function of an *actuator*).

The generation of electric signals from other physics quantities (here acoustic ones) is not subject of this book because this subject belongs in the field of transducers, neither is the use of electric signals to drive optical, mechanical, thermal, or other devices which would be covered by a book on actuators.

When one starts breaking down this system into *devices* (hand sets, antennas, computers, etc.), one needs a much greater electronic knowledge to understand their function. Within these devices, there are electronic *circuits* and these are built from electronic *components*. To understand the function of electronic components you finally will need physics to answer your questions. So we have a hierarchy from the simple to the very complex as shown in Figure 1.2. Depending on where you stand in this hierarchy a different approach is appropriate. In this book, we mainly deal with circuits and networks and in some cases with electronic devices. We are not going to indulge in the physical properties of electronic components which are covered in a large number of books and manuals. However, we need the knowledge of basic electrical properties to understand circuits. More complex systems require specialized knowledge and cannot be dealt with in a general way.

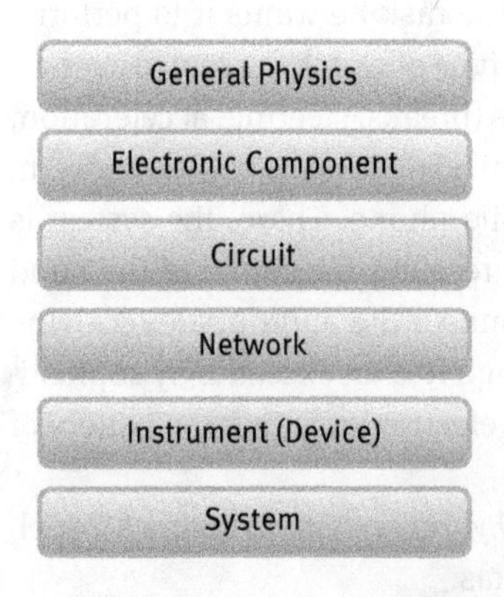

Fig. 1.2. Hierarchy in electronics. In this book, we mainly deal with circuits and networks and in some cases with electronic devices. However, we need the knowledge of basic electrical properties of electronic components to understand circuits.

1.2 Basics of electricity

The central physical quantity in electronics is electric (signal) power $p(t)$. It is not current $i(t)$ or voltage $v(t)$, even if these quantities are the practical basis for determining the electric power according to (1.1)

$$p(t) = v(t) \times i(t) \tag{1.1}$$

meaning that at each moment t the electric power p is the product of voltage v times current i. As we are dealing with electronic principles, we will not deal with high power electronics, nor with very low power electronics. These fields require specific practical knowledge which is outside the scope of this book.

Following a generally accepted practice

- capital letters are used for the symbols of quantities that do not change in time t (current I, voltage V, power P) and
- lower case symbols for quantities that vary in time ($i(t)$, $v(t)$, $p(t)$), i.e., they are functions in time.

Problem

?

1.1. Satisfy yourself that for all electronic instruments that come to your mind power is needed to fulfill the task of this instrument (data transmission, display, computing, mobile phone, etc.). Name electric instruments which do not need auxiliary power to fulfill their task.

Even if electronic measurements involve primarily voltage or current, such devices actually measure power. However, they are destined either to be used as voltage meter or as current meter. Thus, for practical purposes, voltage and current are the basic physical quantities that count in electronics. Both quantities can be derived from one natural phenomenon, the existence of the electrical charge Q. Actually, there are two types of charge, the negative charge and the positive charge. Charge is always the attribute of a particle and consequently quantized. The ordinary electron has one elementary negative charge unit. Positive charge in electronics is the result of missing negative charge, e.g., when an atom has lost a (valence) electron.

The unit of electric power is 1 W(att) which is 1 J(oule)/s(econd). The unit of the electric charge Q is 1 C(oulomb). It takes charge of about 6.2×10^{18} electrons to give 1 Coulomb of negative charge. Moving charge is called current i, i.e., current is charge per time. The unit of current is 1 A(mpere) which is 1 C/s.

The energy (power $p(t)$ times time t) contained in a charge is called voltage v (or potential difference). Its unit is 1 V(olt) and 1 V = 1 J/C. For electronic purposes, voltage can be viewed at as the potential difference between two points constituting an electromotive force that can displace electric charge, i.e., it can enable a current flow. In circuit analysis, the polarity of voltage sources can be assumed at will. When easily feasible it will be so chosen that it conforms to the assumed direction of current or vice versa.

A basic property of electric charge is that it does not vanish, it is conserved. This is not only so in particle physics but is essential for *Maxwell's equations* which govern the propagation of electromagnetic signals in space and time. In particular, they show how electric and magnetic fields are interwoven into each other. In circuit theory, we do not have to exploit fully the beauty of these equations. Because of the relatively small size of circuits usually there is no need to pay heed to the distance (space) covered by the electronic signal. Signal propagation time is in most cases not considered. (Remember: light covers about 30 cm in 1 ns.) In *steady-state* electronics, not only the space but also the time dependence is disregarded.

Consequently, we will first concentrate on the easiest cases, namely those where the electric quantities are constant in time and space. These are commonly called DC (direct current) responses. Later on, we will allow changes in time, i.e., we will consider frequency (steady-state) responses and transient responses. Only in special cases we will cover the propagation of electric charge in space.

1.2.1 Current flow

As there exist two types of electric charge, it became necessary to discriminate between them. They were called positive and negative charge rather arbitrarily. Charges of the same kind repel each other, opposite charges attract each other. If there is a surplus of one type of charge carriers at one terminal and a deficiency at the other, we deal with a current (or voltage) source. If as a result of this unbalance, charge is moved between the two terminals, electric current flows.

Conventional current flow is (for historical reasons) from a positive potential to a less positive (negative) potential.

However, electrons, which are the usual carriers of the electric charge, move the opposite direction (being attracted by the positive potential). Actually, one may *assume* the direction of current flow at one's will as long as one maintains consistency all over the electrical network. If current flows in the assumed direction, it is positive. If the analysis yields a negative current, it just means that you should have made the opposite assumption of a positive current.

Let us start out with *one-ports*, i.e., devices with two terminals. It makes sense to choose the direction of current flow into a one-port such that it is from the higher potential to the lower potential (see Figure 1.3). Thus, the direction of the assumed current flow in a dissipative element, i.e., an element that absorbs power, makes the dissipated power positive, as it should be. However, current delivered by a source has a sign opposite to the sign of its voltage resulting in a negative power; a source does

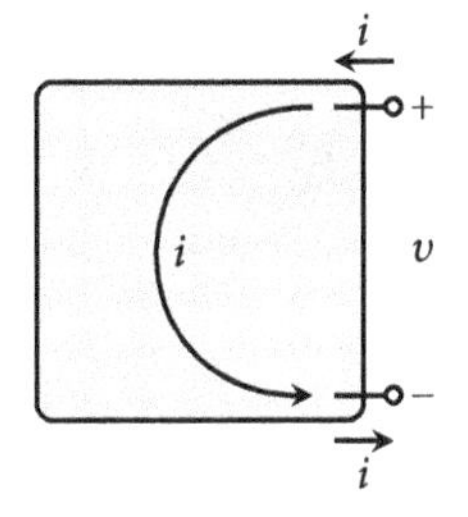

Fig. 1.3. The current flow in a one-port is defined from the terminal of higher potential to the terminal of lower potential. The power $p(t) = v(t) \times i(t)$ of the element is positive for dissipative (passive) elements and negative for active elements.

not consume but supplies power. For that reason, sources are called *active elements*. However, when charging an accumulator (battery), the direction of current is reversed so that power is consumed when it is loaded, i.e., such a source behaves dissipative and is in this operational state a *passive* element.

Many devices function irrespective of the actual direction of the current flow. Devices that respond differently to currents of different direction are called *polarized devices*. The arrow in a symbol of a polarized device indicates the direction of the conventional current flow.

Problem

?

1.2. (a) How would you find out that in an accumulator being loaded power is dissipated?
(b) Find out for yourself, why you may charge accumulators but not regular batteries.

1.3 Helpful basic procedures

1.3.1 Prefixes of units

Electrical quantities cover a huge range of values. To make the numbers better readable prefixes to the units have been introduced, in steps of 10^3. When doing electronics it is unavoidable to get familiar with these prefixes (Table 1.1). In addition, we need the prefix d (deci), i.e., 10^{-1}.

1.3.2 Circuit diagrams

A schematic diagram is the symbolic representation of an electronic network using circuit elements. Actually, it just represents the model of the electronic circuit. It uses idealized circuit elements that only to some extent have the same properties as the physical devices, i.e., the electronic components used in the actual circuit. Just two simple examples: connecting lines in a diagram have no electrical properties to be considered. On the other hand, in particular when dealing with high-frequency cir-

Table 1.1. Commonly used prefixes to electric units.

Name	Abbreviation	Factor
Exa	E	10^{18}
Peta	P	10^{15}
Tera	T	10^{12}
Giga	G	10^{9}
Mega	M	10^{6}
Kilo	k	10^{3}
Milli	m	10^{-3}
Micro	μ	10^{-6}
Nano	n	10^{-9}
Pico	p	10^{-12}
Femto	f	10^{-15}
Atto	a	10^{-18}

cuits, there might be components in the diagram indicating circuit intrinsic (natural) properties that are not presented by lumped components in the circuit.

It should be clear that the location of a component in a circuit diagram has no relationship to the location of its physical counterpart in the actual circuit. We can rearrange components in the schematic as long as the connections between components stay the same as in the circuit.

It is very good practice to have a clear layout in circuit diagrams and to arrange the elements in such a way that the quiescent current flows from top to bottom, i.e., the voltages get smaller and smaller toward the bottom line.

Linearization is, like in other sciences, the basis of easy understanding of electronic circuits. In addition, there are very helpful fundamental laws that alleviate life, e.g., the concept of closed current loops, the existence of a node or the principle of feedback. By restriction to *small-signal response*, nonlinearities can be avoided.

It is easy to bring an electronic circuit, shown in a schematic diagram, to life if the circuit is simple. A lot of know-how (and experience) is needed to do so with demanding circuits. Such practical designs are sometimes a piece of art.

1.3.3 Linearization

Usually, only linearized science is considered because of the easier access to it for the human mind. The same is true for our approach to electronics. At the beginning, we restrict ourselves to linear electronics so that the essentials are not fogged up by

complicated mathematics. The simplest electronic components are linear one-ports. Ideally, there exist three basic types of electronic one-ports
- active elements (sources),
- dissipative elements, and
- reactive elements.

Although these one-ports can be collated to actual electronic components, we will wait to do that until later when we have introduced *duality*. Thus, the amount of detailed knowledge can be reduced by a factor of two. Besides, we will restrict ourselves to linear one-ports which allow the basic understanding of practically all circuits. However, it is not too early to stress one fact that is easily forgotten when overwhelmed by many new insights:

Any electronic component retains its intrinsic properties independent of its (momentary) use, unless it is broken.

1.3.4 Duality

Although duality has been known for more than a century, it is rarely applied despite the fact that it reduces the number of circuits and of laws governing electronics by half, and can be a great help in understanding the circuits. From Maxwell's equations, the symmetry between electric field and magnetic field is obvious. This translates into the duality between voltage and current, i.e., for each component, circuit, equation (law) that deals with current there is an equivalent component, circuit, equation (law) with the voltage as characteristic property.

Current and voltage are equivalent electric quantities, they are dual to each other.

For current, the three basic one-ports have the following properties:

$$\begin{array}{ccc} \text{active element} & \text{dissipative element} & \text{reactive element} \\ i(t) = i_S(t) & i(t) = G \times v(t) & i(t) = C \times \dfrac{dv(t)}{dt} \\ \text{independent of } v(t) & & \end{array} \tag{1.2}$$

with the source current i_S, the conductance G (unit: 1 S(iemens)) and the capacitance C(unit: 1 F(arad)). From duality follows

$$\begin{array}{ccc} \text{active element} & \text{dissipative element} & \text{reactive element} \\ v(t) = v_S(t) & v(t) = R \times i(t) & v(t) = L \times \dfrac{di(t)}{dt} \\ \text{independent of } i(t) & & \end{array} \tag{1.3}$$

with the source voltage v_S, the resistance R (unit: 1 Ω (Ohm)), and the inductance L(unit: 1 H(enry)). Duality makes i_S into v_S, G into R, and C into L. Therefore, the number of basic relations describing linear circuit elements is reduced from six to three.

Among above relations

$$v(t) = R \times i(t) \tag{1.4}$$

is known as *Ohm's law*. It provides a linear relation between voltage v and current i.

Ohm's law is the *basic relation of* linear electronics.

From (1.2) and (1.3) it is clear that the conductance G is dual to the resistance R. As both the conductance and the resistance are realized by the same physical element, called resistor, this element is self-dual. This self-duality of the resistor is the reason why a conductor is rarely considered as physical element. However, for duality reasons it is

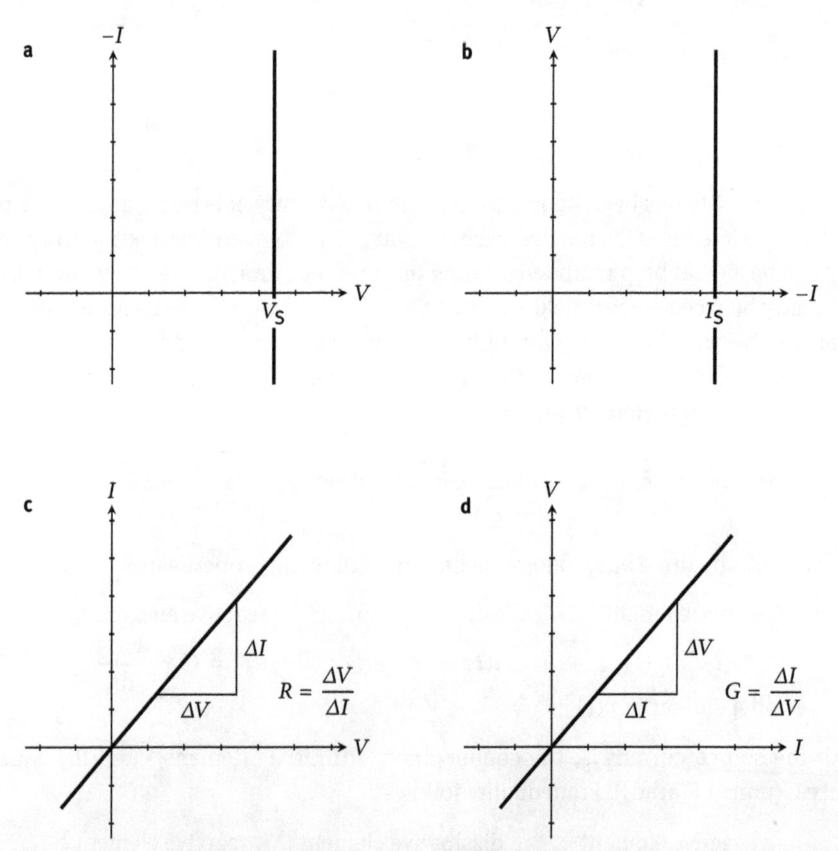

Fig. 1.4. Characteristics of an ideal voltage source (a), an ideal current source (b), a linear resistance (c), and a linear conductance (d). Note that in order to match the physical direction of the currents, the currents' signs are inverted in the characteristics of the active elements (sources).

Table 1.2. Essential features of the current–voltage duality.

	Dualities	
Electrical quantity	Current	Voltage
Field	Magnetic	Electric
Constituent	Electric charge	Magnetic flux
Load	Conductance	Resistance
Complex load	Admittance	Impedance
Zero load	Short-circuit	Open-circuit
Neg. impedance i-v-c Characteristic	N-shaped	S-shaped
Reactive component	Capacitor	Inductor
Kirchhoff's laws	First	Second
Source	Current	Voltage
Theorem	Norton	Thevenin
Structure	Node	Mesh (loop)
Alignment (Superposition)	In parallel	In series
Sensing (Measuring)	In series	In parallel

essential to accept the concept of having resistance and conductance side by side. The dual version of Ohm's law $i(t) = G \times v(t)$ is often disregarded because the physical element representing the conductance is the resistor. Another self-dual element is the switch (Section 2.3.2.1).

Although just one parameter describes a linear device, we are introducing the term *characteristic* already at this point. A characteristic of an electronic element is the graphical presentation of the relation between two variables. A linear relation obviously results in a linear characteristic, as shown in Figure 1.4 for several linear relations between current and voltage. For the beginning, we do not consider signals that change in time. Therefore, the following discussions disregard reactive elements.

Table 1.2 summarizes the essential features of duality between current and voltage.

Observe, that the (real) load is a special case of the complex load. Consequently, the term *impedance* is more general than *resistance* and *admittance* more general than *conductance*. Thus, the terms resistance and conductance are only used when it is important to stress the difference.

1.3.5 Electronic elements with memory effect

From the fact that current is electric charge in motion directly follows that

$$\mathrm{d}q = i \times \mathrm{d}t = C \times \mathrm{d}v ,$$

and by duality

$$d\Phi = v \times dt = L \times di\,.$$

From Ohm's law we have

$$dv = R \times di\,.$$

Thus, the three basic electronic components R, L, and C which we introduced in Section 1.3.4 can be defined in a different way. In 1971, it looked as if Leon Chua has discovered a missing link among these three basic electronic components. Into this scheme of the electronic proponents voltage, current, electric charge, and magnetic flux

$$\begin{aligned} dq &= C \times dv \\ dv &= R \times di \\ di &= (1/L) \times d\Phi \end{aligned}$$

the relation

$$d\Phi = M \times dq$$

fits perfectly well. The up to then unknown property M was called *memrisistivity*, and it was assigned to a component named *memristor*. From the identity formed by these four relations one gets

$$M = \frac{L}{RC}\,,$$

i.e., memrisistivity has the dimension of a resistance. Memristivity is a resistance with memory effect, i.e., its value depends on the amount of current that has flown through it. Much more recently, such a device has actually been built. Later, also capacitors with memory effect have been built. This suggests that components with memory effect fit into some other scheme than suggested, i.e., the suggestion of a missing link now appears in a different light. Components with memory effect have not been used much if at all. For that reason, we are going to pay no further heed to them.

1.3.6 Relativity of attributes in electronics

The attributes small, large, slow, fast, etc., of signals or parameters are practically always *relative*, i.e., in comparison to something specific. For instance, if an output impedance is called small it means that it is much smaller than the impedance of its burden, the *load impedance*, and a small load impedance means that it is a heavy burden to the source, i.e., it is much smaller than the output impedance of the source.

A special case is the *small-signal*. It means that a specific circuit behaves linearly toward this signal. It could be a voltage signal of 100 V or of 10 mV, i.e., a specific signal may be considered as small-signal for one circuit, but not for the other.

Unfortunately, scholars of electronics give too much weight to rules of thumb. Many electronic terms can only be viewed as such, e.g., if a battery (a voltage source) is burdened by a load with an impedance (much) smaller than its own impedance it behaves like a current source. As this is a very rare situation, it is justified to call a battery a voltage source. However, one should always be open to the fact that there will be situations in which the generally used term may be misleading.

Problem

?

1.3. Is a resistance value of 100 Ω small or large?

2 Static linear networks

A functional arrangement of electronic elements is an electronic circuit. An assembly of linear circuits forms a linear network. At this moment we disregard any time dependence in the behavior of electronic circuits. We just consider the momentary status of the electric variables current and voltage. We could state as well that we are only considering the DC (direct current) behavior of the networks.

2.1 One-ports

An electronic element with two terminals (leads) is called *one-port*. To understand its properties, we need to know the voltage $v(t)$ between the terminals (across the one-port) and the magnitude and direction (sign) of the current $i(t)$ through the one-port (Figure 1.3).

2.1.1 Discrete ideal one-ports

Practical electronic one-port components are, e.g.,
- batteries as sources of electric power,
- *resistors* (dissipative components) representing their ideal property conductance G or resistance R, and the reactive components,
- *capacitors* representing very well the capacitance C, and
- *inductors*, representing not quite as well the inductivity L.

Using ideal (linear) components helps us to concentrate on the essentials. Table 2.1 summarizes the six ideal linear one-ports.

Exchanging current with voltage according to duality halves the number to three ideal linear one-ports. One must realize that one can choose freely either the *current* one-ports or the *voltage* one-ports when trying to understand a circuit. Usually, one of these choices will give an easier approach to the understanding of the property of a circuit. With some experience, it will be quite obvious which to choose, e.g., in a loop the current will have more importance, in a parallel arrangement the voltage.

When trying to understand a circuit one is free to choose the voltage or the current version; one of them will be more appropriate for the problem at hand.

For the beginning, we restrict ourselves to sources and dissipative elements, i.e., to four elements.

Table 2.1. Family of ideal linear one-ports and their symbols as used in circuit diagrams.

Sources	Dissipative components	Reactive components
current source	conductor	capacitor
voltage source	resistor	inductor

2.1.1.1 Merging two ideal one-ports

Obviously, one can arrange two one-ports either in *parallel* or in *series*. One-ports arranged in parallel will have the identical voltage across them, one-ports arranged in series will have the identical current flow through them. Consequently, some combinations of one-ports will be contradictory, and some are meaningless in view of the intrinsic properties (1.2) and (1.3) of ideal sources. An ideal current source delivers its current independent of what the load is, i.e. which other one-port is arranged in series.

Combining a one-port with an ideal current source can only be done sensibly in parallel.

Problem

?

2.1. Combining an ideal current source in series with another one-port:

(a) Place alternately an ideal voltage source, a resistor, or an ideal current source in series to an ideal current source and find out, why none of these combined one-ports makes sense.

(b) Why is there no sense in arranging any one-port in series to an ideal current source? (How does the augmented one-port act differently from the ideal one?)

Considering that an ideal voltage source maintains it voltage independently of what one connects its leads to, one gets the dual version:

Combining a one-port with an ideal voltage source can only be done sensibly in series.

There is no sensible way of combining an ideal voltage source with an ideal current source, i.e. one cannot add a voltage to a current which nobody would suggest anyway. One can only combine ideal sources of the same kind. Generalizing one gets:

Voltages are *added* in series, currents in parallel.

Problem

2.2. Combining an ideal voltage source in parallel with another one-port:

(a) Place alternately an ideal voltage source, a resistor, or an ideal current source in parallel to an ideal voltage source and find out, why none of these combined one-ports makes sense.

(b) Why does it not make sense to arrange any one-port in parallel to an ideal voltage source? (How does the augmented one-port act differently from the ideal one?)

This leads to another rule of duality:

A serial arrangement of one-ports is dual to the parallel arrangement of their dual counterpart.

Usually, electronic engineers appear to prefer working with voltages rather than with currents. Why is that so? To have direct access to current one must place the sensing device (the meter) into the loop, i.e. one must open the loop in which the current flows because one must arrange the meter in series. A voltage measurement can be done in parallel, i.e. one can just put the terminals of the voltmeter to the two spots between which the voltage should be measured. The development of current probes has lessened the importance of voltage measurements because the current is measured by the magnetic field accompanying it. In that case, the loop stays intact.

Direct sensing (measuring) of voltages is done in parallel, of currents in series.

Based on the fact that ideal sources are unchangeable in their output signal the following two principles can be used for simplifying circuits.

1. Circuit components that do not affect the functionality of a circuit may be removed from this circuit.
2. Under the presupposition that a voltage between two points (a current through a junction) is fixed, i.e., it has an invariable value, then for the analysis of the network, an ideal source can be used to replace that part of the circuit that provides the invariable electrical signal.

2.1.2 Real linear sources

An ideal source is characterized by its property to maintain at its output its electrical signal independent of the load, i.e.
- the ideal current source will deliver the current i_S into any circuit connected to it, and
- the ideal voltage source will maintain v_S at its leads independent of the load connected to it.

However, there is a practical catch in these definitions. Obviously, a current source cannot feed current into an open-circuit (having $G_L = 0$), and correspondingly a voltage source cannot maintain its voltage if it is short-circuited (with $R_L = 0$). This brings about another rule of duality:

Short-circuits are dual to open-circuits.

As short-circuits and open-circuits do occur, it is necessary to modify our notion of sources. Including a dissipative element into the source one-port will make such sources immune against these above-mentioned forbidden situations. These are called *real sources* rather than *ideal sources*.

In Figure 2.1, each of the two ideal sources is combined with a dissipative element resulting in real (linear) sources. Their properties are considerable closer to real sources of electricity (like a battery). For obvious reasons, the internal dissipative elements must be arranged as shown in Figure 2.1. It must be arranged in parallel to the ideal current source and in series to the ideal voltage source.

Without a load, the real and the ideal source deliver the identical electrical quantity. But now it is feasible to short-circuit a voltage source, and to leave the current source open-circuited.

Real sources play an important role in the analysis of linear circuits because both *Thevenin's* theorem, and *Norton's* theorem, its dual counter-part, makes use of them. The common formulation is as follows:

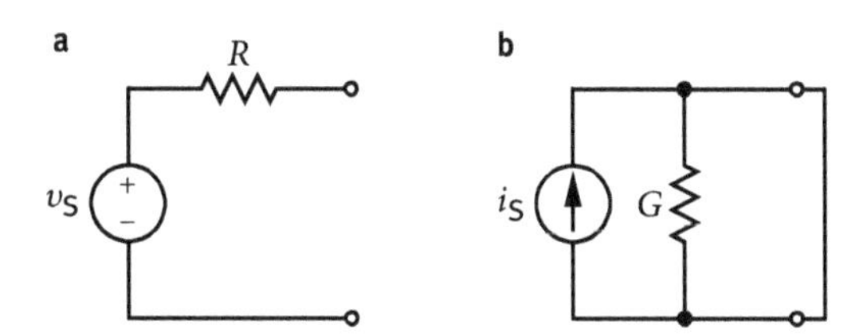

Fig. 2.1. Real linear sources, a combination of an ideal voltage source with a resistor R (a), and an ideal current source with a conductor G (b).

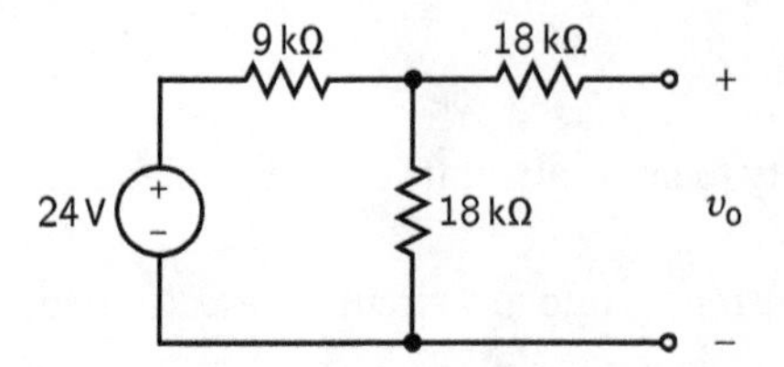

Fig. 2.2. Linear circuit from Problem 2.5.

Any combination of linear one-ports, i.e. any linear electric network is between two arbitrarily chosen points, electrically equivalent to a real linear source, by choice either a linear voltage source (Thevenin) or a linear current source (Norton).

These theorems only state that such a replacement can be done, but not, how it is done. Two linear dependences are identical if two independent parameters of them agree. There are three parameters that usually can be accessed easily between the two points in the circuit: open-circuit voltage, short-circuit current, and impedance. Which pair is the easiest to use cannot be said but must be guessed to reduce the calculational effort.

Problems

2.3. Apply Thevenin's theorem to a real linear current source. What is the difference in behavior between a real linear voltage source and a real linear current source?

2.4. A linear network has a voltage of 10 V between two terminals. When these terminals are short-circuited a current of 10 mA flows through this short-circuit. Replace this linear network between these terminals according to Norton's theorem.

2.5. Replace the linear circuit of Figure 2.2 at the output terminals
(a) by a real voltage source (Thevenin's theorem), and
(b) by a real current source (Norton's theorem).
(c) Determine v_o in both cases.

It depends on the problem at hand whether the preferred electrical quantity is voltage or current, i.e. whether a current or a voltage source is used for the substitution of a linear network.

2.1.3 Adding electrical signals (Kirchhoff's theorems)

Kirchhoff's first law is based on the conservation of electric charge. It can be formulated as follows

The sum of all currents flowing into a *node* (*junction*) equals the sum of currents flowing out of this node.

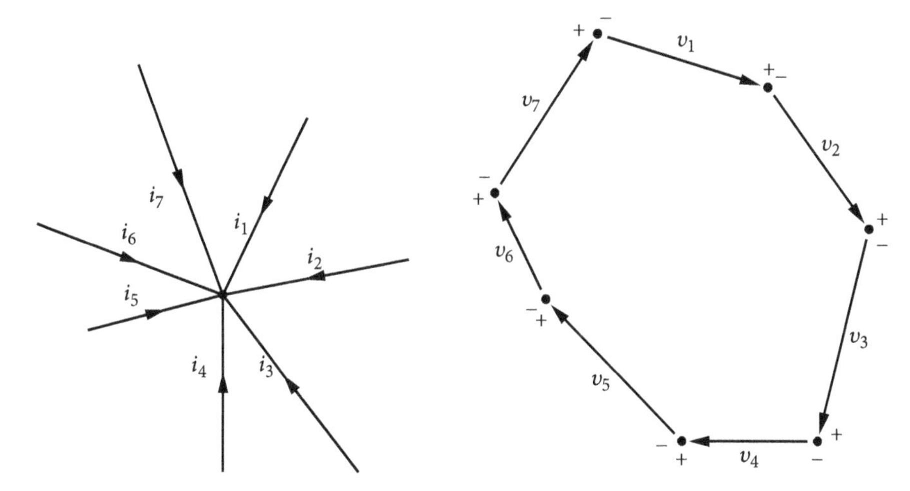

Fig. 2.3a. Explaining Kirchhoff's first law.

Fig. 2.3b. Explaining Kirchhoff's second law.

Considering that current is a signed quantity according to the assumed direction of flow (Figure 2.3a) this law (for an n-fold node) may be written as

$$\sum_{k=1}^{n} i_k(t) = 0. \tag{2.1}$$

Duality delivers instantly Kirchhoff's second law (Figure 2.3b)

$$\sum_{k=1}^{n} v_k(t) = 0. \tag{2.2}$$

The sum of all voltages around a loop equals zero.

To arrive at this simple equation, it was taken advantage of the fact that voltage like current are signed quantities.

A loop is the dual counterpart of a node.

This is nothing really new because we came already across the duality between short-circuit and open-circuit as special cases of node and loop. (As a short-circuit is the smallest reasonable node with $k = 2$, an open-circuit is for duality reasons the smallest loop.) The following prescription, already given before, is confirmed by Kirchhoff's laws. As already shown, voltages are added in series, currents in parallel. This is demonstrated in Figure 2.4.

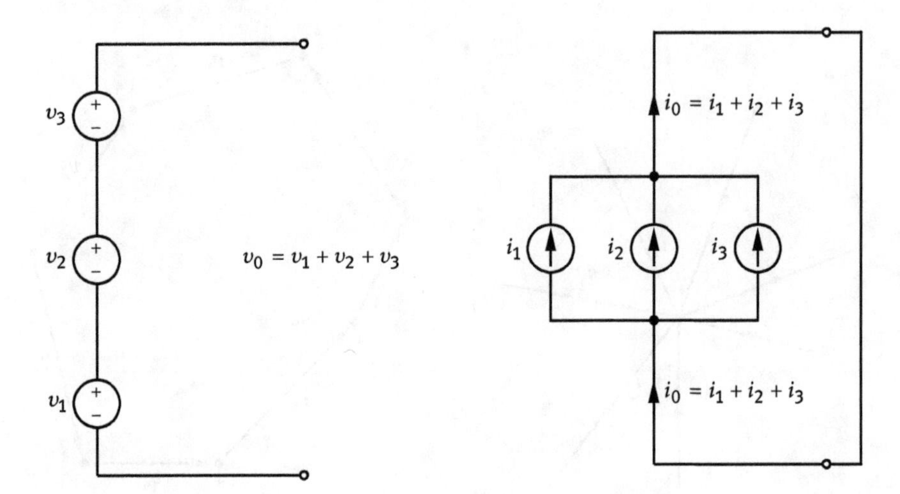

Fig. 2.4a. Adding voltages.

Fig. 2.4b. Adding currents.

2.1.3.1 Pairs of basic one-ports in parallel

Formally there are six different pairs of sources and dissipative elements to be considered (Figure 2.5). Having two current sources in parallel manifests a very important property. As each of the sources delivers their specific current independent of the load,

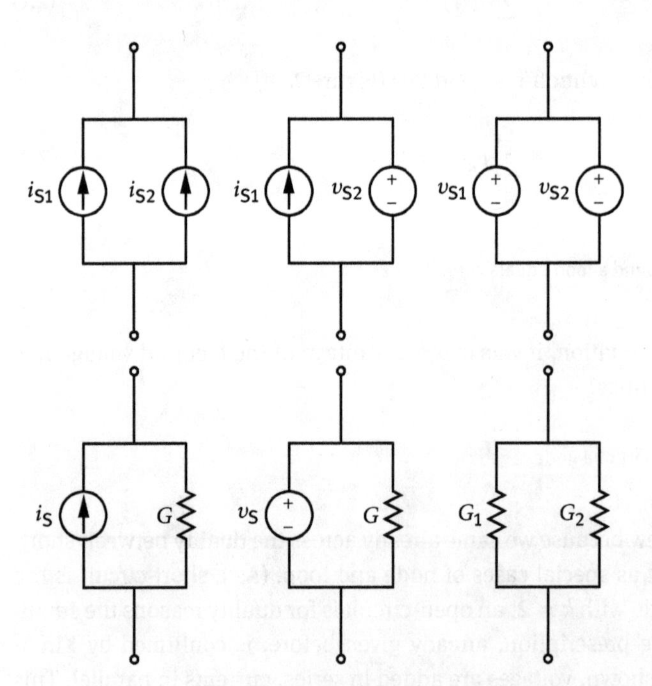

Fig. 2.5. Six possibilities of arranging sources and dissipative elements in parallel.

the current of one source *cannot* flow through the other source, i.e.

the conductance of an ideal current source is zero.

This finding is consistent with the slope of the characteristic of an ideal current source in Figure 1.4.

Two ideal current sources can be replaced by one ideal source with a total source current i_S

$$i_S = i_{S1} + i_{S2} \tag{2.3}$$

Having two ideal voltage sources in parallel is in contradiction to the definition that an ideal voltage source maintains the voltage at its terminals independent of the load. There cannot be two (different) voltages at the common terminals of the combined one-port. This fact has practical implications. Without specific precautions, battery cells (which in many cases come close to ideal voltage sources) may not be wired in parallel.

A combination of an ideal voltage source in parallel to an ideal current source makes the current source redundant. The voltage at the common terminals is determined by the voltage of the voltage source and is independent of the current delivered by the current source. The same ruling applies to a resistor in parallel to an ideal voltage source.

As discussed above, a conductance in parallel to an ideal current source constitutes a "real" current source. Applying Thevenin's theorem, it may be replaced by a real voltage source.

Finally, one can place two dissipative elements in parallel. In such a case, it is more favorable to view them as conductors because the two conductances G_1 and G_2 can easily be combined to one.

$$G = G_1 + G_2 \tag{2.4}$$

2.1.3.2 Pairs of basic one-ports in series

All that is needed is to translate above section by means of duality. See Figure 2.6 for the dual counterparts of arrangements shown in Figure 2.5. There remain only three sensible arrangements: the real voltage source and the serial connection of two voltage sources and of two resistors. The two ideal voltage sources in series can be replaced by one source with

$$v_S = v_{S1} + v_{S2} \tag{2.5}$$

The two resistors in series can be replaced by a single resistor R

$$R = R_1 + R_2 \,. \tag{2.6}$$

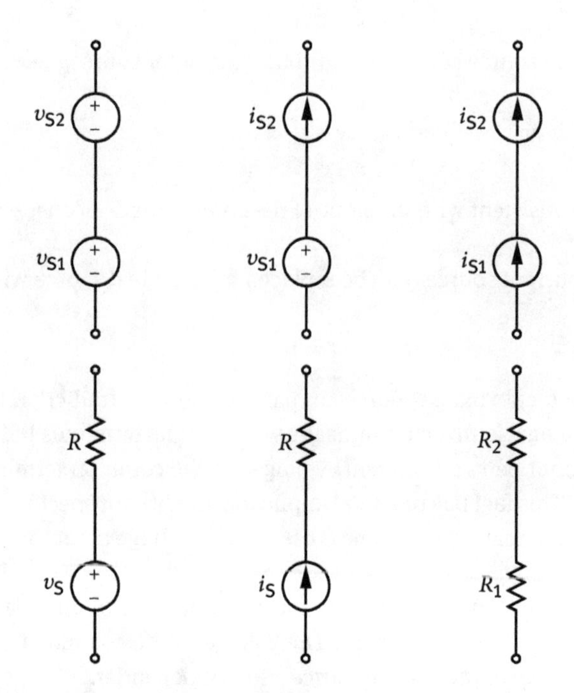

Fig. 2.6. The six dual counterparts of the arrangements shown in Figure 2.5.

However, the most important finding is that

> the resistance of an ideal voltage source is zero.

This finding is consistent with the slope of the characteristic of an ideal voltage source in Figure 1.4.

Example 2.1 (Analyzing dual circuits based on voltages and currents).

(a) An ideal voltage source with the voltage v_S is burdened by two resistors R_1 and R_2 in series. The voltage v_1 is across R_1, v_2 across R_2. Then

$$v_S = v_1 + v_2$$

according to Kirchhoff's second law. After adding R_1 and R_2 to the combined resistor R one gets for the current (Ohm's law, Section 1.3.4)

$$i_S = \frac{v_S}{R}.$$

(b) An ideal current source with the current i_S is burdened by two conductors G_1 and G_2 in parallel. The current i_1 flows through G_1, i_2 through G_2. Then

$$i_S = i_1 + i_2$$

according to Kirchhoff's first law. After adding G_1 and G_2 to a combined conductor G one gets (inverse Ohm's law, Section 1.3.4)

$$v_S = \frac{i_S}{G}.$$

2.1.3.3 Several one-ports in series

Loading a real voltage source with a resistor R_L constitutes a loop of an electric network (Figure 2.7). It is essential to realize that the *same* current flows through *all* elements of this loop in one direction. The two resistors in series can be joined together to give $R = R_S + R_L$. Applying Ohm's law yields the loop current $i_L = v_S/R$. The maximum current that can be drawn from this real voltage source is called *short-circuit current* i_{sc} and is obtained by making the load resistance zero (a short-circuit)

$$i_{sc} = \frac{v_S}{R_S}. \tag{2.7}$$

Arranging n one-ports in series has as a consequence that the current i in the loop is identical in all one-ports. If, aside from the voltage source, all components are resistors, such an arrangement is called voltage divider. The voltage drop v_j at each resistor R_j is according to Ohm's law $i \times R_j$, the total voltage drop, $v_t = i \times \sum_{j=1}^{n} R_j$. Therefore, the voltage v_j at each resistor R_j is a fraction of the total voltage v_t

$$\frac{v_j}{v_t} = \frac{R_j}{\sum_{j=1}^{n} R_j}. \tag{2.8}$$

For the case $n = 2$, i.e. two resistors in series, this becomes the basic *voltage divider equation*

$$\frac{v_1}{v_t} = \frac{R_1}{R_1 + R_2}. \tag{2.9}$$

Problems

?

2.6. An ideal current source of 10 mA feeds a resistor of 1 kΩ in series to a resistor of 10 kΩ.

(a) Give the current through the second resistor.

(b) Give the voltage across the first resistor.

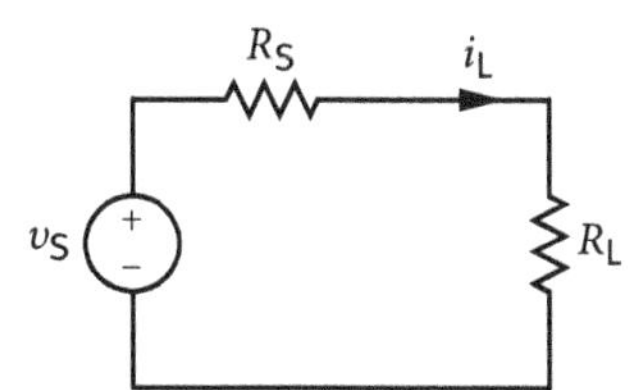

Fig. 2.7. A real voltage source with load resistance R_L.

2.7. An ideal voltage source of 10 V, an ideal current source of 10 mA, and a resistor of 10 kΩ form a loop. Give the voltage across the resistor.

2.8. A chain of five resistors of 1 kΩ each, lies in parallel to an ideal voltage source of 8 V.
(a) simplify the circuit by combining the upper four resistors.
(b) Determine the voltage across the bottom resistor.
(c) The divided voltage is used as a real voltage source. Apply Thevenin's theorem and find the source voltage and impedance.
(d) Replace the real voltage source of (c) by a real current source using Norton's theorem.

2.1.3.4 Several one-ports in parallel

All that is needed is to translate above section by means of duality. Note that in this case the *voltage* across each component is common. The currents through the conductors G_j are proportional to their values yielding, after exchanging voltage and current and resistors and conductors, exactly the same equations for current division as before for voltage division.

Problems

2.9. Convert Figure 2.7 to a parallel configuration by applying duality.

2.10. Write down the equations for current division, using resistors instead of conductors.

2.11. Short-circuit current and duality
(a) Give the short-circuit current i_{sc} of a real current source with a source current i_S.
(b) Apply duality to (a).

2.12. An ideal voltage source with v_S is shunted by four resistors, having an impedance of 0.1 kΩ, 1 kΩ, 10 kΩ, and 100 kΩ, respectively.
(a) Give the voltage across the 1 kΩ resistor.
(b) Give the total current delivered by the source.

2.13. An ideal current source of 10 mA is shunted by an ideal voltage source of 10 V and a 10 kΩ resistor.
(a) Give the current through the resistor.
(b) Give the voltage across the resistor.

2.1.3.5 Superposition theorem

In any linear network both the voltage and the current response of any linear element to more than one *independent* source equals the (algebraic) sum of the responses to each source acting alone, with all other sources substituted by their impedances.

The superposition theorem presupposes a *linear* network.

Recommendation:

- Current sources contained in a loop should be converted to voltage sources applying Thevenin's theorem. The voltage sources not under consideration must be replaced by their impedances. Then the individual voltage contribution can be obtained by voltage division between the impedance of the voltage source under consideration and the sum of all the other impedances.
- Voltage sources connected at a node should be converted to current sources applying Norton's theorem. The current sources not under consideration must be replaced by their admittances. Then the individual current contribution can be obtained by current division between the admittance of the current source under consideration and the sum of all the other admittances.

In circuit analysis a voltage source (e.g., a power supply) that is independent of the signal source may be replaced by a short-circuit, i.e. *both terminals of a grounded power supply act as signal ground.*

This means, in practical applications, that the signal path through a voltage power supply must not have any appreciable parasitic impedance for the signal (from the wiring etc.).

Example 2.2 (Replacing a voltage divider by a real voltage source using Thevenin's theorem). Some (input) voltage v_i is reduced by voltage division to (the output voltage) v_o. To this end, it is applied to two resistors R_1, and R_2 arranged in series. Like any other linear network, this network may be replaced by a real voltage source.

- The quantity of interest is v_o. In our case it is the voltage across R_2. Therefore, this voltage must be retained in the simplified circuit.
- A linear network, represented by a linear u–i-function (i.e. a straight line), needs two independent quantities for its description. One may choose between three convenient ones: open-circuit voltage, short-circuit current, and (output) resistance (conductance).
- One choice is obvious: the variable of interest is v_o which is the open-circuit voltage.

Applying Kirchhoff's second law we get $v_i = v_1 + v_2$. From Ohm's law, we get $v_1 = i \times R_1$ and $v_2 = i \times R_2$. Thus, $v_i = i \times (R_1 + R_2)$, and the voltage division is

$$\frac{v_o}{v_i} = \frac{v_2}{v_i} = \frac{i \times R_2}{i \times (R_1 + R_2)} = \frac{R_2}{R_1 + R_2} \tag{2.10}$$

As we need the value of the resistor R of the real voltage source, we choose the (output) resistance as the second independent property. Let us have a look at Figure 2.8.

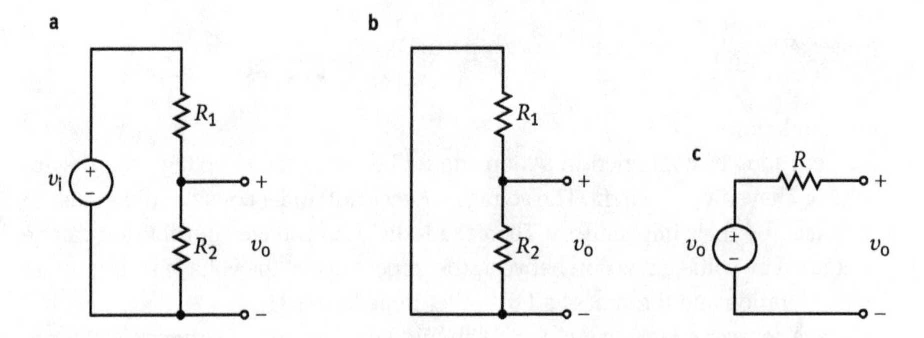

Fig. 2.8. (a) Circuit diagram of the voltage division. (b) Parallel configuration of R_1 and R_2 when viewed from the output. (c) The real voltage source.

In Figure 2.8a, the voltage v_i is symbolized by an ideal voltage source of that value. According to the superposition theorem this source is equivalent to a short-circuit when viewed from the output as shown in Figure 2.8b. Consequently, R_1 and R_2 are in parallel. And for parallel resistors their conductance (their inverse) is added to give the total conductance

$$\frac{1}{R} = \frac{1}{R_1} + \frac{1}{R_2} = \frac{R_2 + R_1}{R_1 \times R_2} \tag{2.11}$$

so that

$$R = \frac{R_1 \times R_2}{R_1 + R_2} \tag{2.12}$$

is obtained. Thus, it is shown that a real voltage source with an ideal voltage source of the value v_o and a resistor R replaces the voltage divider in any regard that deals with its output (Figure 2.8c).

The main purpose of this exercise is to stress the importance of the voltage divider equation and the equation describing the parallel configuration. Probably, anybody who deals with electronic networks knows these equations by heart to spare time. Replacing a current divider according to Norton's theorem by a real current source is dual to above approach.

Problems

2.14. Do partial electric signals in nonlinear networks add up (linearly) to the total one?

2.15. Is Thevenin's theorem more important than Norton's?

2.16. Two real voltage sources ($v_{S1} = 2.4$ V, $R_{S1} = 1$ kΩ, $v_{S2} = 1.2$ V, $R_{S2} = 1$ kΩ) supply in parallel a load resistor of $R_L = 100$ Ω. Calculate all three currents.

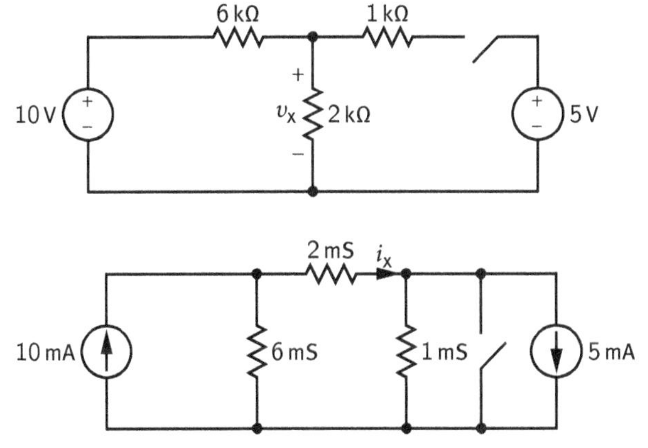

Fig. 2.9. Linear circuit from Problem 2.17.

Fig. 2.10. Linear circuit from Problem 2.18.

2.17. Find the values of v_x for both switch positions in Figure 2.9 after simplifying the 10 V voltage divider circuit using Thevenin's theorem and using the superposition theorem.

2.18. Find the values of i_x for both switch positions in Figure 2.10, applying Norton's theorem to the 10 mA-current divider circuit and using the superposition theorem.

Example 2.3 (Combined voltage signal). Figure 2.11a shows a combination of two ideal voltage sources loaded by a resistor of 1 kΩ and Figure 2.11b shows the resulting voltage at the resistor. One source delivers a quiescent voltage $V_{op} = 6$ V and the other source delivers a voltage v_s that switches between −1 V and +1 V resulting in a combined voltage between 5 V and 7 V as shown in Figure 2.11b. The current through the resistor is the superposition of i_S, a current switched between −1 mA and +1 mA, with the quiescent current I_{op} of 6 mA.

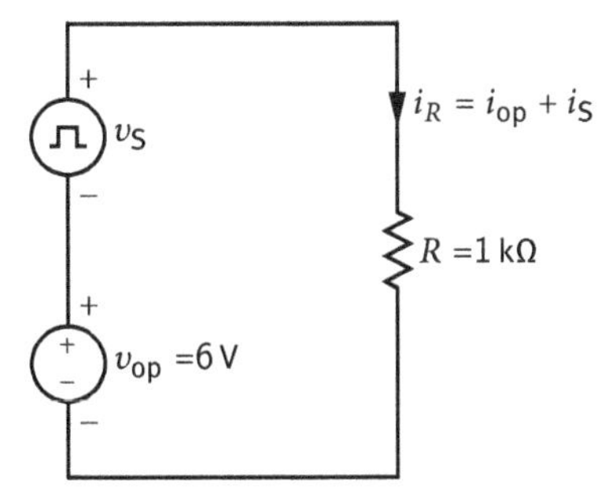

Fig. 2.11a. Superposition of two voltage sources.

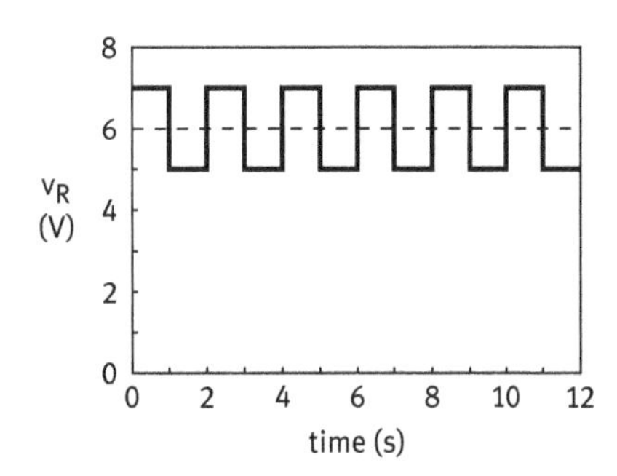

Fig. 2.11b. The resulting voltage across the resistor R.

Let us look backward from the result of the example. The output signal shown in Figure 2.11b is moderately complicated and dealing with it as a whole is not straightforward. However, taking advantage of the superposition theorem makes life much easier. Splitting the output voltage into two voltages, transforms one complicated task into two easy tasks. This is a generally obeyed practice. The quiescent voltage V_{op} is the operating point voltage of the device (resistor) and the changing voltage v_S is considered as signal. So we have an operating point of the resistor of (V_{op} = 6 V, I_{op} = 6 mA) or more compact (6 V, 6 mA) and a signal voltage v_S of ±1 V.

Problem

2.19. Avoid bipolar signals for the signal source. Find the two obvious combinations of voltage sources that would also give the voltage pattern shown in Figure 2.11b.

2.1.4 Special linear one-ports (meters)

The measurement of an electric signal consumes energy independent of the applied method. The advancement of instrumentation has made the minimum amount of energy needed for a measurement small, but it is not nil. Consequently, a measuring device loads an electric circuit which means that (in most cases) it behaves like a resistor (disregarding reactive behavior at the moment). This behavior is called *input resistance* (or more general *input impedance*). The power dissipated in this resistance is taken from the circuit which is "loaded" by the measuring device. Consequently, the meter measures the loaded condition of the circuit for which it must be corrected if, as usual, the unloaded condition is of interest.

Like any measuring device, electric meters measure their own response to a signal applied to their input. To generate a useful output (by means of an indicator, a display etc.), some power is needed. Obviously, the minimum power that can be measured must be higher than the power consumed by the instrument. Developing *active* meters, i.e. meters involving amplifiers, the sensitivity was much increased as compared to passive meters.

As the answer of an electric meter depends on the power dissipated in its input resistance R_i, it should be clear that, independent of the type of calibration, each meter can, in principle, be used for the measurement of voltage and current by applying Ohm's law for the conversion. Voltmeter and ammeter are essentially the same, just as resistor and conductor. To avoid excessive loading, a versatile voltmeter will have a high input resistance compared to the resistance of the circuit under investigation, an ammeter a high input conductance. Otherwise, correcting for the loading might introduce excessive uncertainty.

?

Problems

2.20. A (passive) meter has an input resistance of 2.5 kΩ and has full-scale reading at an input power of 4 μW. By adding one one-port, this meter can be made to be used as voltmeter for measuring voltages up to 100 V. Which arrangement is necessary and what are the essential properties of this one-port?

2.21. A (passive) meter has an input conductance of 0.4 mS and has full scale reading at an input power of 4 μW. By adding one one-port, this meter can be made to be used as an ammeter for measuring currents up to 1 A. Which arrangement is necessary and what are the essential properties of this one-port?

2.22. An active voltmeter has in the 1 V range a burden current of less than 10 pA, i.e. it draws from the circuit to be measured at most that much current. The reading, when measuring the output voltage of a real voltage source having an internal resistance of 0.1 Ω, is 1.000 000 00 V. Is there a need to correct the reading for the loading of the circuit?

2.23. An active ammeter has in the 0.1 μA range a burden voltage of 55 mV, i.e. at full scale it draws from the circuit to be measured a power of 5.5 nW. The reading when measuring the output current of a real current source having an internal conductance of 1 μS is 1.000 00. Is there a need to correct the reading for loading the circuit?

Example 2.4 (Measurement of the resistance of a linear resistor by measuring simultaneously voltage and current). According to Ohm's law the resistance value R_x of a linear resistor is given by $R_x = V_x/I_x$. Therefore, a simultaneous measurement of V_x and I_x is needed. Figure 2.12 shows the two dual configurations that may be used. In Figure 2.12a the actual current through the resistor is measured, whereas the voltage is measured across the series arrangement of the resistor with the ammeter. Consequently, the unknown current $i_x = i_m$, the measured current. From the ratio of v_m/i_m,

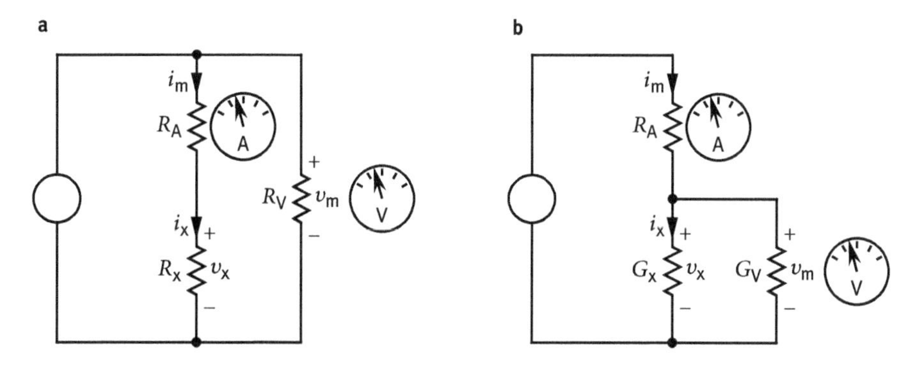

Fig. 2.12. Simultaneous measurement of the current and the voltage through an unknown component: (a) with the amperemeter in series, (b) with the voltmeter in parallel.

the sum of the unknown resistor R_x with the resistance R_A of the ammeter is obtained $v_m/i_m = R_x + R_A$. Consequently, the value of the unknown resistor is $R_x = v_m/i_m - R_A$. The correct value is smaller than the measured ratio suggests; it must be corrected by subtracting the resistance of the ammeter.

The arrangement in Figure 2.12b is dual. Exchanging the current (meter) with a voltage (meter), and resistance with conductance, and paying attention to the duality of parallel and serial configuration, results in equations with an identical structure.

Problems

2.24. Correction factors

(a) Give the correction factor for the unknown resistance rather than unknown conductance, as obtained by duality for the configuration in Figure 2.12b.

(b) Which of the two configurations in Figure 2.12 requires a smaller correction factor under the assumption that the resistor $R_x = 1/G_x = 10\,\text{k}\Omega$ with $R_A = 100\,\Omega$, and $R_V = 250\,\text{k}\Omega$? Give the correction factor of the measured value for either configuration in per cent.

2.25. The voltage division g_v of a voltage divider (consisting of two $200\,\text{k}\Omega$ resistors in series) is measured with an active voltmeter ($R_V = 10\,\text{M}\Omega$) using the 10 V range. The voltage is provided by a voltage source with negligible output impedance. The measured total voltage $v_i = 10.000\,0\,\text{V}$, the measured divided voltage $v_o = 4.950\,9\,\text{V}$ (the uncertainties of these two measurements are assumed to be negligible).

(a) Give the value for the amount of voltage division g_v corrected for the loading by the voltmeter.

(b) Explain why the (absolute) calibration of the voltmeter has no impact on the result.

2.26. A voltage divider consisting of a $10\,\text{k}\Omega$ resistor in series to a $20\,\text{k}\Omega$ resistor divides a voltage of 20 V. At the output, there is a $20\,\text{k}\Omega$ resistor in series to an ammeter represented by its (linear) input impedance R_A.

(a) Replace the linear network by a real current source (Norton's theorem).

(b) What is the maximum value for R_A to ensure a correction smaller than 1%?

(c) Give the current i_A under the condition of (b).

(d) Give the current i_A for $R_A = 0$.

Caution: The display (indicator) of a meter shows, at best, the electric power dissipated in the input of the meter. Such a result must be corrected for the burden that the meter is for the circuit. However, it cannot be corrected if the signal that the meter measures is different from what it is supposed to be. In doubtful cases, the use of an oscilloscope is unavoidable.

2.1.5 Simplifying circuit diagrams by fusing one-ports

It helps visually oriented people in understanding circuits when the assumed (or even better actual) quiescent current flows from top to bottom. Therefore, the circuit elements in most circuit diagrams should be vertically oriented. Often it pays to redraw the circuit diagram arranging the elements in such a way that it is intuitively clear in which direction the current flows. Thus, the highest positive voltage will be at the top, the lowest (negative) voltage at the bottom. In cases where no negative voltage is present the *ground*, the reference point, at which the voltage is assumed to be zero, will be at the bottom. If this point is connected to earth potential, it might be called *earth* instead.

Redundant components (e.g., elements in series to a current source or parallel to a voltage source) should be left out as well as functionless ones (e.g., one-ports with only one terminal connected). Elements of the same kind in series or in parallel may be (properly) combined to one element as long as no relevant information gets lost. Depending on the problem at hand the source can be converted to the proper kind (applying Thevenin's or Norton's theorem, respectively). If components are arranged in parallel a current source is appropriate if arranged in series a voltage source.

2.1.6 *i*–*v*-Characteristics of nonlinear one-ports

Properties of electronic components are displayed by means of their characteristics. As these properties depend on ambient conditions, these should be given as parameters. Besides, it is assumed that the component is located in free air, i.e. that the surroundings do not interact.

The most important parameter (particularly with semiconductor components) is temperature, and the frequency of the signal at which the property in question was determined. In addition, the kind and pressure of the atmosphere, the magnetic field, humidity, and other ambient properties might play a role.

Characteristics given in data sheets usually are typical ones. In that case, their variance will be given, too.

For one-ports, the *i*–*v*-characteristic, the relation between the current flowing through the component and the voltage across it, is most important. Only in special cases the characteristic can be described by an equation. Such a case is the forward characteristic of a junction diode that, over several decades of current gives an exponential dependence of the current on the voltage (Figure 2.13).

Two elements having the same electrical characteristic are electronically indistinguishable, i.e. the behave the same way.

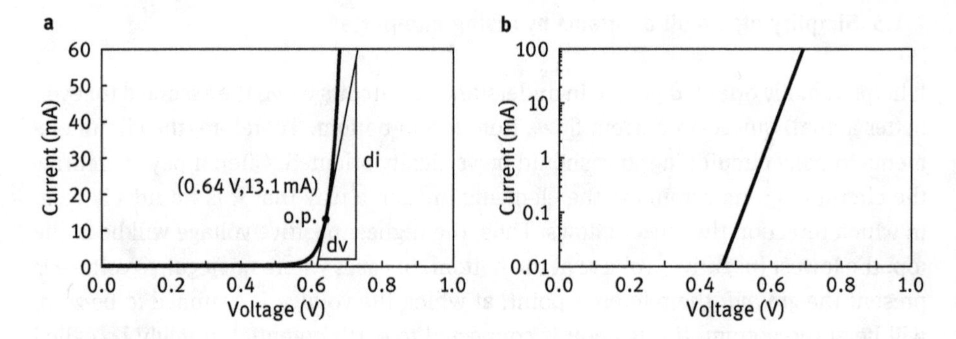

Fig. 2.13. Forward characteristic of a junction diode: (a) "normal" presentation with operating point, (b) semilogarithmic presentation.

In a circuit an electronic component will usually have a bias, i.e. a voltage–current pair constituting the *operating point*. Without signal, the voltage across the component and the current through it is given by these *quiescent* values.

Problem

2.27. Give the operating point of a passive (Section 2.2.4) electronic component that is not connected to any power supply.

In Figure 2.13a an operating point (o.p.) is shown together with the tangent through it. The slope of this tangent usually depends on the position of the operating point. The ratio $\mathrm{d}v/\mathrm{d}i$ is called (small-signal) *impedance* Z of said component

$$Z = \left.\frac{\mathrm{d}v}{\mathrm{d}i}\right|_{\text{o.p.}} . \tag{2.13}$$

In view of duality, we introduce, in addition, the ratio $\mathrm{d}i/\mathrm{d}v$ which we call (small-signal) *admittance* Y

$$Y = \left.\frac{\mathrm{d}i}{\mathrm{d}v}\right|_{\text{o.p.}} . \tag{2.14}$$

The impedance (and admittance) of *linear* components is obviously independent of the operating point so that we arrive at

$$Z = \frac{v}{i} , \quad \text{and} \quad Y = \frac{i}{v} . \tag{2.15}$$

This relation we know as Ohm's law (and its reverse) with the resistance replaced by the more general term impedance, and the conductance by admittance. In the future, we will refrain, whenever appropriate, from using the term resistance (and conductance) and will be using the more general terms impedance (and admittance) instead.

Keep in mind that

any practical *i–v*-characteristic is, exactly speaking, *nonlinear*.

An operating point may not even temporarily exceed a maximum voltage or current (Section 2.3.1). Even before these limits are reached, it is plausible that the voltage-to-current ratio of a linear device will change, i.e. that it will become nonlinear, before the device is destroyed by too high power, current, or voltage. So, components with ideal linear characteristic do not exist, they will be linear just in a limited operating range, called *small-signal operating range* (or small-signal *dynamic* range). Such a range could be as large as several hundred (even thousand) volts or as small as, say, 10 mV.

The nonlinearity of a "linear" component (e.g., resistor) is an unwanted byproduct. However, there are several nonlinear components with very useful properties. Some such *i–v*-characteristics are sketched in Figure 2.14.

In many applications, the property of a diode is best described as that of a current check valve. Current is allowed to flow in one direction, but not (or nearly so) in the other. The allowed direction is called *forward* direction indicated by the arrow in the

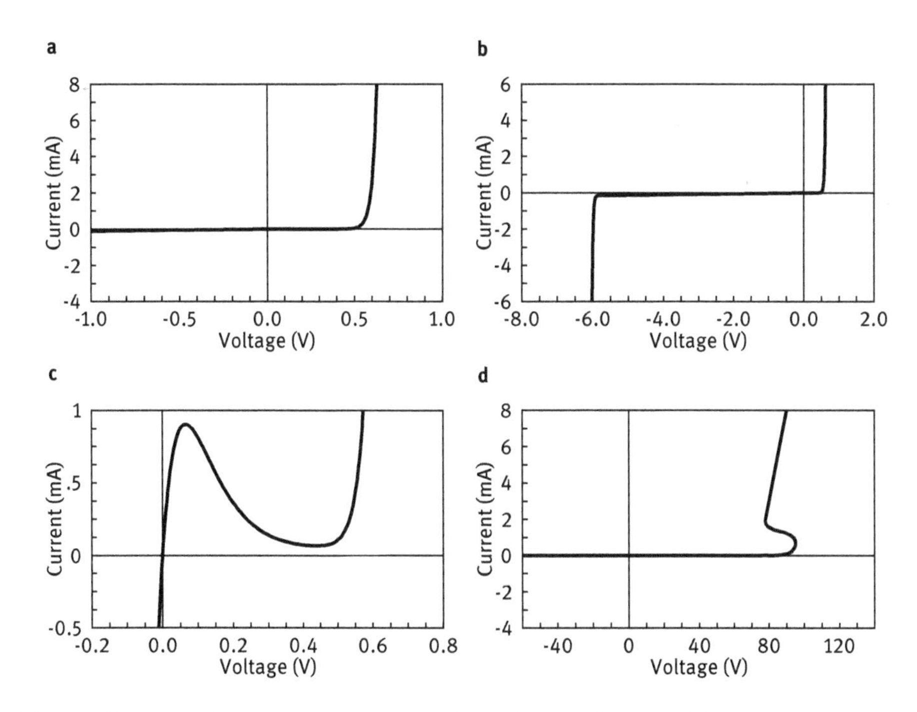

Fig. 2.14. Typical i-v-characteristics of some nonlinear elements: (a) diode, (b) Zener diode with $v_Z = 6.0$ V, (c) N-shaped, and (d) S-shaped.

symbol, the current in the *reverse* direction i_r is usually very small depending on the technology used to produce the diode.

With Zener diodes advantage is taken of the (very) low impedance occurring at a characteristic reverse voltage v_Z. If the operating point is set there, the voltage does not change much with current, as required for a voltage source. Consequently, such Zener (or breakdown) voltages can be used to establish a reference voltage for a power supply.

The two characteristics in Figures 2.14c and 2.14d have a portion with negative impedance. There

$$Z = \left.\frac{\mathrm{d}v}{\mathrm{d}i}\right|_{\text{o.p.}} < 0 \,. \tag{2.16}$$

In addition, they are dual to each other, i.e. exchanging v with i converts one into the other. Any (simple) characteristic with a portion of negative impedance must either be S- or N-shaped because the (more or less) linear region with negative impedance must be connected to the origin (0 V, 0 mA). One-ports with negative impedance may be used for the amplification of signal power (Section 2.2.5), and therefore, are called *active* one-ports (devices). There exist also dynamic negative impedances (Section 2.4.3.2), i.e. circuits in which negative impedance is brought about by (positive) feedback action (Section 2.5.4.1).

2.1.7 Linearizing one-ports (small-signal behavior)

The reasons for linearization were outlined in Section 1.3.3. Linearization is done by approximating (some part of) an i–v-characteristic by a straight line. By specifying the required closeness between the curve and its linear approximation, a range both for the voltage and the current is defined, in which this approximation is valid. Signals within this range are called small-=signals, independent of their actual size.

Any signal may be decomposed into the quiescent component (operating point values) and a variable *signal* (Section 2.1.3.5). If this variable signal is a small-signal, then one can operate with the linear impedance Z making life easy. In that case, in a model circuit the nonlinear element is replaced by Z under the tacit assumption that the bias of the circuit is such that the correct operating point is set. Thus, we are dealing with a linear circuit.

Problem

2.28. A current of 0.25 mA flows through some nonlinear element if a voltage of 0.5 V is applied. When this voltage is −10 V, the current is 0.002 mA. What is the (small-signal) impedance of this one-port?

In Figure 2.15 the characteristics of Figure 2.14 are linearized in a way that they resemble the original ones. In the case of the diode, a moderately complicated model may

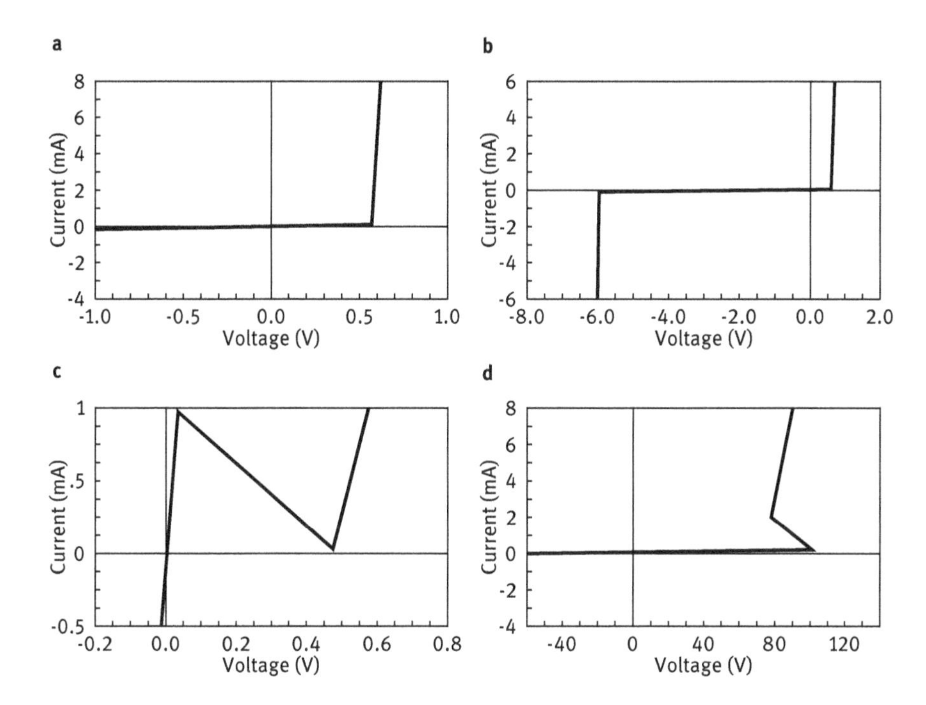

Fig. 2.15. Linearization of the characteristics of Figure 2.14.

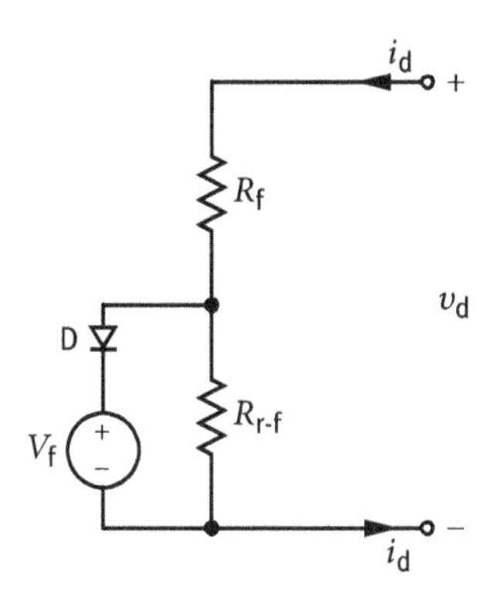

Fig. 2.16. Example of a linear model of a diode; $R_{r-f} = R_r - R_f$.

be applied using an ideal diode D, a voltage source V_f, and two resistors R_f and R_{r-f} (see Figure 2.16).

Example 2.5 (Linear model of a junction diode). A simple linear model of a junction diode needs three parameters, the forward impedance R_f, the reverse impedance R_r and the forward voltage V_f, plus an ideal current check valve, an ideal diode D. The values of these parameters vary according to the technology of the diode: $R_f = 10^0$ to $10^1\ \Omega$, $R_r = 10^5\ \Omega$ to $10^7\ \Omega$, and $V_f = 0.3$ V to 0.7 V. If the ideal diode does not conduct, it has the admittance $Y_D = 0$ S, and the voltage across the ideal diode is given by

$v_d \times (R_r - R_f)/R_r - V_f$, and the impedance of the one-port $Z_d = v_d/i_d = R_r$. For voltages $v_d \geq V_f/(1 - R_f/R_r)$ the ideal diode is conducting having an impedance of $Z_D = 0\,\Omega$ and the impedance of the one-port is $Z_d = R_f$. Thus, at the point $v_d = V_f/(1 - R_f/R_r)$ and $i_d = V_f/(R_r - R_f)$ the slope of the characteristic changes from $1/R_f$ to $1/R_r$. Observe that the voltage source used in this model cannot be used as active source. It does not supply power because the reverse-biased ideal diode prohibits the current to flow out of the source. It just sinks current maintaining its voltage.

? **Problem**

2.29. The characteristic of a Zener diode is well described by a straight line intersecting the abscissa at the reverse breakdown voltage V_Z. From the slope of this line, the (constant) small-signal impedance Z_Z (or r_Z) is obtained. With $V_Z = 5.96$ V, an operating point on the abscissa (5.96 V, 0.00 mA) is defined. With $Z_Z = 2.5\,\Omega$, the operating point at any current can be calculated within this model. Find the parameters of the small-signal model for the following circuit at the given operating point: A Zener diode with the above properties is (reverse) biased by a voltage source of $V_s = 13.5$ V and a series resistor of 390 Ω. The load resistor that shunts the Zener diode has 390 Ω, as well. Replace the linear network by a real voltage source (Thevenin's theorem) expressing the model properties by means of another real voltage source. Then calculate the current in this loop, and the voltage at the terminals of the Zener diode model yielding the actual operating point.

2.1.8 Graphical methods for dealing with nonlinear elements

It is tiresome if not impossible to deal with nonlinear characteristics. If a linearization cannot be employed there are general ways to deal with nonlinear elements if the rest of the network is linear. As a first step we apply Thevenin's theorem and replace the linear network by a real voltage source with V_S and R_S. Thus, we have the situation depicted in Figure 2.17a.

The current i_{ne} is given by the voltage difference across the source resistor R_S:

$$i_{ne} = \frac{V_S - v_{ne}}{R_S} \quad \text{or}$$

$$i_{ne} = -\frac{v_{ne}}{R_S} + \frac{V_S}{R_S}$$

which is the equation of a straight line, the so-called *load line* because R_S functions as a load for the nonlinear element. This line is easily constructed by the zero voltage point with $i_{ne0} = V_S/R_S$ and the zero current point with $v_{ne0} = V_S$. Independent of the characteristic of the nonlinear element, its operating point must lie on this load line. The actual operating point is the intersection of the (unknown) characteristic with the

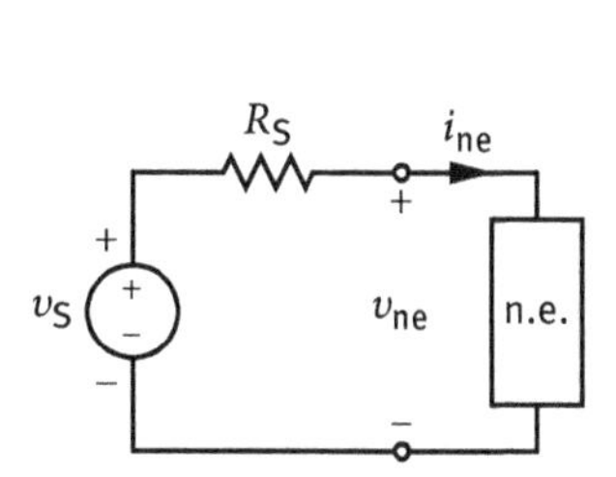

Fig. 2.17a. A nonlinear element at the output of a real voltage source.

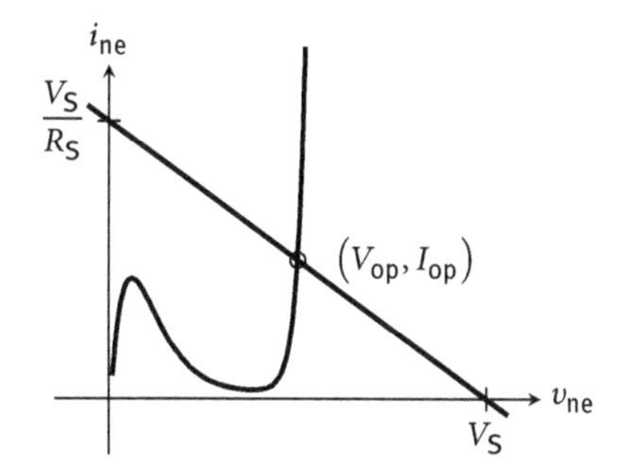

Fig. 2.17b. The operating point must lie on the load line independent of the characteristic of the nonlinear element. As an example the characteristic of a tunnel diode is shown.

load line. In Figure 2.17b the load line is shown. As an example the characteristic of a tunnel diode is given, too.

Thus, all possible operating points must lie on the load line, i.e. the two-dimensional manifold of possible operating points has been reduced to a one-dimensional one. If, in addition, the maximum allowable power (Section 2.3.1) of the nonlinear element is known, then this information can be added by way of the maximum-power hyperbola. The equation $P = I \times V$ is that of an equilateral hyperbola. If the operating point is outside the maximum-power hyperbola then the nonlinear element would be destroyed by too high a temperature (Section 2.3.1). Therefore, if the load line does not intersect the maximum-power hyperbola, one can be assured that the power dissipated in the nonlinear element will not exceed the maximum allowed power independent of the actual characteristic. As pointed out in Section 2.2.4.1 the maximum power of an operating points is in the middle of the dynamic range. There the hyperbola has its closest approach to the load line.

Problems

?

2.30. The linear network in Figure 2.18 is intended to feed any (nonlinear) one-port.
(a) Determine the maximum power that this circuit can deliver into any one-port.
(b) Draw the load line into an i–v diagram.
(c) If the characteristic of the one-port is such, that the current of the operating point is 10 mA, how large is the power dissipated in that one-port?

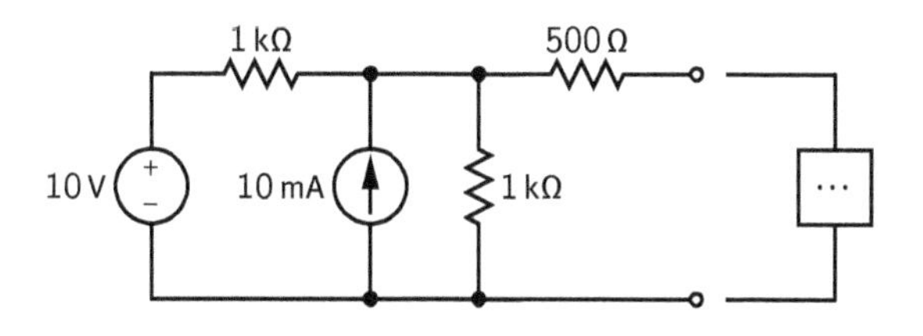

Fig. 2.18. Network of Problem 2.30.

2.31. To protect a nonlinear element with unknown characteristic but a known maximum power rating of 0.1 W a resistor of 1 kΩ is inserted between the element and the voltage source.

(a) What is the maximum allowed voltage $V_{S\,max}$ that may be applied without damaging any nonlinear element?

(b) What is the worst case maximum power dissipated in the 1 kΩ resistor?

2.2 Two-port network models

Electrical networks that modify signals can be described by two-ports having an input and an output port. The modification can embrace the amplitude or/and the time behavior. Any two-port may be used with input and output exchanged. This is reflected in the so-called *two-port convention* defining the sign of the currents and the voltages at the two-port (Figure 2.19). In this book, we use the subscripts "i" to indicate input variables and "o" to designate output variables. In some instances (e.g., two-port matrices) the alternative method of using subscripts 1 and 2, respectively, makes more sense.

The symmetry between (assumed) output and input is important. In many cases there will be an obvious (natural) input/output.

Two-ports may be arranged in a chain of two or more (even an infinite number of) two-ports connected in a *cascade* (with each output port connected to the input port of the following).

Problem

2.32. There will be a step-up transformer feeding a high-voltage transmission line and a step-down transformer to feed the utilities. Viewing transformers as two-ports, is it possible to recognize which port is the natural input port?

2.2.1 Active two-ports (dependent signal sources)

Two-ports in which the output signal can contain more signal power than is dissipated in the input are called *active* two-ports. Such *amplification* can be symbolized by the

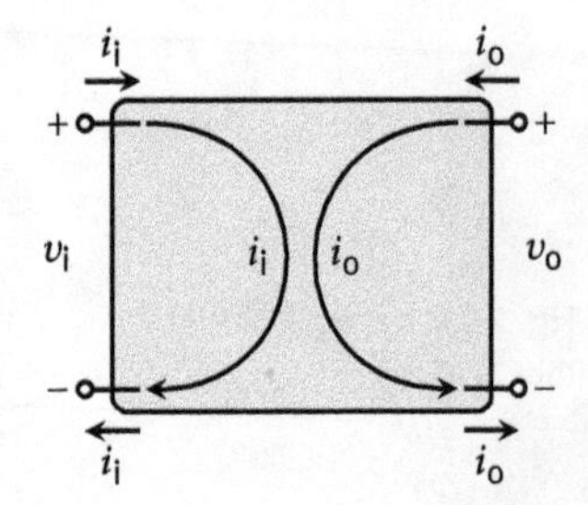

Fig. 2.19. Definition of the signs of the electrical signals at a two-port.

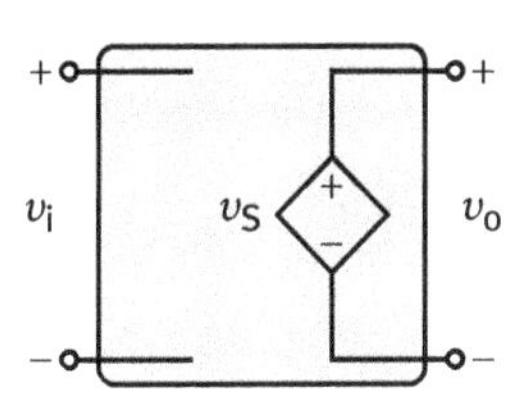

Fig. 2.20a. Voltage-controlled voltage source with $v_S = f(v_i)$.

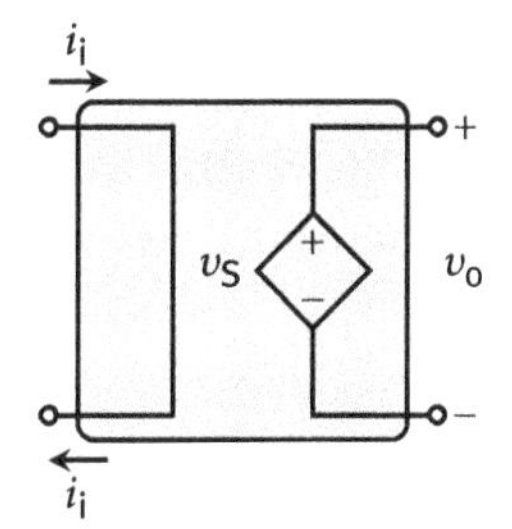

Fig. 2.20b. Current-controlled voltage source with $v_S = f(i_i)$.

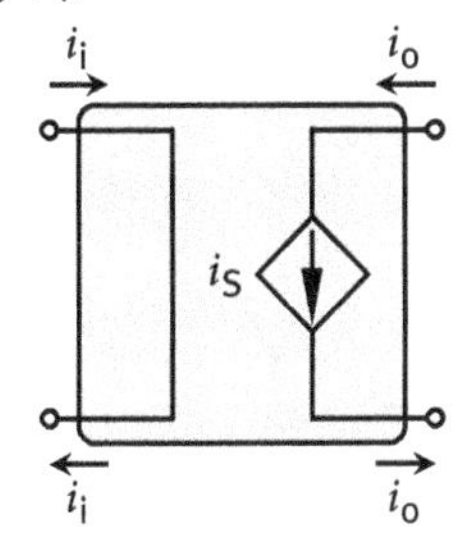

Fig. 2.20c. Current-controlled current source with $i_S = f(i_i)$.

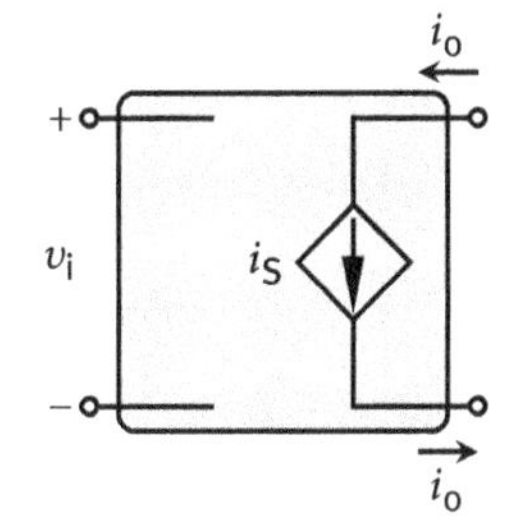

Fig. 2.20d. Voltage-controlled current source with $i_S = f(v_i)$.

use of a *controlled* electric source inside the two-port. Figure 2.20 summarizes the four types of controlled sources.

By applying repeatedly the principles of duality on a voltage-controlled voltage source with $v_S = f(v_i)$ the three other types are easily arrived at: voltage-controlled current source with $i_S = f(v_i)$, current-controlled current source with $i_S = f(i_i)$, current-controlled voltage source with $v_S = f(i_i)$, thus defining the four (forward) transfer parameters:

- voltage gain $g_v = v_o/v_i$,
- transadmittance (mutual conductance) $g_m = i_o/v_i$,
- current gain $g_i = i_o/i_i$, and
- transimpedance (mutual resistance) $r_m = v_o/i_i$.

In view of the basic symmetry of a two-port, there will be *two* controlled sources in any practical two-port. One for the forward transfer, and one for the reverse transfer, according to the choice of the input port. The reverse transfer takes care of the *internal feedback* of the output signal to the input signal.

2.2.2 Small-signal models of two-ports

To describe voltage and current both at the input and output, one needs four independent parameters. The best presentation depends on the application. Depending on the type of transfer parameter there is a choice of four linear models of two-ports. Of course, the parameters of one model can be converted into those of any other model. When connected to other two-ports, there is an optimum model depending on the configuration. (Remember: Instead of the term resistor we use the more general word impedance, and instead of conductor the more general term admittance.)

2.2.2.1 Forward transfer is voltage gain (g-parameters)

The relation between the four variables of such a two-port (Figure 2.21) is given by

$$i_\mathrm{i} = g_\mathrm{i} v_\mathrm{i} + g_\mathrm{r} i_\mathrm{o} \tag{2.17a}$$

$$v_\mathrm{o} = g_\mathrm{f} v_\mathrm{i} + g_\mathrm{o} i_\mathrm{o} \tag{2.17b}$$

with

$$g_\mathrm{i} = \left.\frac{i_\mathrm{i}}{v_\mathrm{i}}\right|_{i_\mathrm{o}=0} \qquad g_\mathrm{r} = \left.\frac{i_\mathrm{i}}{i_\mathrm{o}}\right|_{v_\mathrm{i}=0} \tag{2.18a}$$

$$g_\mathrm{f} = \left.\frac{v_\mathrm{o}}{v_\mathrm{i}}\right|_{i_\mathrm{o}=0} \qquad g_\mathrm{o} = \left.\frac{v_\mathrm{o}}{i_\mathrm{o}}\right|_{v_\mathrm{i}=0} \tag{2.18b}$$

where

g_i is the input admittance with an open-circuit at the output,
g_r is the reverse current gain with a short-circuit at the input,
g_f is the forward voltage gain with an open-circuit at the output,
g_o is the output impedance when there is a short-circuit at the input.

Observe that each of the parameters has a different dimension because the fixed parameter for the input variables is the output current whereas it is the input voltage for

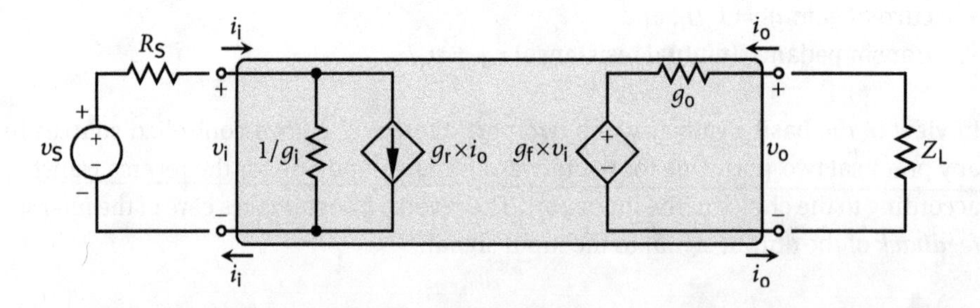

Fig. 2.21. Inverse hybrid parameter two-port with an independent voltage source v_S, source resistance R_S at the input and load impedance Z_L at the output as well as two dependent sources describing the forward and reverse transfer of the two-port.

the output variables. This mixing is reflected in the term hybrid. The full name of this parameter choice is *inverse hybrid parameters* because of the *reverse* current gain.

The g-parameters are the optimum option of two-port parameters when two two-ports are connected in a parallel–series configuration as done in parallel–series feedback (Section 2.4.3).

Problem

?

2.33. A two-port described by the following g-parameters is loaded by a 10 kΩ resistor: $g_i = 1\,\text{mS}$, $g_r = 0$, $g_f = 110$, and $g_o = 100\,\text{k}\Omega$.
(a) Determine the current gain.
(b) Give the output impedance.

2.2.2.2 Forward transfer is transadmittance (y-parameters)

The relation between the four variables of such a two-port (Figure 2.22) is given by

$$i_i = y_i v_i + y_r v_o \tag{2.19a}$$

$$i_o = y_f v_i + y_o v_o \tag{2.19b}$$

with

$$y_i = \left.\frac{i_i}{v_i}\right|_{v_o=0} \qquad y_r = \left.\frac{i_i}{v_o}\right|_{v_i=0} \tag{2.20a}$$

$$y_f = \left.\frac{i_o}{v_i}\right|_{v_o=0} \qquad y_o = \left.\frac{i_o}{v_o}\right|_{v_i=0} \tag{2.20b}$$

where
y_i is the input admittance when there is a short-circuit at the output,
y_r is the reverse transadmittance with a short-circuit at the input,
y_f is the forward transadmittance with a short-circuit at the output,
y_o is the output admittance when there is a short-circuit at the input.

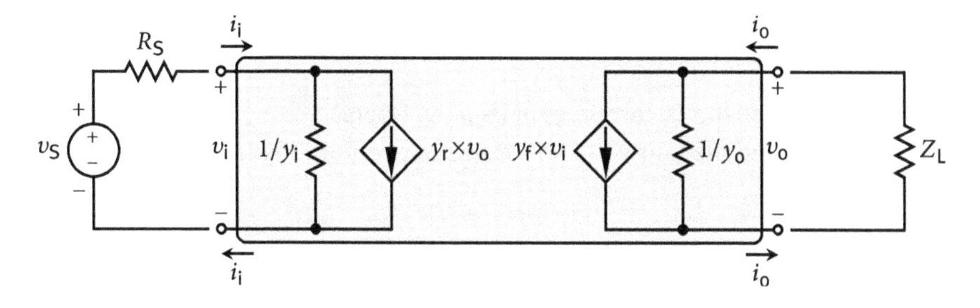

Fig. 2.22. Admittance-parameter two-port with an independent real voltage source v_S at the input and load impedance Z_L at the output as well as two dependent current sources describing the forward and reverse transfer of the two-port.

All y-parameters are admittances with the unit 1 S. The y-parameters are the optimum option of two-port parameters when two two-ports are connected in a parallel–parallel configuration as used in parallel–parallel feedback (Section 2.4.3).

Problem

2.34. A two-port described by the following y-parameters is loaded by a 100 kΩ resistor: $y_i = 1\,\text{mS}$, $y_r = 0\,\text{mS}$, $y_f = 1\,\text{mS}$, $y_o = 0.01\,\text{mS}$.
(a) Determine the voltage gain.
(b) Give the output impedance.

2.2.2.3 Forward transfer is transimpedance (z-parameters)

The relation between the four variables of such a two-port (Figure 2.23) is given by

$$v_i = z_i i_i + z_r i_o \tag{2.21a}$$

$$v_o = z_f i_i + z_o i_o \tag{2.21b}$$

with

$$z_i = \left.\frac{v_i}{i_i}\right|_{i_o=0} \qquad z_r = \left.\frac{v_i}{i_o}\right|_{i_i=0} \tag{2.22a}$$

$$z_f = \left.\frac{v_o}{i_i}\right|_{i_o=0} \qquad z_o = \left.\frac{v_o}{i_o}\right|_{i_i=0} \tag{2.22b}$$

where
z_i is the input impedance with open-circuit output,
z_r is the reverse transimpedance with open-circuit input,
z_f is the forward transimpedance with open-circuit output,
z_o is the output impedance with open-circuit input.

All z-parameters are impedances with the unit 1 Ω. The z-parameters are the optimum option of two-port parameters when two two-ports are connected in a series–series configuration as used in series–series feedback (Section 2.4.3).

2.2.2.4 Forward transfer is current gain (h-parameters)

The relation between the four variables of such a two-port (Figure 2.24) is given by

$$v_i = h_i i_i + h_r v_o \tag{2.23a}$$

$$i_o = h_f i_i + h_o v_o \tag{2.23b}$$

with

$$h_i = \left.\frac{v_i}{i_i}\right|_{v_o=0} \qquad h_r = \left.\frac{v_i}{v_o}\right|_{i_i=0} \tag{2.24a}$$

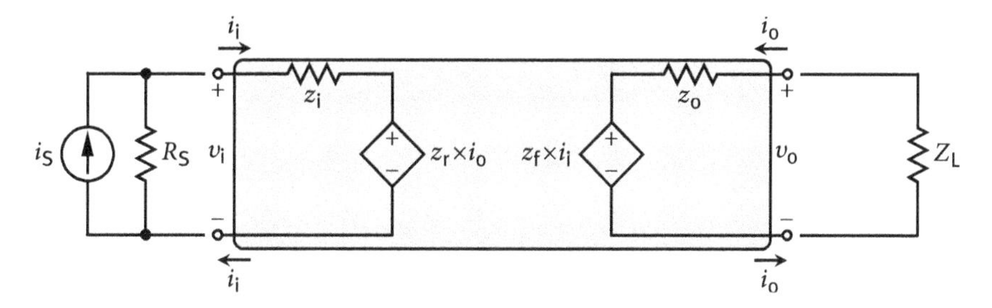

Fig. 2.23. Impedance parameter two-port showing an independent real current source i_S at the input and the load impedance Z_L at the output as well as two dependent voltage sources describing the forward and reverse transfer of the two-port.

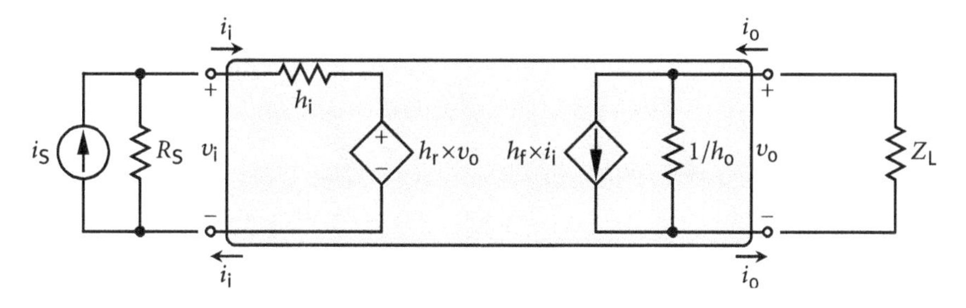

Fig. 2.24. Hybrid-parameter two-port with an independent real current source i_S, load impedance Z_L, and two dependent sources.

$$h_f = \left. \frac{i_o}{i_i} \right|_{v_o=0} \qquad h_o = \left. \frac{i_o}{v_o} \right|_{i_i=0} \tag{2.24b}$$

where
h_i is the input impedance when there is a short-circuit at the output,
h_r is the reverse voltage gain with an open-circuit at the input,
h_f is the forward current gain with a short-circuit at the output,
h_o is the output admittance with open-circuit input.

Observe that the mixing of the independent variables (voltage at the input, current at the output) results in the name *hybrid parameters* (characterized by the *forward* current gain). The h-parameters are the optimum option of two-port parameters when two two-ports are connected in a series–parallel configuration as used in series–parallel feedback (Section 2.4.3).

2.2.2.5 Summary

Which kind of parameter family should be chosen, depends on the configuration of the other two-ports or on the kind of transfer parameter that is desired. Alone from the fact that the input parameters are given under the condition $v_o = 0$ (short-circuit at the output), or $i_o = 0$ (open-circuit at the output) it is evident

that the input properties depend on the load impedance. In analogy the output properties depend on the source impedance.

2.2.3 Transforming a two-port into a one-port

The characteristic property of a one-port is just one parameter, its impedance, relating a change in current to its change in voltage. However, linear two-ports are described by two coupled linear equations, relating input voltage and current to output voltage and current.

However, when loading the output of a two-port with the impedance Z_L, the relation between v_o and i_o

$$v_o = Z_L \times (-i_o) \tag{2.25}$$

reduces the number of free parameters to one. Therefore, the situation at the input can be expressed by the input impedance Z_i which depends on the four two-port parameters and Z_L. Using the z-parameters (that are appropriate for this case) the following relation is easily obtained from Figure 2.23:

$$v_i = Z_i \times i_i = \left(z_i - \frac{z_r \times z_f}{Z_L + z_o} \right) \times i_i \,. \tag{2.26}$$

Because Z_i depends on gain (the transimpedances z_f and z_r), it is a *dynamic impedance*. We will deal some more with dynamic impedances in Section 2.4.3.2.

2.2.4 Amplification (power gain)

Sloppily, the transfer parameters are called *gain*. However, it is difficult to visualize transimpedance and transadmittance as "gain" because gain is usually thought to be a dimensionless ratio. This leads us back to the beginning of Chapter 1. There we stressed the importance of power over current and voltage. Consequently, what counts is power gain g_p which can be calculated using any of the four transfer parameters. Only amplifiers that can provide power gain are of any value.

Power gain is a measure of signal transfer from the two-port input to its output. Therefore, it is a function of its parameters and its load Z_L. It is given by the ratio of power p_L dissipated in the circuit load over the power p_i dissipated in the input of the circuit

$$g_p = \frac{p_L}{p_i} \,. \tag{2.27}$$

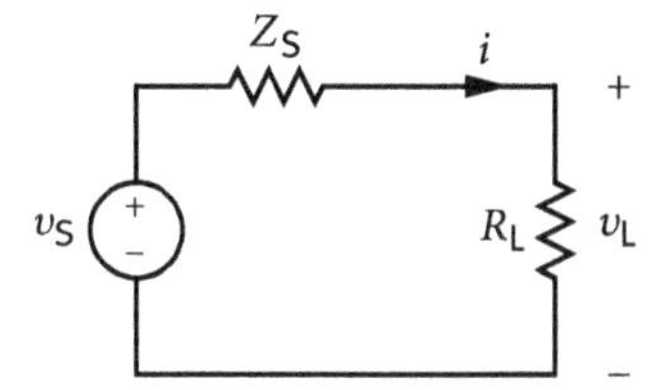

Fig. 2.25. A load in series to a real voltage source.

It is obviously dependent on Z_L and independent of the source impedance Z_S. A two-port where g_p cannot exceed unity is called a *passive element*. Only *active elements* can provide power amplification, i.e. $g_p > 1$. Applying Ohm's law we arrive at the following relation for the power:

$$p = v \times i = i \times R \times i = i^2 \times R \tag{2.28}$$

and for the power gain

$$g_p = \frac{p_L}{p_i} = \frac{i_o^2 \times Z_L}{i_i^2 \times Z_i} = g_i^2 \times \frac{Z_L}{Z_i} . \tag{2.29}$$

The current gain g_i is readily obtained from (2.21b) and Figure 2.23 as

$$g_i = \frac{i_o}{i_i} = -\frac{z_f}{z_o + Z_L} \tag{2.30}$$

yielding with (2.26)

$$g_p = \frac{z_f^2}{z_o + Z_L} \times \frac{Z_L}{z_i \times (z_o + Z_L) - z_r z_f} . \tag{2.31}$$

An amplifier does not amplify power, but amplification is a controlled conversion of power from a power supply into signal power.

2.2.4.1 Optimum power transfer (power matching)

Under which condition will there be the maximum power transfer from a source with a given source impedance? Figure 2.25 illustrates the situation.
The load power in the circuit in Figure 2.25 is given by

$$p_L = v_L \times i = v_L \times \frac{v_S - v_L}{Z_S} = Y_S \times (v_L \times v_S - v_L^2) \tag{2.32}$$

With partial differentiation one gets

$$\frac{\partial p_L}{\partial v_L} = Y_S \times (v_S - 2v_L) \tag{2.33}$$

From equating the result with zero

$$v_S - 2v_L = 0 \tag{2.34}$$

the final result for maximum power transfer is obtained:

$$v_L = \frac{v_S}{2} \tag{2.35}$$

This result is called *power matching* because the power in the source impedance equals that in the load. If the load impedance and the source impedance are resistors, the maximum power transfer from the source to the load is accomplished if the values of the two impedances are equal. In such a case, the term *impedance matching* is justified.

Problems

2.35. Power matching

(a) Find in Figure 2.26 the current through R_x in dependence of its value.

(b) At which value of R_x is the dissipated power at its maximum and what is this maximum power?

2.36. Power matching

(a) Determine in Figure 2.27 the value of R_x so that its dissipated power is maximal.

(b) What is the maximum power that can be delivered by this circuit?

2.37. The maximum power delivered into a one-port occurs when half of the source voltage lies across the one-port. What is the maximum power p_{Smax} that can occur across the source resistance, i.e., for any load, when compared to the maximum power of the load one-port p_{Lmax}?

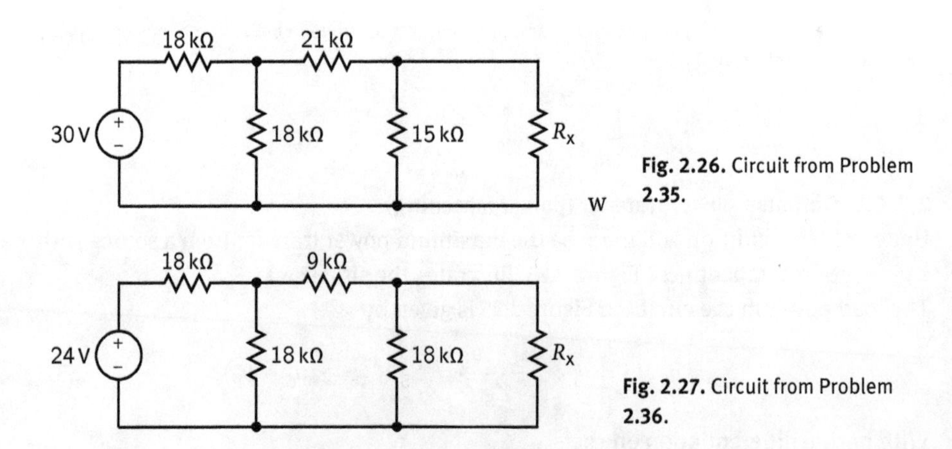

Fig. 2.26. Circuit from Problem 2.35.

Fig. 2.27. Circuit from Problem 2.36.

2.2.4.2 Amplification over several stages

Cascading two or more amplifiers makes it necessary to take a multiple product of the power gains. This multiple product can be converted into a multiple sum by taking logarithms of the power values. Consequently, a logarithmic measure of power gain was introduced, the (dimensionless) Bel (B)

$$g_p = \log \frac{p_L}{p_i} \text{ (B)} . \tag{2.36}$$

This measure is not convenient for smaller gains so that normally the 10 times smaller unit deciBel (dB) is used.

$$g_p = 10 \times \log \frac{p_L}{p_i} \text{ (dB)} \tag{2.37}$$

The unit dB is also used to quantify current and voltage gains. To conform to the definition of power gain, it is necessary to take into account the quadratic dependence of power on current (2.28) or voltage. This quadratic dependence introduces a factor of 2 in the logarithmic presentation, i.e.

$$g_v = 20 \times \log \frac{v_L}{v_i} \text{ (dB)} \tag{2.38}$$

The same formalism is used for attenuation (gain$<$ 1) which can be recognized by the negative sign. A (power) attenuation to 1/2 is a -3 dB attenuation ($\log(2.0) = 0.303\ldots$). However, voltage attenuation to 1/2 is a -6 dB (voltage) attenuation. Table 2.2 might come handy when converting ratios to its logarithmic equivalent (in dB).

Table 2.2. Potentially helpful numbers when dealing with gains.

Logarithmic gain (dB)	p_L/p_i	v_L/v_i i_L/i_i
40	10 000	100
20	100	10
10	10	$\approx$ 3.16
6	$\approx$ 4	$\approx$ 2
3	$\approx$ 2	$\approx$ 1.41
1	$\approx$ 1.26	$\approx$ 1.12
0	1	1
-1	$\approx$ 0.79	$\approx$ 0.89
-3	$\approx$ 0.5	$\approx$ 0.71
-6	$\approx$ 0.25	$\approx$ 0.5
-10	0.1	$\approx$ 0.32
-20	0.01	0.1
-40	0.0001	0.01

Problem

2.38. A voltage attenuation is given by −3 dB. What fraction is that?

2.2.5 One-port used as two-port

A linear one-port element with negative impedance $-Z$ is placed into a two-port. This two-port has a load impedance of Z_L. Its input signal comes from a linear network that, according to Norton's theorem, has been replaced by a real current source with a source admittance Y_S. At this point, we switch from impedances to admittances and arrive at the circuit of Figure 2.28 where the signal current source is shunted by the source admittance Y_S, the admittance of the component with negative impedance $-Y$, and the admittance of the load Y_L. The maximum power a source can deliver to a load is with admittance matching (Section 2.2.4.1). Thus, the maximum power is

$$p_{L,max} = \frac{i_S^2}{Y_S} \times \frac{1}{4} \tag{2.39}$$

This is the maximum available power from the generator. The actual power depends on the current division due to the value of Y_L

$$p_L = \frac{i_L^2}{Y_L} = i_S^2 \times \frac{Y_L}{(Y_S + Y_L)^2} \tag{2.40}$$

By considering the shunting negative admittance, the current division changes affecting the denominator of the above equation to

$$p_L = \frac{i_L^2}{Y_L} = i_S^2 \times \frac{Y_L}{(Y_S + Y_L - Y)^2} \tag{2.41}$$

By dividing this power by the maximum power that is available from the source, we arrive at a power gain g_p of

$$g_p = \frac{p_L}{p_{L,max}} = \frac{4 \times Y_S \times Y_L}{(Y_S + Y_L - Y)^2} \tag{2.42}$$

If the sum of the positive admittances is made equal to the magnitude of the negative admittance $-Y$ the denominator becomes zero and the power gain infinite. The degree

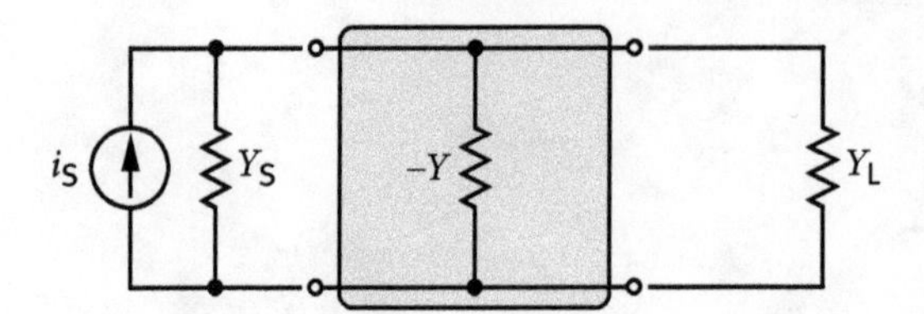

Fig. 2.28. One-port with negative impedance used as a two-port.

of agreement between the sum $Y_S + Y_L$ and $|Y|$ will determine the gain. For $|Y| > Y_S + Y_L$, the power gain would be negative indicating that the circuit is unstable under this condition. At an operating point with smaller absolute admittance $|Y|$ this instability is circumvented. Due to the nonlinearity of the i-v-characteristic, such an operating point will always be available.

Elements that can provide power gain are called *active elements*. As we could show, one-ports qualify if they contain negative impedance. Practical devices with negative impedance are gas discharge devices (glow lamps), and semiconductor components (e.g., the old-fashioned tunnel diode). Clearly, there is no source of electric power in such one-ports. As signal amplification means conversion of some electric power into signal power (Section 2.2.4) there is no need of such a power source.

Problem

?

2.39. In the circuit of Figure 2.28 the source impedance is 600 Ω and the load Z_L is a resistor of 100 Ω.

(a) What is the minimum absolute value of the negative impedance of the active element to ensure that the power gain $g_p > 1$?

(b) Give the value of the voltage transfer g_v under above condition.

2.2.6 Three-terminal element used as two-port

Traditional active elements:
- vacuum triode,
- bipolar junction transistor (BJT),
- junction-gate field-effect transistor (JFET), and
- metal-oxide-semiconductor field-effect transistor (MOSFET)

are three-terminal elements. When used as amplifier they, inevitably, become two-ports with input and output port. To investigate how this is achieved we need to use one of them as an example. At the risk of being coined "old-fashioned," we decided on the bipolar junction transistor because its properties are far off ideal ones, so that all four parameters must be used for its description, i.e. none of them may be disregarded.

There are six ways of arranging a three-terminal device in a two-port. Taking into account that one of the three terminals is the "natural" input lead that must be connected to the input there remain three qualified configurations to be considered. Each of these configurations is named after that electrode of the three-terminal device that is common to input and output, e.g., for a bipolar junction transistor
- common-emitter circuit,
- common-base circuit, and
- common-collector circuit.

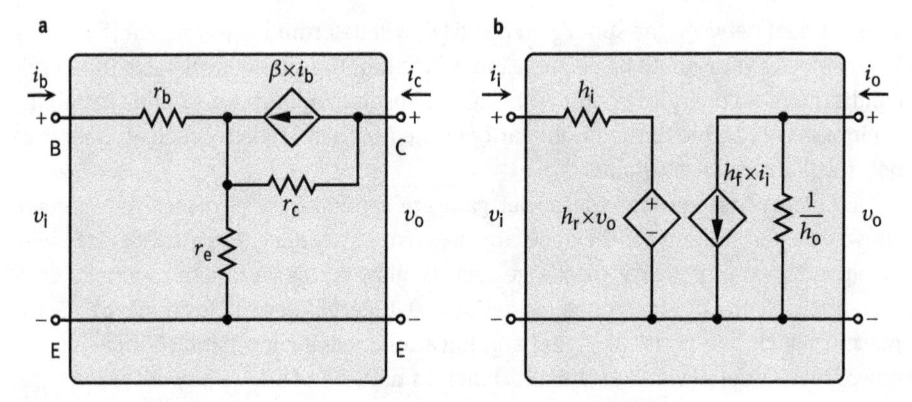

Fig. 2.29. The T-model of a three-terminal device placed into a two-port (a) collated with a two-port using h-parameters (b).

Actually, only the configuration with the highest power gain must be considered (e.g., the common-emitter circuit, Figure 2.31). The other two can be understood as feedback circuits, as will be discussed in detail later (Section 2.4.3.5). Nowadays, three terminal devices are only used in special applications (e.g., high-frequency oscillators, Section 4.2.1.7) because using ready made integrated amplifiers is in most cases more convenient, cheaper and gives more reliable results.

The appropriate model of a three-terminal device is the T-model. Figure 2.29a shows how a small-signal T-model of a bipolar transistor with its four components (parameters) placed into a two-port. One input terminal is connected to one output terminal, establishing a *ground* terminal. This fact gives rise to alternative names, e.g., grounded-emitter circuit instead of common-emitter circuit. Figure 2.29b shows a two-port with its four h-parameters. The dependent source in the T-model is a current-controlled current source. Consequently, the hybrid parameter presentation of the two-port is chosen because the characteristic transfer parameter h_f is the short-circuit current gain. The current i_b into the electrode B (which is the natural input) controls the current source ($\beta \times i_b$). It is at the same time the input current i_i of the two-port. The electrode common to output and input is E, it would be called *common-E circuit*. The h-parameters are obtained from the four T-model parameters as

$$h_i = r_b + r_e \frac{1 + \beta}{1 + r_e/r_c} \tag{2.43}$$

$$h_r = \frac{r_e}{r_c + r_e} \tag{2.44}$$

$$h_f = \frac{\beta r_c - r_e}{r_c + r_e} \tag{2.45}$$

$$h_o = \frac{1}{r_c + r_e} \, . \tag{2.46}$$

On the other hand, the four T-parameters can be expressed by the h-parameters as follows:

$$r_\mathrm{e} = \frac{h_\mathrm{r}}{h_\mathrm{o}} \tag{2.47}$$

$$r_\mathrm{b} = h_\mathrm{i} - \frac{h_\mathrm{r} \times (h_\mathrm{f} + 1)}{h_\mathrm{o}} \tag{2.48}$$

$$r_\mathrm{c} = \frac{1 - h_\mathrm{r}}{h_\mathrm{o}} \tag{2.49}$$

$$\beta = \frac{h_\mathrm{r} + h_\mathrm{f}}{1 - h_\mathrm{r}} \tag{2.50}$$

$$\tag{2.51}$$

From Figure 2.29 we get

$$v_\mathrm{i} = r_\mathrm{b} i_\mathrm{i} + r_\mathrm{e} \times (i_\mathrm{i} + i_\mathrm{o}) , \tag{2.52}$$

and

$$v_\mathrm{o} = r_\mathrm{c} \times (i_\mathrm{o} - \beta i_\mathrm{i}) + r_\mathrm{e} \times (i_\mathrm{i} + i_\mathrm{o}) . \tag{2.53}$$

At the output we have

$$v_\mathrm{o} = -i_\mathrm{o} \times Z_\mathrm{L} , \tag{2.54}$$

and for a signal applied to the output (with $v_\mathrm{S} = 0$)

$$v_\mathrm{i} = -i_\mathrm{i} \times Z_\mathrm{S} . \tag{2.55}$$

From (2.53) and (2.54) we get for the current gain g_i of the two-port burdened with an impedance Z_L

$$\begin{aligned} g_i &= \frac{i_\mathrm{o}}{i_\mathrm{i}} \\ &= -\frac{r_\mathrm{e} - \beta r_\mathrm{c}}{Z_\mathrm{L} + r_\mathrm{e} + r_\mathrm{c}} \\ &= \frac{\beta}{1 + \dfrac{r_\mathrm{e} + Z_\mathrm{L}}{r_\mathrm{c}}} - \frac{1}{1 + \dfrac{r_\mathrm{c} + Z_\mathrm{L}}{r_\mathrm{e}}} . \end{aligned} \tag{2.56}$$

With $r_\mathrm{c} \gg r_\mathrm{e}$ one gets $g_i \approx \beta$ for very small Z_L (short-circuit), and $g_i = 0$ for obvious reasons, when the load is an open-circuit.

The input impedance Z_i is obtained from (2.52) and (2.56) as follows:

$$Z_i = \frac{v_i}{i_i}$$
$$= r_b + r_e \times (1 + g_i) \tag{2.57}$$
$$= r_b + r_e \times \frac{(\beta + 1) \times r_c + Z_L}{Z_L + r_e + r_c}$$
$$= r_b + r_e \times \left(\frac{\beta + 1}{1 + \frac{r_e + Z_L}{r_c}} + \frac{1}{1 + \frac{r_c + r_e}{Z_L}} \right) \tag{2.58}$$

For an open-circuit at the output, this equation degenerates to $Z_i = r_b + r_e$ as can be easily recognized by looking at Figure 2.29a. For a short-circuit one gets

$$Z_i = r_b + r_e \times \frac{\beta + 1}{1 + \frac{r_e}{r_c}} \tag{2.59}$$

and with $r_c \gg r_e$ it is $Z_i \approx r_b + r_e \times (\beta + 1)$. Again the latter result is obvious when scrutinizing Figure 2.29a. The input impedance is the base resistor r_b in series to the *dynamic impedance* of the emitter resistance $r_e \times (i_o + i_i)/i_i = r_e \times (g_i + 1)$ (As mentioned before, impedances depending on gain are called dynamic impedances, Section 2.4.3.2).

The output impedance Z_o is obtained from (2.52) and (2.53) as

$$Z_o = \frac{r_e(r_b + Z_S)}{r_e + (r_b + Z_S)} + r_c \times \frac{r_e(1 + \beta) + (r_b + Z_S)}{r_e + (r_b + Z_S)} . \tag{2.60}$$

An open-circuit input does not allow any input current to flow. Consequently, the output impedance degenerates to a series connection of r_c and r_e. No dynamic action is involved. A short-circuit input reduces the impedance to r_c.

Due to its small value we disregard the reverse transfer, the reverse voltage gain (h_r). Observe that the forward parameters (Z_i and g_i) depend on the load impedance, whereas, the backward parameters (Z_o and $g_{v\mathrm{rev}}$) depend on the source impedance.

Two-ports are transparent in either direction – they do not isolate input from output.

Figure 2.30 shows the dependence of input impedance, current gain, voltage gain, and power gain on the load impedance and the dependence of the output admittance on the source admittance. For completeness, the corresponding dependences for the common-collector and common-base configurations are included, as well, even if these two configuration can be explained by feedback action as will be shown later (Section 2.4.3.5).

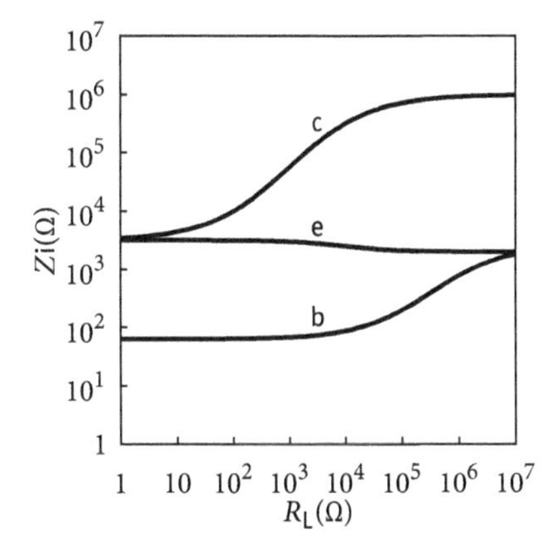

Fig. 2.30a. Dependence of input impedance on the load impedance for the common emitter, base and collector circuits.

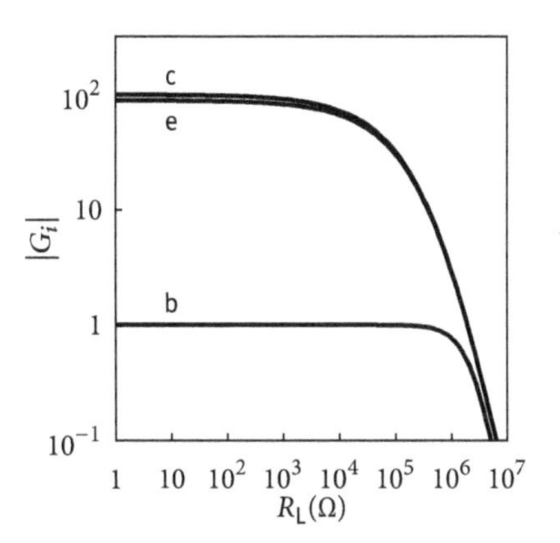

Fig. 2.30b. Dependence of the absolute values of current gain on the load impedance for the common emitter, base, and collector circuits.

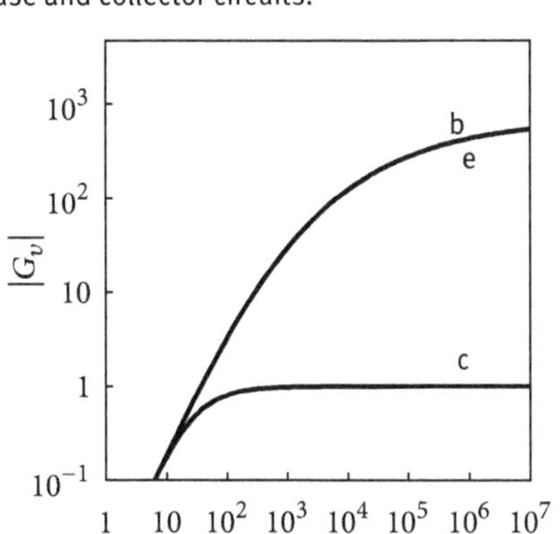

Fig. 2.30c. Dependence of the absolute values of voltage gain on the load impedance for the common emitter, base, and collector circuits.

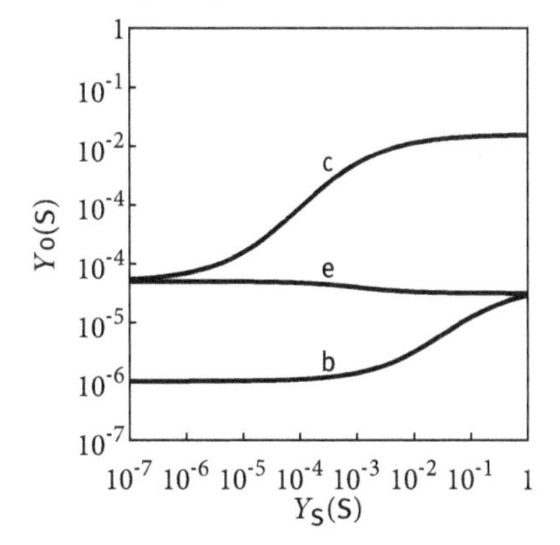

Fig. 2.30d. Dependence of output admittance on the source admittance for the common emitter, base, and collector circuits.

Some features of these figures deserve closer scrutiny. In the case of the common-emitter circuit, the output admittance is practically independent of the source admittance, and the input impedance is independent of the load impedance.

In the case of the common-collector circuit the output admittance is, in a central region, proportional to the source admittance and the input impedance is, in a central region, proportional to the load impedance. The voltage gain g_{vc} of the common-collector circuit is for small load impedances $g_{vc} = g_{ve}/(1 - g_{ve}) \approx -1$ because of the negative feedback action (Section 2.4.3.5) with a closed-loop gain of g_{ve}. The current gain g_{ic} of the common-collector circuit is for geometric reasons for not too small load admittances $g_{ic} = -(g_{ie} + 1)$.

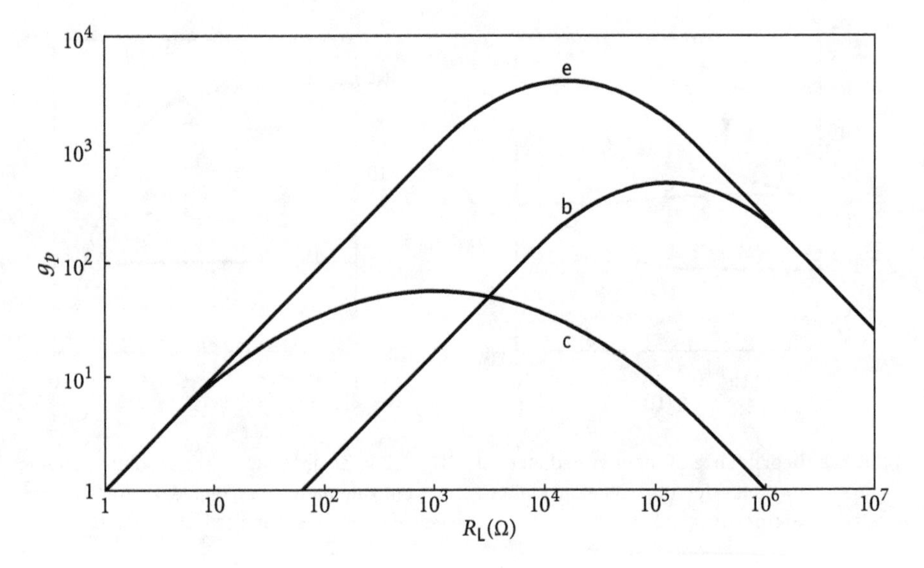

Fig. 2.31. Dependence of power gain on the load impedance for the common emitter, base, and collector circuits.

Applying duality we find that the current gain g_{ib} of the common-base circuit is for small load impedances $g_{ib} = g_{ie}/(1 - g_{ie}) \approx -1$ because of the negative feedback action with a closed-loop gain of g_{ie}. The voltage gain g_{vb} of the common-base circuit is for geometric reasons for not too small load resistances $g_{vb} = -(g_{ve}+1)$. And finally, the common-emitter circuit offers for all burdens the highest power gain (Figure 2.31). This feature qualifies it to be the basic transistor circuit, with the other two just local feedback variants. Consequently, we do not spend any special effort with the common-base and common-collector circuits but refer to Section 2.4.3.5 and later.

Using any three-terminal component of the other technologies gives, qualitatively, the same answer, i.e, the common-cathode and the common-source circuits are basic, the other two in each family can be explained by feedback action.

2.2.7 Passive two-ports

A two-port that does not contain an active element is a passive two-port. Passive two-ports cannot provide power gain. In Section 2.1.3, we arranged resistors in series (and in parallel) to discuss voltage (current) division. Such dividers as shown in Figure 2.32 are the simplest passive two-ports with more than one component.

Whereas active two-ports, in general, are unidirectional, i.e. it is self-evident which port is the input port, passive two-ports, usually, are bidirectional. In addition, the feed-through from the output to the input can be very severe. Therefore, it is wise

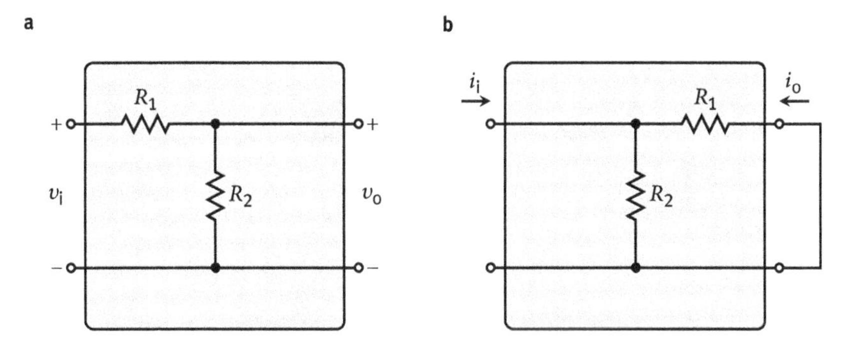

Fig. 2.32. (a) Voltage divider and (b) current divider presented as two-ports.

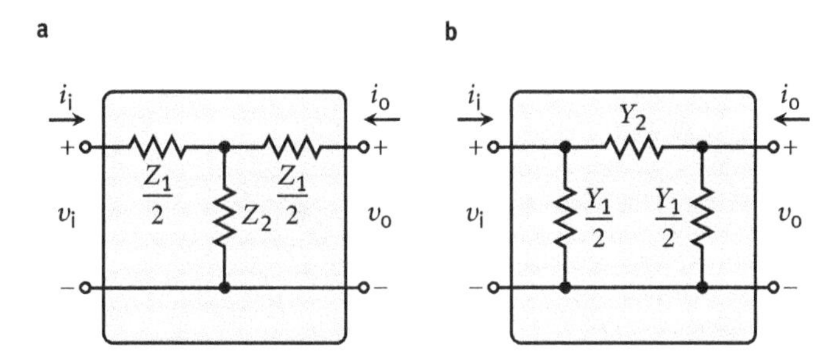

Fig. 2.33. (a) T- and (b) Π-section of attenuators.

to investigate the property of a passive two-port with actual burden, i.e. to include the load into all considerations.

A voltage divider inside a two-port is called *attenuator*. Returning to our statement that electric power is the ultimate electrical variable of interest it should not surprise that attenuators, usually, are made to divide (attenuate) electric power, which is, of course, accompanied by voltage (and current) division.

2.2.7.1 Attenuators

If the transfer through a two-port has a power gain of less than 1, this is called *attenuation* or *loss*. Obviously, attenuation can be achieved with passive devices. In Section 2.1.3.4, we have arranged two dissipative one-ports in series to form a voltage divider network. Attenuators are needed when the source signal is too large to be properly handled by the consecutive device. Presently we place the two impedances into a two-port as shown in Figure 2.32a which is also called *L-section* of an attenuator. Without load its voltage attenuation equals that of the *T-section* (with $Z_1/2 = R_1$ and $Z_2 = R_2$) shown in Figure 2.33a in which another impedance is added to make the two-port symmetric.

A Π-section, which is dual to the T-section, is shown in Figure 2.33b. With a short-circuit at the output it behaves like the current divider of Figure 2.32b. Cascading an infinite number of identical T-sections will result in an input impedance which is called *characteristic impedance* of a (single) section. Making the horizontal impedances equal to $Z_1/2$ and calling the vertical one Z_2 the characteristic impedance Z_{ch} is given by

$$Z_{ch} = \frac{Z_1}{2} \times \sqrt{1 + \frac{4Z_2}{Z_1}}. \tag{2.61}$$

Sometimes the alternative equation using the input impedance Z_{ioc} with open output and Z_{isc} with short-circuit at the output comes handy

$$Z_{ch} = \sqrt{Z_{ioc} \times Z_{isc}}. \tag{2.62}$$

The (power) attenuation or loss is obtained from (2.21a) and (2.21b) using the z-parameters of the T-section (which are symmetric)

$$z_i = z_o = \frac{Z_1}{2} + Z_2, \text{ and}$$

$$z_r = z_f = Z_2$$

by

$$g_p = \frac{Z_2^2}{(Z_2 + \frac{Z_1}{2} + Z_{ch})} \times \frac{Z_{ch}}{(\frac{Z_1}{2} + Z_2) \times (Z_{ch} + Z_2 + \frac{Z_1}{2}) - Z_2^2}. \tag{2.63}$$

Problems

2.40. A T-section of an attenuation ladder has $Z_1 = 80\ \Omega$ and $Z_2 = 60\ \Omega$.
(a) Find the characteristic impedance.
(b) Give the voltage attenuation v_o/v_i per section in dB if the load equals the characteristic impedance.

2.41. A T-section of an attenuation ladder has $Z_1 = 360\ \Omega$ and $Z_2 = 400\ \Omega$.
(a) Find the characteristic impedance.
(b) Give the voltage attenuation v_o/v_i per section in dB if the load equals the characteristic impedance.

To achieve both a given characteristic impedance and at the same time a specific attenuation per section (expressed by $k = \sqrt{1/g_p}$, the square root of the inverse power gain), the following relations can be used:

$$\frac{Z_1}{2} = Z_{ch} \times \frac{k-1}{k+1} \tag{2.64}$$

and

$$Z_2 = Z_{ch} \times \frac{2k}{k^2 - 1}. \tag{2.65}$$

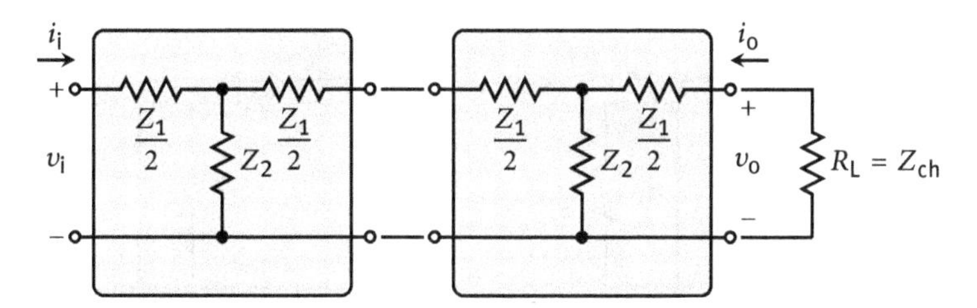

Fig. 2.34. Cascaded T-sections: the attenuation is additive if given in dB; the input impedance equals Z_{ch} when terminated with a load resistance $R_L = Z_{ch}$.

Any number of sections with identical characteristic impedance may be cascaded if the load impedance equals the characteristic impedance. This way the attenuation values (negative, in dB) of each section can be added to yield the total attenuation (in dB). In Figure 2.34, two attenuator sections with the same characteristic impedance are cascaded, and the output is terminated with the characteristic impedance so that the input impedance equals the load impedance. Attenuator sections that are not loaded with their characteristic impedance will attenuate but they will not have the characteristic impedance as input impedance nor will they provide the nominal attenuation that is given just for the matched load impedance case.

2.2.7.2 Nonlinear passive two-ports (clipping)

A nonlinear passive two-port contains at least one nonlinear component. For simplicity reasons, we concentrate on diodes (Section 2.1.6). Let us first get the multiple diode cases done. Putting diodes in series, results in equal currents in all of them. If they are forward biased, there will be no real problem. In general, the voltage values of each operating point will differ considerably unless the diodes are very well matched. Even if the input voltage is not evenly divided between the components, the difference will not be dramatic. If they are reverse biased, the situation is completely different. The reverse current is extremely low (the diode does not "conduct"), but the characteristics are so much different that it is not unusual that nearly all voltage is across one of the diodes and there is practically none across the other. Consequently, putting two diodes in series is not the way to increase (double) the maximum reverse voltage allowed on the two devices.

Putting two diodes in parallel has just the opposite problems. The reverse currents can easily be added in parallel whereas the forward currents will be distributed between the two components very unevenly.

Problem

?

2.42. The forward characteristic of a diode resembles that of a(n ideal) voltage source, the reverse characteristic that of a(n ideal) current source. Verify the above findings by comparing the characteristics of a diode with that of ideal sources.

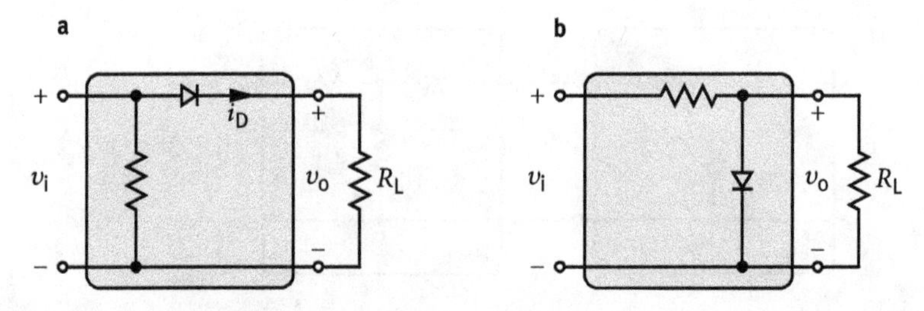

Fig. 2.35. Signal modification by diode: (a) series configuration with $i_L = i_D$ (b) parallel configuration with $v_L = v_D$.

If diodes must be used in parallel, then the characteristic must be changed in a way that the forward characteristics do not differ much. This can best be done by having for each of them a (high enough) resistor in series.

Although any parameter of a two-port can be nonlinear, nonlinearity is usually allotted to the transfer parameter. It is a good exercise to visualize various output signals after being nonlinearly modified in (simple) nonlinear two-ports. Observe that there are usually twin configurations, e.g. Figure 2.35, with the nonlinear element in series or in parallel to the output (load).

Figure 2.36a shows a resistor R in series to a serial combination of a diode with an (ideal) voltage source of V_S and a load resistor R_L at the output. Even if this circuit is rather simple, it is good practice to simplify it. As viewed from the diode, R and R_L are in parallel. Therefore, the circuit can be simplified as shown in Figure 2.36b. The new input voltage $v_{iTh} = v_i \times R_L/(R+R_L)$, and the new series resistor $R_{Th} = R \times R_L/(R+R_L)$. It is important, that v_o, the quantity of interest is still available that it does not get veiled. When the (biasing) voltage V_S of the constant voltage source is zero and the input voltage v_i such that the diode is forward biased (i.e. positive), input current will be short-circuited by the diode and the output voltage v_o will be zero (or the value of V_S, respectively).

If v_i is reversed, the diode is an open-circuit (it may be left out) and the output voltage is $v_o = v_{iTh} = v_i \times R_L/(R+R_L)$ (voltage division, Section 2.1.3.4). Reversing the polarity of diode and voltage source gives the analogous behavior for input voltages of the opposite polarity. Figure 2.36c and 2.36d show the transfer characteristics for both polarities.

Problems

2.43. Reverse in Figure 2.36a only the polarity of the constant biasing voltage source. How does the transfer characteristic look?

2.44. Construct the transfer characteristic, if two anti-parallel diodes, each connected to a suitable constant voltage source, are placed inside the two-port.

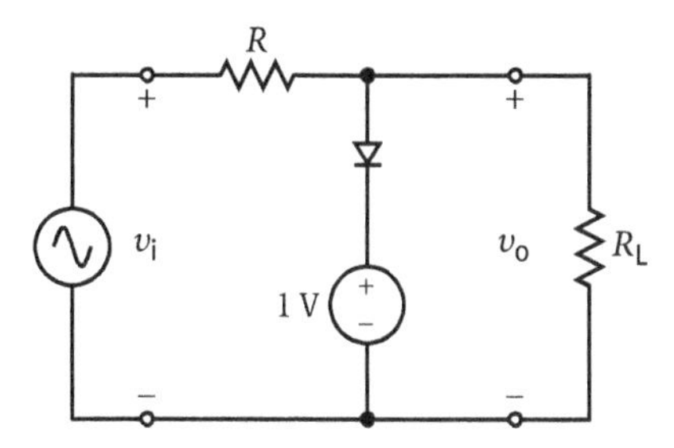

Fig. 2.36a. Diode clipping circuit with the diode parallel to the output.

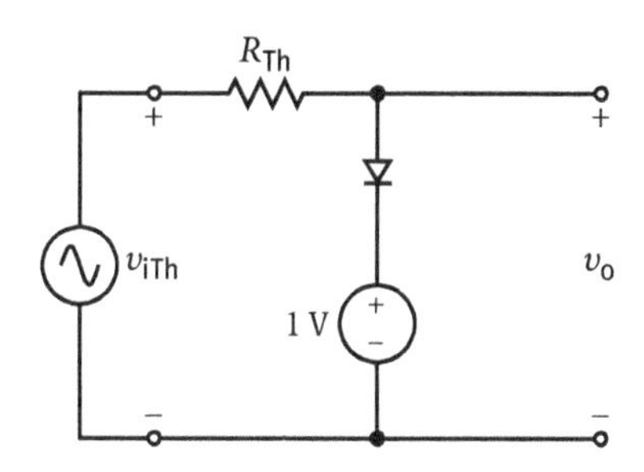

Fig. 2.36b. Simplified circuit diagram by combing the parallel resistors.

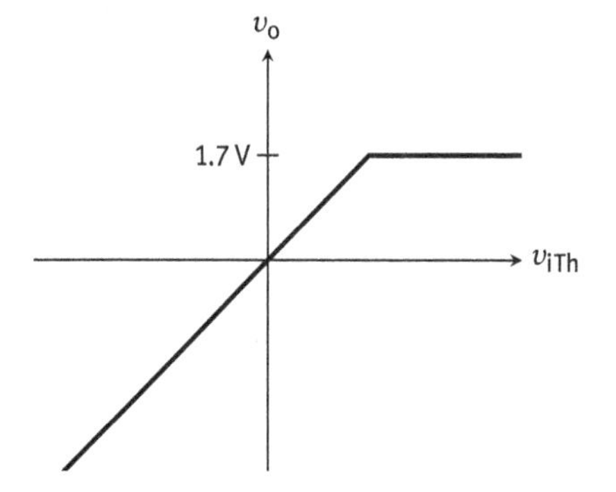

Fig. 2.36c. Transfer characteristic under the assumption of a diode forward voltage of 0.7 V.

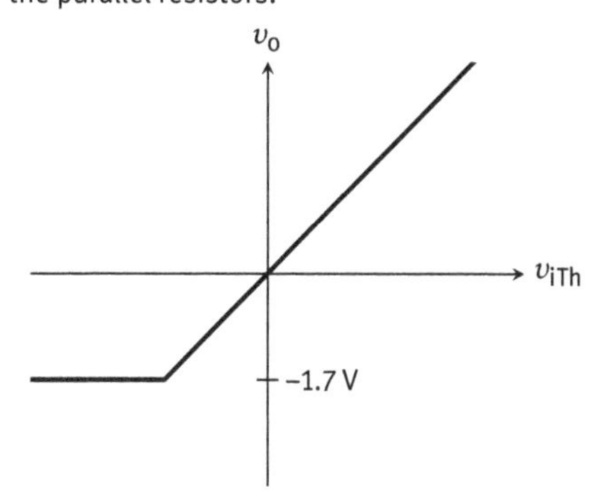

Fig. 2.36d. As c) after reversing the polarity at the diode and the constant voltage source.

Voltage limiting or *clipping circuits* can also be designed by having the clipping diode in series to the load resistor. In this case, the constant voltage source, biasing the diode, is a real source so that its impedance can be combined with the load resistor reducing the number of elements to be considered. Figure 2.37b shows the simplified circuit diagram and Figure 2.37c the voltage transfer characteristic. Reversing the diode, the constant (biasing) voltage, and the polarity of the input signal would result in the transfer characteristic shown in Figure 2.37d.

Problems

?

2.45. Serial clipping diode.

(a) Is the diode in Figure 2.37a forward biased if the input signal voltage is larger than the effective bias voltage?

(b) What is the minimum output voltage?

2.46. Serial clipping diode.

(a) Is the diode in Figure 2.37a forward biased if the input signal voltage is zero?

(b) Calculate the input threshold voltage at which the diode starts conducting.

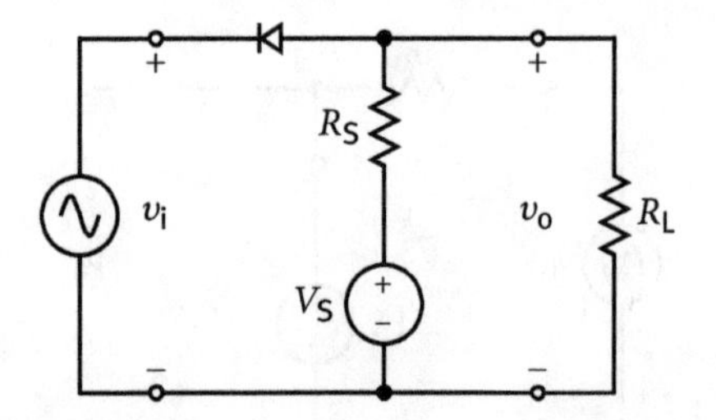

Fig. 2.37a. Diode clipping circuit with the diode in series to the output.

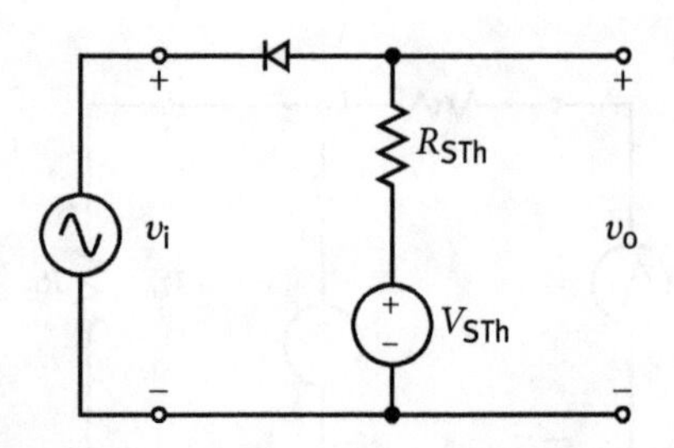

Fig. 2.37b. Simplified diode clipping circuit using a series diode.

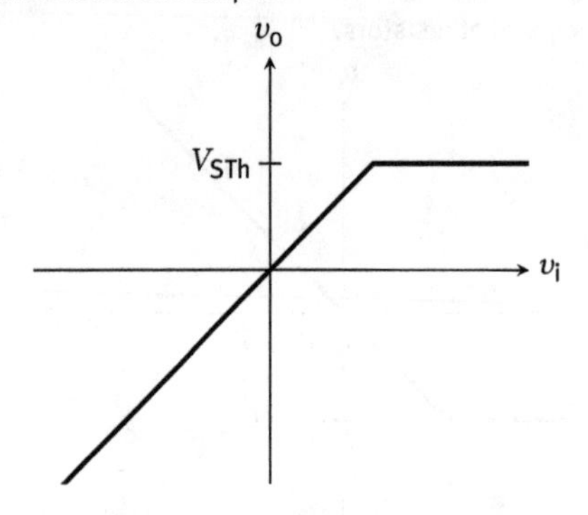

Fig. 2.37c. Transfer characteristic of the circuit shown in (a).

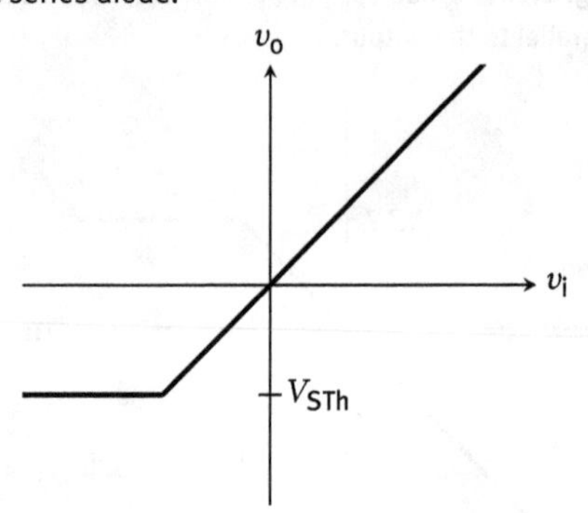

Fig. 2.37d. As (c) after reversing the polarity at the diode and of the constant voltage source.

Another way of clipping signals (limiting voltages) with passive components is to use Zener diodes. In Figure 2.38a, a circuit using one Zener diode is shown; in Figure 2.38b two Zener diodes are arranged in series, with one of them reverse biased limiting both voltage polarities symmetrically.

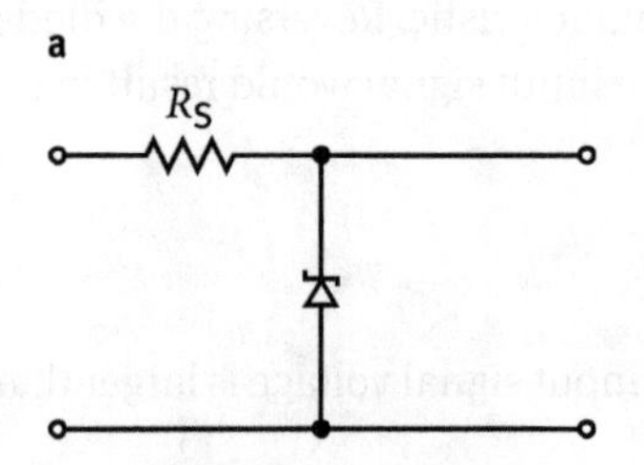

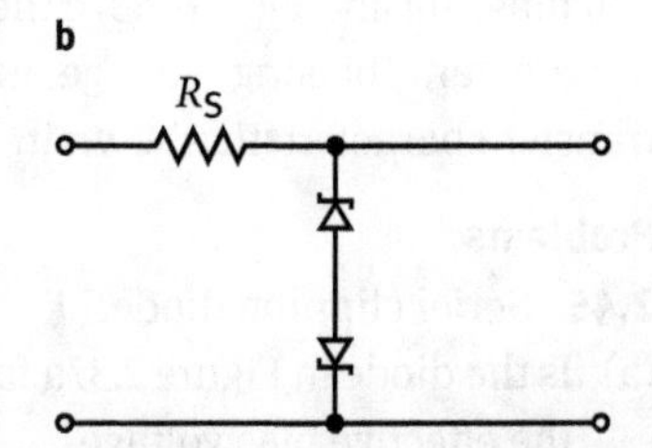

Fig. 2.38. Voltage limiter circuits. (a) Using one Zener diode and (b) having two oppositely biased Zener diodes in series.

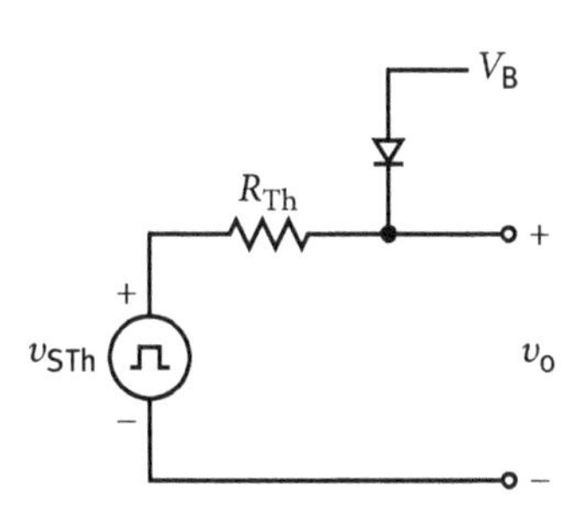

Fig. 2.39a. Diode discriminator with parallel diode (simplified circuit).

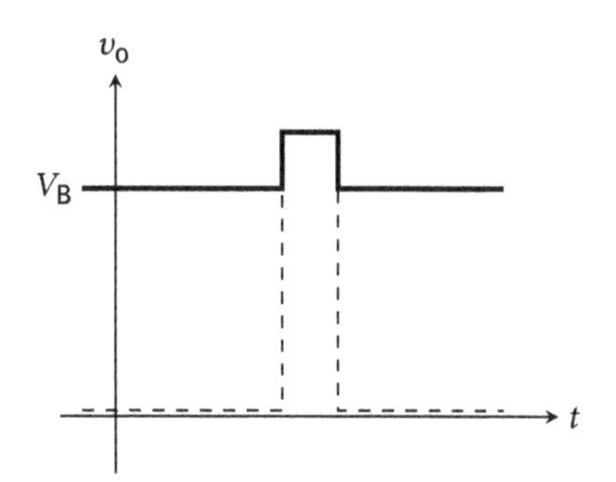

Fig. 2.39b. Output signal if the signal is larger than the bias voltage.

Problem

2.47. Construct the voltage transfer functions for the two cases shown in Figure 2.38 assuming a Zener voltage of 6.0 V each.

Reversing the diode in Figure 2.36 results in a *parallel-diode discriminator*. Figure 2.39a shows the simplified circuit with the parallel resistors R_L and R_S combined to $R_{Th} = R_L \times R_S/(R_L + R_S)$ and the signal amplitude v_S reduced to its effective value $v_{STh} = v_S \times R_L/(R_L + R_S)$. Without signal, the output voltage v_o equals the bias voltage V_B. Only when the effective signal amplitude v_{STh} is larger than V_B, the diode is reverse biased and the portion of v_{STh} that is above V_B gets transmitted. This is shown in Figure 2.39b.

Inverting the diode in Figure 2.37a yields the *serial-diode discriminator*. In this case, the effective bias voltage is $V_{BTh} = V_B \times R_L/(R_L + R_S)$ and $R_{Th} = R_L \times R_S/(R_L + R_S)$. Without signal the output voltage v_o equals the effective bias voltage V_{BTh}. Only when the signal amplitude v_S is larger than V_{BTh} the diode is forward biased and transmits the portion of v_S that is above V_{BTh}. This is shown in Figure 2.40b.

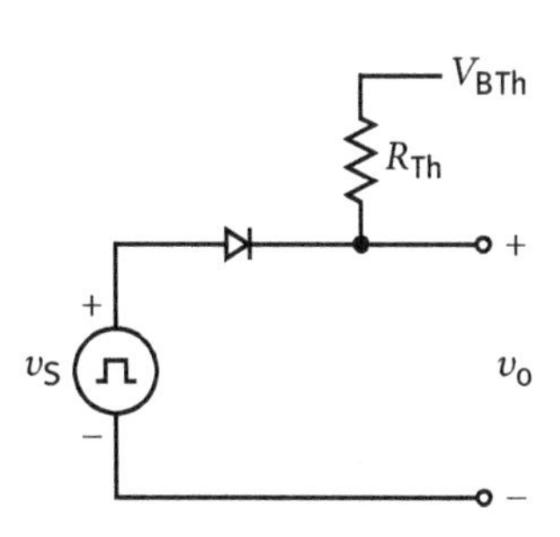

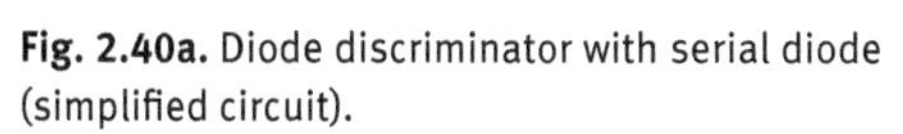

Fig. 2.40a. Diode discriminator with serial diode (simplified circuit).

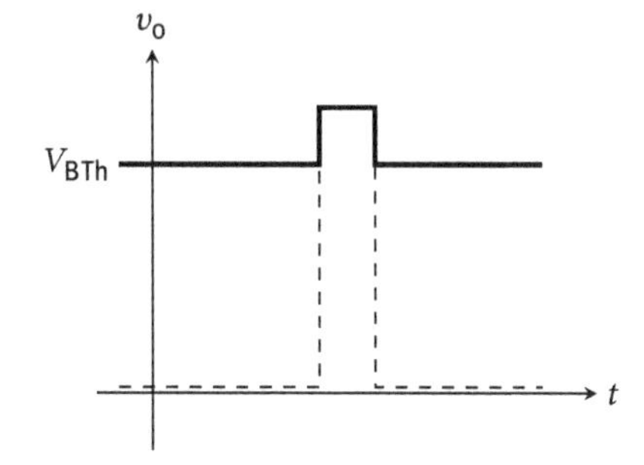

Fig. 2.40b. Output signal if the signal is larger than the effective bias voltage.

2.3 Real active two-ports (amplifiers)

Even if the scope of this book is not the construction of electronic circuits their design requires the knowledge of the operational limits of electronic components. The value pair of voltage and current applied to an electronic component at any instant is called (instantaneous) *operating point* of that device. The operating point without a signal is called quiescent or just operating point. It is established by biasing that device. Linear passive devices (resistors, capacitors, inductors, and transformers) do not need a bias, they are operational without a bias. The properties of nonlinear components change with changing operating point. In these cases, it is essential to bias these components correctly.

The instantaneous operating point can be split into two components, the quiescent value (set by biasing) and the superimposed value from the signal.

There is no general advice of how to best set the operating point. The following considerations will enter into the decision:
- idle (quiescent) power of the component
- choice of the optimum dynamic range
- optimization of parameters (e.g., higher currents will usually allow a faster response; also the small-signal gain will depend on the operating point)
- stability of operation
- avoiding or taking advantage of nonlinearities.

2.3.1 Maximum limits for voltages and currents

The characteristic of an ideal resistor obeying Ohm's law pretends to be a straight line from minus infinity to plus infinity. Obviously no such device can exist. Solid-state properties and gas discharge properties will limit the maximum possible voltage applied to a component because of voltage breakdown. In addition, there exists a maximum temperature up to which a device will function. Only in rare cases this will be the melting temperature. It would rather be the ignition temperature, the recrystallization temperature, etc. As it would be inconvenient to have a temperature measuring device attached to each component, not only the maximum temperature is given as limit, but also the electric power that would bring about that temperature when dissipated in the device which is located in free air of standard atmospheric pressure at 25 °C. The balance between the quantity of heat produced by the dissipated power and the heat loss to the surroundings fixes the device's temperature within the thermal time constant which is of the order of 10 ms. However, as no material is ideally uniform, the device may be overheated locally so that current limits are given, too.

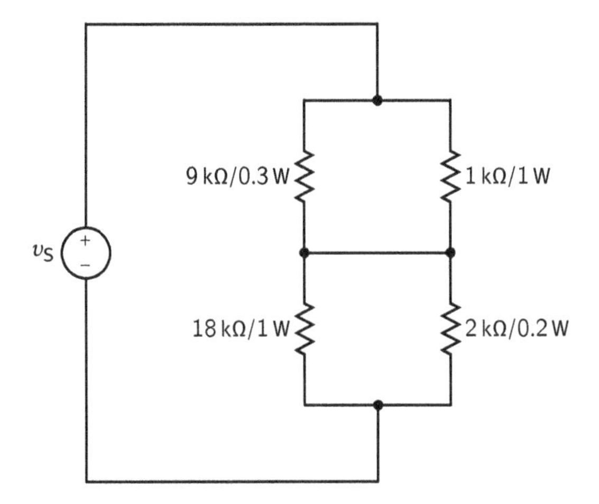

Fig. 2.41. Resistor network from Example 2.6.

The operational limit of electronic devices is primarily the maximum allowed temperature.

Example 2.6 (Maximum power of resistors in a network). Figure 2.41 shows a network consisting of one voltage source and four resistors. For the resistors not only the impedance values but also the power ratings are given. The question to answer is what the maximum voltage v_{Smax} is so that none of the resistors gets a power overload.

As a first step an inspection of the circuit should reveal whether there is the possibility to simplify it. As the question concerns all the resistors, it is not allowed to combine resistors to make the circuit simpler. However, there is one simplification that can be applied. Both voltage dividers divide in the ratio 1 to 2 so that, ideally, no current will be flowing through the connection in the middle. It is superfluous and may be left out. In general, connections that do not carry any current may be removed from the circuit. The same would be true for components with only one terminal connected unless they function as antenna. In the present case, it is useful to translate the maximum allowed power into maximum allowed current. Using $p = i^2 \times R$ we get the following ratings:

9 kΩ /0.3 W:	5.8 mA
18 kΩ /1.0 W:	7.5 mA
1 kΩ /1.0 W:	31.6 mA
2 kΩ /0.2 W:	10.0 mA

Considering that nine times more current flows through the right branch it is clear that the resistor with 2 kΩ/0.2 W is the component that limits the maximum allowed voltage to $v_{Smax} = 30$ V.

Problem

2.48. The load of a real voltage source ($V_S = 110$ V, $R_S = 1$ kΩ/0.2 W) consists of a potentiometer (20 kΩ/1 W) with its sliding contact loaded by a resistor of 20 kΩ/1 W.

(a) At which position of the sliding contact is the (current) load of the potentiometer the largest?

(b) Which of the three resistive components is most likely to be overloaded, i.e. will be thermally destroyed.

2.3.2 Characteristics of two-ports

As detailed in Section 2.2.2, four parameters are needed to fully describe a linear two-port. For real, i.e. nonlinear two-ports such parameters would be the result of linearization at a specific operating point. To fully describe all properties of a nonlinear two-port the dependence of each parameter on the input and output variables must be given. In Figure 2.42 the dependences of the four hybrid parameters of a bipolar junction transistor in the common-emitter configuration are given as an example.

For convenience the four dependences are shown in a combined figure of four quadrants, one for each two-port property. On the right-hand side the reverse properties are shown, on the left-hand side the forward properties. As the hybrid parameters are defined for a fixed output voltage and a fixed input current the appropriate one of these variables is used as parameter.

The upper right quadrant is the output quadrant showing the dependence of the output admittance $h_o = i_o/v_o$ at selected input currents i_i on v_o and i_o. The upper left quadrant is the forward transfer quadrant. There the dependence of the short-circuit current gain h_f on i_i and i_o is given for *one* fixed output voltage v_o. The lower left quadrant is the input quadrant. There the dependence of the input impedance h_i on v_i and i_i at *one* fixed output voltage v_o is displayed. The lower right quadrant shows the dependence of the reverse transfer parameter h_r at fixed input currents i_i on v_o and v_i.

The only characteristic that is linear over most of its range is that of the (forward) current gain. The input characteristic resembles that of the forward characteristic of a junction diode. The reverse voltage gain in the lower right quadrant is so small that we do not deal with it.

The most important quadrant is the output quadrant. For that reason it is often the only quadrant shown, e.g., in data sheets. There are three regions to be considered. The largest is the *linear* region where the individual characteristics are more or less flat indicating a small output admittance, i.e. a high output impedance. Such a behavior is expected for a current amplifier as which a bipolar junction transistor is usually regarded. For small output voltages v_o the characteristics become steep indicating a low output impedance. This region is called region of *saturation* and is not used for

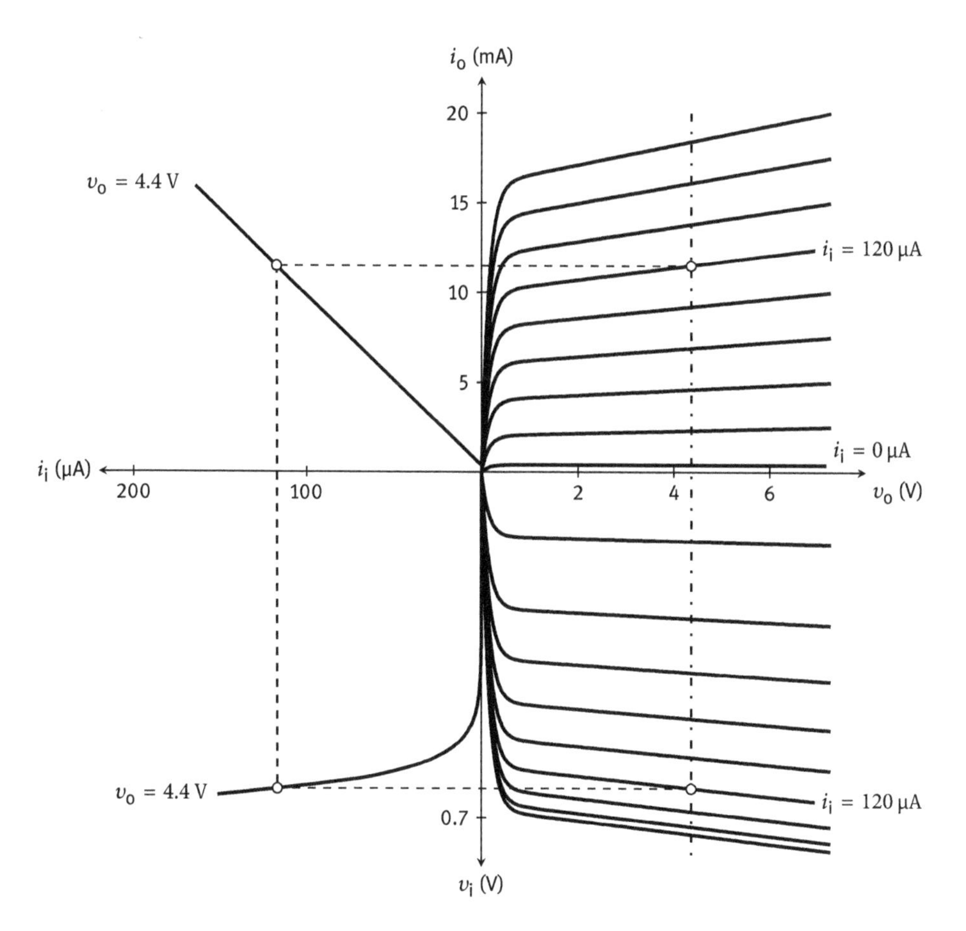

Fig. 2.42. Characteristics of a two-port based on hybrid parameters. Upper right quadrant: output impedance, upper left quadrant: (forward) current gain (single characteristic only), lower left quadrant: input impedance (single characteristic only), lower right quadrant: reverse voltage gain.

linear amplification. The region below the characteristic with $i_i = 0$, i.e. when a very small current flows out of the two-port is called the *cut-off* region.

For completeness the correct definitions of these regions which are specific to bipolar junction transistors are given in Table 2.3. To understand this table, we must know that in a transistor there are two junctions, the base–emitter (B-E) and the base–collector (B-C) junction that behave just like junction diodes, i.e. they may be forward or reverse biased.

The last line of Table 2.3 lists a forth region which cannot be seen in the output quadrant. Because of the symmetry between collector and emitter (as far as the junctions are concerned) a transistor may be used reverse, i.e. with the collector terminal used as if it were the emitter terminal and vice versa. Thus, an additional linear region is obtained whereas the other two remain the same (as is obvious from the Table).

Table 2.3. Regions of operation in bipolar junction transistors.

B-E Junction bias	B-C Junction bias	Region
Forward	Reverse	Forward-linear
Forward	Forward	Saturation
Reverse	Reverse	Cut-off
Reverse	Forward	Reverse-linear

The importance of the cut-off and the saturation regions lies in the low operating power for operating points in those two regions. In the saturation region the voltage is low, in the cut-off region the current. Therefore, bipolar junction transistors qualify as switches, e.g., for applications in binary circuits (Chapter 5).

2.3.2.1 (Electronic) switches

An ideal switch has zero resistance when closed (i.e., when it is in the ON position), zero conductance when open (i.e., when it is in the OFF position), and has no limits on voltage or current rating. It can change instantly from the off state to the on state, and vice versa. Although a switch is a linear one-port and therefore an analog element it is often used as an example of a binary element because its resistivity has two distinctly different states. An example of the use of a switch is given in Figure 2.43.

When the switch is open, no current flows (the current is in the low state L) and the voltage across the switch is v_S (the voltage is in the high state H). When the switch is closed, no voltage will be across the switch (the voltage is in the low state L) and the current is $i_S = v_S/R_L$ (the current is in the high state H). This behavior is summarized in Table 2.4.

The term *truth table* is borrowed from the field of *binary logic*. With it all output situations for all combinations of input variables are given.

Truth tables with the electronic states L and H give a *unique* description of binary *circuits*. Chapter 5 deals in more detail with truth tables.

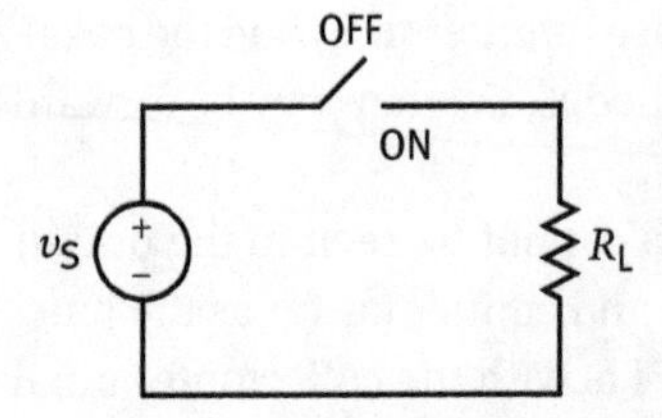

Fig. 2.43. Connecting a load resistor R_L to an ideal voltage source by means of a simple ON/OFF switch.

Table 2.4. Truth table of the electrical variables *across* the ON/OFF switch of Figure 2.43.

Switch position	Voltage (value/symbol)	Current (value/symbol)
ON ($Z = 0$)	0 V / L	$i_S = \frac{v_S}{R_L}$ / H
OFF ($Y = 0$)	v_S / H	0 mA / L

Table 2.4 shows that a ON/OFF switch is self-dual. If it is OFF (open-circuit), the conductance is zero, the voltage H and the current L. If it is ON (short-circuit), the resistance is zero, the current H and the voltage is L.

Mechanical switches have metallic contacts. Due to their elasticity they undergo mechanical oscillations with a frequency on the order of 10^{-1} kHz when switched, resulting in an intermittent contact. This behavior is called *bouncing*. This contact bounce (also called *chatter*) is common to mechanical switches and relays. When the contacts strike together, their momentum and elasticity act together to cause a mechanical oscillation. Instead of a clean transition from zero to full conductance multiple sequential contacts result. Using mercury-wetted contacts eliminates contact bounce. In binary electronics (Chapter 5) bouncing generates spurious signals which must be suppressed by appropriate circuits (e.g., monostable multivibrators, Section 4.1.2.2).

Regular switches are actuators and not pure electronic devices. However, a relay is an electrically operated switch.

Disregarding semiconductor power switches, we concentrate on electronic switches in regular circuits that can be realized by three-terminal components. These components can be so biased that their impedance is either H or L. For bipolar junction transistors these regions of operation are called cut-off and saturation, respectively (Table 2.3). For FETs they are called cut-off and linear. In demanding applications MOSFETs are the best choice for an electronic switch because

- no current flows into the gate (very high input impedance),
- its ON resistance r_{ds}(ON) is typically less than 1 Ω, and
- the cut-off resistance is that of a reverse biased PIN diode (e.g., very high).

Changing the gate voltage of a FET will alter the channel resistance r_{ds}(ON). In this mode the FET operates as a variable (a *voltage-controlled*) resistor. Then the FET operates in the *linear mode* or *ohmic mode*, i.e. the drain current is proportional to the drain voltage v_{ds}. Note that negligible drain current ($i_d \approx 0$) is not only achieved in the cut-off regime, but it can also be achieved in the ohmic regime with $v_{ds} = 0$.

Problems

2.49. Why is a switch self-dual?

2.50. Which family of three-terminal elements has the best switching properties?

2.51. What effect has bouncing of a switch on the signal output?

2.3.3 Setting the operating conditions (biasing)

By establishing predetermined voltage–current pairs at the input (and the output) of an electronic circuit appropriate operating points are set. Such a point is also known as the bias point, quiescent point, or Q-point if no input signal is applied. However, one can also speak of instantaneous operating points in which case the momentary impact of the signal is included.

Biasing of integrated circuits is straightforward by adhering to the voltage power supply limits given in the data sheets. The circuit design takes care of the biasing of the individual components. In high-speed applications, it is important to provide bypass capacitors for the voltage supplies to improve the supply quality. A (tantalum) electrolytic bypass capacitor of at least 1 μF should be placed within 1 cm of each component's power supply pin to ground. In addition, a ceramic capacitor of at least 10 nF should be placed as close as possible from the component's power supply pins to ground. These capacitors serve as charge reservoir for the device during fast switching.

Lumped components, like three-terminal active devices need to be biased according to their specific needs. In all cases the operating point must be stabilized. As the operating point at the output strongly depends on that of the input, stabilizing efforts should concentrate on the input operating point. The first stage of a multistage system must be stabilized for analogous reasons. The best way of controlling the stability of an operating point is to apply (negative) feedback (Section 2.4.1). Current feedback (i.e. a parallel configuration) at the input stabilizes the current value of the operating point, voltage feedback (serial configuration) the voltage value. Other means are less effective and were chosen in the early years when it was to expensive to build feedback circuits with a high enough closed-loop gain.

2.3.4 Classification of amplifiers according to the operating point

Figure 2.44b shows the load line in the field of the output characteristics of an amplifier. In a class A amplifier the operating point is situated in the middle of this line. Signals as large as half the output range can be accommodated in either direction. However, the operating point ($V_{1/2}$, $I_{1/2}$) is at the position of the highest power consumption (Section 2.2.4.1), i.e. without a signal the maximum power is consumed. Moving

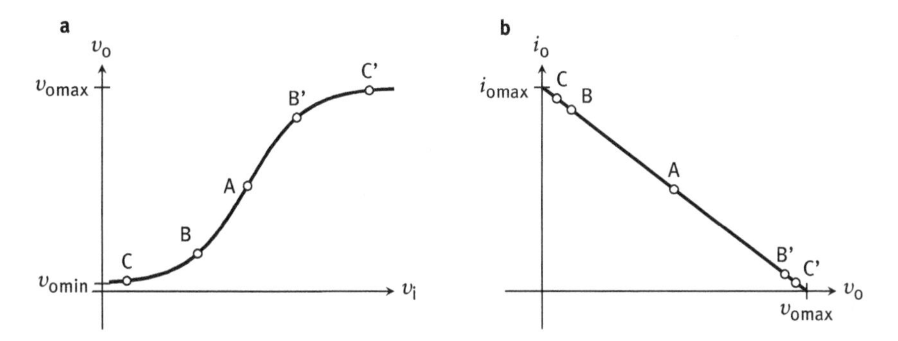

Fig. 2.44. Positions of the operating point of class A, B (and B'), and C (and C') on a voltage transfer characteristic (a), and on the load line in the output field (b).

the operating point either way makes the quiescent power P_{qui} smaller

$$P_{qui} = I_{op} \times V_{op} \leq I_{1/2} \times V_{1/2} \tag{2.66}$$

If signals of only one polarity must be amplified, a position at either end of the output range is better. The quiescent power becomes practically zero and the maximum (unipolar) signal is practically twice as big. This would be a class B amplifier. Complementary semiconductor electronics allows the combination of two complementary B amplifiers to amplify bipolar signals. The quiescent power is close to zero, the signal power can be maximal.

If the operating point is outside the dynamic range of the amplifier, this is called a class C amplifier. Then the lower part of an input signal is used up to put the momentary operating point into the dynamic range so that only that part (if any) of the signal above a certain threshold (as determined by the position of the quiescent operating point) is amplified. Such amplifiers are used as *threshold amplifiers* or *biased* amplifiers.

2.3.4.1 Biased amplifiers (long-tailed pair)

A threshold amplifier is a class C amplifier with a linear transfer characteristic in its dynamic range. Figure 2.45 shows such voltage transfer characteristic.

As long as the input voltage v_i is below the threshold v_{ithr} the gain of the amplifier is zero, the output voltage v_o is that of the quiescent output operating point. Above the cut-in point v_{ithr} a linear characteristic indicates a constant voltage gain g_v. By applying feedback (which is only operative within the dynamic range!) the potential nonlinearity of the amplifier near threshold is much reduced (Section 2.4.1.3).

At this occasion, we want to introduce a powerful combination of two three-terminal amplifying components, called long-tailed pair as shown in Figure 2.46a. The term "long-tail" supposedly comes from the high-impedance resistor connecting the emitters to ground resembling a long tail.

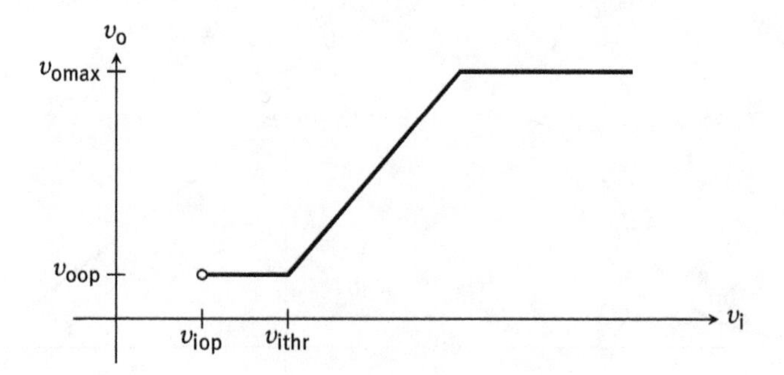

Fig. 2.45. Ideal transfer characteristic of a biased amplifier.

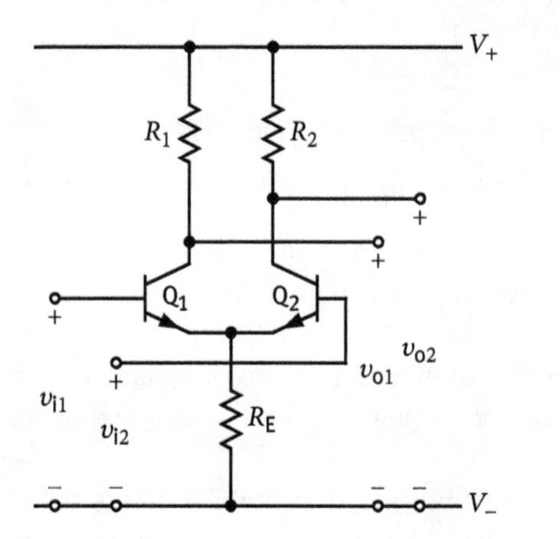

Fig. 2.46a. Basic long-tailed pair circuit with two bipolar NPN transistors.

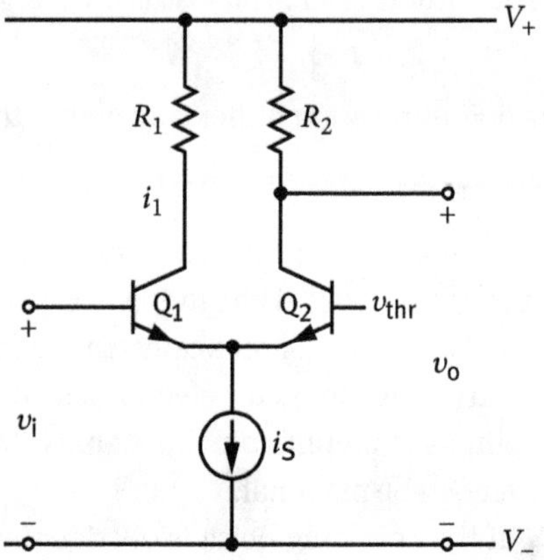

Fig. 2.46b. Long-tailed pair circuit used as a biased amplifier with current source replacing the high impedance resistor R_E.

This circuit constitutes the basic version of the fully differential amplifier (Section 2.3.5.1). As discussed there, a differential amplifier is used for amplifying any difference between the signals applied to its inputs. If the pair of transistors are identical and balanced, a common-mode signal at the inputs will *not* cause a significant signal at the output. Thus, the output of this circuit (which is taken across the collectors) for a common-mode signal would be zero.

A difference between the base signals will be amplified by the transistors, and will result in an output signal proportional to the difference between the two signals. The differential voltage gain of this circuit is high, whereas its common-mode gain is low. Thus this circuit is the standard input circuitry of operational amplifiers (Section 2.3.5.2).

Let us switch to Figure 2.46b where the circuit is improved by substituting the high impedance resistor by a current source. Such a current source can be realized by an

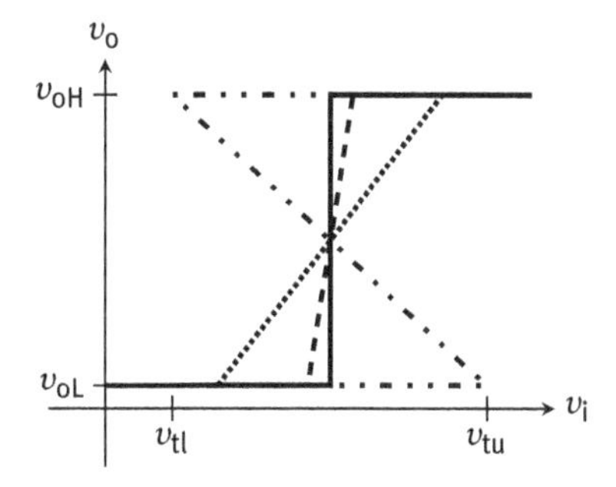

Fig. 2.47a. Family of transfer functions; amplifier, comparator, comparator with stable positive feedback, comparator with unstable positive feedback (Schmitt trigger).

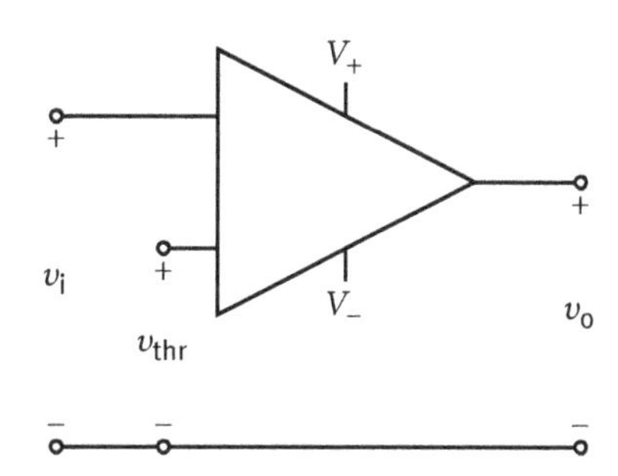

Fig. 2.47b. A bare operational amplifier acting as comparator.

active current source circuit (Sections 2.5.6.1 and 2.5.6.2). Besides just one of the input terminals is used, the other one is grounded (signal-wise). Connecting a pair of FETs at the sources would be at least as good.

The difference between a fixed (threshold) voltage v_{thr} and the input signal v_i is amplified but only if both transistors have at least a class B operating point. If v_i is too small, Q_1 does not conduct and all the current i_S of the constant current source flows through Q_2 resulting in a constant output voltage of $V_+ - R_2 \times i_S$. As soon as Q_1 opens the current i_1 flowing through Q_1 will show up as an increase v_o of the output signal $v_o = i_1 \times R_2$ until $i_1 = i_S$ at which point all the current will flow through Q_1 and the output voltage v_o will coincide with V_+. This asymmetric version of a long-tailed pair is a cascade of a common-collector circuit (Q_1) and common-base stage (Q_2).

2.3.4.2 Comparators

Increasing the gain of a threshold amplifier so much that the linear input range is less than about 10^{-3} V has as consequence that the operating point at the output is either in the lower L or in the higher H saturation region and hardly in the linear region at all. By increasing the gain, e.g., by positive feedback (Section 2.4.2), this gap with linear operation can be made very small. As a consequence small variation in the input signal (or the reference voltage) will be sufficient to make the comparator switch from L to H or vice versa. By increasing the positive feedback to a closed-loop gain ≥ 1 this gap is closed producing a hysteresis in the transfer function (Section 4.1.2). This is illustrated in Figure 2.47a. In Figure 2.47b such a comparator is symbolized by a bare operational amplifier.

Any standard operational amplifier with a well-balanced difference input and a very high gain can act in open-loop configuration as comparator and can, therefore, be used in applications with moderate requirements. When the noninverting input is at a higher voltage than the inverting input, the high gain of the op-amp causes the output to saturate at the highest positive voltage (v_H) it can supply. When the voltage at the noninverting input is below that of the inverting input, the output saturates at the most negative voltage (v_L) it can supply.

Actually, most devices sold as comparators are Schmitt triggers (Section 4.1.2.1). As shown in Figure 2.47a their transfer function has a hysteresis, a result of unstable positive feedback, i.e. there does not exist a region with linear amplification. Thus, it is ensured that the output is either L or H.

2.3.5 Amplifiers as two-ports

An amplifier may have (at least) two terminals for supplying the power, two terminals as input port, and two as output port. Thus, starting out with six terminals we arrive at the following family of amplifier configurations:

- fully differential amplifiers (push-pull differential amplifiers, with six leads),
- differential amplifiers (with five leads),
- push-pull amplifiers (with five leads),
- simple (four-terminal) amplifier, and
- active three-terminal devices.

All amplifiers need (at least) two terminals supplying the power for the operation. However, the input and output ports which are independent in fully differential amplifiers may degenerate by grounding one terminal of either the output or the input or both. The resulting family of amplifiers is displayed in Figure 2.48. By using the remaining output terminal for the supply of the operating power the simple amplifier degenerates further to a three-terminal amplifier.

A *standard operational* (or *operative*) *amplifier* is a differential amplifier with voltage gain as characteristic transfer parameter. For our discussion we will degenerate said standard operational amplifier by grounding the inverting input terminal arriving at a simple voltage amplifier. This will be our standard building block for an amplifier as symbolized in Figure 2.49.

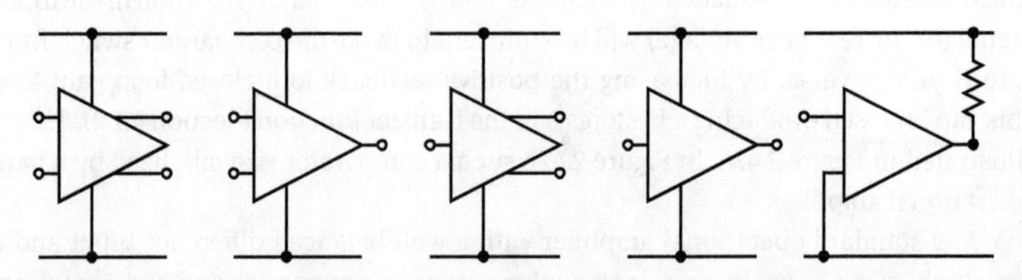

Fig. 2.48. The five steps in the degeneration of a fully differential amplifier to an active three-terminal device.

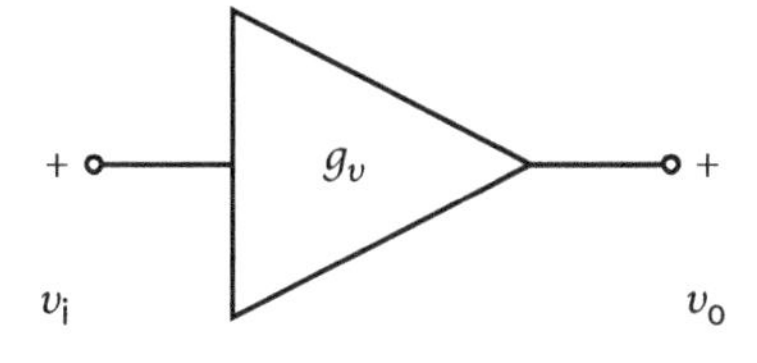

Fig. 2.49. Symbol of a standard voltage amplifier.

2.3.5.1 Fully differential amplifier

In Figure 2.50, two identical standard amplifiers are so arranged that the voltage v_o at the output is proportional to the difference of the input voltages v_{i1} and v_{i2}. With $v_{o1} = v_{i1} \times g_{v1}$ and $v_{o2} = v_{i2} \times g_{v2}$ the voltage difference v_o at the output is given by $v_o = v_{o1} - v_{o2} = v_{i1}g_{v1} - v_{i2}g_{v2}$ and with $g_{v1} = g_{v2} = g_v$ one obtains

$$v_o = (v_{i1} - v_{i2}) \times g_v. \tag{2.67}$$

If the voltage gain of the two amplifiers is equal, this arrangement delivers an output signal that is proportional to the difference of the input signals. Such an arrangement has two drawbacks:

- it suffers from a small dynamic range because the power supply voltage V_+ must be larger than the product of each input voltage and the voltage gain, i.e. $V_+ > v_{ix} \times g_{vx}$, and
- the *common mode* input signal v_{iCM} is amplified in the same way as the differential signal resulting in a common mode output signal $v_{oCM} = g_v \times v_{iCM}$. (The common mode voltage is the mean of the two voltages)

Both disadvantages are overcome in the arrangement shown in Figure 2.51.

Because of negligible admittance, the current source constitutes a *virtual open-circuit*, allowing the potential v_{float} at its top to float without any change in the current flow. At the same time it supplies the operating currents for the two amplifiers, one-

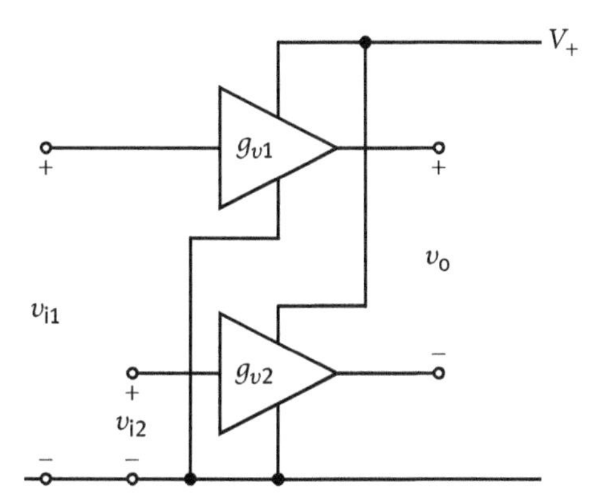

Fig. 2.50. Difference amplifier made of two standard voltage amplifiers.

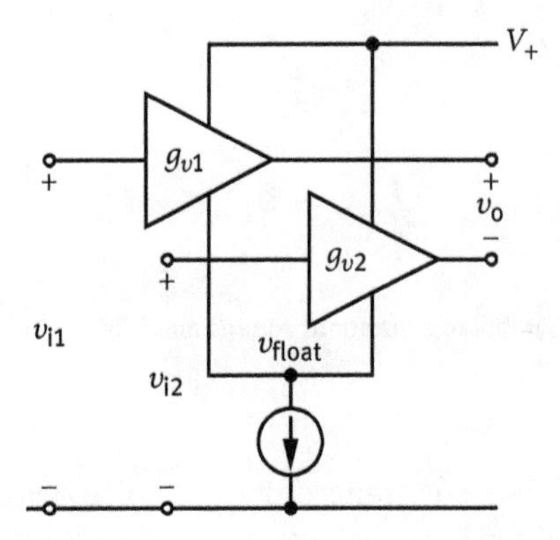

Fig. 2.51. Model of a fully differential voltage amplifier.

half to each (assuming components with identical electrical properties). The decisive point is the following: Changes in the two amplifiers are only possible if their individual operating current changes by the same amount but in the opposite direction. This only happens when (small) voltage signals of opposite sign are applied to the inputs of the individual amplifiers, as is the case. Because of the floating voltage v_{float}, the input voltages v_{ix} are reduced to effective input voltages $v_{ixeff} = v_{ix} - v_{float}$. The floating voltage v_{float} equals the common mode input voltage v_{iCM}, making $|v_{i1eff}| = |v_{i2eff}|$ and of opposite sign.

$$v_{iCM} = v_{float} = \frac{v_{i1} + v_{i2}}{2} \tag{2.68}$$

Substituting in above exercise v_{ix} by $(v_{ix} - v_{float})$ one gets $v_o = (v_{i1} - v_{i2}) \times g_v$ as before. However, contrary to before the common mode output voltage v_{oCM} equals now the common mode input voltage v_{iCM} so that the dynamic range is greatly increased because the common mode input voltage does not get amplified. The common mode gain is 1!

For this ideal model the *common-mode-rejection ratio* (CMRR), i.e. the ratio of the signal gain g_v to the common mode gain, g_{vCM} equals g_v. Such a circuit, when built, has two weak spots:

- The current delivered by the current source is somewhat dependent on the common mode input voltage.
- The two voltage amplifiers would not have identical properties, making the common mode voltage at the output different from that at the input.

In addition, the parameters are subject to drift. Any changes in the parameters of the two amplifiers would have no effect, as long as they are equal. Changes due to temperature changes can be minimized by *thermally coupling* the two input elements, so that a temperature change is common to both. Thus, *thermal drift* is, to a good part, suppressed by common-mode rejection. *Drift* due to aging (microscopic structural changes in semiconducters) can be an issue, too. Other reasons for a change in the (common-mode) output voltage are changes in the supply voltage, in the burden and electronic noise.

The most important application of a fully differential amplifier is as input stage of operational amplifiers.

Problem

?

2.52. A fully differential amplifier has a common-mode gain of $g_{v\mathrm{CM}} = 1$ and a signal gain of g_v. How much is the dynamic range larger than it would be if the CMRR = 1?

2.3.5.2 Operational amplifier

The majority of feedback circuits are based on *operational* (or *operative*) *amplifiers.* Their input stages are usually fully differential *voltage* amplifiers. The aim in the design of such an amplifier is to come as close as possible to the ideal (static) properties of a voltage amplifier:

- zero input conductance,
- zero output impedance, and
- infinite voltage gain.

Practical devices that are moderately complex circuits can only approximate this ideal model. The practical limitations of operational amplifiers can, more or less, be ignored when they are used with negative feedback to perform as *operation amplifiers* (Section 2.5.1).

How close does a practical operational amplifier come to its ideal model?

Finite voltage gain Although high gains can be achieved ($> 10^6$) it is not infinite. High open-loop gains will require special precautions to maintain amplifier stability.

Nonlinear transfer function If the voltage gain is not constant in the signal range of interest, the output voltage will not be accurately proportional to the difference between the input voltages. This effect can be disregarded, except for saturation effects which strongly reduce the loop gain, by using ample negative feedback. The maximum output voltage will be slightly less than the power supply voltage. Therefore, only some limited input voltage can be handled.

Finite common mode rejection ratio The ideal common-mode-rejection ratio (CMRR) is infinite. In applications with a noteworthy common-mode input voltage (e.g., in noninverting operation amplifiers, Section 2.5.2) operational amplifiers with high CMRR must be used.

Nonzero input admittance Although very low values can be obtained using FETs the admittance is not zero. As a consequence input bias current (typically 10 nA for bipolar and several pA for CMOS circuits) flows into the inputs. Unfortunately, this current is mismatched between the inverting and noninverting inputs inducing the need of an input off-set current to balance the output. This fact results in an *input offset* voltage/current making the output voltage zero for a zero input signal.

Nonzero output impedance To achieve low output impedance a relatively high quiescent current of the output stage is required. But even then the output impedance is not zero. In addition, the amount of current (and power) that can be drawn at the output is limited. Most operational amplifiers are designed to drive load resistances down to 2 kΩ.

Temperature effects (drift) The values of all parameters depend on temperature. Temperature drift of the input offset voltage is particularly important.

The impact of nonideal properties of an operational amplifier on the performance of an inverting operation amplifier is investigated in Example 2.12 in Section 2.5.1.

2.4 Static feedback

In Section 2.2.1, we have come across the reverse transfer in two-ports. It takes care of the effect of the output signal on the input signal. In general language this would be called feedback action of the output on the input.

Self-regulating mechanisms are wide-spread, e.g. in biology. All senses in living creatures are designed to provoke a response depending on the signal received by the senses. Also evolution is supposed to take advantage of feedback. Feedback is a general phenomenon found nearly anywhere in human every-day life. For example, driving a car makes use of the visual feedback of relevant information concerning traffic, the lay-out of the road, etc. The existence of feedback in human relations is obvious. Teaching thrives on feedback from students. Also the trial-and-error method takes fully advantage of some feedback action. Numerous feedbacks in economics and finance are known, e.g., if the gasoline price at the gas-filling station depends on supply and demand, raising the price reduces the demand resulting in a lower price. Without further deliberation it should be clear that the information fed back can decrease or increase the quantity under consideration, e.g., the gasoline price may go up or down, dependent on the information received. This is called positive and negative feedback.

In electronic engineering, feedback is one of the most powerful tools available. In mixed systems feedback is used to control mechanical, optical, thermal and other physical processes.

Figure 2.52 depicts a symbolic general feedback loop. Although this is a very common presentation of feedback it is not at all typical. Its merit lies in the presence of a loop which pleases visually oriented people. In electronics, as we will show later, it is not always possible to isolate a (geometric) feedback "loop." Besides, in every-day life it would not be unusual to have several feedback "loops" working in parallel. However, this figure is helpful in defining some important properties of feedback.

Let the *forward* transfer property of the main element A be A, i.e.

$$s_{\mathrm{oA}} = A \times s_{\mathrm{iA}} \,,$$

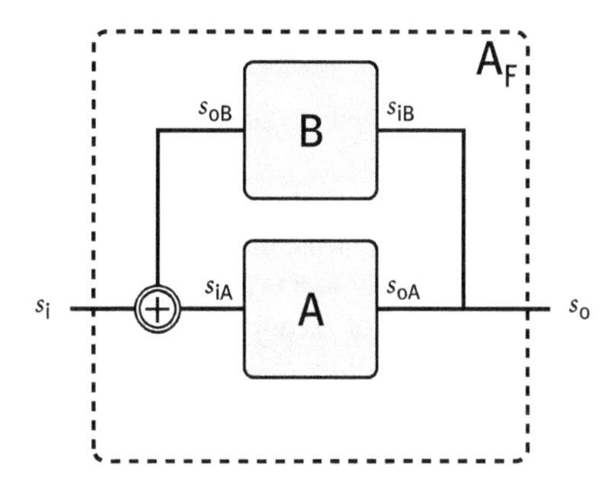

Fig. 2.52. Simple feedback loop with two elements.

the "reverse" transfer property of the feedback net B be B, i.e.

$$s_{oB} = B \times s_{iB} .$$

No other properties are assigned to these elements. In nearly all cases of electronic feedback A would contain at least one operational *active* element and B only passive ones. Hence the symbol A.

When combining these two elements as done in Figure 2.52 the input of B is derived from the output of A, and the output of B must be added to the input signal s_i to add up to s_{iA}:

$$\begin{aligned} s_{iB} &= s_{oA} \\ s_{iA} &= s_i + s_{oB} \\ s_o &= s_{oA} . \end{aligned}$$

The union of these two elements results in a new element with (internal) feedback, called A_F. Let us call its (forward) transfer property A_F, with

$$A_F = \frac{s_o}{s_i} = \frac{s_{iA} \times A}{s_{iA} - s_{iA} \times AB} = \frac{A}{1 - AB} . \tag{2.69}$$

Obviously, the combined element A_F is bidirectional, it has in addition a reverse transfer property B_F

$$B_F = \frac{B}{1 - BA} . \tag{2.70}$$

Even if the individual elements A and B transfer only into their forward direction, the element with feedback transfers either way. This fact by itself makes it clear that feedback does not change the properties of the element A, but results in a new element A_F with properties different from the single element (without feedback).

Applying feedback to an element does not change the properties of said element but gives rise to a novel system with different properties.

However, observe that A_F is of the same nature as A. This can be recognized in (2.69) by the fact that they have the same dimension. The dimensionless product AB is called *forward* (closed-) *loop gain*, whereas BA is called *reverse* (closed-) *loop gain*. The closed-loop gains are the characteristic properties of each feedback loop. Until later we concentrate on the forward properties. There are three cases to be considered:

- $AB > 0$: positive feedback,
- $AB = 0$: no feedback, and
- $AB < 0$: negative feedback, making $A_F < A$.

Actually, we will see that the term $1 - AB$ showing up in the denominator which is called (forward) *return difference* really defines the feedback properties.

Problems

2.53. Determine the (forward) transfer property A_F of the three feedback configurations shown in Figure 2.53.

2.54. By applying negative feedback of the amount B to an active element with the property A the transfer property A_F should be made ½.
(a) How can this be done?
(b) What is peculiar with this solution?

2.55. How does feedback affect the intrinsic properties of elements inside the feedback loop?

2.56. How do you determine whether a feedback is positive or negative?

2.4.1 General accomplishments by negative feedback

For *negative* feedback ($AB < 0$), it follows from (2.69) that $|A_F| < |A|$. This means that the transfer value is reduced by negative feedback. From (2.69) one gets

$$A_F = \frac{1}{\frac{1}{A} - B} . \tag{2.71}$$

Therefore, the transfer property B must have the inverse dimension of A as already evidenced by the fact that the product $A \times B$ is dimensionless. For $|A| \gg 1/|B|$ above equation, is reduced to

$$A_F \approx -\frac{1}{B} . \tag{2.72}$$

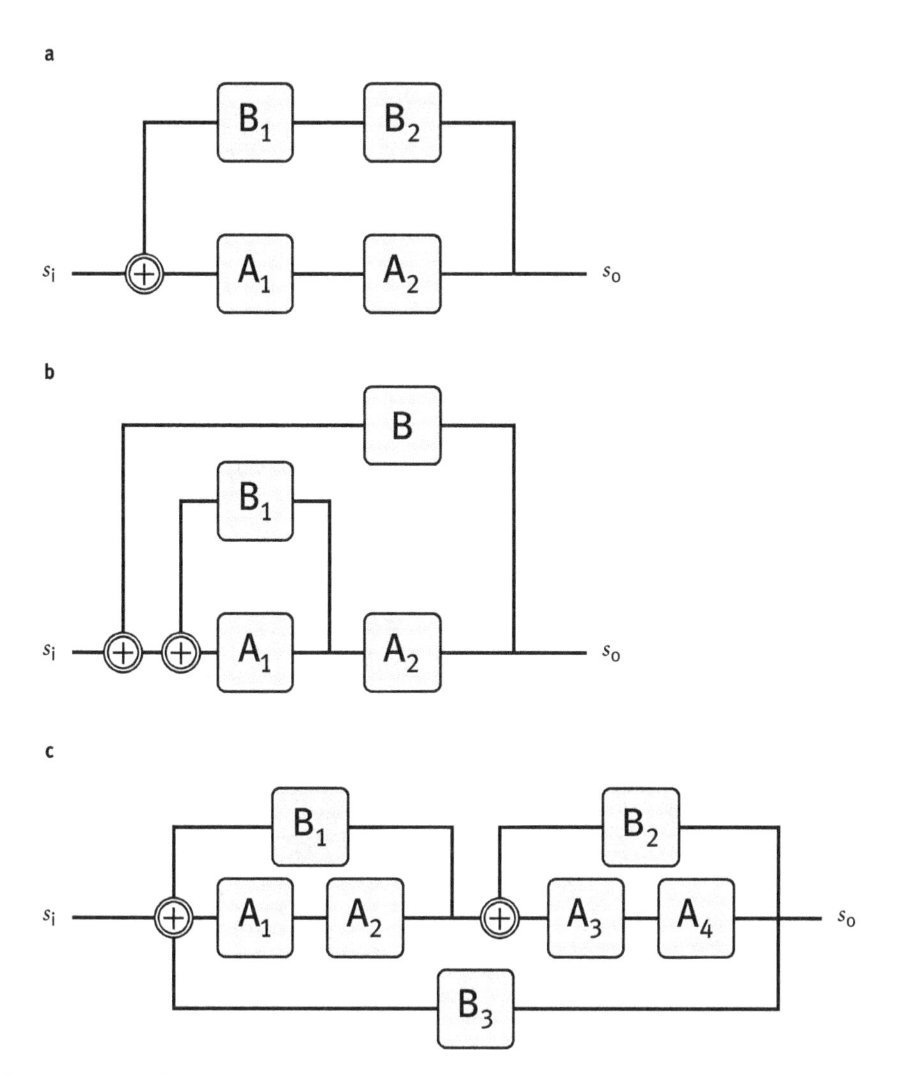

Fig. 2.53. Feedback configurations from Problem 2.53.

Thus, for a closed-loop gain $|AB| \gg 1$, the transfer value A_F becomes de facto independent of A. It is practically entirely dependent on the inverse of the transfer value B of the feedback element. With B made of stable and linear elements there are the following consequences for the quality of A_F as compared to A:

- the stability of is improved,
- the linearity of the transfer function is improved, and
- noise from inside the element A is reduced as much as the signal.

2.4.1.1 Improving the stability

By partial differentiations of (2.69) we obtain

$$\frac{\partial A_F/A_F}{\partial A/A} = \frac{1}{1+|AB|} \tag{2.73}$$

$$\frac{\partial A_F/A_F}{\partial B/B} = \frac{|AB|}{1+|AB|} \approx 1\,. \tag{2.74}$$

Above findings can be expressed in words:
- instabilities (i.e. changes) in A are reduced by the return difference,
- instabilities in B are fully effective.

The de facto sole dependence of A_F on B is the most important principle used in the design and fabrication of linear amplifiers.

2.4.1.2 Feedback over two stages

Feedback over two stages can either be accomplished by two local feedback loops or by one global feedback (Figure 2.54). Which of these arrangements is the more stable, has the least dependence on A under the precondition that the value of the total transfer quantity is the same in both cases and that the active elements are identical?
- Using local feedback over two identical stages in cascade reduces the transfer value per stage to $\sqrt{A_F}$. Having two cascading stages doubles the dependence of the system output on (identical) changes of the two A

$$\frac{\partial A_F}{\partial A} = \frac{2}{A^2} \times A_F\,. \tag{2.75}$$

- Cascading two active elements and using one global feedback with the property B makes the return difference to $1 + A^2B$ giving a dependence of the system output on (identical) changes of the two A of

$$\frac{\partial A_F}{\partial A} = \frac{2}{A^2} \times \frac{A_F^2}{A}\,. \tag{2.76}$$

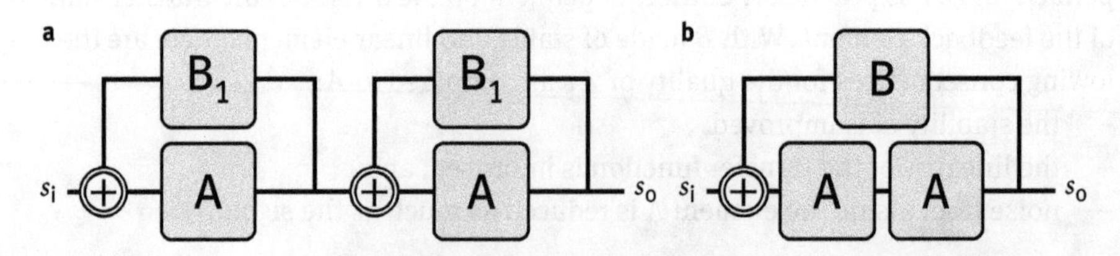

Fig. 2.54. Feedback over two stages. (a) Accomplished by local feedback loops and (b) by one global feedback loop.

Since for negative feedback A_F/A is less than 1, the global feedback depends less on changes of A than the local feedbacks.

The best feedback results are obtained by making the feedback loop as large as feasible.

2.4.1.3 Improving the linearity

No transfer function is truly linear because at some point the maximum output signal is reached (Figure 2.55). The deviation from a linear response is a function $\epsilon(s_o)$ of the size of the output signal s_o, i.e.

$$s_o = A \times s_i - \epsilon(s_o) = s_i \times \left(A - \frac{\epsilon(s_o)}{s_i}\right) . \tag{2.77}$$

With negative feedback (making up for the reduced gain by increasing the input value by the return difference so that $s_{oF} = s_o$)

$$s_{oF} = s_i \times (1 + |AB|) \times \left(A - \frac{\epsilon(s_{oF})}{s_i \times (1 + |AB|)}\right) . \tag{2.78}$$

Thus for $s_o = s_{oF}$ the deviation ϵ from linearity is reduced by the amount of the return difference.

2.4.1.4 Improving the noise immunity

It is obvious that noise at the input cannot be separated from a signal and, therefore, has the same fate as the signal. Therefore, noise generated in the element A or picked

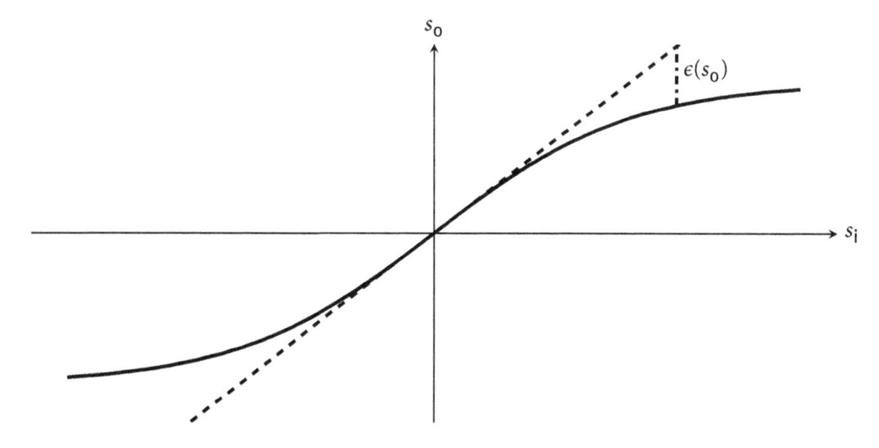

Fig. 2.55. Response function and ideal linear response of an amplifier as well as deviation $\epsilon(s_o)$ from linear response as a function of the output signal s_o.

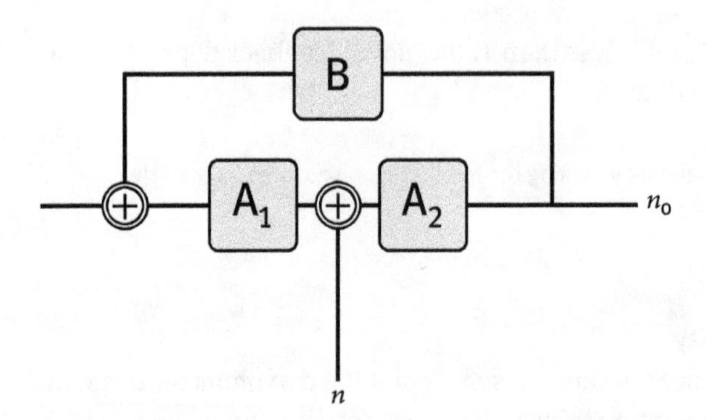

Fig. 2.56. Feedback network with internal source of noise n in amplifier A.

up by it must be converted into equivalent input noise. In Figure 2.56 the position of the noise source is indicated by the appropriate splitting of A into A_1 and A_2 with $A_1 \times A_2 = A$. Some noise n generated directly after A_1 has the same effect as an input noise $n_i = n/A_1$. This is a property of A_1 and consequently independent if there is feedback or not. The noise signal n_o at the output is given in the no-feedback case by

$$n_o = n \times A_2 \tag{2.79}$$

and in the feedback case by

$$n_{oF} = n \times \frac{A_2}{1 + |AB|} \,. \tag{2.80}$$

Internal noise is reduced by the return difference. However, as the signal is reduced by the same amount the *signal-to-noise ratio* stays the same – it is not affected by feedback.

Observe that when one applies this finding to amplifier feedback, it is important to compare the same kind of transfer characteristic for the signal and the noise. Otherwise a comparison does not make sense.

2.4.2 Static positive feedback

With a positive loop gain the feedback equation

$$A_F = \frac{A}{1 - AB} \tag{2.81}$$

resembles that of the infinite geometric series:

$$s = \sum_{k=0}^{\infty} ar^k = \frac{a}{1-r} = a + ar + ar^2 + ar^3 + ar^4 + \cdots \tag{2.82}$$

As k goes to infinity, the absolute value of r must be less than one for the series to converge. Consequently stable positive feedback operation requires that the closed-loop gain AB be less than 1.

Problem

?

2.57. Make it plausible to yourself why stable positive feedback behaves like an infinite geometric series.

What has been said on the effect of negative feedback can be taken over for positive feedback taking into account the opposite sign of the (closed) loop gain. Stability and linearity get worse. Where is then the benefit of stable positive feedback? From (2.81) it should be clear that the transfer value A_F can be made arbitrarily large. For this reason there can be some benefit in applying stable positive feedback locally.

2.4.3 Static feedback in circuits

In electronics, both the active element A and the feedback element B are two-ports. As a consequence there are four possible feedback arrangements of the two elements (Figure 2.57). The input terminals can be in parallel or in series and the output terminals as well. Based on the geometric arrangements of the inputs and the outputs of the two-ports (see also Section 2.2.2) we arrive at the following four types of feedback:
- parallel–parallel feedback,
- series–parallel feedback,
- parallel–series feedback, and
- series–series feedback.

There exist four alternative names for these feedback configurations based on the effective electrical quantities. In this case, the names reflect the feedback action, namely from output to input. As voltage is sensed in parallel and added in series, whereas current is sensed in series and added in parallel the above types of feedback are called
- voltage–current feedback,
- voltage–voltage feedback,
- current–current feedback, and
- current–voltage feedback.

Sometimes just the input is considered and only the terms current feedback and voltage feedback is used. Although the closed-loop gain AB is a dimensionless ratio, it is for the two current-feedback cases a current gain and for the two voltage-feedback cases a voltage gain.

Although (closed) loop gains are dimensionless they are either a current gain or a voltage gain. Consequently, one speaks in short of *current* or of *voltage feedback*, respectively.

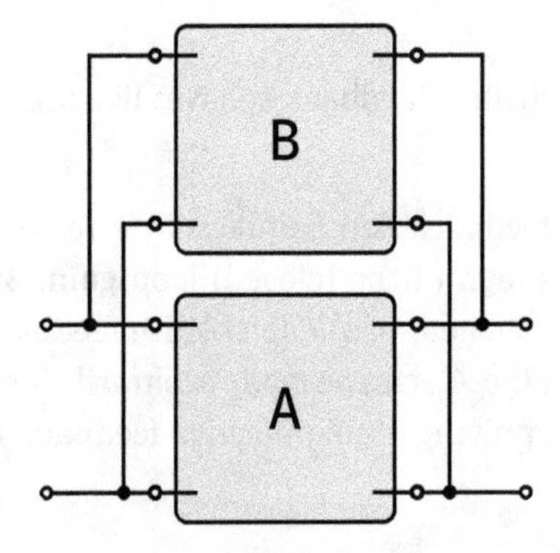

Fig. 2.57a. Parallel–parallel feedback.

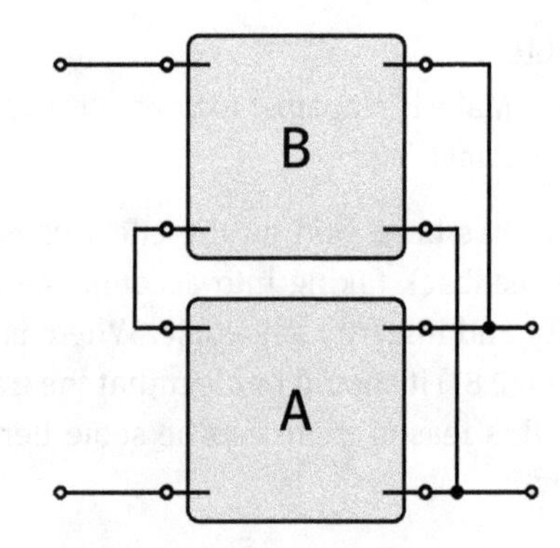

Fig. 2.57b. Series–parallel feedback.

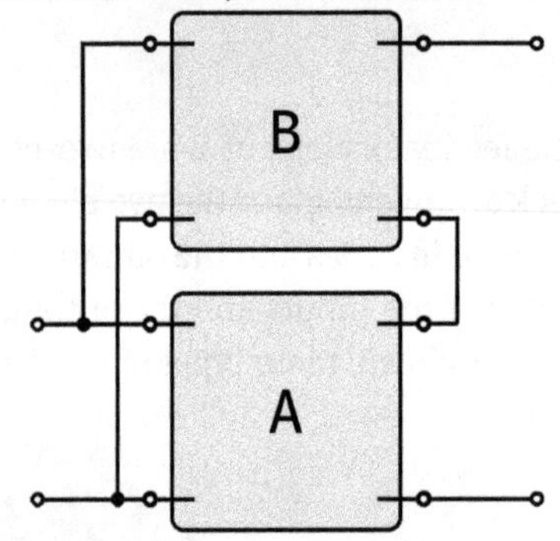

Fig. 2.57c. Parallel–series feedback.

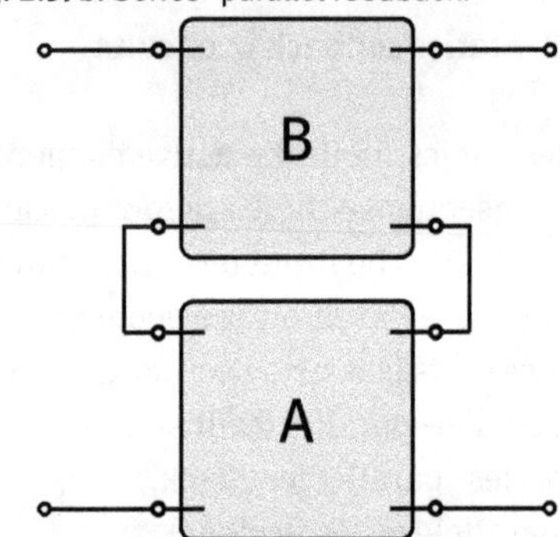

Fig. 2.57d. Series–series feedback.

The type of forward transfer of each of these four configurations is *characteristic* for them, i.e. feedback acts directly *only* on them. They are, in preserved sequence,
– r_{m}, the transresistance (or more general transimpedance),
– g_v, the voltage gain,
– g_i, the current gain,
– g_{m}, the transconductance (or more general transadmittance).

These transfer parameters have been discussed in detail in Section 2.2.1.

Determining the closed-loop gain is not quite straightforward because the loop must be opened at some point to provide access to the output and input variable. To this end the identical loading of the (open) output must be established as existed in the closed case. Consequently, it is advisable to choose the point at which the loop is broken such that the loading has little effect (low output impedance, high input impedance). By now it should be clear that the closed-loop gain determines the property of a feedback circuit. Therefore,

circuit elements that have no impact on the closed-loop gain do not take part in the feedback action.

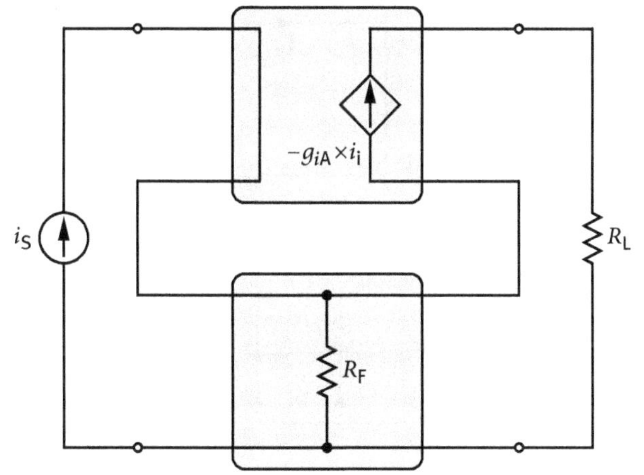

Fig. 2.58. Feedback circuit from Problem 2.58 and Problem 2.59.

Problems

2.58. Feedback is applied to an ideal current dependent current source with a current gain g_{iA} according to Figure 2.58.
(a) What is the current gain g_{iF} of the feedback circuit?
(b) What is the voltage gain g_{vF} of the feedback circuit?

2.59. The output of a feedback two-port lies in series to the output of the active two-port (Figure 2.58).
(a) Does the load resistor take part in the feedback action?
(b) Does shunting the load resistor by a short-circuit cancel the feedback action?

2.4.3.1 Impedances of two-ports with external feedback

How are the input and output impedances of the two-port affected when a feedback two-port is added in any of the configurations shown in Figure 2.57? Exactly speaking, the answer should be: Not at all! What is actually meant is, how do the impedances of the (new) two-port A_F with feedback differ from the original naked two-port A?

To arrive at a valid comparison some basic precautions must be taken. Firstly, it should be clear that the primary two-port A does not change any of its properties because of the added feedback. The primary two-port together with the feedback two-port can be viewed at as a new two-port with internal feedback. And it can be expected that the properties of this two-port with feedback will differ from the properties of the two-port without feedback. Secondly, whenever a valid comparison is made, it is absolutely necessary to make it under "identical" conditions. As the two-ports differ the conditions cannot be really identical but care must be taken that crucial data will be the same. For example, in Figure 2.30 it can be seen that the forward properties (input impedance and forward gain) depend on the load, i.e. the impedance at the output, whereas the reverse properties (output impedance and reverse gain) are functions of the source impedance, i.e. the impedance at the input.

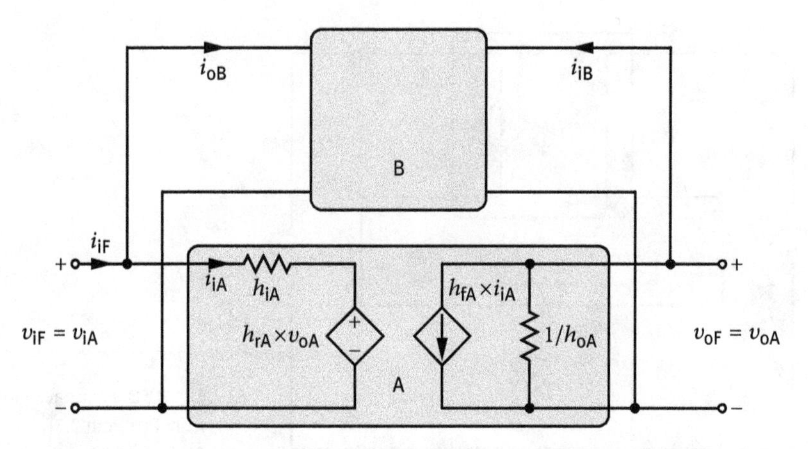

Fig. 2.59. Parallel–parallel feedback applied by a passive two-port B to an active two-port A.

By now, it should not surprise that a circuit responds to the actual impedance which is not necessarily the static impedance, e.g., as given in data sheets. Consequently, any comparison of circuits must be made *under the same electrical burden* to make results comparable.

Taking advantage of duality it suffices to investigate just one configuration at the input and one at the output. Figure 2.59 shows two two-ports in parallel–parallel configuration with hybrid parameters.

The input impedance Z_{iA} of A is given as

$$Z_{iA} = \frac{v_{iA}}{i_{iA}} = \frac{h_{iA}i_{iA} + h_{rA}v_{oA}}{i_{iA}} = h_{iA} + h_{rA} \times Z_{iA} \times g_{vA} \tag{2.83}$$

after inserting the factor $Z_{iA} \times i_{iA}/v_{iA}$ which is identical to 1 and using $g_v = v_{oA}/v_{iA}$. The input impedance Z_{iF} of the arrangement with feedback is given as

$$Z_{iF} = \frac{v_{iF}}{i_{iF}} = \frac{h_{iA}i_{iA} + h_{rA}v_{oA}}{i_{iA} \times (1 + \frac{i_{oB}}{i_{iA}})} = \frac{h_{iA} + h_{rA} \times Z^*_{iA} \times g^*_{vA}}{1 - AB} \tag{2.84}$$

with Z^*_{iA} and g^*_{vA} the values for equal electrical burden, and

$$\frac{-i_{oB}}{i_{iA}} = AB \tag{2.85}$$

the forward (closed-) loop gain. The compact answer for a parallel configuration for which the admittance comes natural is

$$Y_{iF} = Y_{iA} \times (1 - AB)\,. \tag{2.86}$$

The (drastic) reduction of the input impedance by *negative* parallel–parallel feedback has made it the working horse of electronic engineers. This configuration provides a node with virtual ground at the input making the circuit quite independent of the common-mode-rejection ratio (CMRR) of the amplifier in use (Section 2.3.5.1). This is the reason behind the rule of thumb "always invert, except when you can't" meaning "have a negative feedback with a parallel configuration at the input to bring about a node where the voltage hardly changes at all, i.e. with a small impedance (toward ground)." A virtual ground may be at any voltage. However, changing the current in the node will not change the voltage due to its very small impedance. Due to the unavoidable nonlinearity (reducing the closed-loop gain) the small impedance will be limited to not too large currents. Nodes with virtual ground are very convenient reference points in circuit analysis due to their practically invariable voltage.

Virtual ground is a node with (very) small impedance so that the voltage swing due to a current swing is negligible in comparison to other voltages in this circuit.

One terminal of an ideal voltage source with the other terminal connected to ground is such a(n ideal) virtual ground. The voltage does not change. It is independent of the current flow.

By applying duality we arrive at the virtual open-circuit. It can be represented by an ideal current source which provides an invariable current, independent of the applied voltage (Section 2.3.5.1). Consequently, the voltage across an ideal current source is said to be *floating*.

A virtual open-circuit is a terminal pair with a very small conductance so that the current swing due to a voltage swing is negligible.

For a serial arrangement the dual answer is

$$Z_{\mathrm{iF}} = Z_{\mathrm{iA}} \times (1 - AB)\,. \tag{2.87}$$

Duality also changes the current loop gain of the parallel configuration into a voltage loop gain of the serial arrangement.

The reverse "input" impedance, i.e. the output impedance, usually, does not get the attention it deserves. Obviously, when viewing from the back, A and B exchange their function, i.e. the subscript A must be replaced by B in (2.84) and the reverse loop gain BA must be used as loop gain.

$$Z_{\mathrm{oF}} = \frac{v_{\mathrm{oF}}}{i_{\mathrm{oF}}} = \frac{h_{\mathrm{iB}} i_{\mathrm{iB}} + h_{\mathrm{rB}} v_{\mathrm{oB}}}{i_{\mathrm{iB}} \times (1 + \frac{i_{\mathrm{oA}}}{i_{\mathrm{iB}}})} = \frac{h_{\mathrm{iB}} + h_{\mathrm{rB}} \times Z^{*}_{\mathrm{iB}} \times g^{*}_{v\mathrm{B}}}{1 - BA} \tag{2.88}$$

with Z^{*}_{iB} and $g^{*}_{v\mathrm{B}}$ the values for identical electrical burden, and

$$\frac{-i_{\mathrm{oA}}}{i_{\mathrm{iB}}} = BA \tag{2.89}$$

the reverse (closed) loop gain. The compact answer for a parallel configuration for which the admittance comes natural is

$$Y_{\text{oF}} = Y_{\text{iB}} \times (1 - BA)\,. \tag{2.90}$$

For a serial arrangement the dual answer is

$$Z_{\text{oF}} = Z_{\text{iB}} \times (1 - BA)\,. \tag{2.91}$$

In a *parallel* configuration *negative* feedback *de*creases impedances, in a *serial* configuration it *in*creases impedances.

Note: Clearly, the forward closed-loop gain AB depends on the impedance at the output whereas the reverse closed-loop gain depends on the impedance at the input. Their appropriate values must be obtained from the actual closed-loop configurations.

? **Problem**

2.60. A floating voltage (e.g., the voltage across an ideal current source) is cut in half. How much does the current change?

2.4.3.2 Other dynamic impedances

The y-parameters are the optimum option of two-port parameters when two two-ports are connected in a parallel–parallel configuration as shown in Figure 2.60. The feedback two-port B contains just one admittance Y. Then $y_{\text{iB}} = y_{\text{oB}} = Y$, and $y_{\text{rB}} = y_{\text{fB}} = -Y$.

Placing an ideal amplifier into the active two-port A simplifies calculations without affecting the essential. We are choosing an *operational amplifier* characterized by

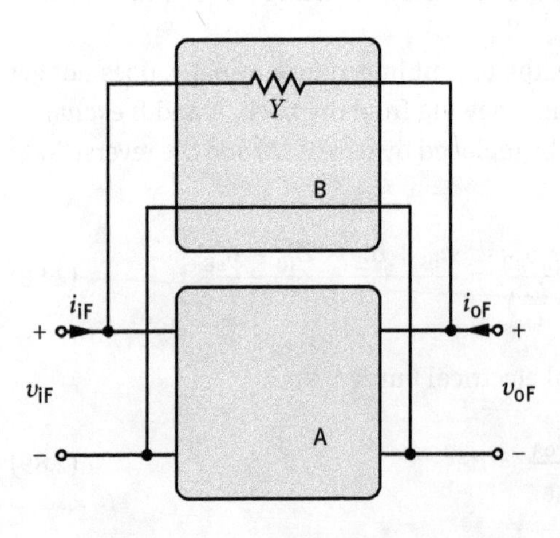

Fig. 2.60. Parallel–parallel negative feedback with one resistor in the feedback two-port.

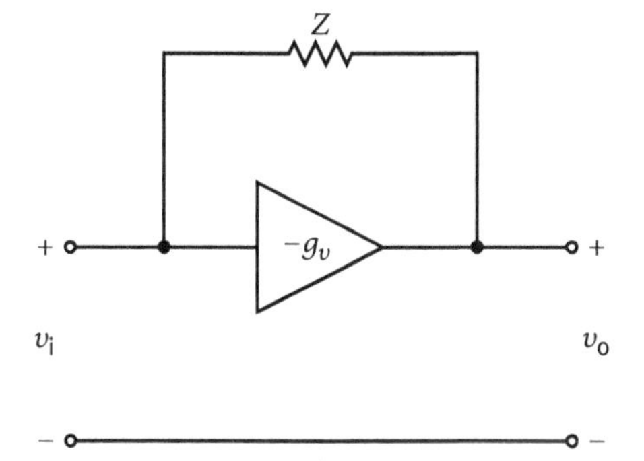

Fig. 2.61. An ideal voltage amplifier with a parallel–parallel feedback.

its voltage gain g_v, an infinite input impedance, zero output impedance and no reverse transfer, i.e. just one nonzero parameter (g_v) is needed for its description.

In the case shown in Figure 2.61 all the input current flows through Y (with $Y = 1/Z$) which results in an output voltage v_o of $v_o = v_i - i_i \times Z$. With $v_o = v_i \times g_v$ the input admittance is given by

$$Y_i = \frac{i_i}{v_i} = Y \times (1 - g_v) . \tag{2.92}$$

The following cases must be considered:

- If $g_v > 1$, it is a case of negative admittance (impedance). A positive input voltage results in an input current flowing out of the circuit.
- If $0 < g_v \leq 1$, such a circuit is called *bootstrap circuit*. The input current is smaller than expected from applying Ohm's law on Z itself. For $g_v = 1$, there is no voltage drop on Z because $v_o = v_i$, i.e. $i_i = 0$ and $Y_{dyn} = 0$.
- If $g_v < 0$, the so-called *Miller effect* increases the admittance of Z accordingly.

This exemplifies how admittances arranged in parallel to an amplifier are made dynamic.

Duality considerations

Without much ado, we can apply duality to the above case (Figure 2.63). The input current of a series–series feedback flows through the feedback impedance Z which is dynamically changed to $Z_i = Z \times (1 - g_i)$. The following cases must be considered:

- If $g_i > 1$, it is a case of negative impedance.
- If $0 < g_i \leq 1$, such a circuit is dual to the bootstrap circuit.
- If $g_i < 0$, the impedance is increased accordingly.

In the previous case, positive feedback (i.e. a positive open-loop gain) is cause of negative *admittance*, in the present case of negative *impedance*. The i-v-characteristic of the *dynamic impedance* has an N-shape (Section 2.1.6). This can easily be found out

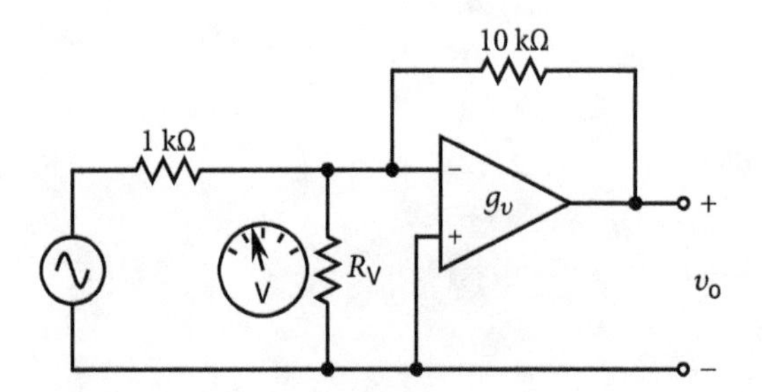

Fig. 2.62. Circuit from Example 2.7.

by starting at the origin. The characteristic through the origin must be that of Z, because for $i = 0$ the amplifier does not work, i.e. $g_i = 0$. With increasing g_i, the slope of the characteristic becomes smaller, becoming zero with $g_i = 1$. Increasing the current gain further causes the negative slope, the negative impedance. At the end of the dynamic range, the current gain decreases. When it becomes one, the characteristic is flat again and from then on the impedance approaches the value of Z again.

The characteristic of the *dynamic admittance* dealt with just before has an N-shape, too, as voltage is used as dependent variable. However, when, as usual, current is used then the characteristic of the dynamic impedance has an S-shape as expected because the S-shape is dual to the N-shape. Above deliberation supports the use of conductance as an independent property and not just as reciprocal resistance.

Example 2.7 (Measuring the voltage across a dynamic impedance). Figure 2.62 shows an inverting operation amplifier. The voltage at the input of the operational amplifier shall be measured. As discussed in Section 2.1.4 the loading of the circuit by the measuring instrument requires a correction of the measured value to obtain the correct value (without loading). If the correction is below a given level, one may disregard it saving time and effort. In the present case this level is assumed to be 1%. What minimal impedance R_V must the voltmeter have so that the correction will be not more than 1%? As a first step the supposedly linear network is replaced by a real voltage source. At this point we note that the absolute voltage of this source need not be determined as we are only concerned with the relative voltage change. At the input of the operational amplifier three impedances are in parallel, the impedance of the 1-kΩ resistor, that of the 10-kΩ resistor and the input impedance of the amplifier. The last, usually, is so high that it may be disregarded. Thus the impedance of the real voltage source is given by the parallel combination of just two impedances. That of the 1-kΩ resistor is just 1 kΩ because it is parallel to the voltage to be measured. (The signal voltage source can be replaced by a short-circuit according to the rules of the superposition theorem.) However, the impedance of the 10-kΩ resistor is dynamically changed as discussed above. The second terminal of this resistor is not connected to ground but to the output of the amplifier that has a voltage gain $g_v = -1000$. Consequently, its impedance as seen from the input is $Z_{10k} = 10\,000/1001$ which is about 10 Ω. Therefore, the impedance of the real voltage source is due to the 1-kΩ resistor in parallel a

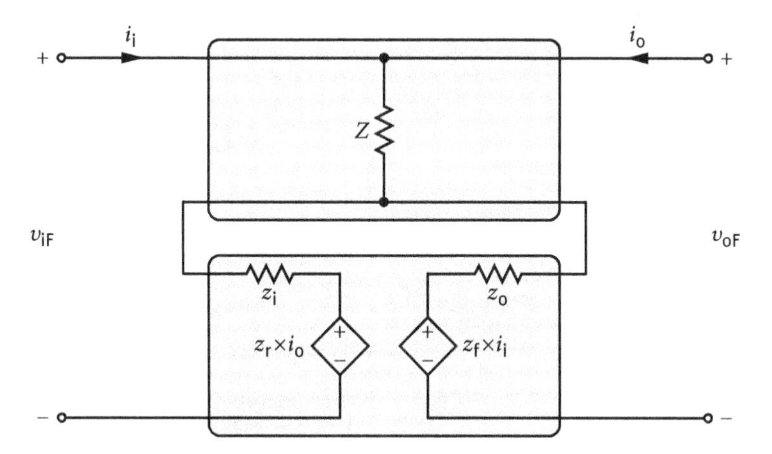

Fig. 2.63. Series–series negative feedback with one conductor in the feedback two-port.

little smaller, namely $Z_{Th} = 9.89\,\Omega$. Thus, a voltmeter with an impedance of $979\,\Omega$ would load the circuit so little that the correction would amount to 1%.

Now let us investigate the conductance of Y in Figure 2.60 viewed from the output. Any signal v_o at the output is necessarily accompanied by a signal $v_i = v_o/g_v$ at the input. Thus the voltage across Y is $v_o - v_o/g_v$. Consequently, the conductance Y_o of Y as viewed from the output is $Y_o = Y \times (1 - 1/g_v)$. We should now understand the difference between the bare conductance Y measured between the two terminals of the component and the two dynamic conductances Y_i and Y_o measured at each port against *ground*. Although the conductance of Y does not change by including it into a feedback loop the feedback action generates dynamic values when measured against the reference (ground).

Using an impedance Z as series–series feedback (Figure 2.63) with an (ideal) current amplifier the *dual circuit* to that of Figure 2.60 is obtained, giving dynamic impedances: $Z_i = Z \times (1 - g_i)$ and $Z_o = Z \times (1 - 1/g_i)$. Although such a circuit has very little importance because current amplifiers are much less common than voltage amplifiers the appropriate configurations are shown in Figure 2.63 as an exercise in duality. In this case z-parameters are optimum because two two-ports are connected in a series–series configuration as shown in Figure 2.63. The feedback two-port contains just one impedance Z. Then $z_{iB} = z_{rB} = z_{fB} = z_{oB} = Z$.

Internal feedback in two-ports

Considering that the reverse transfer constitutes an internal feedback in an active two-port we come across dynamic impedances without an external feedback loop. E.g., in Section 2.2.5 we found that there the short-circuit input impedance h_i depends on the current gain β of the T-model, i.e. the impedance h_i is dynamic.

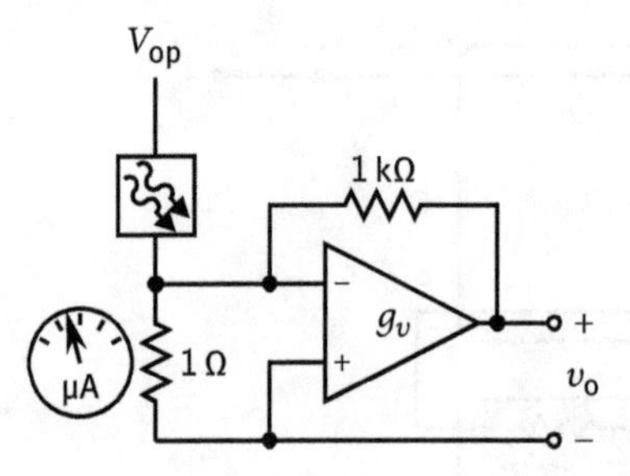

Fig. 2.64. Circuit from Problem 2.63 (g_v = 1000).

Problems

2.61. Does a resistor change its value, i.e. change its impedance when inserted into a feedback loop?

2.62. Investigate the impedance of the following one-port: the port with the voltage v is shunted by a resistor R and in parallel to it is a real *dependent* voltage source with a source resistor R_S and a voltage of $v_S = v \times g_v$.
(a) Determine the (dynamic) impedance of the one-port.
(b) Under which condition is the admittance $Y = 1/Z$ zero?

2.63. In Figure 2.64 the dark current of a photo cell is measured by an ammeter having an impedance of 1 Ω. The reading is 1 μA.
(a) Apply the loading corrections to find the actual dark current.
(b) What output voltage do you expect with the ammeter removed?

2.4.3.3 Transfer properties of two-ports with external feedback

To compare the transfer property with and without feedback the same basic precautions must be taken as discussed before. In particular, it should be remembered that the primary two-port in question does not change any of its properties because of the added feedback, and that the isolated active two-port must have the same burden to make results comparable.

Applying duality to the output and/or the input it suffices to investigate just one of the four configurations into detail. We choose the most popular configuration, the parallel–parallel feedback (using a voltage amplifier). Contrary to the recommendation in Section 2.2.2.2 to use for this configuration admittance parameters in calculations with small-signal parameters, we use the inverse hybrid parameters for the active two-port representing best the character of the operational amplifier (voltage amplifier). The feedback two-port is described by its serial feedback resistor R (Figure 2.65).

Even if the transimpedance is the characteristic transfer parameter of this configuration we start with comparing the voltage gain g_v for which the active two-port is

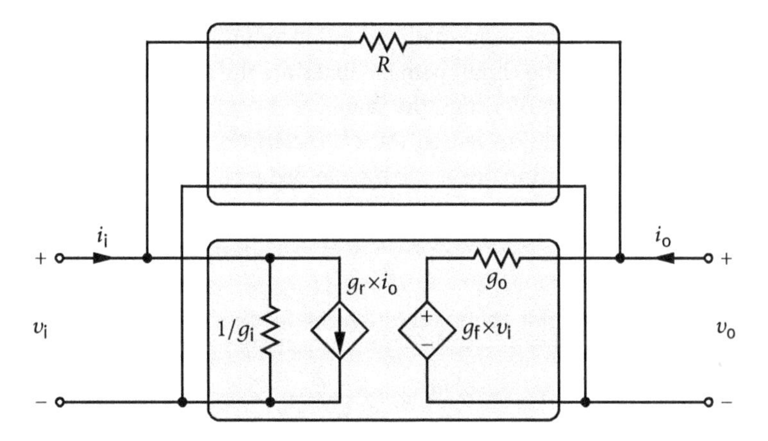

Fig. 2.65. Parallel–parallel feedback with a feedback two-port B applied to an active two-port A. Inverse hybrid parameters are chosen to comply with the voltage gain of the amplifier.

specified. Without feedback ($1/R = 0$) we get

$$g_v = \frac{v_o}{v_i} = \frac{v_i \times g_f \times \dfrac{Z_L}{g_o + Z_L}}{v_i} = \frac{g_f}{1 + \dfrac{g_o}{Z_L}} \, . \tag{2.93}$$

With feedback (finite impedance of R) the output is loaded by Z_L^* consisting of the parallel configuration of Z_L and the (dynamic) impedance Z_{dynR} of the feedback resistor R as viewed from the output. As discussed above

$$Z_{\mathrm{dynR}} = \frac{R}{1 - \dfrac{1}{g_v}} \tag{2.94}$$

so that the equivalent load impedance Z_L^* becomes

$$Z_L^* = \frac{1}{\dfrac{1}{Z_L} + \dfrac{1 - 1/g_v}{R}} \, . \tag{2.95}$$

Thus the decrease of the voltage gain when connecting the feedback resistor is solely the effect of additional loading of the amplifier and not due to feedback action.

As a stringent test, make g_o in (2.93) equal zero. Such a choice makes the active two-port closer to ideal so that an improved feedback circuit can be expected if at all. In that case the loading of the active two-port has no effect and consequently the voltage gain of the two two-ports, with and without feedback, is *identical.*

Besides, all these considerations are completely unnecessary if you just stick to a simple fact (stressed before). Any electronic component retains it properties (at a given operating point, i.e., with the same loading!) independently of its use. Both the input and the output voltage of the circuit with feedback are the same as without feedback, therefore, their ratio necessarily stays the same.

Negative feedback only decreases the *characteristic* transfer parameter. For a parallel–parallel configuration this is the transimpedance r_m

$$r_m = \frac{v_o}{i_i} = Z_i \times \frac{i_i}{v_i} \times \frac{v_o}{i_i} = Z_i \times g_v \,. \tag{2.96}$$

The voltage gain g_v does not change through the feedback action, as discussed, so that

$$r_{mF} = Z_{iF} \times g_v = \frac{Z_i}{1 - AB} \times g_v \tag{2.97}$$

yielding

$$r_{mF} = \frac{r_m}{1 - AB} \,. \tag{2.98}$$

The return difference is the factor by which negative feedback reduces the *characteristic* transfer parameter. This was expected from the general properties of feedback (Section 2.4.1).

Feedback does not act directly on any noncharacteristic transfer parameter.

Example 2.8 (Series–series feedback as source of the common-mode rejection in a long-tailed pair). An essential ingredient of practically all operational amplifiers with differential input is an input stage based on a long-tailed pair. For convenience, we use bipolar junction transistors as active elements. Figure 2.66 shows the circuit diagram, modified insofar, as the resistor R_E that forms the long tail is split up in two resistors with the value of $2R_E$ each, connected in parallel.

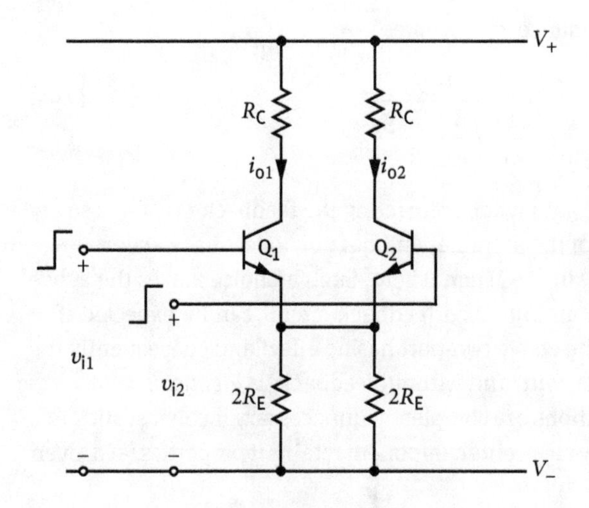

Fig. 2.66. Common mode signal applied to a long-tail pair.

Thus we have two symmetric transistor stages Q_1 and Q_2, with equal collector R_C and emitter resistors $2R_E$. It is obvious that a signal applied to the bases of both transistors will result in the *same voltage at the emitters* of both transistors if the pairs of transistors and resistors have identical properties each. As the voltages are equal no current will flow through the connection between the resistors so that this connection may be removed without disturbing the values of the electrical variables.

The resistor $2R_E$ at the emitter of the transistor performs a negative series–series feedback. The output current builds up a voltage by flowing through $2R_E$ that counteracts the input voltage so that we have a current–voltage feedback also called series–series feedback. As we know from Section 2.4.3.1 such a feedback changes the input impedance from Z_{iA} to $Z_{iF} = Z_{iA} \times (1 - AB)$.

With $g_{vA} = g_{iA} \times R_C/Z_{iA}$ and $g_{vF} = g_{iA} \times R_C/Z_{iF}$ the *common-mode-rejection ratio* becomes

$$CMRR = \frac{g_{vA}}{g_{vF}} = (1 - AB) \tag{2.99}$$

with $AB = -(\beta + 1) \times 2R_E/Z_{iA}$. In practical circuits a constant current source with its very high output impedance replaces R_E providing optimal $CMRR$.

Problems

?

2.64. Series–series feedback

(a) Does a series–series feedback act directly on the current gain?
(b) Does a series–series feedback act directly on the transimpedance?
(c) Does a series–series feedback act directly on the voltage gain?
(d) On which transfer property does a series–series feedback act directly?

2.65. The voltage gain in an amplifier varies by 40%. Which kind of negative feedback must be applied using how much closed-loop gain to reduce the variations to 1%?

2.66. The transfer function of a voltage amplifier deviates by up to 12% from linearity. Which kind of feedback must be applied using how much closed-loop gain to reduce the nonlinearity to less than 1%?

2.67. In Figure 2.67, the feedback loop of a parallel–parallel feedback can be opened by means of a switch. How does the voltage gain change when this switch is closed terminating the open-loop condition?

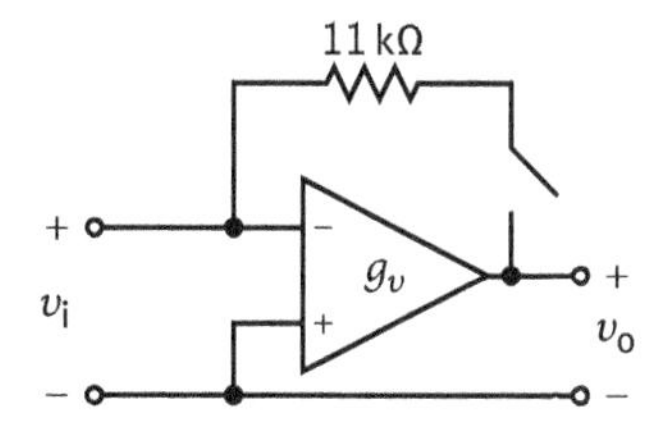

Fig. 2.67. Feedback loop of Problem 2.67 ($g_v = 1000$).

Table 2.5. Negative feedback action on two-port properties.

Forward quantities		
Input configuration	Serial	Parallel
Shared variable	Input voltage	Input current
Forward loop gain	$AB = -v_{oB}/v_{iA}$	$AB = -i_{oB}/i_{iA}$
Input impedance	$Z_{iF} = Z_{iA} \times (1 - AB)$	
Input conductance		$Y_{iF} = Y_{iA} \times (1 - AB)$
(Forward) transfer	$A_F = A/(1 - AB)$	
Characteristic gain	s/p: g_v – s/s: g_m	p/s: g_i – p/p: r_m
Reverse quantities		
Output configuration	Serial	Parallel
Shared variable	Output current	Output voltage
Reverse loop gain	$BA = -i_{oA}/i_{iB}$	$BA = -v_{oA}/v_{iB}$
Output impedance	$Z_{oF} = Z_{iB} \times (1 - BA)$	
Output conductance		$Y_{oF} = Y_{iB} \times (1 - BA)$
Reverse transfer	$B_F = B/(1 - BA)$	
Characteristic gain	s/p: g_i – s/s: r_m	p/s: g_v – p/p: g_m

2.4.3.4 Summary of feedback actions on two-ports with external feedback

Because each of the four two-port properties are differently affected in the four basic feedback configurations, Table 2.5 summarizes how they are affected.

Remember: With *negative* feedback impedances become smaller if the configuration is in parallel, larger when in series.

If in doubt what kind of feedback is acting, here are simple tests to find out.

Configuration at the input:
If a short-circuit at the input makes the feedback ineffective then you have current, i.e. parallel feedback at the input, otherwise voltage, i.e. serial feedback at the input.

Configuration at the output:
If a short-circuit at the output does not affect the feedback action then you have current, i.e. serial feedback at the output, otherwise voltage, i.e. parallel feedback at the output.

Example 2.9 (All four types of feedback configurations in one circuit). Figure 2.68 shows a circuit in which, depending on the choice of the input and output terminals all four feedback configurations can be found. As we already know, the easiest way to

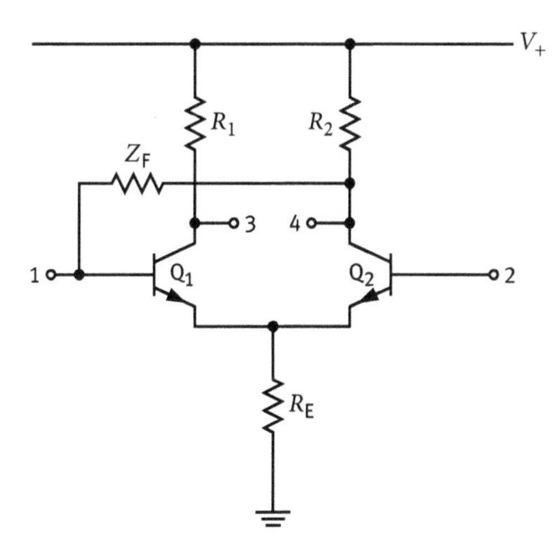

Fig. 2.68. Long-tailed pair with two inputs and two outputs can be used to demonstrate all four feedback configurations.

Table 2.6. Identification of the four feedback configurations in Figure 2.68.

Input#	Output#	Type of feedback	Charact. transfer function
1	4	Parallel–parallel	transimpedance
1	3	Parallel-in series	Current gain
2	4	In series–parallel	Voltage gain
2	3	In series-in series	Transadmittance

recognize whether the feedback is in parallel or in series at a port of the active element is to apply a short-circuit at that port. If this short-circuit nullifies the feedback action, one has a parallel configuration. Applying this recipe to the circuit of Figure 2.68 we get the answers listed in Table 2.6.

If a short-circuit at a port nullifies the feedback action, parallel feedback is present at that port.

Problem

2.68. Verify the findings of Table 2.6 for the four feedback variants.

?

2.4.3.5 Feedback in circuits with three-terminal components

In Section 2.2.6, it was pointed out that the configuration that gives the highest power gain among the three-terminal components is the *basic* circuit, whereas the other two are parallel–series or series–parallel feedback configurations of the basic circuit. Therefore, according to Figure 2.31, the common-emitter circuit using bipolar junction transistors (or the common-source circuit with field effect transistors) is basic.

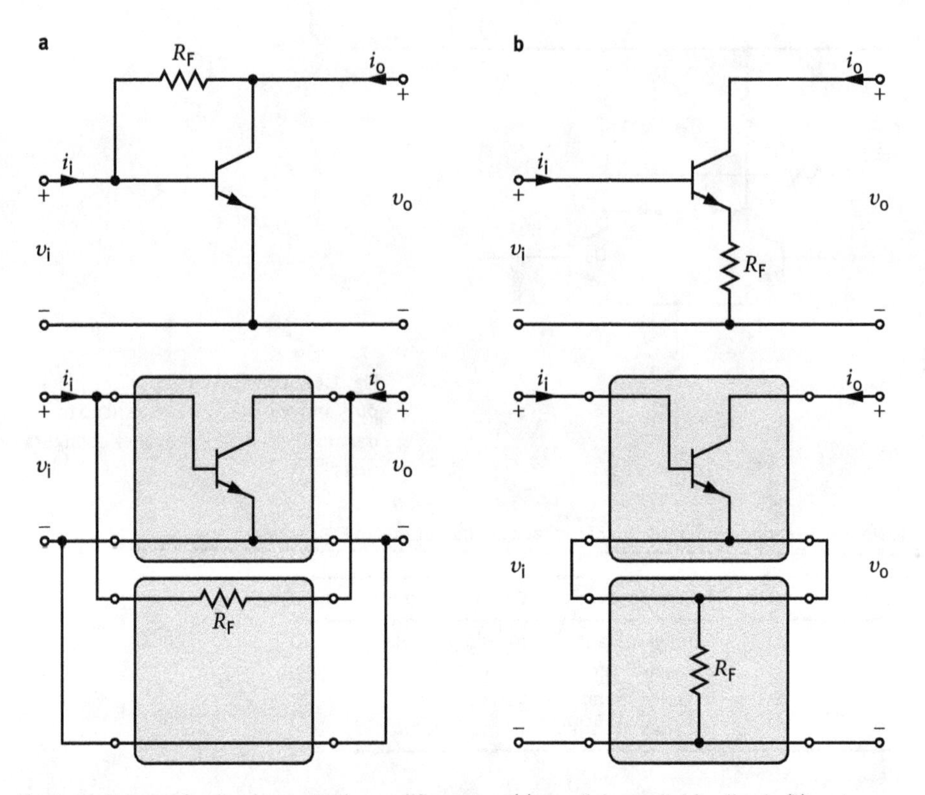

Fig. 2.69. External feedback at a single amplifier stage: (a) parallel–parallel feedback, (b) series–series feedback (and the corresponding two-port circuit models).

Figure 2.69 shows the two obvious feedback configurations with a bipolar junction transistor. In Figure 2.69a, the feedback resistor R_F is from the collector to the base, establishing a parallel–parallel feedback. In Figure 2.69b, the feedback resistor R_F (usually called R_E) leads from the emitter to ground giving rise to a series–series feedback.

As we will show a series–parallel feedback applied to a common-emitter circuit coincides with the common-collector circuit. Applying duality will spare us the effort to prove that the common-base circuit is a common-emitter circuit with parallel–series feedback.

For this exercise it is important that these two facts mentioned before were understood.

- Just from looking at a two-port feedback configuration its symmetry with regard to input and output is obvious (see Figure 2.19). Such an arrangement is bidirectional. Therefore, there is a *forward*-loop gain AB (which, usually, is called just loop gain) and a *reverse*-loop gain BA.
- Whenever a valid comparison is made, it is absolutely necessary to make it under "identical" conditions. Therefore, care must be taken that crucial data will be

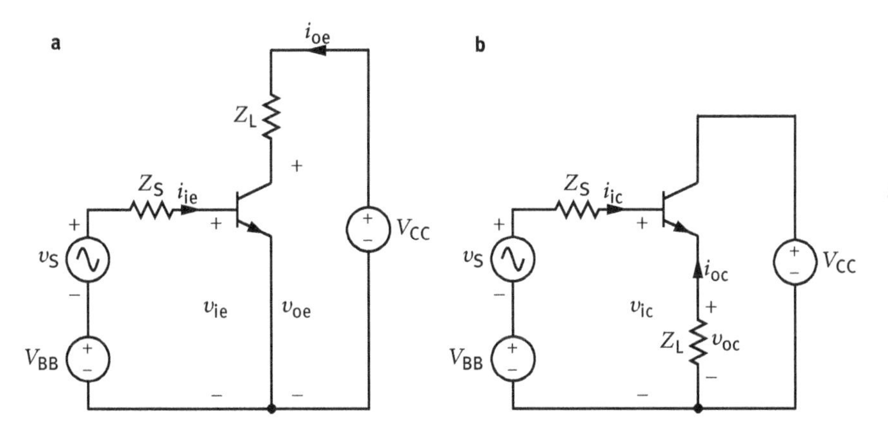

Fig. 2.70. Common-emitter circuit (a) and common-collector circuit (b) with signal voltage v_S, source impedance Z_S and load impedance Z_L. The transistor's operating point is set by the power supply (modelled by V_{BB} and V_{CC}). Note that the signs of input and output quantities are chosen according to the two-port convention.

the same, in particular, that the electrical burden is the same in both cases. Forward properties are burdened by the load impedance and the impedance of the feedback network at the output, reverse properties are burdened by the source impedance and the impedance of the feedback network at the input.

The correct presentation of the output properties of feedback arrangements requires that the reverse loop gain is *not* disregarded.

Example 2.10 (Maximum series–parallel feedback of a common-emitter circuit yields the common-collector circuit). In Figure 2.70b the collector terminal is connected both to the source and the load impedance, i.e. it is grounded. According to the superposition theorem the unconsidered voltage source – here the voltages V_{CC} and V_{BB} of the power supplies – should be replaced by a short-circuit, each. The output voltage of the common collector-circuit is the same as for the common-emitter circuit.

$$v_{oc} = v_{oe}$$

and the input voltage v_{ic} is

$$v_{ic} = v_{ie} - v_{oe} .$$

The output current is given by

$$-i_{oc} = i_{oe} + i_{ie} ,$$

the input current by

$$i_{ic} = i_{ie} .$$

A series configuration at the input makes the (forward) loop gain to a voltage gain therefore we must investigate how much the voltage gain g_{vc} is reduced by the feedback

$$g_{vc} = \frac{v_{oc}}{v_{ic}} = \frac{v_{oe}}{v_{ie} - v_{oe}} = \frac{\frac{v_{oe}}{v_{ie}}}{1 - \frac{v_{oe}}{v_{ie}}} = \frac{g_{ve}^*}{1 - g_{ve}^*} = \frac{g_{ve}^*}{1 + |g_{ve}^*|}. \tag{2.100}$$

From the general feedback theory we know that negative feedback reduces the characteristic transfer parameter by the return difference

$$g_{vc} = \frac{g_{ve}^*}{1 + |AB|}$$

so that the forward loop gain AB is obtained as $AB = g_{ve}^*$. The asterisk denotes that the value under actual load conditions is used.

All of the output voltage is fed back, i.e. the maximum possible feedback (with passive elements) is applied. Any (reasonable) active element would have a rather high voltage gain so that the voltage gain of a common collector circuit is only marginally smaller than 1, e.g., with $|g_{ve}^*| = 100$ the common collector circuit voltage gain g_{vc} would be 0.99. Common collector/drain circuits are called *emitter/source follower* because a voltage gain of about one means that the (small-signal) output voltage (at the emitter/source) is identical with (i.e. it follows) the input voltage (at the base/gate). Using operational amplifiers the corresponding circuit is called *voltage follower* (Section 2.5.2.1).

The input impedance is obtained from

$$Z_{ic} = \frac{v_{ic}}{i_{ic}} = \frac{v_{ie} - v_{oe}}{i_{ie}} = \frac{v_{ie} \times (1 - \frac{v_{oe}}{v_{ie}})}{i_{ie}} = Z_{ie}^* \times (1 - g_{ve}^*) = Z_{ie}^* \times (1 + |AB|). \tag{2.101}$$

Again, the result conforms with the general equation.

Negative series–parallel and series–series feedback provides high input impedances.

The reverse properties can be investigated by using the output as input and replacing the (ideal) signal generator by a short-circuit following the recipes of the superposition theorem (Section 2.1.3.5). Because of the parallel configuration at the output the reverse loop gain BA has the dimension of a current gain.

The parallel configuration pleads for the use of an admittance rather than impedance, therefore, the input admittance Y_{iB} of the feedback two-port should be used

$$Y_{iB} = \frac{1}{Z_S + Z_{ie}^*}. \tag{2.102}$$

Using $v_{oc} = -v_{oe} = v_{ic} - v_{ie} = i_{ie} \times Z_S - i_{ie} \times Z_{ie}^*$ and $i_{oc} = i_{ie} - i_{oe}$ with $v_{ic} = i_{ic} \times Z_S = i_{ie} \times Z_S$ the output admittance Y_{oc} is obtained as

$$Y_{oc} = \frac{i_{oc}}{v_{oc}} = \frac{-i_{ie} - i_{oe}}{i_{ic} \times Z_S + v_{ie}} = \frac{1 + g_{ie}^*}{Z_S + Z_{ie}^*} = Y_{iB}^* \times (1 + BA) \tag{2.103}$$

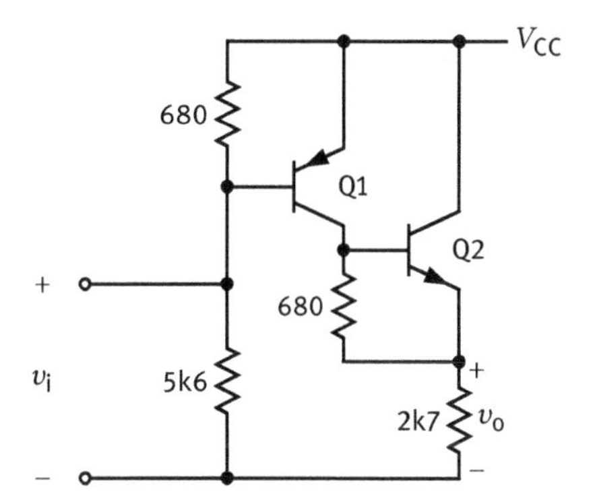

Fig. 2.71. Two-stage amplifier with a dynamically increased load resistance for the transistor Q1.

in agreement with our general findings on feedback actions on impedances.

Negative parallel–parallel and series–parallel feedback provides low output impedances.

Example 2.11 (Bootstrapping with a common collector circuit). In Figure 2.71, a circuit with two transistor stages Q1 and Q2 is shown. Q1 is a common-emitter circuit, Q2 a common-collector circuit. The passive load resistance of Q1 (if the 680 Ω resistor is not connected to the emitter) is 680 Ω in series to 2.7 kΩ, i.e. 3.38 kΩ, parallel to the input impedance of Q2 of roughly 270 kΩ, i.e. about 3.3 kΩ. The voltage gain g_v of Q1 would be low being about proportional to the load impedance. By connecting the 680 Ω resistor to the emitter of Q2 (which acts as an emitter follower with an assumed voltage gain of $g_v = 0.99$) the dynamic impedance of the 680 Ω resistor is $Z_{680} = 680/(1 - 0.99) = 68$ kΩ (bootstrap effect, Section 2.4.3.2) so that the load impedance of Q1 becomes 54 kΩ increasing the voltage gain of Q1 (and of the two-stage amplifier) by roughly a factor of 16.

By using a voltage follower (Section 2.5.2.1) instead of the emitter follower the voltage gain would be much closer to 1 so that the dynamic increase of a resistor due to the bootstrap effect can be much more dramatic.

Problems

?

2.69. Common-Base Circuit

(a) Apply the principles of duality on Example 2.10 and the circuit of Figure 2.70b and get the corresponding equations for the parallel–series feedback, i.e. for the common-base circuit.

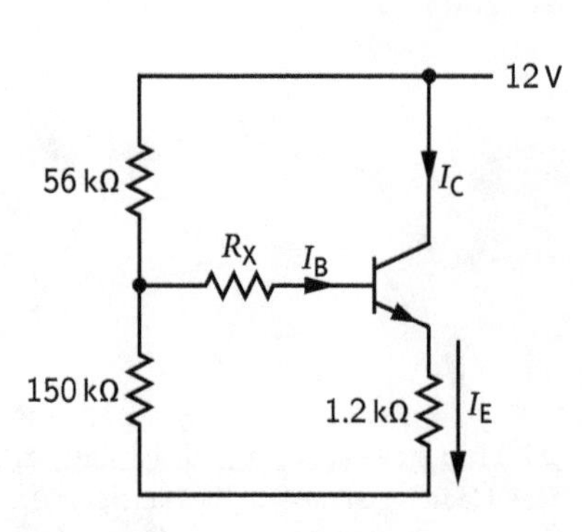

Fig. 2.72. Circuit from Problem 2.70.

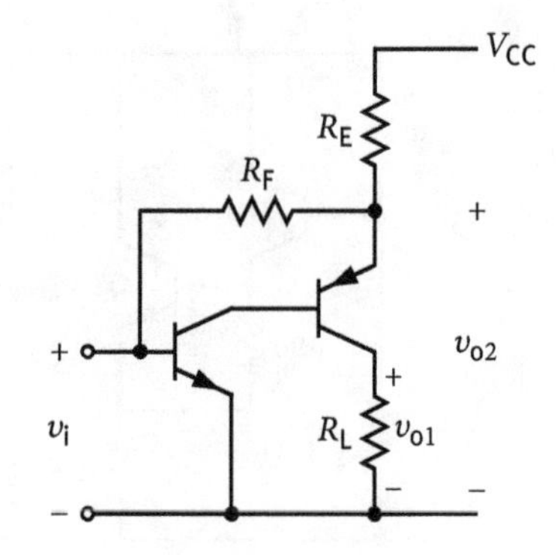

Fig. 2.73. Circuit from Problem 2.71.

(b) Verify the equations by straight calculations with the currents and voltages given in Figure 2.70a.

2.70. Analyze the circuit of Figure 2.72 using Thevenin's theorem and taking advantage of the knowledge that R_E is dynamically enhanced and do the following:
(a) Give the value of R_x when $I_E = 5\,\text{mA}$, $g_i = I_C/I_B = 99$, and $V_{BE} = 0.699\,\text{V}$.
(b) Give the complete (i.e. the input and the output) operating point of the transistor.

2.71. Answer the following questions concerning the circuit of Figure 2.73 for both outputs designated by v_{o1} and v_{o2}.
(a) What kind of feedback exists?
(b) Name the characteristic transfer parameter.
(c) Give attributes (low, medium, and high) to the values of the input and the output impedances.

2.5 Operation amplifiers

Amplifiers with external feedback are called *operation amplifiers* if the characteristic transfer parameter is de facto independent of the transfer parameter of the active device.

Although all four types of active elements with their four types of transfer properties would qualify, usually just *operational amplifiers* (Section 2.3.5.2) are considered. Those are voltage amplifiers with differential input. As such they have high input impedance and low output impedance. As a high forward closed-loop gain (Section 2.4.1) is required, the voltage gain should be as high as possible. In many cases, the essentials of a feedback circuit can be understood using an ideal operational amplifier with input conductance and output resistance zero and an infinite voltage gain.

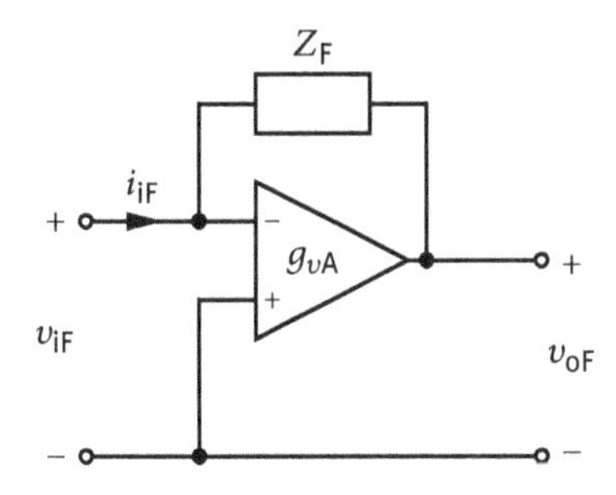

Fig. 2.74. Inverting operation amplifier with general feedback network.

2.5.1 Inverting operation amplifiers

If parallel–parallel feedback with an impedance Z_F is applied to an operational amplifier in the inverting configuration (i.e. the noninverting input is grounded) then the configuration of amplifier and feedback circuit is called *inverting* operation amplifier. Such an operation amplifier is shown in Figure 2.74.

Problem

2.72. Is the (closed-) loop gain of an inverting amplifier with parallel–parallel feedback under normal circumstances (i.e. using passive one-port elements in the feedback loop) positive or negative?

To simplify life we now assume that the operational amplifier has an input admittance of $Y_{iA} = 0$ and an output impedance of $Z_{oA} = 0$. Besides, the voltage gain g_{vA} should be as high as possible. The two following simple facts make an analysis of this circuit a straightforward exercise:

- From (2.92) we know that the input impedance of this circuit is $Z_{iF} = Z_F/(1-g_{vA})$, i.e. with $-g_{vA} \gg 1$ it can be very low establishing a virtual ground (Section 2.4.3.1).
- Because of $Y_{iA} = 0$ (or very small when compared to Y_{iF}) all of the input current i_{iF} flows through the feedback network so that the output voltage v_{oF} is given by $v_{oF} = -i_{iF} \times Z_F$. As i_{iF} flows into the feedback network the voltage at the other end of the network (at the amplifier output) must be negative, in agreement with the inverting mode of the amplifier.

This answer is in perfect agreement with the findings of the general feedback theory where we found that the transfer property of a system with negative feedback becomes independent of that of the active element if a very high (closed-) loop gain is present (2.72). Thus, an inverting operation amplifier is an amplifier with transimpedance $-Z_F$ as characteristic transfer property.

Figure 2.75 depicts the most common use of an inverting operation amplifier. The resistance R_S of a real voltage source (e.g., obtained by Thevenin's theorem) is cascaded with an operation amplifier using a resistor R_F in a parallel–parallel feedback configuration. Because of the node with the virtual ground, the input current is $i_{iF} =$

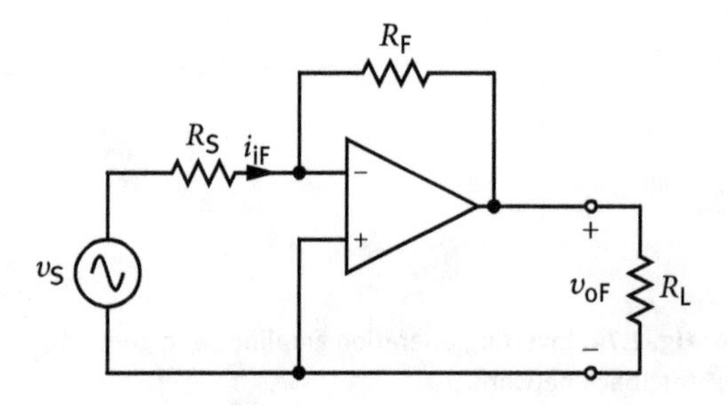

Fig. 2.75. Inverting operation amplifier, e.g., combination of a real voltage source with an inverting operation amplifier.

v_S/R_S and consequently the output voltage

$$v_{oF} = -\frac{i_{iF}}{Y} = -i_{iF} \times R_F = -\frac{v_S}{R_S} \times R_F\,, \quad \text{and} \tag{2.104}$$

$$\frac{v_{oF}}{v_S} = -\frac{R_F}{R_S}\,. \tag{2.105}$$

Some people who have problems in dealing with currents, include R_S into their feedback arrangement which should not be done, alone for the fact that this resistance is outside the feedback loop, and quite self-evidently elements outside the feedback loop cannot have any direct influence on properties of the circuit with feedback. Nevertheless, such an arrangement is called inverting (operation) (voltage) amplifier. If $R_F = R_S$, then $v_{oF}/v_S = -1$ and the signal of the source is just inverted. Such an arrangement is called *inverter*.

Circuit elements that lie outside the feedback loop have no direct influence on the feedback action.

The correct interpretation of the inverting operation amplifier within the framework of feedback circuits is as follows: Because of the virtual ground the resistor R_S acts as a voltage-to-current converter with the transadmittance $1/R_S$ whereas the cascaded operation amplifier works as a current-to-voltage converter with the transimpedance $-R_F$ so that the combined transfer property, voltage gain from source to output, becomes $-R_F/R_S$.

Problem

2.73. An operational amplifier has an input impedance $Z_{iA} = 10\,\text{k}\Omega$ and an open-loop voltage gain of $g_{vA} = -200$. The output impedance is $Z_{oA} = 0\,\Omega$. By connecting the output to the inverting input with a resistor of $10\,\text{k}\Omega$ feedback is established.

(a) What kind of feedback is this?

(b) How much does the voltage gain g_{vF} of the feedback circuit differ from g_{vA} the voltage gain of the amplifier?

Example 2.12 (Ideal vs. real operational amplifier). The circuit of an inverting operation amplifier (Figure 2.76) uses a very unsophisticated operational amplifier. The

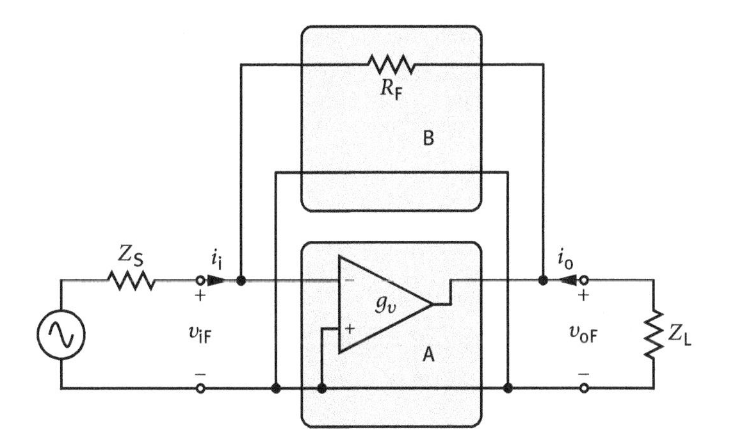

Fig. 2.76. Parallel–parallel feedback applied to a differential amplifier.

Table 2.7. Hybrid parameters of the two two-ports of the feedback loop in Figure 2.76.

	Two-port A		Two-port B
h_{iA}	$Z_{iA} = 20\,\text{k}\Omega$	h_{iB}	$R_{FB} = 50\,\text{k}\Omega$
h_{rA}	0	h_{rB}	1
h_{fA}	$-g_v \times Z_{iA}/Z_{oA}$ $= 5{\times}10^3 \times 2{\times}10^4/25$	h_{fB}	-1
h_{oA}	$1/Z_{oA} = 40\,\text{mS}$	h_{oB}	0
ΔA	$h_{iA}h_{oA} - h_{rA}h_{fA} = 800$	ΔB	$h_{iB}h_{oB} - h_{rB}h_{fB} = 1$

parameter values of this operational amplifier and of the feedback two-port are given in Table 2.7. The impedance Z_S of the source is 1 kΩ and that of the load Z_L is 1 kΩ, as well. With these numbers we get

- the forward (closed-) loop gain $AB = i_{oB}/i_{iA} = -1950.7$,
- the reverse (closed-) loop gain $BA = i_{oA}/i_{iB} = -192\,514$,
- the input admittance $Y_{iF} = Y_{iA} \times (1 - AB) = 1/10.248$ S,
- the output admittance $Y_{oF} = Y_{iB} \times (1 - BA) = 1/0.265$ S.

The loaded voltage gain g_v^* is smaller than the open-loop gain $g_v = -5000$ because of the voltage division between Z_{oA}, and Z_L shunted by the dynamic impedance of R_F as seen from the output, so that $g_v^* = -4876$.

Finally, the characteristic transfer parameter, the transimpedance $r_{mF} = r_{mA}/(1-AB) = Z_{iA} \times g_v^*/(1 - AB) = -49.963$ V/mA. In the ideal case, it is expected to be the inverse of the transadmittance of the feedback two-port $A_F = 1/B = -\frac{1}{1/(50\,\text{k}\Omega)} = -50$ V/mA.

Table 2.8 compares the ideal values with the values obtained with this unsophisticated operational amplifier. In particular, the characteristic transfer parameters agree rather nicely. Up-to-date operational amplifiers resemble their ideal counterpart much closer. Both the input impedance and the open-loop voltage gain are about two orders of magnitude higher. Thus operation amplifiers using these devices will approach the ideal behavior even more closely.

Table 2.8. Comparison of the ideal values with realized values of the amplifiers from Example 2.12.

Type	Ideal value	Actual value
	Operational amplifier	
$1/Z_{iA}$	0	1/(20 kΩ)
Z_{oA}	0	25 Ω
$-1/g_{vA}$	0	1/5000
	Inverting operation amplifier	
Z_{iF}	0	10.25 Ω
Z_{oF}	0	0.26 Ω
$-1/g^*_{vF}$	0	1/4876
$-1/AB$	0	1/1951
$-1/BA$	0	1/192 514
r_{mF}	−50 kΩ	−49.96 kΩ

2.5.1.1 Summing amplifier

The low-impedance node of the inverting amplifier is an ideal spot for summing n currents i_k, resulting in an output voltage v_{oF} of

$$v_{oF} = -R_F \times \sum_{k=1}^{n} i_k . \tag{2.106}$$

If the currents i_k come from voltage sources v_{Sk} with equal resistances R_S, then one gets

$$\sum_{k=1}^{n} i_k = \frac{1}{R_S} \times \sum_{k=1}^{n} v_{Sk} \tag{2.107}$$

and

$$v_{oF} = -\frac{R_F}{R_S} \times \sum_{k=1}^{n} v_{Sk} . \tag{2.108}$$

With $R_F = R_S$ this becomes

$$v_{oF} = -\sum_{k=1}^{n} v_{Sk} . \tag{2.109}$$

Thus the name summing amplifier (Figure 2.77).

2.5.1.2 Nonlinear amplifiers

As the nature of R_S and R_F do not enter into above considerations we may switch to the generalized values Z_S and Z_F. Considering that the task of R_S is voltage-to-current conversion and that of the operation amplifier current-to-voltage conversion, it should be clear that instead of either or both of the resistors in Figure 2.75 nonlinear elements may be used.

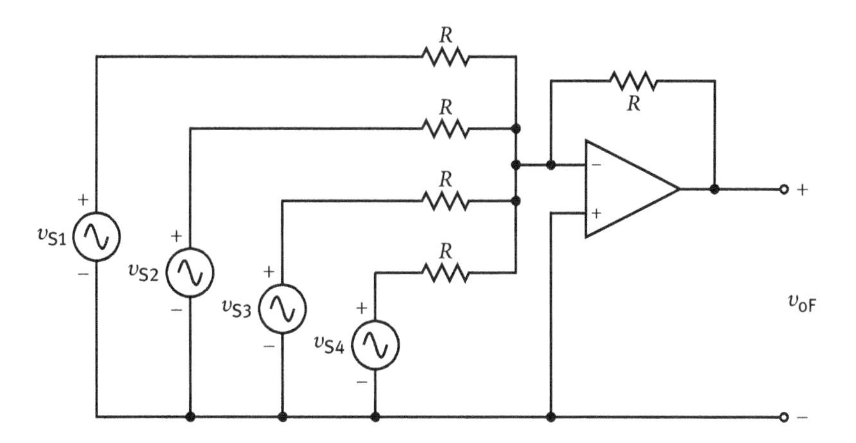

Fig. 2.77. A summing amplifier.

In the forward direction, the current I_D of a diode is, over several decades, given by

$$I_D = f_1 \times \exp(f_2 \times V_D)\,. \tag{2.110}$$

Using a diode instead of R_S gives an input current that depends exponentially on v_S and consequently an output voltage v_o that depends exponentially on v_S providing an exponential amplifier.

On the other hand, replacing only R_F by a diode produces an output voltage v_o that is the logarithm of i_i which is $i_i = v_S/R_S$, so that v_o would be logarithmically dependent on v_S and a logarithmic amplifier is obtained. The nonlinear characteristic need not be realized by a single component but may be the result of a sophisticated circuit composed of several or many (nonlinear) components.

The quality of the nonlinear conversion depends solely on the shape and stability of the characteristic of the nonlinear component. In sensitive applications thermal changes of the characteristic can be strongly reduced by keeping the nonlinear component at a fixed temperature (by means of a thermostat).

Cascading a current-to-voltage converter with a voltage-to-current converter results in a current-to-current converter, i.e., a current amplifier (Section 2.5.6.2).

2.5.1.3 Active voltage clipping (voltage limiter)

Figure 2.78 shows how diode limiters (Section 2.2.7.2) can be incorporated into a feedback loop. The feedback resistor R_F is shunted by a diode. If the diode is reverse biased it has nearly zero conductance and is, therefore, functionless. However, a positive input signal to the inverting input makes the output voltage negative, forward-biasing the diode. As the differential input voltage stays necessarily very small (output voltage divided by voltage gain) the output voltage is the voltage at the noninverting input mi-

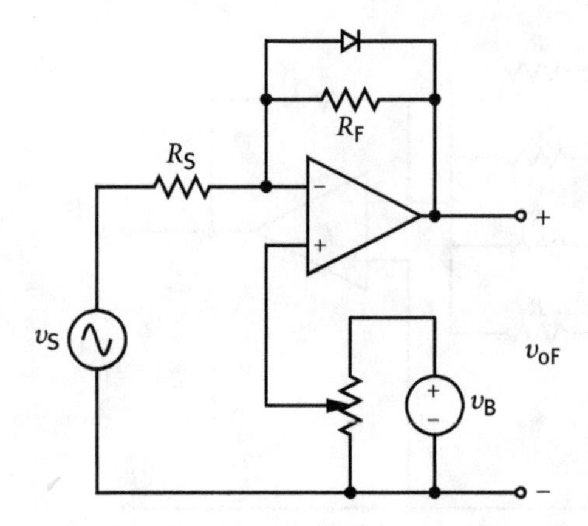

Fig. 2.78. A voltage limiter accomplished by active voltage clipping.

nus the forward voltage of the diode. It cannot be less, i.e. the output voltage is limited to that voltage.

By raising the voltage at the noninverting input (e.g., by means of a potentiometer as shown in Figure 2.78) the voltage at the inverting input is raised by the same amount and consequently the output voltage as well. This allows setting the voltage limit according to the need.

2.5.2 Noninverting operation amplifiers

Operational (or *operative*) amplifiers, i.e. differential amplifiers with high input impedance and low output impedance, and high gain are also used for operation amplifiers with series–parallel feedback. Such an operation amplifier is shown in Figure 2.79.

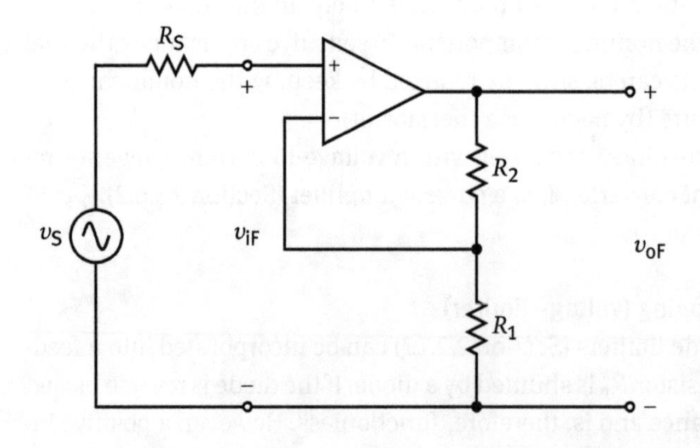

Fig. 2.79. Noninverting operation amplifier with standard resistive feedback network.

The following two simple facts make an analysis of this circuit a straightforward exercise.

- From (2.87) we know that the input impedance Z_{iF} of this circuit is $Z_{iA} \times (1 - AB)$. With $|g_{vA}| \gg 1$ the (negative) forward loop gain which is also a voltage gain due to the series configuration at the input, is very large, too, making the input impedance Z_{iF} very large (1 MΩ to 1 TΩ) even for nonideal input impedances Z_{iA}.
- Because of the high gain ($|g_{vA}| \gg 1$) the differential input voltage at the operational amplifier is essentially zero.

With this information (negligible input admittance and negligible differential input voltage) we get from the voltage division (in Figure 2.79) $v_{iF} = v_{oF} \times R_1/(R_2 + R_1)$ the voltage gain g_{vF} (the characteristic transfer parameter)

$$g_{vF} = \frac{v_{oF}}{v_{iF}} = 1 + \frac{R_2}{R_1} . \tag{2.111}$$

Again the characteristic transfer property is solely determined by the property of the feedback circuit (voltage divider) as required for operation amplifiers (Section 2.4.1).

2.5.2.1 Voltage follower

Figure 2.80 depicts a special case of a noninverting operation amplifier. With $R_2 = 0$ the voltage gain g_{vF} in above equation becomes 1, i.e. the output voltage v_{oF} follows the input voltage v_{iF}. (The value of R_1 does not matter so that R_1 may be left out.) This circuit is called voltage follower. Actually g_{vF} is smaller than 1, namely

$$g_{vF} = \frac{1}{1 + \frac{1}{g^*_{vA}}} . \tag{2.112}$$

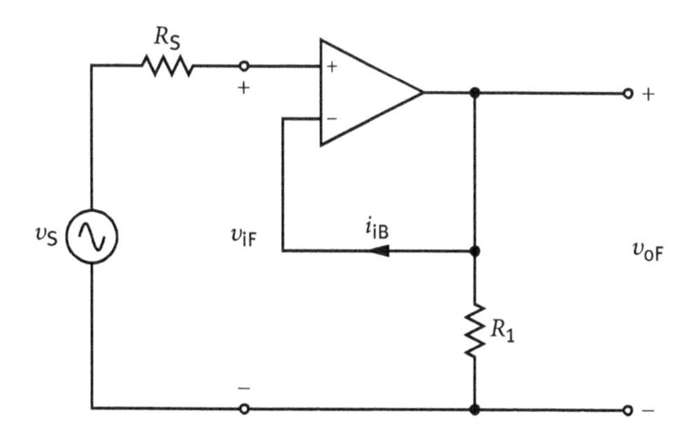

Fig. 2.80. Noninverting operation amplifier with $R_2 = 0$, i.e. with a voltage gain $g_{vF} = 1$ (voltage follower).

This difference is due to the (very small) differential input voltage which is the output voltage divided by the voltage gain g^*_{vA} of the amplifier under actual burden as indicated by the asterisk.

However, there is another factor to be considered. As mentioned before (Section 2.3.5.1) the suppression of the common-mode signal in amplifiers with differential input is not perfect. As $v_{iCM} \approx v_{iF}$ and the contribution of the common mode signal to the output signal is suppressed by the common mode rejection ratio $CMRR$, g_{vF} becomes

$$g_{vF} = \frac{1}{1 + \frac{1}{g^*_{vA}}} \times \left(1 + \frac{1}{CMRR}\right). \tag{2.113}$$

Is an amplifier with a voltage gain of (less than) 1 an amplifier at all? The answer is yes, if the power gain $g_p > 1$. From (2.29) it is clear that with $g_v = 1$, $Z_{iF} > Z_L$ is necessary to achieve a power gain $g_p > 1$.

Comparing equation (2.112) with the general equation (2.69) using the (forward) loop gain results in a perfect agreement when equating AB with g^*_{vA}:

$$g_{vF} = \frac{1}{1 + \frac{1}{g^*_{vA}}} = \frac{1}{\frac{1}{g^*_{vA}} + \frac{AB}{g^*_{vA}}} \tag{2.114}$$

Thus, the forward closed-loop gain is a voltage gain as required by the serial configuration at the input $AB = g^*_{vA}$.

Example 2.13 (Importance of reverse loop gain). The calculation of the output impedance of a voltage follower (Figure 2.80) requires the knowledge of the reverse loop gain. Because of the parallel configuration at the output the reverse-loop gain is a current gain, $BA = i_{oA}/i_{iB}$. The input current i_{iB} into the feedback network flows through the input impedance Z_{iA} of the amplifier and the impedance R_S of the source to ground

$$Z_{iB} = Z_{iA} + R_S. \tag{2.115}$$

The reverse closed-loop gain BA equals

$$BA = \frac{i_{oF}}{i_{iB}} = \frac{i_{oA}}{-i_{iA}} = -g^*_{iA}. \tag{2.116}$$

Thus, the output admittance Y_{oF} is given as

$$Y_{oF} = Y_{iB} \times (1 - BA) = Y_{iB} \times \left(1 + g^*_{iA}\right) = \frac{1}{Z_{iA} + R_S} \times (1 + g^*_{iA}). \tag{2.117}$$

In a simplistic view, disregarding the reverse loop gain, one would expect

$$Y_{oF} = Y_{oA} \times (1 + g^*_{vA}) \tag{2.118}$$

which is wrong because

$$Y_{oA} \neq \frac{1}{Z_{iA} + R_S}, \text{ and} \tag{2.119}$$

the forward loop gain g^*_{vA} differs from the reverse loop gain g^*_{iA}, e.g., Table 2.8.

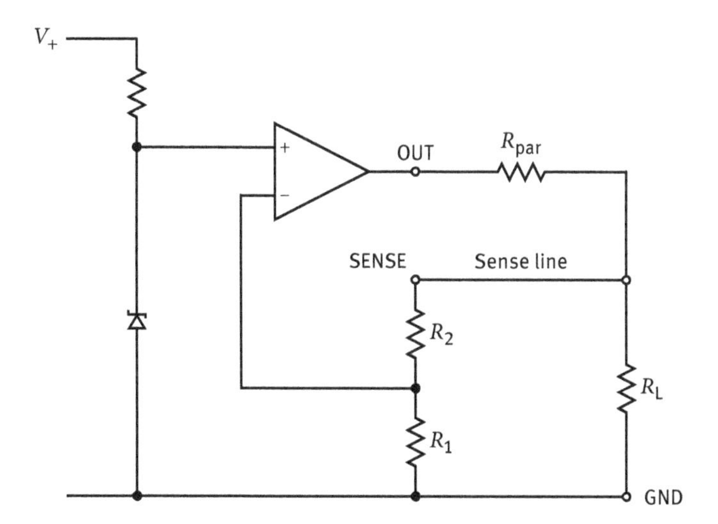

Fig. 2.81. Voltage power supply with sense line: noninverting amplifier with a correction for the signal loss in the output line (*remote sensing*).

2.5.2.2 Linear voltage power supplies (power amplifiers)

An ideal voltage source delivers a constant voltage through an insignificant impedance. So it is not surprising that the output of an operation amplifier with serial-parallel feedback is a good choice for supplying a voltage. All that is needed is a reference voltage at the noninverting input. By varying the feedback resistor the gain can be adjusted allowing a variation of the output voltage. The reference voltage is usually obtained from a reference (Zener) diode which is so constructed that at a given working point the temperature coefficient (i.e. the dependence of the reference voltage on temperature) is effectively zero. The (very) high input impedance of the amplifier is only a small burden to the reference voltage circuit so that the operating point of the reference diode is not affected by the loading by the power amplifier.

Figure 2.81 depicts the case of a power supply with a sense line allowing remote sensing. At higher currents the voltage drop in the supply line of the load due to parasitic resistances R_{par} may not be negligible, i.e. the load does not receive all the voltage provided by the output of the amplifier. In that case the voltage drop can be compensated for by raising (automatically) the output voltage of the amplifier by that amount. For that purpose a sense line to the load is necessary, so that the actual voltage at the load is fed back, rather than the output voltage of the amplifier.

The name *power supply* suggests already that such a voltage source is expected to deliver high output power (i.e. high output current). As we have learnt a standard class A amplifier consumes the maximum of power in the idle condition. But it allows that the output current may be increased or decreased as needed. From a general voltage source it is expected both to deliver current and to sink current. The latter is in agreement with the superposition theorem which states that a voltage source is a short-circuit for external currents.

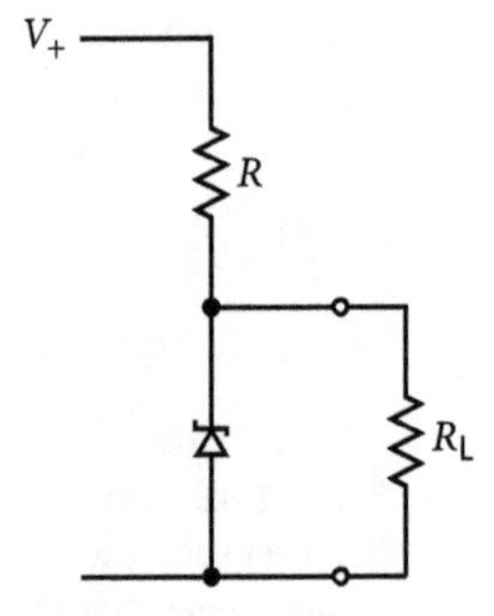

Fig. 2.82. Voltage stabilization by means of a Zener diode in series to a resistor R providing shunt regulation.

Power considerations might require that a class B or B' amplifier be used, i.e. the operating point is at the edge of the dynamic range, either allowing the current to increase, i.e. to deliver current as done by *series regulators*, or to decrease, i.e. to sink current as done by *shunt regulators*. A simple demonstration of shunt regulation is given in Figure 2.82. There a Zener diode that maintains its voltage will sink current fed into the output.

If you need both delivering current and sinking current, the output stage of the amplifier should be a push-pull device with complementary semiconductor power components, with class B operating points. The N-type would be used for the series regulator, the P-type for the shunt regulator.

Commercial power supplies, if at all linear, are very sophisticated having several additional features: Possibly a preregulator that minimizes the power dissipated in the output stage of the amplifier by maintaining a low operating voltage of that stage, an output characteristic with constant voltage (cv) and constant current (cc; making the device safe against short-circuits), and provisions against *thermal*, *current*, and *power* overload. The cc mode does not only allow limiting the output current to a preselected value but makes it possible to use voltage power supplies in parallel in contradiction to the behavior of ideal voltage sources. In that case the lowest voltage of the supplies will be provided in the cv mode with the other supplies in the cc mode supplementing additional current needs.

A linear regulated power supply excels the competitors with regard to regulating properties and fast response to changes of line or/and load. Its recovery time after such changes is smaller than in supplies using other techniques. Its circuit simplicity provides a very effective solution coupled with high reliability, sufficient power in most applications with stable regulation and little noise.

2.5.2.3 Precision half-wave rectifier

Across a diode there will be a voltage drop, called forward voltage v_f when used in the forward direction (see, e.g., Figure 2.13). However, at very small forward voltages a diode will have a rather high impedance. In the circuit of Figure 2.83 the voltage drop

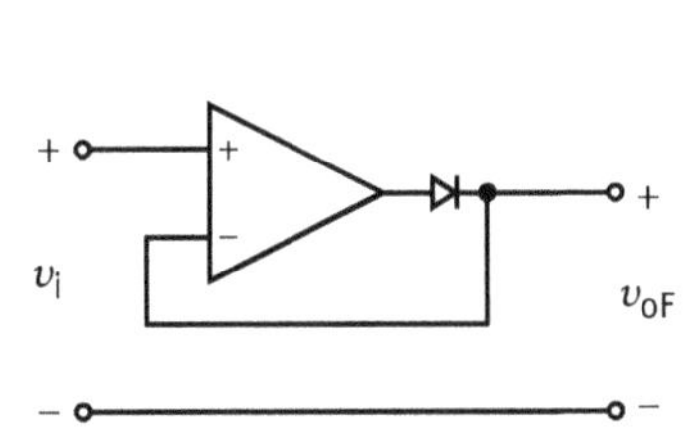

Fig. 2.83a. Basic precision half-wave rectifier.

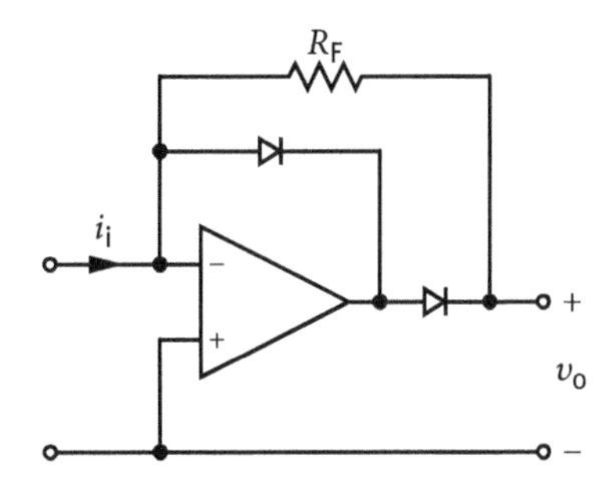

Fig. 2.83b. Improved precision half-wave rectifier.

at the diode is minimized by feedback action. Observe that

- v_{oA} differs from v_{oF},
- due to the diode action no output current will flow for $v_{oA} < 0$ so that v_{oF} will be zero under this condition,
- the feedback loop is open if no output current flows,
- because of the required high voltage gain g_{vA} the differential input voltage v_{iA} has to be small (if the amplifier is in the operational mode), and
- for the last reason the feedback action must increase v_{oA} as compared to v_{oF} so that the latter voltage is $v_{oF} + v_{iA} = v_{iF}$, i.e. $v_{oF} \approx v_{iF}$.

Feedback effectively removes the forward voltage of diodes making it possible that even small (positive, in this circuit) signals can pass the diode.

Problem

2.74. The precision rectifier of Figure 2.83a will have zero output voltage v_o for negative input voltage v_i. Explain why it is possible that under these conditions the differential input voltage is not necessarily small!

2.5.3 Difference operation amplifier

The circuit in Figure 2.84 is a composite of an inverting and a noninverting amplifier. Assuming linearity (small-signals) we can apply the superposition theorem and compose the response of the circuit to the two voltage sources from the responses to the individual sources. To this end we replace first the ideal voltage v_{S2} and then v_{S1} by a short-circuit.

When we shorten source 2 the circuit becomes a cascade of a series resistor R_{S1} with an inverting operation amplifier. The voltage gain g_{v1} of this arrangement is (see Section 2.5.1)

$$g_{v1} = -\frac{R_F}{R_{S1}} . \tag{2.120}$$

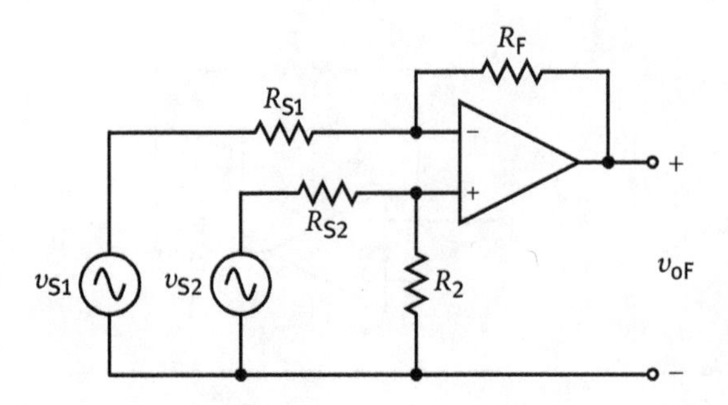

Fig. 2.84. Difference operation amplifier.

However, there are two resistors which are not present in Figure 2.75. They are in parallel to each other and in series to the input impedance. Because of the required high input impedance of the operational amplifier these resistors do not really affect the small-signal operation of the inverting amplifier.

Shortening voltage source 1 makes the circuit to a cascade of a voltage divider and a noninverting operation amplifier. The voltage v_{S2} is first divided to the fraction $R_2/(R_{S2} + R_2)$ which is then amplified by the noninverting operation amplifier with $(R_F + R_{S1})/R_{S1}$. Under the condition that

$$\frac{R_F}{R_{S1}} = \frac{R_2}{R_{S2}} \tag{2.121}$$

the combined output voltage v_{oF} is given by

$$v_{oF} = (v_{S2} - v_{S1}) \times \frac{R_F}{R_{S1}} , \tag{2.122}$$

it is proportional to the difference of the two source voltages. For this reason this circuit is called *difference amplifier* (or ambiguously differential amplifier).

There are two main contributors that influence the quality of the resulting output voltage (see Section 2.3.5.1)

- common mode contributions, and
- asymmetry (offset) contributions.

The common mode input signal $v_{iCM} = (v_{i1} + v_{i2})/2$ adds to the output signal an (amplitude dependent) component $v_{oCM} = v_{iCM} \times g_v^*/CMRR$, and because of the unavoidable asymmetry of the two input stages an offset voltage (and current) at the input is needed to balance the output to zero if the input voltages are zero. The latter requires that the impedances of the circuit as seen from the two inputs are equal.

Figure 2.85 shows a network for subtracting signals that avoids common mode problems. Subtracting is reached by summing signal 1 with the inverted signal 2. Both amplifiers have virtual ground (Section 2.4.3.1) so that common mode problems do not occur. This arrangement does not only require an additional amplifier but also the asymmetry of the two input channels limits its application.

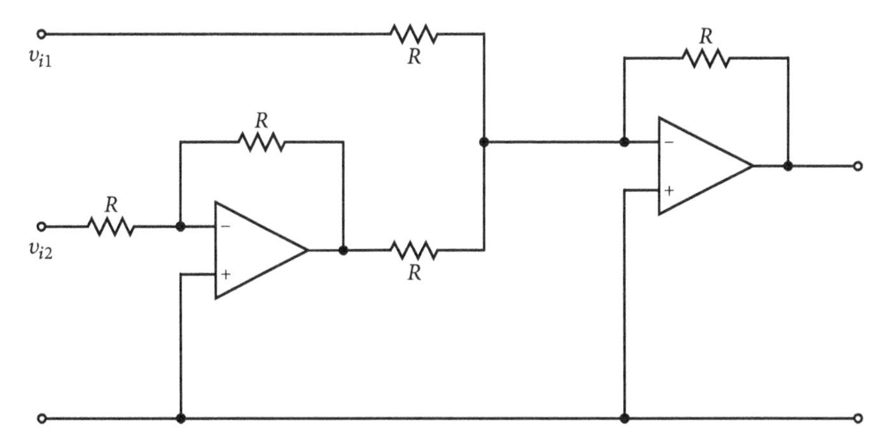

Fig. 2.85. Composite difference amplifier.

2.5.4 Operation amplifiers with positive feedback

All equations containing the (closed-) loop gain are valid for positive feedback as well, considering that now the loop gain is positive. Therefore, the return difference is $1 - |AB|$ or $1 - |BA|$, respectively. When the loop gain is positive but less than 1 the return difference becomes < 1 and the properties of a two-port are affected in a way just opposite to negative feedback, i.e. all advantages of negative feedback for amplifiers are reversed to disadvantages. Only the increase of the transfer parameter can be beneficial. In special cases it pays to increase the gain by a local positive feedback to increase the overall loop gain of a negative feedback loop that extends over several stages. The main application of positive feedback is in oscillator design with which we will deal in Chapter 4. In tailoring dynamic negative impedances, stable positive feedback is also invaluable.

Problem

2.75. Under which condition is an amplifier with positive feedback stable?

2.5.4.1 Negative impedance converter (NIC)

Figure 2.86 shows an operation amplifier configuration that produces negative impedances dynamically. There is negative feedback to the inverting input and positive feedback to the noninverting input. Assuming that the operating point of the circuit provides a stable condition the analysis is much helped by the fact that the input voltage of the operational amplifier is practically zero $v_{iA} \approx 0$. For that reason $v_{iF} = v_{oNIC}$. On the other hand $v_{oNIC} = v_{oA} \times R_L/(R_2 + R_L)$ and $v_{oF} = v_{oA} = v_{iF} - i_{iF} \times R_1$. Using these equations we get

$$Z_{iF} = \frac{v_{iF}}{i_{iF}} = \frac{v_{oNIC}}{i_{iF}} = -R_L \times \frac{R_1}{R_2}, \tag{2.123}$$

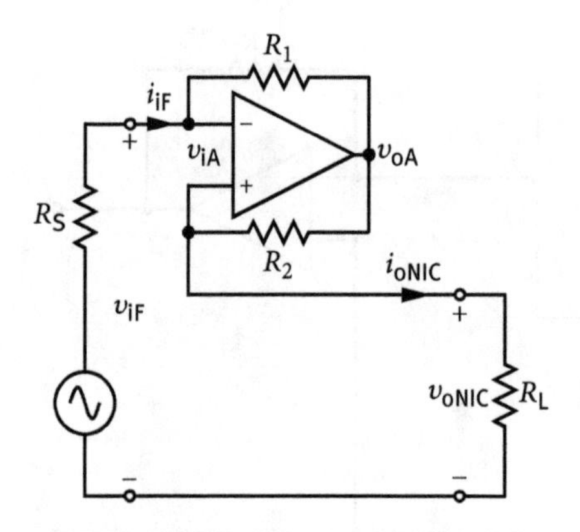

Fig. 2.86. Negative impedance converter based on negative and positive feedback with parallel configurations at the output of the amplifier.

and in analogy

$$Z_{oNIC} = \frac{v_{oNIC}}{i_{oNIC}} = \frac{v_{iF}}{i_{oNIC}} = -R_S \times \frac{R_2}{R_1} . \tag{2.124}$$

Alternatively, the output impedance can be obtained by calculating the dynamic impedance of R_2 (Section 2.4.3.2)

$$Z_{oNIC} = \frac{R_2}{1 - g_{vF}} = \frac{R_2}{1 - \frac{R_1 + R_S}{R_S}} = -R_S \times \frac{R_2}{R_1} . \tag{2.125}$$

To have the operation amplifier in a stable condition the signal provided by positive feedback to the noninverting input *must at no moment* be stronger than that of the negative feedback acting on the inverting input.

The fractions of v_{oF} that are fed back can be obtained by voltage division of v_{oF} $(= v_{oA})$

$$v_{iF} = v_{oA} \times \frac{R_S}{R_1 + R_S} \geq v_{oNIC} = v_{oA} \times \frac{R_L}{R_2 + R_L} . \tag{2.126}$$

With $R_1 = R_2$ the stability condition degenerates to $R_S \geq R_L$ which means that the system is stable with $1/R_S = 0$ (open-circuit) or with $R_L = 0$ (short-circuit). In other words, the input i-v-characteristic has an S-shape, the output i-v-characteristic an N-shape.

Problems

2.76. Show that the load line of a short-circuit has only one intersection with an N-shaped i-v-characteristic.

2.77. Show that the load line of an open-circuit has only one intersection with an S-shaped i-v-characteristic.

2.5.4.2 Applications of negative impedances

Negative impedances (static or dynamic ones) may be used to counteract the effect of positive impedances of the same kind. The best known example is the *undamping* of a resonant circuit (e.g., Example 4.2 in Section 4.2.1.1). There, the effect of a resistive component in the circuit is cancelled by a negative dynamic resistance. In Section 2.4.3.2 we have found that for parallel–parallel feedback the dynamic admittance of a feedback admittance Y_F becomes at the input $Y_{iF} = Y_A \times (1 - g_v)$. For $g_v > 1$ we get a negative admittance with an N-shaped i-v-characteristic, i.e. the feedback is stable when the circuit is short-circuited at the input.

The dual arrangement with a feedback impedance Z_F in a serial-serial feedback circuit gives a dynamic impedance of $Z_i = Z_F \times (1 - g_i)$. With $g_i > 1$ negative impedances with an S-shaped i-v-characteristic are obtained. Such circuits are stable with a open-circuit at the input.

Standard operational amplifiers are voltage amplifiers. For them the parallel–parallel configuration comes naturally (high input impedance, low output impedance). In this case, conductances or capacitances in the feedback loop can produce negative conductances or capacitances at the input. Thus the (positive) admittance of a current source can be compensated so that the source is not loaded but delivers effectively all of its current into the amplifier circuit. Negative resistance finds use at the output of amplifiers for loudspeakers. In this case, *microphonics* of the loudspeakers are damped optimally. Negative capacitance at the input of amplifiers is very helpful in amplifiers for electrophysiological applications because it compensates unavoidable (stray-)capacitances in the sample to be investigated.

2.5.5 Dissecting the term gain

As discussed before, the main purpose of electronics is to transfer (signal) power into a load. Usually, the circuit transferring the power is symbolized by a two-port with two input variables and two output variables. Electric power is best determined through measurements of voltage or/and current. Therefore, there are four options to describe the transfer property of a two-port (Section 2.2.2). The usual favorite is the ratio between output voltage v_o and input voltage v_i, called voltage gain g_v. The following discussion will be based on the voltage gain as transfer property. The other three behave in a dual way.

Voltage gain depends on the operating point. Therefore, it should be presented by a characteristic rather than a number. Besides each device has its own individual characteristic. Therefore, it is necessary to work with typical characteristics. As a byproduct this requires that only such circuits should be designed and built that are not sensitive to the actual characteristic.

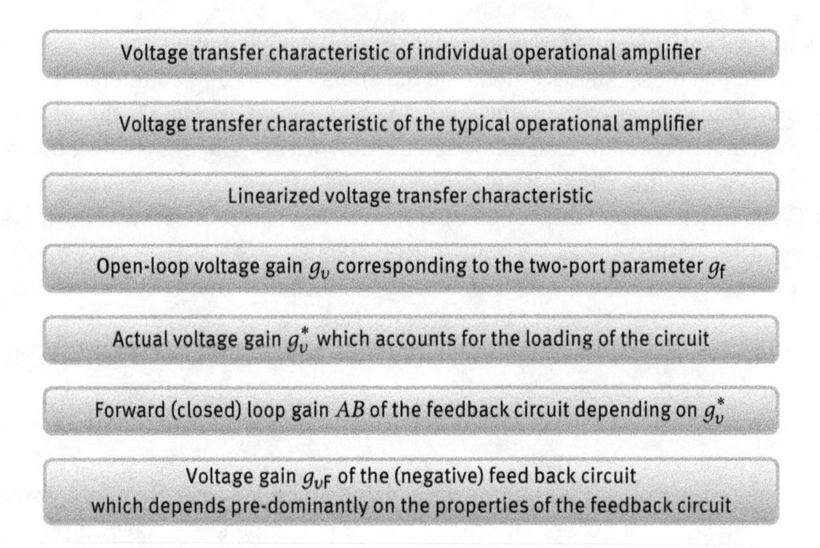

Fig. 2.87. Hierarchy of the voltage transfer in a series–parallel feedback circuit.

Modern circuit design is aimed at becoming independent of the individual characteristics of the active electronic components.

Thus, the first step is to find the typical voltage transfer characteristic and to linearize it so that it can be expressed by a single parameter, the voltage gain g_v as found in data sheets. As this parameter should be usable for all loads, it is given under *open-circuit* condition and is called *open-loop gain*. Thus, it is identical with the linear two-port parameter g_f (Section 2.2.2.1).

However, in an actual circuit the voltage gain with the actual burden is of particular interest, symbolized in this book by g_v^*.

Applying series–parallel feedback introduces a new type of voltage gain v_{oF}/v_{iF}, the voltage gain of the amplifier with feedback which, in the case of negative feedback, is primarily determined by the feedback circuit and not by the transfer property g_v of the active two-port. Furthermore, the voltage gain g_v^* is part of the forward loop gain AB of this feedback circuit. This hierarchy is sketched in Figure 2.87.

In addition there are two "odd" voltage gains to consider.

- It is generally accepted (bad) practice to make the inverting operation amplifier with transimpedance as its characteristic transfer property to a voltage amplifier by including the source impedance R_S arriving at a *source-voltage gain* which is defined as $g_{vS} = v_{oF}/v_S$.
- Then there is the noise gain. Not quite unexpectedly, the feedback circuitry may act differently on amplifier noise which originates from inside the circuit than on a signal from outside. An example is given in Figure 2.88.

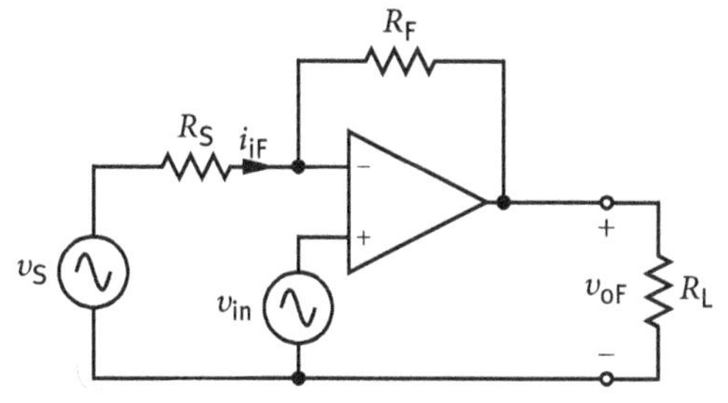

Fig. 2.88. Negative parallel–parallel feedback with internal noise.

For a source signal this is a transimpedance amplifier with a voltage gain from source to output of $-R_F/R_S$. Noise in the amplifier is usually expressed by a noise voltage source at the input of said amplifier. For a noise signal the signal source v_S is equivalent to a short-circuit (due to the superposition theorem) so that for the noise signal v_{in} a serial-parallel feedback configuration is in force with a noise voltage gain g_{vn} of

$$g_{vn} = 1 + \frac{R_F}{R_S} . \tag{2.127}$$

Problems

2.78. Why is it not unexpected that a feedback circuit may behave differently to external and internal signals?

2.79. A change in the biasing (of the operating point) acts very similar to noise generated in the active element. What kind of feedback configuration is responsible for the counter reaction of the circuit on such changes?

2.5.6 Current amplifiers

A current mode operational amplifier is basically an active device with low input impedance, high current gain, and high output impedance. None of the three-terminal components provide these characteristics by themselves. As a minimum, a two-stage structure is required, consisting of a common base input stage providing a low input impedance and a common-emitter output stage providing high current gain and a high output impedance. Thus, a transimpedance input stage is followed by a transadmittance output stage. The current gain is the product of the transimpedance and the transadmittance.

If a differential output is required, a differential long-tailed pair is an obvious choice for the output stage. One input to the long-tailed pair is used for the signal the other input is connected to a constant bias voltage source.

Problem

2.80. Name one reason why voltage amplifiers are preferred over current amplifiers.

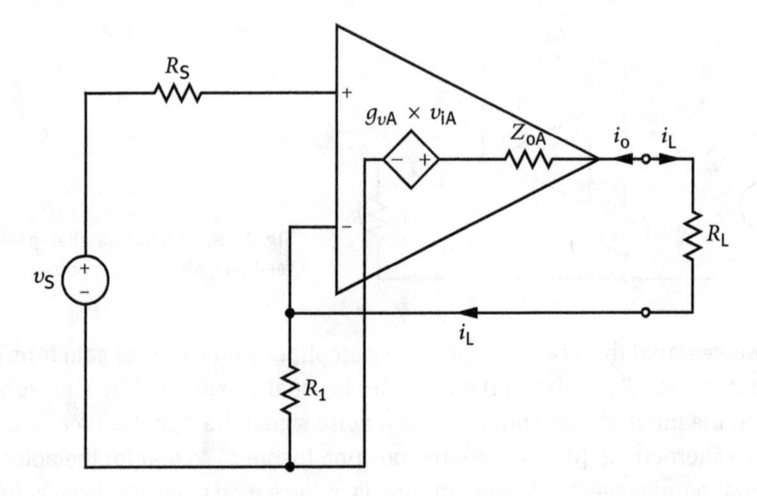

Fig. 2.89. Active current source based on an operational amplifier. The load current i_L is a high impedance output current.

2.5.6.1 Current sources

An ideal current source has zero conductance, just dual to the voltage source. Consequently the output configuration of an active current source must be dual to that of an active voltage source. There, the low output impedance is obtained by negative parallel feedback at the output. Therefore, negative serial feedback must be applied.

The common base circuit is a parallel–series feedback circuit of a common-emitter circuit. Therefore, it qualifies too. Standard operational amplifiers with their low output impedance are not well suited for a series feedback at the output because that feedback depends on the output current, not voltage. Nevertheless, the circuit shown in Figure 2.89 provides an active current source based on a conventional operational amplifier.

This circuit resembles that of the noninverting operation amplifier except for the resistor R_2 that is replaced by the load resistance R_L. When calculating the (output) impedance of the circuit (at the position of the load resistor) the signal source may be replaced by a short-circuit (its impedance, due to the superposition theorem). Then for a current i_L through the current source the following relations are found (assuming infinite input impedance of the amplifier):

$$v_{iA} = i_L \times R_1 \,,$$

and with $v_o = v_L$ and $i_o = -i_L$ we get

$$v_o = i_o \times R_1 \times |g_{vA}| + i_o \times Z_{oA} + i_o \times R_1 \,,$$

yielding

$$Z_{oL} = \frac{v_L}{i_L} = Z_{oA} + R_1 \times (1 + |g_{vA}|) \,. \tag{2.128}$$

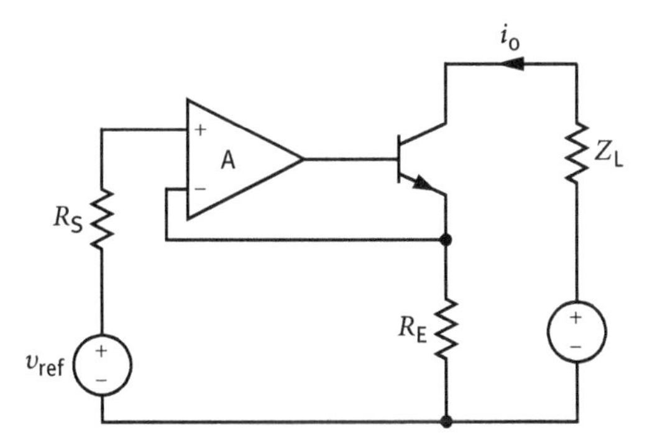

Fig. 2.90. An operational amplifier aided current source.

Thus, the impedance of this current source is effectively the dynamically increased impedance of R_1 (Section 2.4.3.2). The disadvantage of this circuit is that the current is not delivered toward ground but against a floating voltage. Note that the load resistance was omitted in this calculation just the current i_L through R_1 was used. Consequently, the value of the load resistance does not matter, i.e. the current is independent of the value of the load resistor, just as required of a current source. With a high voltage gain g_{vA}, Z_{oL} could well be many megohms.

Problem

2.81. What is the disadvantage of the circuit in Figure 2.89?

Example 2.14 (Output impedance of an operational amplifier aided current source). The circuit of Figure 2.90 delivers a highly constant output current i_o which is the collector current i_c of the transistor. In this case the output impedance of a common-emitter circuit with negative series–series feedback by the resistor R_E is enhanced by an operational amplifier A. In the forward direction this amplifier assures that the voltage at the emitter stays constant at the value of the reference voltage v_{ref}. Consequently, the emitter current i_e is constant and thus, the output current i_o. With $i_{RE} = v_{RE}/R_E = (v_{ref} - v_{iA})/R_E$ one gets with $v_{iA} \ll v_{ref}$

$$i_o = i_c = i_e \times \frac{\beta}{\beta + 1} = \frac{v_{ref}}{R_E} \times \frac{\beta}{\beta + 1} = \frac{v_{ref}}{R_E} \times \frac{1}{1 + \frac{1}{\beta}} \tag{2.129}$$

Changes of v_{iA} and β are strongly suppressed so that a stable output current is guaranteed.

The output impedance is the output impedance of the common-emitter circuit in series to the (dynamic) impedance of R_E. The effective source impedance (between the base and the emitter of the transistor) is obtained by

$$Z_S = Z_{oA} + R_E \times (1 + |g_{vA}|), \tag{2.130}$$

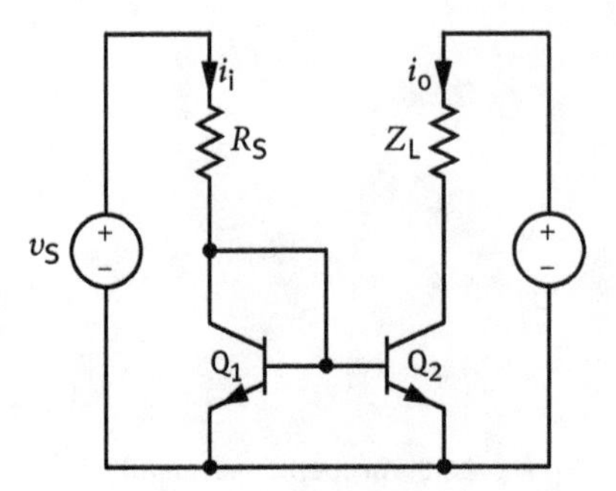

Fig. 2.91. A current mirror using two NPN bipolar transistors with a resistor R_S providing the input current i_i from the source voltage v_S.

i.e. it is very high. Therefore, the equation for the output impedance of common-emitter circuits with very high source impedances applies

$$Z_{oe} = \frac{r_c}{\beta + 1}. \tag{2.131}$$

Together with the impedance of R_E the total output impedance is obtained as

$$Z_o \approx \frac{r_c}{\beta + 1} + R_E \times \frac{\beta + 1}{\beta} \tag{2.132}$$

2.5.6.2 Current mirror

A current mirror reverses the direction of an input current. According to the two-port convention, this means that the current gain $g_i = 1$. A current mirror is characterized by

- its current gain,
- its output impedance,
- its *compliance voltage*, the minimum voltage across the output necessary to make the mirror work correctly.

The dynamic voltage range of the mirror is called the *compliance range*.

The basic circuit of a current mirror is shown in Figure 2.91. A voltage applied to the base-emitter junction of Q_1 is at the same time the input voltage of Q_2 with its collector current as output quantity. Q_2 acts as an *exponential voltage-to-current converter*. The circuit that includes Q_1 can be viewed at as a negative feedback circuit. This interpretation looks rather far-fetched so that it deserves some scrutiny. In Figure 2.92 the first stage of the current mirror is put into the two two-ports of a parallel–parallel feedback (Section 2.5.1). The situation is extremely unusual as the active element is in the feedback two-port and the wire connecting collector with base rests in the two-port A (see Figure 2.92). As primarily the (closed) loop gain is essential for the feedback action it does not matter in which two-port the active element is situated so that our findings on feedback (Section 2.4.1) are valid without question. At high enough (closed-) loop gain the transfer of the feedback circuit is (entirely) the inverse of that of the feedback

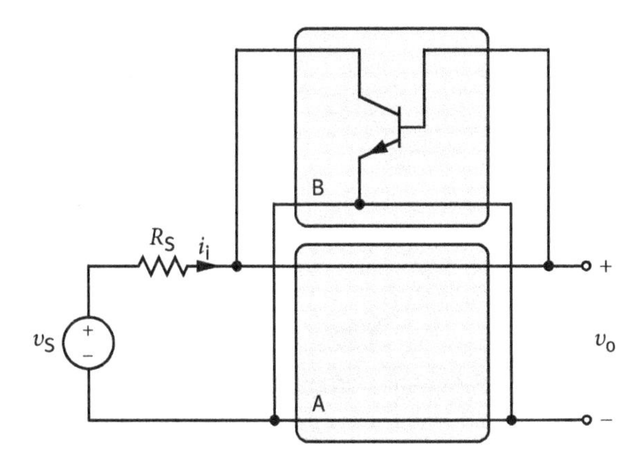

Fig. 2.92. Input stage of the simple current mirror from Figure 2.91 redrawn as feedback circuit.

network. Therefore, the first stage has the inverse transfer property of a transistor, i.e. its transfer function is a *logarithmic current-to-voltage* converter. Cascading a *logarithmic current-to-voltage converter* (Q_1) with an *exponential voltage-to-current converter* (Q_2) results in a current amplifier with gain 1 if the transfer functions of Q_1 and Q_2 are matched. Exactly speaking it does not matter which shape the transfer function has. If the two elements have equal voltage-to-current transfer functions, the circuit will inverse the transfer function of the first stage so that the product of both will result in a current gain $g_i = 1$. Considering the geometric asymmetry of the two stages one gets for equal current gains β of the transistors

$$g_i = \frac{i_{c2}}{i_{c1} + 2 \times i_b} = \frac{\beta_2}{\beta_1 + 2} = \frac{1}{1 + \frac{2}{\beta}} \approx 1 . \tag{2.133}$$

The compliance voltage, the lowest output voltage that results in correct mirror behavior, is at $V_o = V_{BE}$ because $V_{CB2} \geq 0$ V keeps Q_2 active. The output impedance Z_o is just the output impedance Z_{oe} of Q_2 which is a common-emitter circuit.

Problem

?

2.82. When the simple current mirror is explained by negative feedback action, the active element is in the feedback two-port. Why do all the equations concerning feedback also apply for such cases?

Example 2.15 (Current mirror aided by an operational amplifier). Although, the circuit in Figure 2.93 resembles that of a simple current mirror, insofar, as the first stage is rather similar, its behavior is distinctively different. Again, there is a two-stage arrangement with the first stage performing a current-to-voltage conversion and a second stage a voltage-to-current conversion. However, in this case there is no need to match the characteristics of the active elements because the conversions are done by the resistors R_{E1} and R_{E2}, respectively. Let us assume that both resistors have the same value R_E. Assuming high input resistance of the operational amplifier ($Z_{iA} \gg R_E$) then all of the input current i_i flows through R_{E1} resulting in an input voltage v_+ at

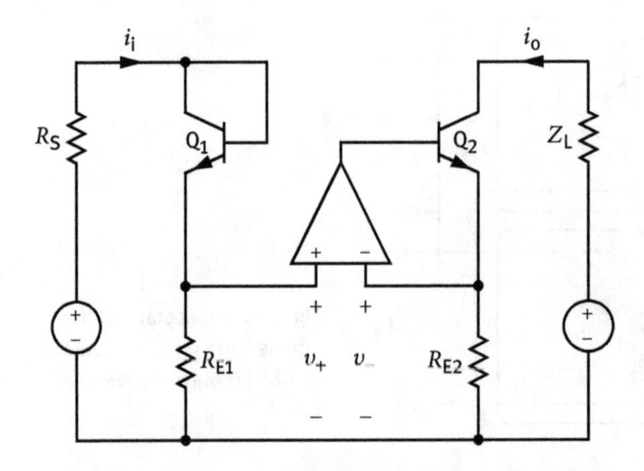

Fig. 2.93. Current mirror with operational amplifier feedback to increase output resistance according to Example 2.14.

the operational amplifier of $v_+ = i_i \times R_{E1}$. The input voltage v_- at the inverting input is $v_- = v_+ - v_{iA}$. From the output voltage v_{oA} of the amplifier one gets its input voltage $v_{iA} = v_{oA}/g^*_{vA}$. As the emitter follower Q_2 has a voltage gain of practically 1, its (small-signal) output voltage equals v_{oA} which at the same time is v_-. Therefore, one gets

$$v_- = v_+ - v_{iA} = i_i \times R_{E1} - \frac{v_{oA}}{g^*_{vA}} = i_i \times R_{E1} - \frac{v_-}{g^*_{vA}}\,, \tag{2.134}$$

and

$$\begin{aligned} v_- \times \left(1 + \frac{1}{g^*_{vA}}\right) &= i_i \times R_{E1} = (i_o + i_{B2}) \times R_{E1} \times \left(1 + \frac{1}{g^*_{vA}}\right) \\ &= i_o \times \left(1 + \frac{1}{\beta_2}\right) \times R_{E1} \times \left(1 + \frac{1}{g^*_{vA}}\right) \end{aligned} \tag{2.135}$$

yielding

$$g_i = \frac{i_o}{i_i} = \frac{1}{\left(1 + \frac{1}{\beta_2}\right) \times \left(1 + \frac{1}{g^*_{vA}}\right)} \approx 1\,. \tag{2.136}$$

This answer is closer to one than that of the circuit in Figure 2.91. Actually these two circuits have little in common except for the input stage and the split of the current gain into a current-to-voltage conversion and a voltage-to-current conversion.

The second stage with the operational amplifier is just the operational amplifier aided current source of the previous section. There, we found an increase of the output impedance Z_{o2} at the collector of Q_2. This increased output impedance is another benefit of this circuit.

? **Problem**

2.83. Name the two benefits of the operational amplifier aided current mirror of Figure 2.93.

2.84. What function has Q_1 in Figure 2.93?

3 Dynamic behavior of networks (signal conditioning)

In the preceding chapters, we were solely interested in the amplitude at a given moment or unchangeable in time (the latter commonly called *direct current* = *DC* behavior). When dealing with signals a new dimension enters – that of time. Signals of natural origin are mainly characterized by their amplitude and their behavior in time. (In sophisticated applications additional signal properties play a role.) Both properties are analog, i.e., to give their true value one must express them by real numbers (with an infinite number of digits). The need of using only a limited number of digits for the quantitative description of any property leads to *digitizing*. A digitized quantity has only values which are multiples of the least significant bit (*binary digit*) *LSB*. As (in most cases) digitizing is done electronically there are four types of electronic signals, depending on which of the two properties (amplitude and time) has been digitized: *analog-analog*, *digital-analog*, *analog-digital*, and *digital-digital*.

Observe that a constant signal frequency is the analogon to a DC signal amplitude: there is no time element, the (analog) value of either does not change in time. Besides, frequency is irrelevant for a DC signal and likewise is amplitude for an intrinsic frequency signal.

From now on we will concentrate on the dynamic behavior of circuits, i.e., the generation, transmission, and conditioning of signals that are analog both with respect to amplitude and time. In other words, we include the time element into our considerations.

The unit of time is the second [s]. The angular frequency ω with the unit [s^{-1}] is the conjugate variable to the time and is, therefore, the basis for the Fourier transform (Section 3.1.1.3).

The advancement of electronics is closely connected to the improved speed of the circuits allowing the handling of shorter and shorter signals providing more and more information in a given time interval. Besides, the higher speed procures new applications that were not feasible before.

For a beginning, we extend the term steady-state signal to signals that vary *regularly* in time. In the mathematical model these are signals that have no beginning and no end. In practice, an electronic device will be turned on (and off) introducing a transient so that an ideal steady-state signal can only be (very well) approximated. Consequently, there does not exist a pure DC signal. Unavoidably, it will be a rectangular signal of significant length.

The mathematical presentation of a steady-state signal $s(t)$ that repeats itself after a time interval T, called *period* or *cycle*, is $s(t) = s(t + nT)$, with n for all integer numbers. The most important steady-state signal is the sinusoidal signal which has a natural period. For a voltage signal it has the form $v(t) = V_{max} \times \sin(\omega t + \varphi)$. V_{max} is the (maximum) *amplitude*, ω the *angular* (or radian) *frequency* and φ the *phase* shift angle (measured in radians) relative to some reference angle. The presentation of both the frequency f by the angular frequency ω (measured in s^{-1}) which is the frequency $f = 1/T$ (measured in Hz) times 2π and the phase shift φ in radians is a requirement because the argument of the sine function must be proportional to 2π (i.e, for practical purposes a fraction of it).

Other common steady-state signals are the square wave, the triangular wave, and the saw-tooth signal.

Problems

3.1. Give the mathematical presentation for a bipolar saw-tooth signal with a period T oscillating between +1 V and −1 V.

3.2. Give the mathematical presentation for a bipolar square-wave signal with a period T oscillating between +1 V and −1 V.

3.3. Give the mathematical presentation for a bipolar triangular wave signal with a period T oscillating between +1 V and −1 V.

3.4. Explain, why one can charge a capacitor with a DC voltage supply even if the admittance of a capacitor equals zero at $\omega = 0$.

3.5. Name two signal properties that are in common to a constant voltage and a constant frequency.

3.1 Decomposition of signals

Reversing Kirchhoff's laws we arrive at the fact that in a linear network any current through a node (or voltage between two points) can be composed of partial currents (voltages).

In Figure 3.1 one simple example is given where the superposition of a unipolar voltage square-wave signal with a constant voltage level is demonstrated. This figure

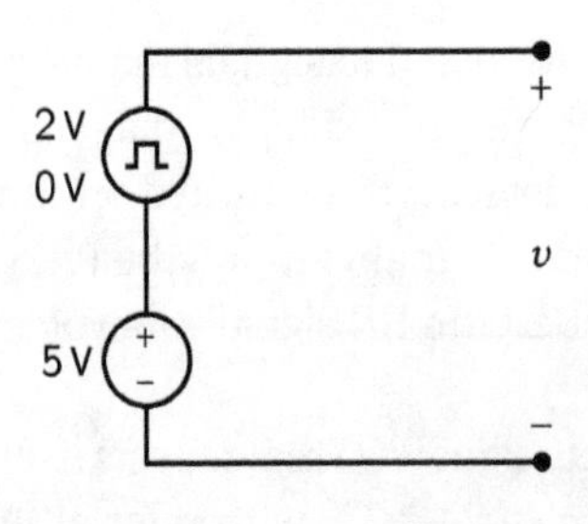

Fig. 3.1a. Superposition of a constant voltage with a unipolar square-wave voltage signal.

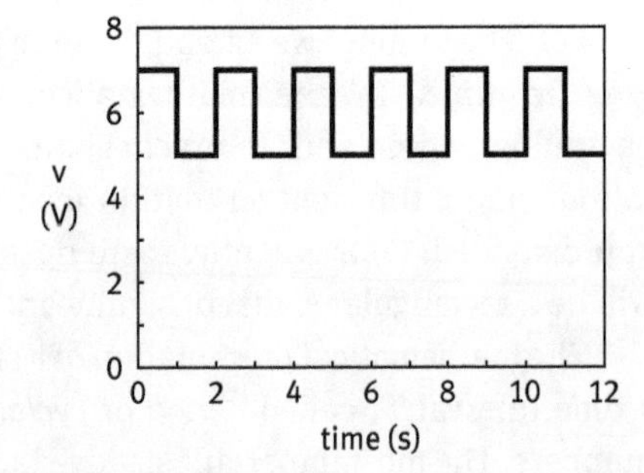

Fig. 3.1b. Combined output of the two voltage sources

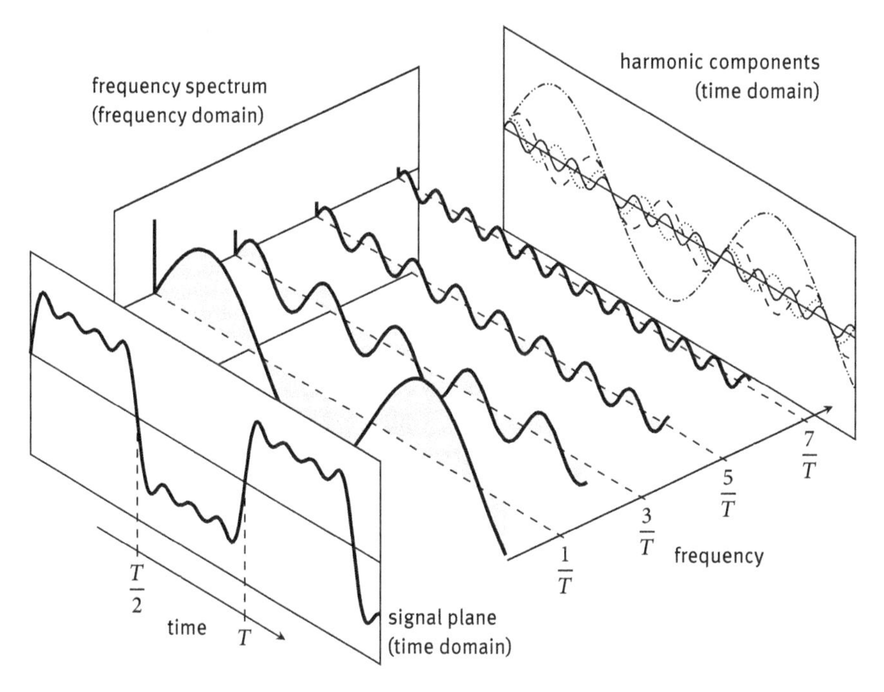

Fig. 3.2a. Decomposition of the signal from (3.1) into harmonic functions (here sine-functions). The front panel shows the signal in the time domain. The individual components are shown behind this panel.

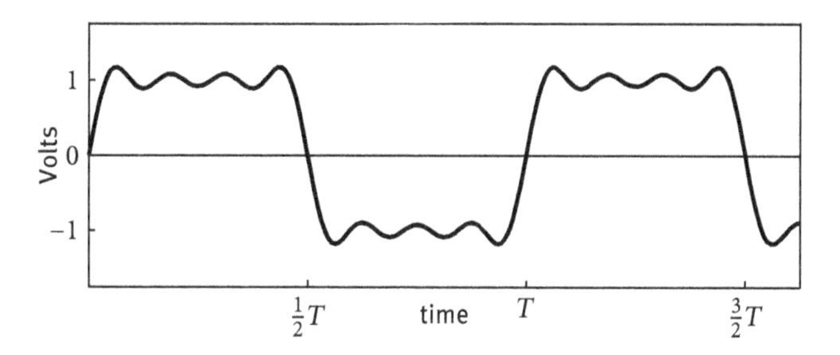

Fig. 3.2b. Signal (time domain) from (3.1).

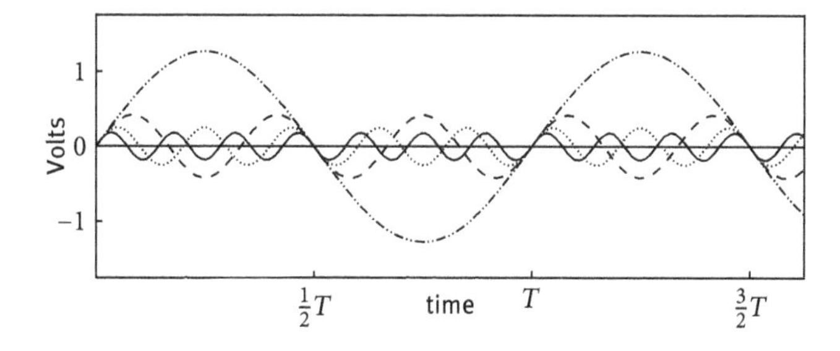

Fig. 3.2c. Signal components (time domain).

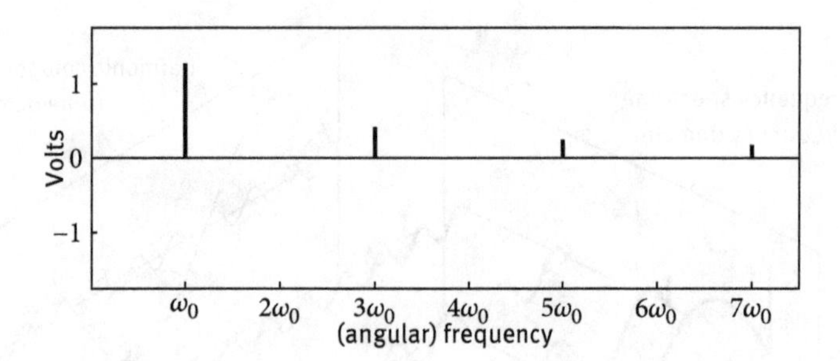

Fig. 3.2d. Frequency spectrum (frequency domain).

may be read in either direction: Adding the output of two voltage sources yields the combined output as shown. The combined output voltage signal can be decomposed into the signals of two appropriate voltage sources. In this case, the decomposition is not unique because the square wave may be assumed to be bipolar, or unipolar in either direction, each requiring a different DC voltage to produce the identical combined signal.

Decomposition into a DC voltage and a bipolar square-wave signal is most convenient for further use. It resembles the decomposition of the electrical variables of the momentary operating point into a quiescent value and a small-signal value (Sections 2.1.3.5, 2.3.3, and 2.3.3).

More to the point is the superposition of steady-state sinusoidal signals, e.g., the following sum

$$v(t) = 1.27\sin(\omega_0 t) + 0.42\sin(3\omega_0 t) + 0.25\sin(5\omega_0 t) + 0.18\sin(7\omega_0 t)[\mathrm{V}]\,. \quad (3.1)$$

As the frequencies are integer multiples of the basic frequency ω_0 one expects a steady-state signal that is repetitive with a period of $T = 2\pi/\omega_0$.

Figure 3.2d shows the four amplitudes of the frequency components in dependence of the angular frequency. This figure is the presentation of the combined signal in the *frequency domain*. Figure 3.2b shows the same signal in the *time domain*, i.e., the result of the addition of the above frequency components.

Reversing above procedure, namely decomposing periodic steady-state signals, i.e., periodic signals with properties that are unchanging in time into their harmonic components is a powerful tool in electronics. The Fourier transform (Section 3.1.1) decomposes periodic time-signals into their conjugate frequency components. Spectrum analyzers (Section 6.8.3) using a Fourier transform algorithm would perform such a decomposition of electrical signals.

Problem

3.6. Which physical quantity is conjugate to time?

In the following sections the relevant equations needed for Fourier analysis are collected. These sections may be skipped without endangering the understanding of any important topic of this book.

3.1.1 Fourier analysis

There are various ways of dealing with signal decomposition. The decomposition of signals into harmonic functions (sine- and cosine-functions) is called *Fourier analysis*. In the following we will explain the basics of Fourier series representation of periodic signals and will then give an introduction into the *continuous Fourier transform* which is a generalization of the Fourier series method. Finally we will give a short overview about how the Fourier transform method assists in solving systems of differential equations.

There exist several different notations for the Fourier series and for the Fourier transform as well. We will present the three standard forms of Fourier series representation (sine- and cosine-sums, cosine-sums with phase shifts and exponential form) using angular frequencies ($\omega = 2\pi \times f$). Normalization factors in the series coefficients will, therefore, be multiples of the fraction $1/T$. If regular frequencies (f) were used instead of angular frequencies, the normalization factors would be multiples of $1/2\pi$ instead of $1/T$.

Problem

?

3.7. Fourier analysis decomposes a periodic time signal into a finite number of components. What are these components like?

3.1.1.1 Fourier series

Every periodic signal $s(t)$ with period T can be written as a(n infinite) sum of sine- and cosine-functions with proper coefficients a_k and b_k:

$$\begin{aligned} s(t) &= \frac{a_0}{2} + a_1 \cos(\omega_0 t) + b_1 \sin(\omega_0 t) + a_2 \cos(2\omega_0 t) + b_2 \sin(2\omega_0 t) + \ldots \\ &= \frac{a_0}{2} + \sum_{k=1}^{\infty} \{a_k \cos(k\omega_0 t) + b_k \sin(k\omega_0 t)\} \end{aligned} \tag{3.2}$$

with

$$\omega_0 = 2\pi \times f_0 = \frac{2\pi}{T} \, . \tag{3.3}$$

The following facts about integration of products of sine- and cosine-functions are the basics for calculation of the coefficients a_k and b_k. For any l and m

$$\int_{t'}^{t'+T} \sin(l\omega_0 t)\cos(m\omega_0 t)dt = 0 \quad (3.4)$$

$$\int_{t'}^{t'+T} \sin(l\omega_0 t)\sin(m\omega_0 t)dt = \begin{cases} 0 & l \neq m \\ \frac{T}{2} & l = m \end{cases} \quad (3.5)$$

$$\int_{t'}^{t'+T} \cos(l\omega_0 t)\cos(m\omega_0 t)dt = \begin{cases} 0 & l \neq m \\ \frac{T}{2} & l = m\,. \end{cases} \quad (3.6)$$

This universal property of harmonic functions is called *orthogonality*. We are not going to give proof of this feature but instead Figure 3.3 illustrates some examples. The highlighted areas equal the integrals from t' to $t' + T$. Since the functions are periodic the lower limit t' for integration can be chosen arbitrarily. Note that when the two harmonic functions differ, the negative areas equal the positive areas. If they do not differ, the area under the function equals $T/2$ (see Figure 3.3c).

Multiplying both sides of (3.2) with $\sin(l\omega_0 t)$ and integrating yields

$$\int_{t'}^{t'+T} s(t)\sin(l\omega_0 t)dt$$

$$= \int_{t'}^{t'+T} \sum_{k=1}^{\infty} \{a_k \cos(k\omega_0 t)\sin(l\omega_0 t) + b_k \sin(k\omega_0 t)\sin(l\omega_0 t)\}\, dt$$

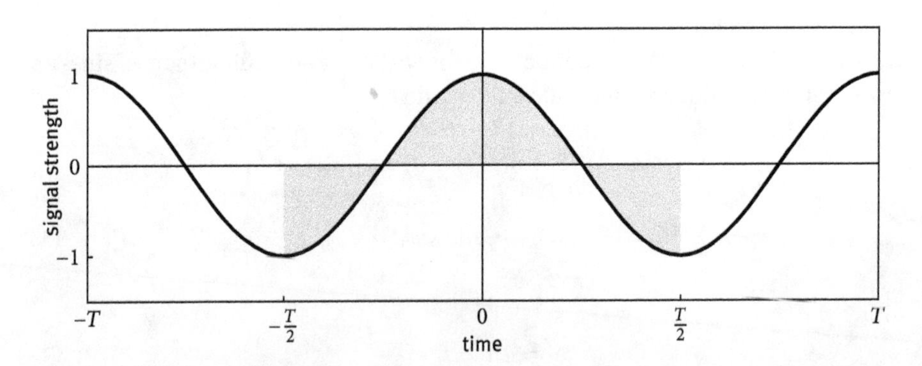

Fig. 3.3a. Graph of the function $\cos(1\omega_0 t)cos(0\omega_0 t)$, i.e., $cos(\omega_0 t)$. The highlighted area between the time axis and the function graph equals the integral from $t' = -T/2$ to $t' + T = T/2$. Observe that the negative portions and the positive areas are equal. Thus the integral is zero.

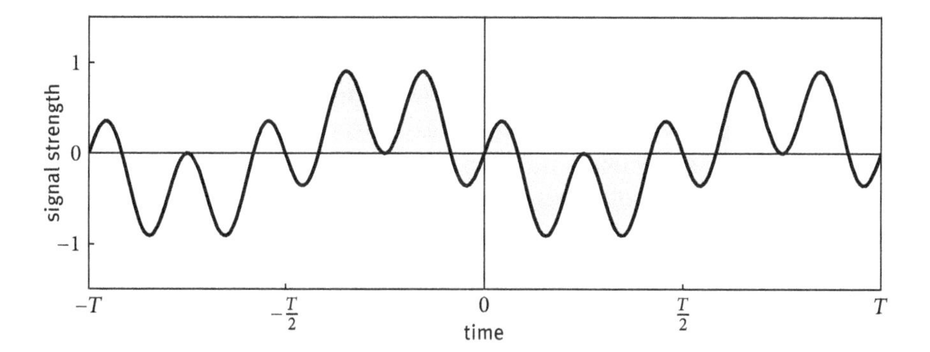

Fig. 3.3b. Graph of the function $\sin(2\omega_0 t)\cos(3\omega_0 t)$.

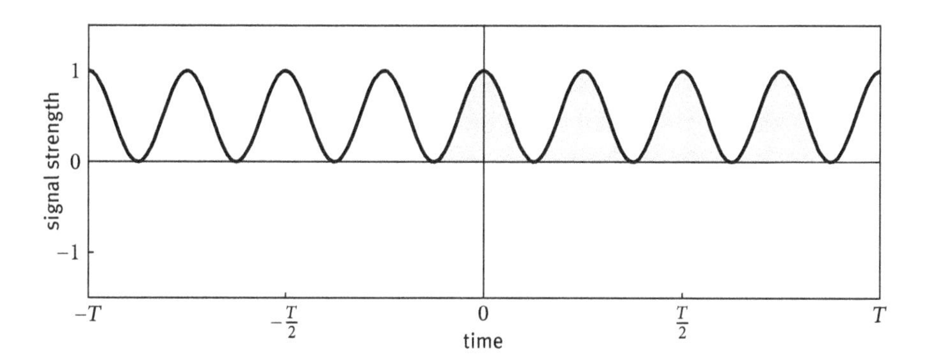

Fig. 3.3c. Graph of the function $\cos(3\omega_0 t)\cos(3\omega_0 t)$. The highlighted area between t-axis and the function's graph equals $T/2$.

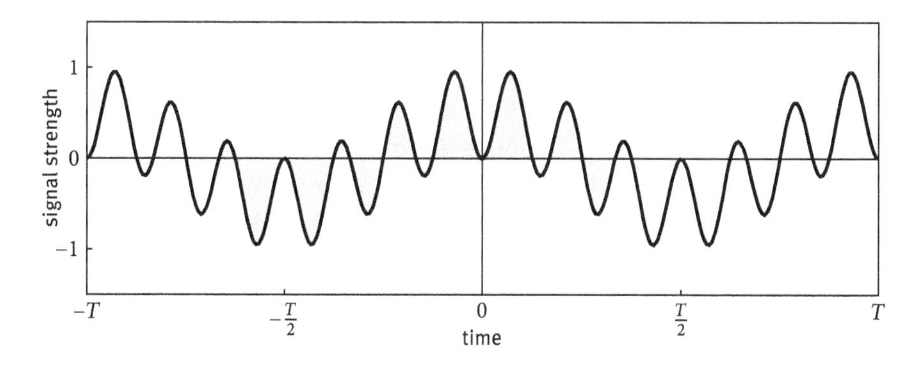

Fig. 3.3d. Graph of the function $\sin(4\omega_0 t)\sin(3\omega_0 t)$.

$$= \sum_{k=1}^{\infty} \left\{ a_k \int_{t'}^{t'+T} \cos(k\omega_0 t) \sin(l\omega_0 t) \mathrm{d}t + b_k \int_{t'}^{t'+T} \sin(k\omega_0 t) \sin(l\omega_0 t) \mathrm{d}t \right\}$$

$$= b_l \times \frac{T}{2} . \tag{3.7}$$

Multiplying (3.2) with $\cos(m\omega_0 t)$ and subsequent integration yields

$$\int_{t'}^{t'+T} s(t) \cos(m\omega_0 t) \mathrm{d}t$$

$$= \int_{t'}^{t'+T} \sum_{k=1}^{\infty} \{a_k \cos(k\omega_0 t) \cos(m\omega_0 t) + b_k \sin(k\omega_0 t) \cos(m\omega_0 t)\} \, \mathrm{d}t$$

$$= \sum_{k=1}^{\infty} \left\{ a_k \int_{t'}^{t'+T} \cos(k\omega_0 t) \cos(m\omega_0 t) \mathrm{d}t + b_k \int_{t'}^{t'+T} \sin(k\omega_0 t) \cos(m\omega_0 t) \mathrm{d}t \right\}$$

$$= a_m \times \frac{T}{2} . \tag{3.8}$$

Thus the equations for the series coefficients are obtained

$$a_k = \frac{2}{T} \times \int_{t'}^{t'+T} s(t) \cos(k\omega_0 t) \mathrm{d}t , \tag{3.9}$$

$$b_k = \frac{2}{T} \times \int_{t'}^{t'+T} s(t) \sin(k\omega_0 t) \mathrm{d}t, \text{ and} \tag{3.10}$$

$$a_0 = \frac{2}{T} \times \int_{t'}^{t'+T} s(t) \mathrm{d}t . \tag{3.11}$$

The limit t' can be chosen arbitrarily and should be chosen so that integration of above terms is easiest. For signals with the property $s(t) = s(-t)$ all coefficients b_k are zero and for signals with $s(t) = -s(-t)$ all a_k are zero.

There are two components for each frequency $k\omega_0$: a sine-component and a cosine-component. All linear combinations of these components $a_k \cos(k\omega_0 t) + b_k \sin(k\omega_0 t)$ can be rewritten by means of the trigonometric angle-sum identity

$$\cos(\alpha - \beta) = \cos(\alpha) \cos(\beta) + \sin(\alpha) \sin(\beta) \tag{3.12}$$

so that

$$a_k \cos(k\omega_0 t) + b_k \sin(k\omega_0 t) = A_k \cos(k\omega_0 t - \varphi_k) , \tag{3.13}$$

with the coefficients following

$$a_k = A_k \cos(\varphi_k) \tag{3.14}$$
$$b_k = A_k \sin(\varphi_k) \tag{3.15}$$
$$A_k = \sqrt{a_k^2 + b_k^2} \tag{3.16}$$
$$\varphi_k = \begin{cases} \arctan(\frac{b_k}{a_k}) & a > 0 \\ \arctan(\frac{b_k}{a_k}) + \pi & a < 0 \\ \frac{\pi}{2} & a = 0, b > 0 \\ -\frac{\pi}{2} & a = 0, b < 0\,. \end{cases} \tag{3.17}$$

Thus, the Fourier series representation of the signal $s(t)$ can be written

$$s(t) = \frac{A_0}{2} + \sum_{k=1}^{\infty} A_k \cos(k\omega_0 t - \varphi_k)\,. \tag{3.18}$$

The numbers A_k form the (discrete) *frequency spectrum* and φ_k the *phase shift spectrum* of the signal $s(t)$.

3.1.1.2 Exponential Fourier series

The following equation that links the exponential function of imaginary numbers to the harmonic functions is called *Euler's identity*. In circuit theory, it is general practice to use $\mathrm{j} = \sqrt{-1}$ as imaginary unit to avoid confusion with the symbol i which is already used for current.

$$\mathrm{e}^{\mathrm{j}\varphi} = \cos(\varphi) + \mathrm{j}\sin(\varphi) \tag{3.19}$$

By means of Euler's identity the harmonic functions can be presented as

$$\sin(\varphi) = \frac{\mathrm{e}^{\mathrm{j}\varphi} - \mathrm{e}^{-\mathrm{j}\varphi}}{2\mathrm{j}} = -\mathrm{j}\frac{\mathrm{e}^{\mathrm{j}\varphi} - \mathrm{e}^{-\mathrm{j}\varphi}}{2} \tag{3.20}$$
$$\cos(\varphi) = \frac{\mathrm{e}^{\mathrm{j}\varphi} + \mathrm{e}^{-\mathrm{j}\varphi}}{2}\,. \tag{3.21}$$

Each of the linear combinations $a_k \cos(k\omega_0 t) + b_k \sin(k\omega_0 t)$ of the Fourier series representation can be rewritten utilizing above identities so that the Fourier series becomes

$$\begin{aligned} s(t) &= \frac{a_0}{2} + \sum_{k=1}^{\infty} \{a_k \cos(k\omega_0 t) + b_k \sin(k\omega_0 t)\} \\ &= \frac{a_0}{2} + \sum_{k=1}^{\infty} \left\{a_k \frac{\mathrm{e}^{\mathrm{j}k\omega_0 t} + \mathrm{e}^{-\mathrm{j}k\omega_0 t}}{2} - b_k \mathrm{j}\frac{\mathrm{e}^{\mathrm{j}k\omega_0 t} - \mathrm{e}^{-\mathrm{j}k\omega_0 t}}{2}\right\} \end{aligned}$$

$$
\begin{aligned}
&= \frac{a_0}{2} + \sum_{k=1}^{\infty} \left\{ \frac{a_k - b_k j}{2} e^{jk\omega_0 t} + \frac{a_k + b_k j}{2} e^{-jk\omega_0 t} \right\} \\
&= \frac{a_0}{2} + \sum_{k=1}^{\infty} \frac{a_k - b_k j}{2} e^{jk\omega_0 t} + \sum_{k=1}^{\infty} \frac{a_k + b_k j}{2} e^{-jk\omega_0 t} \\
&= c_0 + \sum_{k=1}^{\infty} c_k e^{jk\omega_0 t} + \sum_{k=-\infty}^{-1} c_k e^{jk\omega_0 t} \\
&= \sum_{k=-\infty}^{\infty} c_k e^{jk\omega_0 t}
\end{aligned}
\tag{3.22}
$$

with the coefficients

$$
c_k = \begin{cases} \frac{a_0}{2} & k = 0 \\ \frac{a_k - b_k j}{2} & k > 0 \\ \frac{a_k + b_k j}{2} & k < 0\,. \end{cases}
\tag{3.23}
$$

Generally, the coefficients c_k can be expressed as

$$
c_k = \frac{1}{T} \times \int_{t'}^{t'+T} s(t) e^{-jk\omega_0 t} \mathrm{d}t\,.
\tag{3.24}
$$

The sequence of (complex) coefficients c_k is called the *(discrete) complex Fourier spectrum*. These coefficients can be expressed by their amplitude and phase terms

$$
c_k = A_k e^{-j\varphi_k}\,,
\tag{3.25}
$$

with

$$
A_k = |c_k| = \sqrt{a_k^2 + b_k^2}
\tag{3.26}
$$

$$
\varphi_k = \begin{cases} \arctan(\frac{b_k}{a_k}) & a > 0 \\ \arctan(\frac{b_k}{a_k}) + \pi & a < 0 \\ \frac{\pi}{2} & a = 0, b > 0 \\ -\frac{\pi}{2} & a = 0, b < 0 \end{cases}
\tag{3.27}
$$

so that the exponential Fourier series becomes

$$
s(t) = \sum_{k=-\infty}^{\infty} A_k e^{j(k\omega_0 t - \varphi_k)}\,.
\tag{3.28}
$$

3.1.1.3 Continuous Fourier transform

The frequencies of the Fourier spectrum of a signal $s(t)$ are integer multiples of the basic angular frequency ω_0 if the signal is repetitive with a period of $T = 2\pi/\omega_0$. For each frequency $\omega_k = k\omega_0$ the corresponding Fourier series coefficient (amplitude) is $c(\omega_k) = c_k$. The difference between two adjacent frequencies is $\Delta\omega = \omega_{k+1} - \omega_k = \omega_0 = 2\pi \times f = 2\pi/T$. Obviously, longer periods yield denser frequency spectra. By letting $T \to \infty$ the frequency spectrum eventually becomes continuous since $\Delta\omega$ converges to 0.

The Fourier series becomes

$$
\begin{aligned}
s(t) &= \lim_{T\to\infty} \sum_{\omega_k=-\infty}^{\infty} \overbrace{\frac{1}{T} \times \int_{-T/2}^{T/2} s(t)\mathrm{e}^{-\mathrm{j}\omega_k t}\,\mathrm{d}t}^{c(\omega_k)} \times \mathrm{e}^{\mathrm{j}\omega_k t} \\
&= \lim_{\substack{T\to\infty \\ \Delta\omega\to 0}} \sum_{\omega=-\infty}^{\infty} \frac{\Delta\omega}{2\pi} \times \int_{-T/2}^{T/2} s(t)\mathrm{e}^{-\mathrm{j}\omega t}\,\mathrm{d}t \times \mathrm{e}^{\mathrm{j}\omega t} \\
&= \lim_{\substack{T\to\infty \\ \Delta\omega\to 0}} \sum_{\omega=-\infty}^{\infty} \frac{1}{2\pi} \times \int_{-T/2}^{T/2} s(t)\mathrm{e}^{-\mathrm{j}\omega t}\,\mathrm{d}t \times \mathrm{e}^{\mathrm{j}\omega t}\,\Delta\omega \\
&= \int_{-\infty}^{\infty} \frac{1}{2\pi} \times \underbrace{\int_{-\infty}^{\infty} s(t)\mathrm{e}^{-\mathrm{j}\omega t}\,\mathrm{d}t}_{\tilde{s}(\omega)} \times \mathrm{e}^{\mathrm{j}\omega t}\,\mathrm{d}\omega\,.
\end{aligned}
\tag{3.29}
$$

This way the Fourier method can be generalized for nonrepetitive signals leading to a *continuous frequency spectrum* instead of a sequence of amplitudes c_k. The spectrum is denoted by a function $\tilde{s}(\omega)$ which is called the *Fourier transform* of $s(t)$:

$$
\tilde{s}(\omega) = \int_{-\infty}^{\infty} s(t)\mathrm{e}^{-\mathrm{j}\omega t}\,\mathrm{d}t\,. \tag{3.30}
$$

The inverse Fourier transform is

$$
s(t) = \frac{1}{2\pi} \times \int_{-\infty}^{\infty} \tilde{s}(\omega)\mathrm{e}^{\mathrm{j}\omega t}\,\mathrm{d}\omega\,. \tag{3.31}
$$

When in applications the frequency f is used instead of the angular frequency ω, the transforms become

$$
\tilde{s}(f) = \int_{-\infty}^{\infty} s(t)\mathrm{e}^{-2\pi\mathrm{j}ft}\,\mathrm{d}t \tag{3.32}
$$

$$
s(t) = \int_{-\infty}^{\infty} \tilde{s}(f)\mathrm{e}^{2\pi\mathrm{j}ft}\,\mathrm{d}f\,. \tag{3.33}
$$

Very often the transformation is denoted

$$\mathcal{F}[s] = \tilde{s} \tag{3.34}$$

$$s = \mathcal{F}^{-1}[\tilde{s}] . \tag{3.35}$$

3.1.1.4 Properties of the Fourier transform

The Fourier transform has specific properties which facilitate the calculation of the transform of a given signal. In practice calculation of the transforms and the inverse transforms is extensively aided by means of tabulated signals and their corresponding transforms. The usage of the following basic properties combined with already known pairs of signals and their transforms can considerably reduce the computational effort. The following properties directly follow from the definitions (3.30) to (3.33).

Linearity

Let a and b be two (complex) numbers and $s_1(t)$ and $s_2(t)$ two signals, then

$$\begin{aligned}\mathcal{F}[a \times s_1(t) + b \times s_2(t)] &= a \times \mathcal{F}[s_1(t)] + b \times \mathcal{F}[s_2(t)] \\ &= a \times \tilde{s}_1(\omega) + b \times \tilde{s}_2(\omega) .\end{aligned} \tag{3.36}$$

Time shift

Let be $s(t)$ a signal and Δt a time difference, then

$$\begin{aligned}\mathcal{F}[s(t - \Delta t)] &= \mathrm{e}^{-j\omega\Delta t} \times \mathcal{F}[s(t)] \\ &= \mathrm{e}^{-j\omega\Delta t} \times \tilde{s}(\omega) .\end{aligned} \tag{3.37}$$

Modulation (frequency shift)

Let $\Delta\omega$ be a frequency shift and $s(t)$ a signal, then

$$\mathcal{F}[\mathrm{e}^{jt\Delta\omega} s(t)] = \tilde{s}(\omega - \Delta\omega) . \tag{3.38}$$

Scaling

With the signal $s(t)$ and any real number a, then

$$\mathcal{F}\left[s\left(\frac{t}{a}\right)\right] = \frac{1}{|a|} \times \tilde{s}\left(\frac{\omega}{a}\right) . \tag{3.39}$$

Differentiation
Let be $s(t)$ a signal, then

$$\begin{aligned}\mathcal{F}\left[\frac{\mathrm{d}}{\mathrm{d}t}s(t)\right] &= \mathrm{j}\omega \times \mathcal{F}[s(t)] \\ &= \mathrm{j}\omega \times \tilde{s}(\omega), \text{ and} \end{aligned} \tag{3.40}$$

$$\begin{aligned}\mathcal{F}[t \times s(t)] &= \mathrm{j} \times \frac{\mathrm{d}}{\mathrm{d}\omega}\mathcal{F}[s(t)] \\ &= \mathrm{j} \times \frac{\mathrm{d}}{\mathrm{d}\omega}\tilde{s}(\omega)\,. \end{aligned} \tag{3.41}$$

Multiplications and convolutions
Let be $s_1(t)$ and $s_2(t)$ signals and $(s_1 * s_2)(t)$ the convolution of them with

$$(s_1 * s_2)(t) = \int_{-\infty}^{\infty} s_1(\tau)s_2(t-\tau)\mathrm{d}\tau\,. \tag{3.42}$$

Then we get

$$\mathcal{F}[s_1(t) \times s_2(t)] = \frac{1}{2\pi}(\tilde{s}_1 * \tilde{s}_2)(\omega) \tag{3.43}$$

$$\mathcal{F}[(s_1 * s_2)(t)] = \tilde{s}_1(\omega) \times \tilde{s}_2(\omega)\,. \tag{3.44}$$

3.1.1.5 Application of the Fourier transform

One of the major applications of the Fourier transform is solving of (partial) differential equations. In fact, this field is the historical origin of the transform. In the following we will briefly outline the standard procedure in dealing with the computation of quantities in electronic circuits. Eventually we will address the problems associated with the procedure.

The circuit in Figure 3.4 contains all four passive one-ports. Consequently, it constitutes an apt example for the application of Fourier analysis.

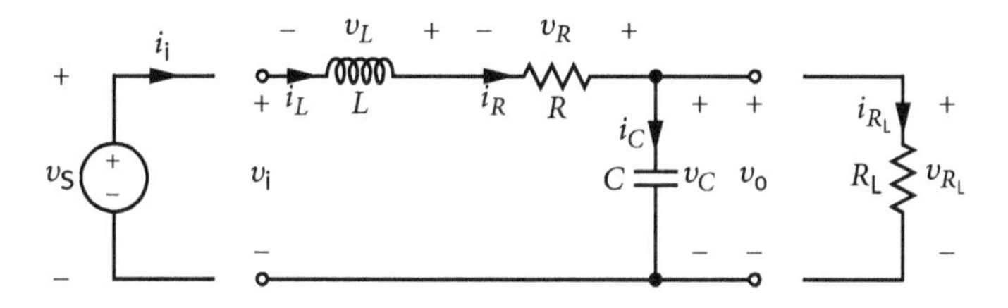

Fig. 3.4. Example circuit for finding an unknown signal $v_o(t)$ by Fourier transform.

Given a (time dependent) signal $v_i(t)$ as input voltage there are nine unknown quantities; the output voltage $v_o(t)$ at the load resistance R_L, the output current $i_o(t)$, the voltage $v_C(t)$ across the capacity C, the current flowing into the capacity $i_C(t)$, the voltage $v_L(t)$ across the inductivity L, the current flowing through the inductivity $i_L(t)$ as well as the input current $i_i(t)$. Clearly, all quantities are time dependent since the input voltage is time dependent.

Thus, we need a set of nine independent equations involving these variables. Three equations are trivially obtained from the quantities' definitions (see Figure 3.4)

$$\begin{aligned} i_i(t) &= i_L(t) \\ i_L(t) &= i_R(t) \\ v_C(t) &= v_o(t)\,, \end{aligned}$$

and two equations can be found by applying Kirchhoff's first (2.1) and second (2.2) law.

$$\begin{aligned} i_o(t) + i_R(t) &= i_C(t) \\ v_o(t) + v_R(t) + v_L(t) &= v_i(t)\,. \end{aligned}$$

Furthermore four equations are obtained from the relations (1.2) and (1.3) between currents and voltages at the basic one-ports

$$\begin{aligned} i_C(t) &= C \times \frac{\mathrm{d}v_C(t)}{\mathrm{d}t} \\ v_L(t) &= L \times \frac{\mathrm{d}i_L(t)}{\mathrm{d}t} \\ v_R(t) &= R \times i_R(t) \\ v_o(t) &= v_{R_L}(t) \\ &= i_{R_L}(t) \times R_L \\ &= -i_o(t) \times R_L\,. \end{aligned}$$

The set of nine equations can be rearranged into two equations with the two unknown variables $v_o(t)$ and $i_i(t)$

$$\begin{aligned} v_i(t) &= v_L(t) + v_R(t) + v_o(t) \\ &= L \times \frac{\mathrm{d}i_L(t)}{\mathrm{d}t} + R \times i_R(t) + v_o(t) \\ &= L \times \frac{\mathrm{d}i_i(t)}{\mathrm{d}t} + R \times i_i(t) + v_o(t) \\ i_i(t) &= i_R(t) \\ &= i_C(t) - i_o(t) \\ &= C \times \frac{\mathrm{d}v_o(t)}{\mathrm{d}t} + \frac{v_o(t)}{R_L}\,. \end{aligned}$$

Thus, we have to deal with two differential equations for $v_\text{o}(t)$ and $i_\text{i}(t)$:

$$v_\text{i}(t) = L \times \frac{\text{d}i_\text{i}(t)}{\text{d}t} + R \times i_\text{i}(t) + v_\text{o}(t)$$
$$i_\text{i}(t) = C \times \frac{\text{d}v_\text{o}(t)}{\text{d}t} + \frac{v_\text{o}(t)}{R_\text{L}} .$$

By applying the properties of the Fourier transform (3.36) and (3.40) the equations are transformed to

$$\tilde{v}_\text{i}(\omega) = L \times \omega\text{j} \times \tilde{i}_\text{i}(\omega) + R \times \tilde{i}_\text{i}(\omega) + \tilde{v}_\text{o}(\omega)$$
$$\tilde{i}_\text{i}(\omega) = C \times \omega\text{j} \times \tilde{v}_\text{o}(\omega) + \frac{\tilde{v}_\text{o}(\omega)}{R_\text{L}}$$

which is a substantial simplification because instead of dealing with differential equations one is just confronted with algebraic equations. This clearly shows one of the most important features of the Fourier-transform procedure – partial differential equations become algebraic equations for the transforms and solving them is easier than solving differential equations.

Substituting $\tilde{i}_\text{i}(\omega)$ in the first equation with the expression from the second equation yields

$$\begin{aligned}\tilde{v}_\text{i}(\omega) &= \left[L\omega\text{j} + R\right] \times \left[C\omega\text{j} \times \tilde{v}_\text{o}(\omega) + \frac{\tilde{v}_\text{o}(\omega)}{R_\text{L}}\right] + \tilde{v}_\text{o}(\omega) \\ &= \left[1 + \frac{R}{R_\text{L}} - LC\omega^2 + \left(\frac{L}{R_\text{L}} + RC\right) \times \omega\text{j}\right] \times \tilde{v}_\text{o}(\omega)\end{aligned}$$

and the the Fourier transform $\tilde{v}_\text{o}(\omega)$ of the previously unknown $v_\text{o}(t)$ is

$$\tilde{v}_\text{o}(\omega) = \frac{1}{1 + \frac{R}{R_\text{L}} - LC\omega^2 + \left[\frac{L}{R_\text{L}} + RC\right] \times \omega\text{j}} \times \tilde{v}_\text{i}(\omega) . \tag{3.45}$$

The standard procedure for finding an unknown $v_\text{o}(t)$ firstly involves computation of the Fourier transform $\tilde{v}_\text{i}(\omega)$ of the given $v_\text{i}(t)$. Subsequently, the system of algebraic equations involving the transform $\tilde{v}_\text{o}(\omega)$ has to be solved. Finally the inverse Fourier transform thereof has to be computed.

However, there are some difficulties in dealing with this procedure. Although the computation of the inverse transform is aided by the application of the properties (3.36) to (3.44) from Section 3.1.1.4 and the knowledge of tabulated transforms, it still remains very involved. Even more problematic is the fact that the Fourier transform does not exist for every signal. Until now we have tacitly assumed that all signals can be transformed. In fact, this is not true. The transforms do only exist for signals where the limits in (3.29) exist by letting $\Delta\omega \to 0$ and $T \to \infty$. It can be shown, that this is the fact for some signals, but unfortunately, for many important signals this is not the case. This issue can be dealt with to some extent by using the *Laplace transform* instead of the Fourier transform (Section 3.1.2).

Last but not least the most serious issue in calculating the behavior of circuits by means of the Fourier method is that one does not gain anything in understanding the circuit. Although the response of a circuit to a signal can be calculated exactly, this does not help in *understanding* the circuit.

Nevertheless, the transform is a very powerful tool. If only harmonic signals (sine- and cosine-functions) are considered, the outlined method allows for direct and easy investigation of the circuits' transfer characteristics. Thus, one can see at a glance how a given circuit affects certain frequency components of an applied signal. Since harmonic functions transform into constants in the frequency domain (possibly with constant phase shift as well) all that must be done is some simple algebra (given the familiarity of complex numbers). There is no need for dealing with differential equations anymore and there is also no need for explicitly performing the transform and the reverse transform as well. With the generalized complex impedances the calculation of circuits' signal behaviors can be done the same way as if it were DC-signals. This tremendously facilitates understanding, as we will see in the remaining sections of this chapter.

3.1.2 Laplace transform

The computation of the Fourier transform of a signal $v(t)$ is done by evaluating the integral

$$\tilde{v}(\omega) = \int_{-\infty}^{\infty} v(t)\mathrm{e}^{-\mathrm{j}\omega t}\,\mathrm{d}t\,. \tag{3.46}$$

In 3.1.1.5 it was already stated that the integral does not converge for all possible signals, i.e., the Fourier transform does not exist. This problem of bad convergence behavior is circumvented by adding a term $\mathrm{e}^{-\sigma t}$ to the integrand. This effectively renders convergence. The resulting transform is called *Laplace transform* and is a function of two variables ω and σ which can both be referred to as one complex variable s with

$$s = \sigma + \mathrm{j}\omega\,. \tag{3.47}$$

The Laplace transformation of a signal $v(t)$ for $t = 0 \ldots \infty$ is denoted as $\mathcal{L}[v]$. We will use $\breve{v}(s)$ for the transformed signal. The Laplace transform has a much broader range of convergence regarding signals than the Fourier transform procedure has. Therefore it is of considerable importance for all kinds of electronic calculations or automated computations of circuit behavior. The (unilateral) Laplace transform is defined

$$\breve{v}(s) = \int_{0}^{\infty} v(t)\mathrm{e}^{-\mathrm{j}\omega t} \times \mathrm{e}^{-\sigma t}\,\mathrm{d}t = \int_{-\infty}^{\infty} v(t)\mathrm{e}^{-st}\,\mathrm{d}t\,. \tag{3.48}$$

The inverse transform is given by

$$v(t) = \lim_{\omega \to \infty} \int_{\sigma - j\omega}^{\sigma + j\omega} \breve{v}(s) \times e^{st} ds \tag{3.49}$$

with a real number σ so that the contour path of the complex integration is in the region of convergence of $\breve{v}(s)$.
Very often the Laplace transformation is denoted

$$\breve{v} = \mathcal{L}[v] \tag{3.50}$$

$$v = \mathcal{L}^{-1}[\breve{v}] \; . \tag{3.51}$$

Of course it is important to use the right notation (i.e., notation of variables) if transformations are used for analysis. Nevertheless, as already outlined in 3.1.1.5, in understanding circuits one will only very seldom have to deal with the calculation of transforms. Using either the Laplace notation $\breve{v}(s)$ or the Fourier notation $\tilde{v}(\omega)$, probably denoted $v(j\omega)$ or even more convenient only v is just a matter of convention and taste in most *every-day* analyses. The consequent application of the *generalized complex impedance* Z releases us from directly (mathematically) dealing with the various transforms.

Problem

?

3.8. In which cases must a Laplace transform be used rather than the Fourier transform?

3.2 Frequency dependent linear one-ports

In Section 1.3.4 the capacitance C was defined by the equation

$$i(t) = C \times \frac{dv(t)}{dt} \; . \tag{3.52}$$

The electronic component capacitor approximates this capacitance closely. For a steady-state sinusoidal voltage signal across a capacitor, the current flow through it is its time derivative, which is the cosine function. Thus, the current precedes the voltage by a quarter cycle (period) which corresponds to a phase shift of −90°. From

$$\hat{i} \times \cos(\omega t) = C \times \frac{d\,(\hat{v} \times \sin(\omega t))}{dt} = C \times \hat{v} \times \omega \times \cos(\omega t) \tag{3.53}$$

follows

$$\hat{Y} = \frac{\hat{i}}{\hat{v}} = \omega \times C \; . \tag{3.54}$$

Therefore, there is not only a phase shift of 90° between current and voltage at a capacitor but also proportionality to the angular frequency ω in the ratio of current to voltage. Putting a resistor (in which no phase shift exists between voltage and current) in series to a capacitor (that has −90° phase shift) into a one-port provides a complex impedance with a phase shift somewhere between 0° and −90° depending on the contribution of each component to the total impedance. Obviously, the behavior of such an impedance cannot be described by a single property any more. After all, there are two different kinds of components involved. It takes a two-dimensional relation to describe this complex impedance.

The polar presentation, which is based on the two properties *magnitude* and *phase shift*, has the advantage that these two properties are measured in an actual circuit. This makes this representation very convenient.

Using the two cartesian coordinates (on the real axis and the imaginary axis) in the complex number plane is an alternative way of presenting an impedance that has two properties. Although this presentation is more abstract, it is preferred because the calculational effort is often smaller. It just requires the handling of complex numbers. It will be the standard presentation of complex quantities in this book. As already mentioned (Section 3.1.1.2), it is general practice to use $\mathrm{j} = \sqrt{-1}$ as imaginary unit to avoid confusion with the symbol i which is reserved for current.

Thus, the complex admittance $Y(\mathrm{j}\omega)$ of a capacitor has the value $Y(\mathrm{j}\omega) = \mathrm{j}\omega C$. A series connection of a capacitor C with a resistor R is just an addition of the two impedances to yield the complex total impedance $Z(\mathrm{j}\omega) = R + \frac{1}{\mathrm{j}\omega C} = R - \frac{\mathrm{j}}{\omega C}$. R is the *real part* $\mathrm{Re}[Z(\mathrm{j}\omega)]$ of $Z(\mathrm{j}\omega)$, and $-\frac{1}{\omega C}$ is referred to as the *imaginary part* $\mathrm{Im}[Z(\mathrm{j}\omega)]$ of $Z(\mathrm{j}\omega)$.

As the inductivity L is dual to the capacity C we apply duality to get the complex impedance of an inductor $Z(\mathrm{j}\omega) = \mathrm{j}\omega L$ and the admittance of a conductor in parallel to an inductor $Y(\mathrm{j}\omega) = G - \frac{\mathrm{j}}{\omega L}$.

From the complex presentation of the impedance $Z = R + \mathrm{j}X$ with the *resistance* R and the *reactance* X, the two parameters magnitude and phase shift of the polar presentation can be derived as follows. The magnitude $|Z|$ presenting the ratio of the voltage peak amplitude $\hat{v}$ across the impedance to the peak amplitude of the current $\hat{i}$ flowing through the impedance is given by

$$|Z| = \sqrt{\mathrm{Re}[Z]^2 + \mathrm{Im}[Z]^2} = \sqrt{R^2 + X^2}\,, \tag{3.55}$$

the phase shift φ, i.e., the phase difference between voltage and current is given by

$$\varphi = \arctan\frac{\mathrm{Im}[Z]}{\mathrm{Re}[Z]} = \arctan\frac{X}{R}\,. \tag{3.56}$$

Combining both presentations one gets

$$Z = R + \mathrm{j}X = \sqrt{R^2 + X^2} \times \mathrm{e}^{\mathrm{j}\arctan\frac{X}{R}}\,. \tag{3.57}$$

When impedances must be added or subtracted it is more convenient to use the cartesian form, but multiplication or division of complex numbers is easier when the polar form is used. Using admittances instead of impedances, when appropriate, might help in avoiding multiplications and divisions.

Problems

?

3.9. What does the symbol j stand for?

3.10. Name the two parameters needed to describe fully infinite sinusoidal time dependences.

3.11. Can any two-dimensional mathematical relation be used for the presentation of infinite sinusoidal time dependences?

3.2.1 Capacitors

From

$$i_C(t) = C \times \frac{dv_C(t)}{dt} \tag{3.58}$$

follows

$$V_C(T) = \frac{1}{C} \times \int_0^T i_C(t)\,dt = \frac{Q_C}{C}\,. \tag{3.59}$$

If a capacitor C is charged with the charge Q_C, there is a voltage V_C across it. It is important to interpret these equations correctly.

- Firstly, the (momentary) current that may flow into a capacitor is unlimited!
- Secondly, if there is no charge in the capacitor the voltage across it is zero. Therefore, it should not surprise that there is a *lag* between current and voltage in a capacitor, called *phase shift* in steady-state applications. The maximum current will flow when the voltage is zero and the maximum voltage occurs when the current is zero. Thus, the phase shift between voltage and current is −90°.

From equation (3.58), we learn that the current i_C through a capacitor is proportional to the time derivative of the voltage v_C across the capacitor. The transfer function of a series capacitor in a two-port (which is a transadmittance; see Figure 3.5a) is that of a *differentiator*. It is proportional to the angular frequency (Figure 3.5b).

From equation (3.59), we learn that by *integration* of the current i_C through a capacitor we obtain the voltage v_C across it. The transfer function of a parallel capacitor in a two-port (which is a transimpedance: see Figure 3.6a) is that of an *integrator*. It is inversely proportional to the angular frequency (Figure 3.6b).

Another relation of importance shows that a capacitor stores electric energy E in the form of electric charge

$$E_Q = \int_0^Q \frac{q}{C}\,dq = \frac{Q^2}{2C} = \frac{C \times V_C^2}{2}\,. \tag{3.60}$$

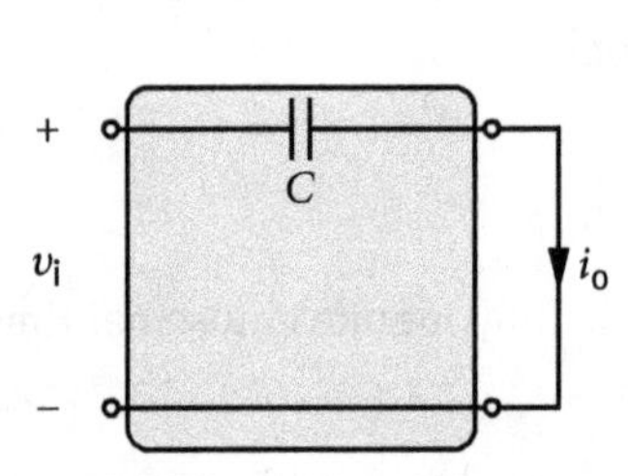

Fig. 3.5a. Series capacitor in a two-port with $R_L = 0$.

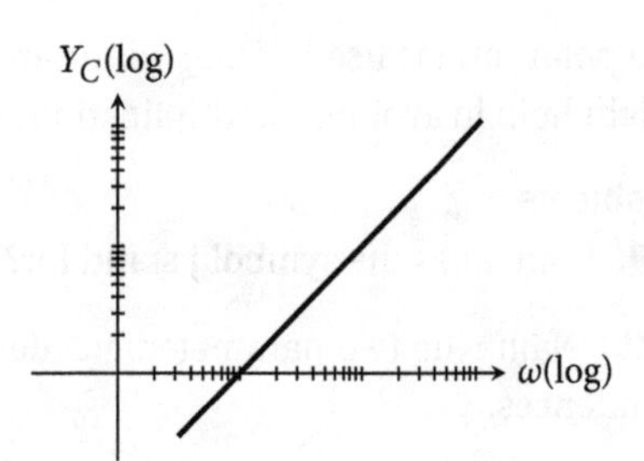

Fig. 3.5b. Frequency dependence of the transadmittance, characteristic of a differentiating circuit.

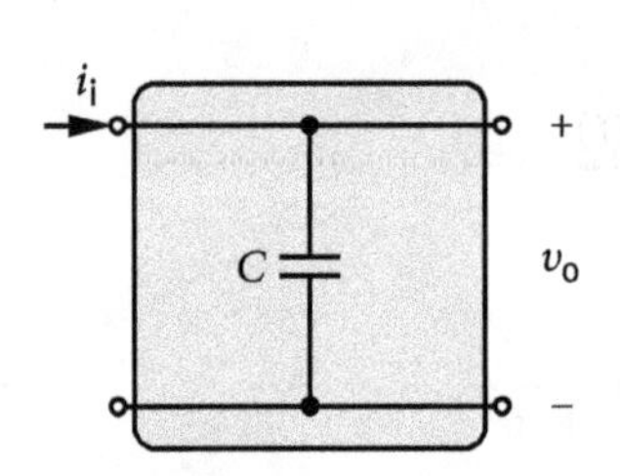

Fig. 3.6a. Parallel capacitor in a two-port with no load ($Y_L = 0$).

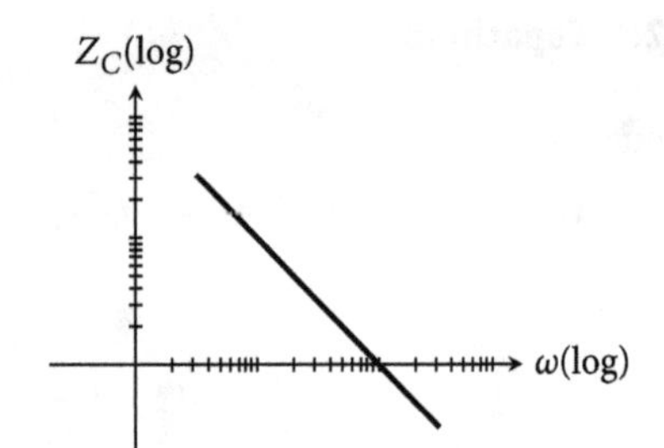

Fig. 3.6b. Frequency dependence of the transimpedance, characteristic of an integrating circuit.

The function of a charged capacitor ist that of a temporary voltage source. In a short enough time interval the voltage stays about the same, independent of the current flow because for high frequencies, i.e., fast changes, the impedance of a capacitor is very low.

In steady-state calculations using complex notation, the above mentioned phase shift of −90° makes the capacitor a purely imaginary admittance $Y_C = j\omega C$, or $Z_C = -\frac{j}{\omega C}$ a purely imaginary negative impedance.

For $\omega = 0$ the admittance is zero, which means that no direct current (DC) can flow through the capacitor. For $1/\omega = 0$ the impedance is zero which means that at high enough frequencies a capacitor acts as a short-circuit.

However, a real capacitor is not a pure capacitance. The leads will have some resistance (and even inductance) as parasitic series impedance associated with them. Besides, the dielectric material inside the capacitor has some conductance dissipating (a very small amount of) electric power. This loss due to this parasitic parallel conductance is expressed by the *loss tangent* $\tan\delta$ which is a parameter of the dielectric material in the capacitor. This tangent is the ratio of the resistive (lossy) component and the reactive (lossless) component of the admittance. The reciprocal of the loss

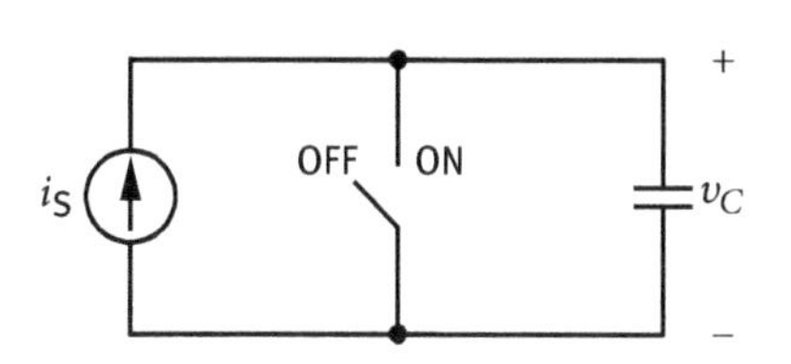

Fig. 3.7a. Charging a capacitor with constant current.

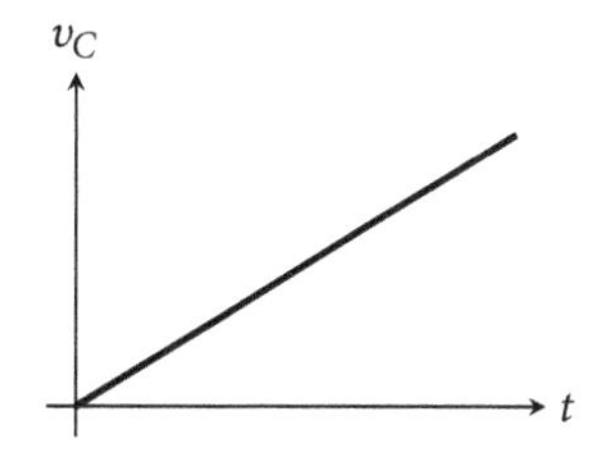

Fig. 3.7b. Time dependence of capacitor voltage.

tangent is the *quality factor Q*

$$Q = \frac{1}{\tan\delta}\,. \tag{3.61}$$

Some properties of various types of capacitors are discussed in Section 4.2.1.2.

Example 3.1 (Charging a capacitor with constant current). In Figure 3.7 a simple switch allows the current of an ideal current source either to flow into a capacitor C (*off* position) or alternatively to ground (*on* position). Let us investigate the behavior of the circuit in Figure 3.7 at an elementary level. In the *on* position of the switch, there will be no voltage $v_C(t)$ across the capacitor C, it is discharged. As soon as the switch goes into the *off* position, the source current i_S will start flowing into the capacitor.

Using the moment of the switching as a time reference, i.e., $t = 0$, we get

$$i_C(t = 0) = i_S\,, \quad \text{and} \quad v_C(t = 0) = 0$$

because it takes charge, i.e., time to build up a voltage across the capacitor. By the current flow the capacitor collects charge increasing the voltage across the capacitor proportionally. From (3.58) one gets $i_c(t)\mathrm{d}t = \mathrm{d}q_C = C \times \mathrm{d}v_C$. With a constant current i_S the charge Q_C and consequently the voltage v_C across the capacitor C increases linearly with time providing a linear voltage ramp (Figure 3.7b).

This linear voltage increase in time can be an unwanted effect in (active) circuits when the amount of current charging a (stray) capacitance is limited due to a too high impedance of the signal source. This effect is called *slewing* and the slew rate (of amplifiers) is typically given in V/μs.

Problem

?

3.12. A capacitor C lies in series to a resistor R shunted by a capacitor C of the same size which are situated at the output of a two-port. This circuit can be viewed at as a complex voltage divider. Give the frequency dependence of the amplitude transfer function and the phase shift transfer function.

3.2.1.1 Stray capacitance

Any two adjacent conductors act as a capacitor. The capacitance will be small unless the conductors are close together for long distances or have large areas. Such a parasitic capacitance, which also occurs inside electronic components is called *stray capacitance*. Stray capacitances make signals leak between otherwise isolated circuits (which is termed *crosstalk*). This can seriously affect the functioning of circuits at high frequencies.

In macroscopic circuits (unavoidable) stray capacitances are on the order of 1 pF, seemingly rather small. However, as with any ordinary capacitance their static value is subject to a dynamic increase if they are part of a parallel feedback loop (Miller effect, Section 2.4.3.2).

Problem

3.13. Think it over. It is advantageous (smaller energy loss!) to transmit electric power over very long distances in the form of direct current rather than alternating current. Why?

3.2.2 Inductors

In Section 1.3.4 we learned that the property that causes a voltage drop that is proportional to a current change is called inductance L which is measured in units of henry (H). The electronic component having this property is called inductor

$$u_L(t) = L \times \frac{\mathrm{d}i_L(t)}{\mathrm{d}t} \tag{3.62}$$

and

$$I_L(T) = \frac{1}{L} \times \int_0^T v_L(t)\,\mathrm{d}t = \frac{\Phi_L}{L}\,. \tag{3.63}$$

The magnetic flux $\Phi = L \times I$ is dual to the electric charge $Q = C \times V$. It is important to interpret these equations correctly. Firstly, the (momentary) voltage that may occur across an inductance is unlimited! Secondly, if there is no magnetic flux in the inductor then there is no current through it. Therefore, it should not surprise that there is a lag between voltage (that builds up the flux) and current in an inductor, which is called phase shift in steady-state applications resulting in a phase shift between voltage and current of 90°. The possibility of high voltage spikes across inductors which can be much higher than the supply voltage and are potentially dangerous to electronic components (having a limited voltage capability, Section 2.3.1) asks for safety provisions like using voltage limiting circuits (Section 2.2.7.2).

Another important relation shows that an inductor stores magnetic energy E_Φ in the form of magnetic flux

$$E_\Phi = \int_0^\Phi \frac{\varphi}{L}\,\mathrm{d}\varphi = \frac{\Phi^2}{2L} = \frac{L \times I_L^2}{2}\,. \tag{3.64}$$

In steady-state calculations using complex notation, the above mentioned phase shift of 90° makes the inductor a purely imaginary impedance $Z_L = j\omega L$. For $\omega = 0$ the impedance is zero. For $1/\omega = 0$ the conductance is zero which means that at high enough frequencies no current can flow through the inductor, it acts as an open-circuit. The high impedance for high frequencies prohibits fast changes of the current, i.e., an inductor is a temporary current source. Its current is independent of the value of the voltage across it.

Problems

?

3.14. An inductor L across the input of a two-port is shunted by a series combination of a resistor R and another inductor L of the same size across the output. This circuit can be viewed at as a complex current divider. Give the frequency dependence of the complex amplitude transfer function.

3.15. A two-port with a conductance of 0.6 S across the input and one of 0.3 S in series to the output has a load of 0.3 S shunted by an inductor of 2 mH. The source is a 9-mA current source with an angular frequency of $\omega = 5 \times 10^3\,\text{s}^{-1}$.

(a) Find out (by thinking) if the following pairs of electrical variables are in phase (whether the phase shift is zero) $v_S - v_o$, $v_o - i_o$, and $v_S - i_S$.

(b) Calculate i_o. (Hint: replace the source plus two-port according to Norton's theorem.)

Example 3.2 (Identification of (exactly) two linear elements which are not accessible). Two linear components (e.g., hidden in a black-box) shall be identified by the amplitude and/or phase response of the black-box. Figure 3.8 shows the arrangement chosen to perform the measurements (with an oscilloscope). From the measurements the following four results are obtained:

1. $g_v(f = 0) = \frac{v_o}{v_i} = 0.50$
2. $g_v(f = 0.1\,\text{MHz}) = 0.71/2$
3. $(\varphi_{vi} - \varphi_{ii})(f = 0.1\,\text{MHz}) = 45°$
4. $g_v(f = 2\,\text{MHz}) = 1.00$

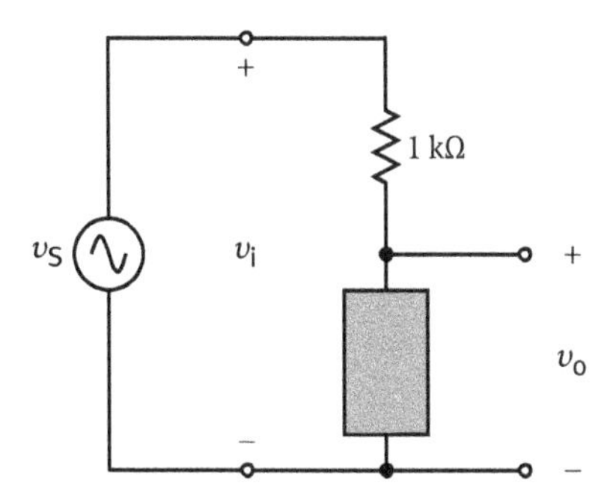

Fig. 3.8. Circuit of Example 3.2.

The increase of the voltage gain from 0 to 2 MHz (results #1 and #4) indicates inductive behavior. Result #4 excludes configurations where a resistor is shunted by an inductor. Thus, we know that a resistor is in series to an inductor.

In this case, it is clear that the value of the resistor equals that of the external resistor, namely, 1 kΩ (from result #1) because at $f = 0$ the impedance of the inductor is zero. The results #2 and #3 are equivalent, i.e, only one of them is necessary. A phase difference of 45° indicates that the real part and the imaginary part of the complex impedance are equal $\omega L = 2\,\text{k}\Omega$. Therefore, the unknown pair of linear one-ports in the box is a resistor of 1 kΩ in series to an inductor of $10/\pi$ mH. The position of the two elements within the black-box is electronically irrelevant. Thermography could find the position of the resistor because the electric power dissipated in it would result in a local temperature increase.

3.3 Time domain vs. frequency domain

As time is the conjugate variable to the angular frequency ω, the frequency response of a circuit may also be specified by parameters describing the time dependence of the response to a step signal. The following properties describe the time behavior of the output signal in response to a step signal at the input (see Figure 3.9).

Step response is the time behavior of the output of a (two-port) circuit when its input signal changes from zero to a flat maximum in a very short time. In binary electronics, this would be a switch from the low state L to the high state H at the input.

Propagation delay is the time difference between the time when the step occurs and the moment when the output response reaches half its final value (the very first time).

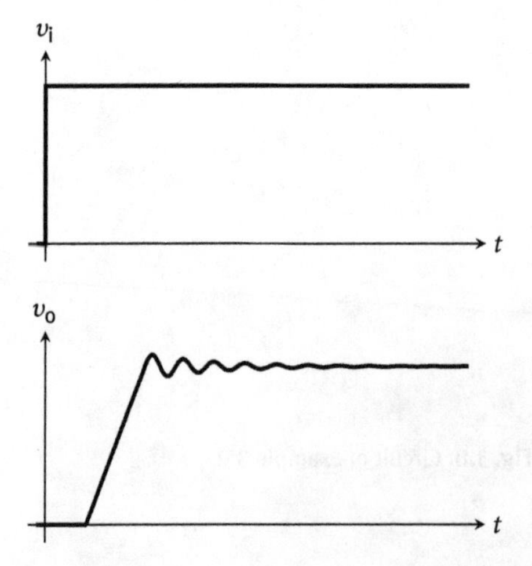

Fig. 3.9. Example of a response of a two-port to a step signal at the input.

Rise time is the time it takes a signal to change from a given low value to a given high value. These values are usually 10% and 90% of the step height. For negative going signals, the term *fall time* is appropriate. When applied to an output signal of a two-port, both parameters depend both on the rise or fall time of the input signal and the characteristics of the two-port.

Steady-state error is the difference between the actual final output value after the circuit reaches a steady state and the one expected for an ideal circuit.

In Figure 3.9 ringing due to the presence of both capacitor and inductor in the circuit is seen which is described by the following properties:

Overshoot is present when the signal exceeds its expected steady-state amplitude.

Peak time is the time it takes the output signal to reach the first peak of the overshoot.

Settling time is the time elapsed between the occurrence of an ideal step at the input to the time at which the amplitude of the output signal has settled to its final value.

As an ideal step signal contains all frequencies, a correct transfer by the (active) two-port would require that the amplitude and the phase of all frequency components are transferred correctly to give an ideal step signal at the output. From an analysis of the actual shape of the output signal, information can be gained *in one step* on the two-port's transfer property at all frequencies.

Working with step functions in the mathematical sense (covering minus and plus infinity) is utterly impractical. Adding two step functions of the same amplitude but opposite sign and delayed by some time interval t_1 results in a single rectangle of length t_1. In a practical application, such a signal of amplitude A would be repeated in time intervals of length T. This periodic rectangular wave has a fixed amplitude A for some interval t_1 (the mark) during the period of length T and has the value zero (the space) for the rest of this period. The length of the mark divided by the period is called *duty cycle d*

$$d = \frac{t_1}{T} \tag{3.65}$$

The *duty cycle* is usually given in percent. A rectangular wave with a duty cycle of 50% is called *square wave*. A duty cycle is also defined for nonperiodic signals. It is the fraction of the total time under consideration in which a device is actively producing a signal. The inverse of the period T is the *repetition frequency* $f = 1/T$ [Hz]. Measurements of the step response are performed with rectangular signals having a long t_1 and a not too high f. Obviously, in practical applications the sharpness of an ideal step can only be approximated, i.e., the rise time of a signal can never be zero.

Problems

?

3.16. A steady flow of 10 V high rectangular pulses with a duty cycle of $d = 0.1$ (= 10%) into a resistor of 1 kΩ heats it up.

(a) Determine the average power dissipated in the resistor.

(b) Under which circumstances must the maximum dissipated power be considered rather than the average to determine whether the dissipated power is above the power rating of the resistor?

3.17. Periodic rectangular signals with a repetition frequency of $f = 10$ kHz and a pulse length $t_1 = 10$ μs are fed into a resistor of 100 Ω.
(a) What is the duty cycle d?
(b) What is the dissipated power in the resistor if the pulse height is 10 V?

3.18. A steady flow of ±1 V high (bipolar) rectangular pulses with a duty cycle $d = 0.01$ (= 1%) into a resistor of 100 Ω heats it up.
(a) Determine the average power dissipated in the resistor.
(b) Explain the difference in dissipative power if the duty cycle is $d = 99\%$.

3.3.1 Voltage step applied to a capacitor

For a change, let us investigate the behavior of the circuit in Figure 3.10 at an elementary level. In the *off* position of the switch (Section 2.3.2.1), the capacitor C will discharge. After equilibrium has been reached, no current will flow and both voltages $v_C(t)$, and $v_R(t)$ will be zero. As soon as the switch is in the *on* position current will flow, and according to Kirchhoff's second law the voltages for each moment t afterward are given by the relation

$$v_S(t) = v_R(t) + v_C(t) .$$

Using Ohm's law we can express the voltage across the capacitor by

$$v_C(t) = v_S(t) - R \times i(t)$$

where $i(t)$ is the current through all elements in the loop. Using the moment of the step (of the switching) as a time reference, i.e., $t = 0$, we get

$$i(t = 0) = \frac{v_R(t = 0)}{R} = \frac{v_S}{R}$$

because it takes charge, i.e., time to build up a voltage across the capacitor. Clearly, in the first moment, all the voltage is across R and thus, the current has above value.

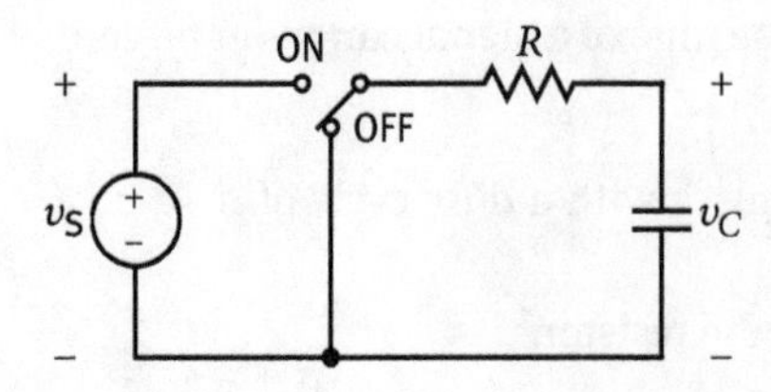

Fig. 3.10. Simulating a voltage step signal to charge a capacitor.

By the current flow the capacitor collects charge increasing the voltage across it which in turn decreases the voltage across R reducing the current so that at any $t > 0$ the current is given by

$$i(t) = \frac{v_S(t) - v_C(t)}{R} .$$

On the other hand, we know (3.58) that

$$i_C(t) = C \times \frac{dv_C(t)}{dt} = \frac{v_S(t) - v_C(t)}{R} ,$$

and that $v_C(t = 0) = 0$, so that solving above differential equation yields

$$v_C(t) = v_S \times (1 - e^{-\frac{t}{RC}}) = v_S \times (1 - e^{-\frac{t}{\tau}}) \tag{3.66}$$

with τ the *time constant* $\tau = R \times C$. Using the relation $v_R(t) = v_S(t) - v_C(t)$ we get

$$v_R(t) = v_S \times e^{-\frac{t}{RC}} . \tag{3.67}$$

The voltage across the capacitor, as the response to a step signal from a real voltage source, is again some kind of step signal, however, with a finite rise time. The rise time t_{rs}, the time difference $(t_{0.9} - t_{0.1})$ between the time when 10% of the final step size is reached, and the time when 90% of the final step size is reached, is easily obtained from

$$t_{rs} = (\ln 0.9 - \ln 0.1) \times \tau = \ln 9 \times \tau \approx 2.2 \times \tau . \tag{3.68}$$

A negative step requires a source with a negative voltage. All above relations are the same except that the term *fall time* t_f is used instead of rise time

$$t_f = (\ln 0.9 - \ln 0.1) \times \tau \approx 2.2 \times \tau . \tag{3.69}$$

As an ideal step signal cannot be realized, it is important to know how the response is to a real step signal with an intrinsic rise time t_{rs}. The answer is that rise times add quadratically

$$t_{rs} = \sqrt{t_{rs1} + t_{rs2} + t_{rs3} + \ldots} . \tag{3.70}$$

This means that contribution to rise times that are at least a factor of 3 smaller than the largest, the dominant rise time, may be disregarded accepting a maximal 5% deviation from the correct value. This quadratic addition is very important because signal generators with finite (well defined) rise times may be used as signal source of step signals.

As the circuit *response* is the relationship between the circuit's output to the circuit's input, it is not surprising that a rise time is also assigned to actual circuits.

The rise time of a circuit, a device, an instrument is the rise time of the response of the system in question to a hypothetical ideal step signal at the input.

This rise time can be obtained from the bandwidth *BW* (Section 3.4.2) of an instrument by $t_{rs} = \frac{2.2}{2\pi \times BW}$.

3.3.2 Charged capacitor

In Section 3.3.1, we have seen that a capacitor gets charged (and discharged) by the current that accompanies a signal flow. If the time constant is short enough, i.e., if the impedance of the signal source is small enough, the charge is removed "instantly" as soon as the signal goes to ground. In this case, the charging may be disregarded and the capacitor behaves just as a passive conductance of $j\omega C$. If, on the other hand, the time constant is long, the signal current charges the capacitor to a level at which, on average, the charging current equals the discharging current. Observe that if the source is a voltage source v_S with an impedance R_S the charging current i_C depends on the voltage v_C across the capacitor

$$i_C = \frac{v_S - v_C}{R_S}$$

which means that only the voltage difference supplies the signal. For this signal, a (linear) capacitor still behaves as a passive conductance of $j\omega C$. This situation can be described by a (quiescent) operating point on the i–v characteristic provided by the charge condition of the capacitor. In Figure 3.11, a capacitor is shown that is charged by a voltage source v_S via a source resistor R_S and shunted by the load resistor R_L. Obviously, some of the source current gets lost by flowing through R_L. What is the best way to analyze this situation? As we are interested only in the electrical variables in connection with the capacitor, we can replace the remaining linear network according to Thevenin's theorem by a real voltage source with v_{Th} and R_{Th}. v_{Th} is the open-circuit

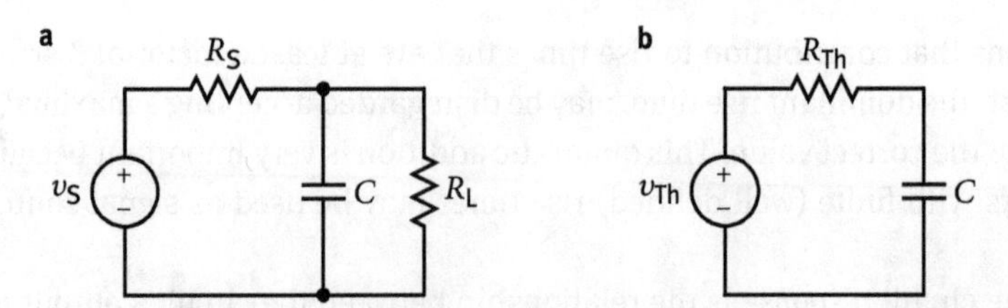

Fig. 3.11. Loading a capacitor by a linear network (a) circuit (b) circuit after applying Thevenin's theorem.

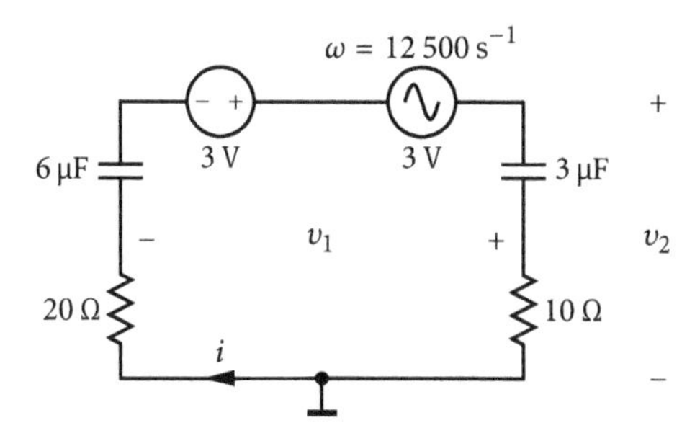

Fig. 3.12. Circuit of Problem 3.20.

voltage of the voltage divider consisting of R_S and R_L

$$v_{Th} = v_S \times \frac{R_L}{R_S + R_L},$$

and R_{Th} is R_S shunted by R_L

$$R_{Th} = R_S \times \frac{R_L}{R_S + R_L},$$

so that v_{Th} is effective instead of v_S and R_{Th} instead of R_S.

The circuits in the following sections either take advantage of the properties of charged capacitors or show how the effect of charging can be minimized.

Problems

3.19. Because of $Y_C = j\omega C$ the conductance of a capacitor at $\omega = 0$ is zero, i.e., no real DC current can flow into a capacitor. Explain why the voltage of a battery that is connected to two (equal) capacitors arranged in series is divided (equally) by this capacitive voltage divider.

3.20. Calculate the amplitude of the following electrical variables from Figure 3.12:
(a) The voltage v_1 across (10 Ω + 20 Ω).
(b) The voltage v_2 across (10 Ω + 3 μF).
(c) The current i in the loop.

3.3.2.1 RC circuits with two different time constants

Some elementary nonlinear components are directional. Their i–v-characteristics depend on the direction of current flow. This becomes obvious, if such components are used to charge and discharge capacitors because the time constants of these two processes may be substantially different.

Example 3.3 (Charging of a capacitor involving a diode). Figure 3.13a presents a circuit with a fall time considerable less than the rise time. A positive voltage step makes the (ideal) diode reversed biased so that the time constant τ_{rs}, effective for the rise time, is $\tau_{rs} = C \times 9\,\mathrm{k\Omega}$. With a capacitor of 5 nF the rise time t_{rs} becomes 99 μs. After equilibrium, the negative voltage step brings the input voltage back to the ground. Now

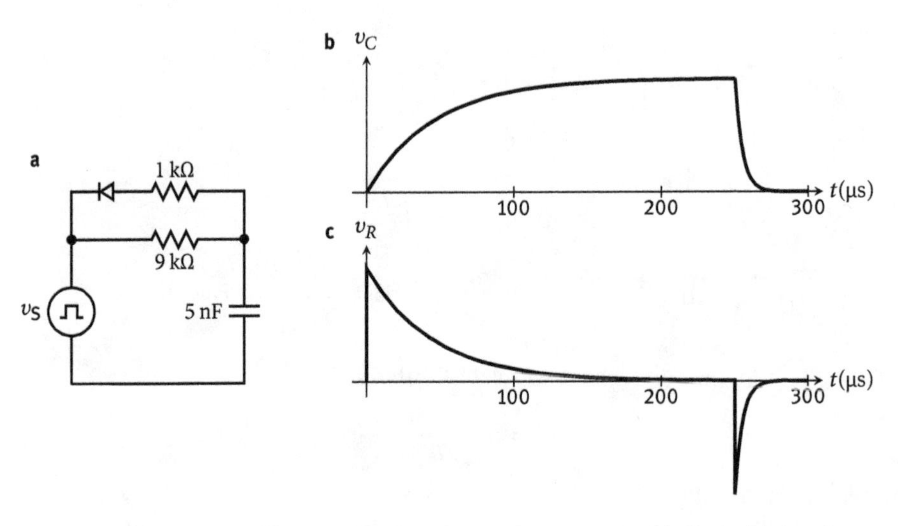

Fig. 3.13. Illustrating a step response with two different time constants: (a) circuit, (b) time dependence of capacitor voltage, and (c) resistor voltage.

the diode is forward biased so that the time constant τ_f, effective for the fall time, is $\tau_f = C \times 0.9\,\text{k}\Omega$. With a capacitor of 5 nF the fall time t_f becomes 9.9 μs, i.e., it is ten times shorter. This is portrayed in Figure 3.13b which shows the time dependence of the voltage across the capacitor. In Figure 3.13c the time dependence of voltage across the resistor(s) is shown. This curve is just the difference between the rectangular input signal and the voltage across the capacitor. The positive and negative spikes in Figure 3.13c reflect the charging and discharging of the capacitor. One might wonder why the areas of these spikes are not equal. After all, the same amount of charge enters and leaves the capacitor. Thus, it takes (much) more time to charge the capacitor than to discharge it as can be seen in Figure 3.13b. If one investigates the time dependence of the current, one will find out that the negative current peak is eleven times as large as the positive one so that in this picture the areas of the spikes (representing the charge) are equal, as expected.

Example 3.4 (Charging of a capacitor involving an emitter follower). The reverse biasing of a diode can have serious consequences when it occurs in amplifier stages. Let us consider the output of an emitter follower (Section 2.5.2.1) that is loaded with a resistor R_E shunted by a capacitor C as shown in Figure 3.14a. When charging the capacitor by means of the emitter current I_E, the transistor is in the amplifying state having a low output impedance of $Z_o \approx r_e = 25\,\text{mV}/I_E$ (with Z_o in Ω and I_E in mA). As seen from the capacitor, the impedance is R_E shunted by Z_o and, therefore, the charging time constant is $\tau_{rs} = C \times R_E \times \frac{Z_o}{R_E+Z_o}$. If v_i falls (suddenly) to zero, the base voltage v_B is zero whereas the emitter voltage v_E is still positive because the capacitor is charged positively. Therefore, the base-emitter diode is reverse biased and the transistor is not

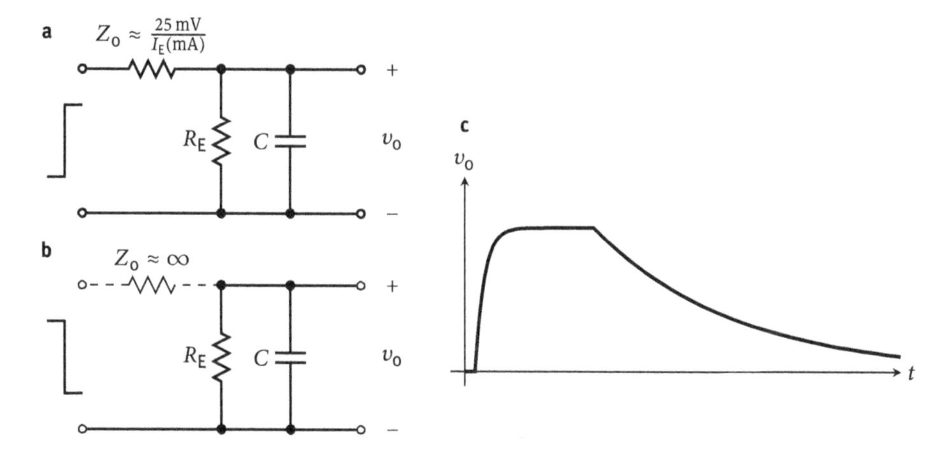

Fig. 3.14. Output situation at an emitter follower with capacitive load: (a) amplifying state of the transistor, (b) transistor in the cut-off state, and (c) resulting response to a fast rectangular input signal.

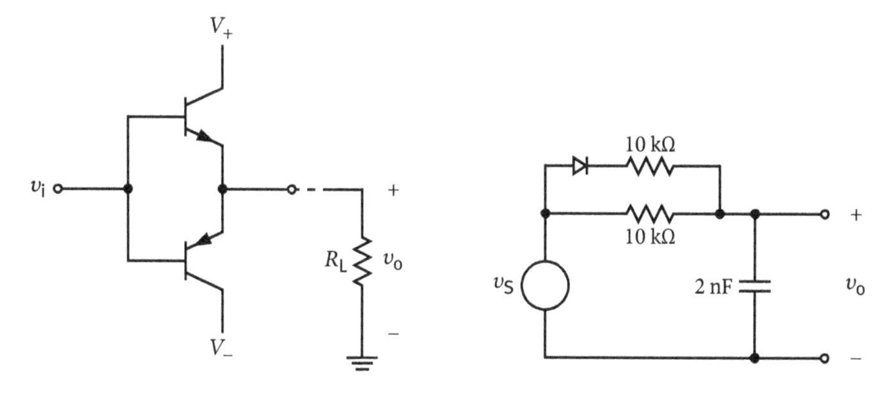

Fig. 3.15. Basic complementary emitter follower.

Fig. 3.16. Circuit of Problem 3.21.

working, it is in the cut-off region. The output impedance of this emitter follower is that of a reverse biased diode, i.e., it is very high, and the time constant commanding the fall time is $\tau_f = C \times R_E$. In Figure 3.14b this behavior is sketched.

By means of complementary emitter followers (Figure 3.15), it can be avoided that the rise and the fall time differ that much. Under no signal condition, the circuit delivers no output current and consequently $v_o = 0$. In this case, the base-emitter bias V_{BE} of both transistors is zero, so that they are in the cut-off region with both transistors not conducting. A positive signal that is not sufficient to bias the upper (NPN) transistor properly (i.e., that is smaller than $\approx$ 0.65 V) will not take the operating point into the amplifier region to provide enough current for the output voltage to follow the input, and the input of the lower (PNP) gets even stronger reverse biased by the positive going input. The same is true for the lower transistor but for a negative going input. Thus, between about ±0.65 V (for technologies based on silicon) of input, the circuit does not work as emitter follower resulting in a kink in the output signal for an input signal crossing from negative to positive or vice versa. This kink is a form of *crossover distortion*. To minimize this distortion it is necessary to bias (e.g., by means of diodes) both transistors in a way that the quiescent operating points of both transistors are in the amplifying region (B-amplifiers Section 2.3.4). Then, the upper emitter follower will deliver current for the full positive swing of the input signal and the lower transistor will sink the current for the full negative swing. In this case charging and discharging of a capacitor shunting the load resistor is done with time constants that are as equal as the output impedances of the two transistors are equal.

Problem

3.21. Investigate the voltage transfer of a negative 1 V rectangular signal of 10 μs length through the circuit of Figure 3.16 (assuming an ideal diode).
(a) Determine the time constants for the negative τ_- and the positive τ_+ slope.
(b) Give the rise time t_{rs} and the fall time t_f of the output signal.
(c) What is the minimum output voltage?

3.3.2.2 Clamping

Let us investigate the idealized circuit in Figure 3.17a. Without further information, there are three obvious properties of this circuit.
- The output voltage v_o cannot go negative.
- No current will flow unless the input voltage is negative.
- The charging of the capacitor (after a current flow) will be such that the output voltage will be positive for a zero input voltage.

Let the input voltage v_i be a square wave oscillating between −3 V and +2 V. During the time with negative amplitude the capacitor gets charged to $V_C = 3$ V, when the signal is positive no current will flow so that the charge of the capacitor does not change.

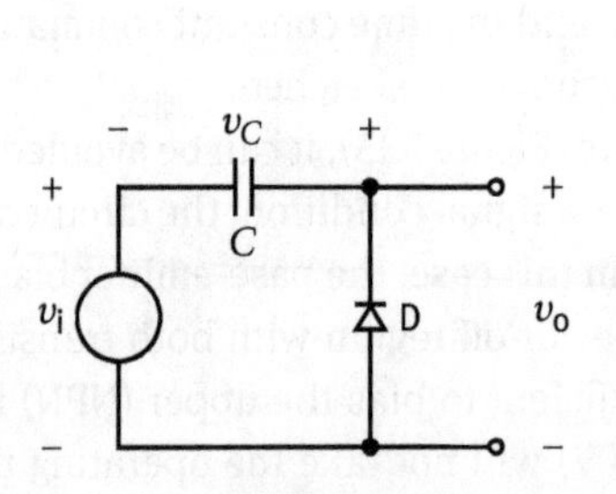

Fig. 3.17a. Idealized diode clamp circuit: an ideal voltage source in series with a capacitor C and an ideal reverse biased diode D.

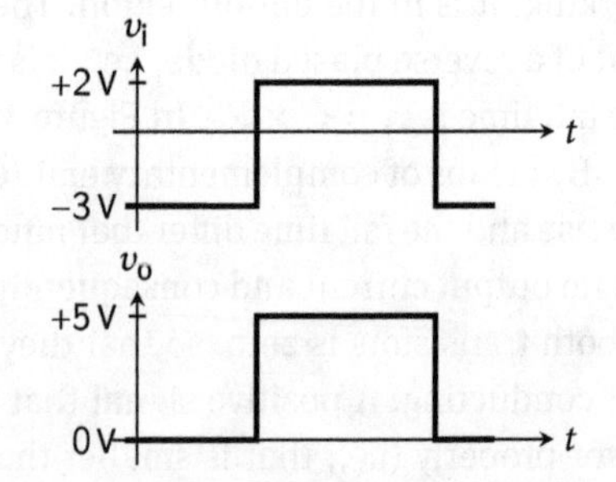

Fig. 3.17b. Input signal (v_i) and output signal (v_o).

After the capacitor is completely charged to 3 V (which is promptly only in the ideal case) the quiescent output voltage is 0 V to which the input signal v_i is added so that the output voltage is a square wave oscillating between 0 V and 5 V.

Problems

?

3.22. Can the clamping property of a capacitor explained by its impedance?

3.23. A DC component will not be transmitted by a capacitor. Which property of a capacitance is responsible for that?

3.24. Why does an output capacitor of a DC-voltage-supply get charged despite the fact that DC-current does not flow into a capacitor?

3.3.2.3 Baseline restoring

In a real-clamping circuit, the diode will be shunted by some resistance, e.g., the load impedance R_L which together with the capacity C makes a finite time constant $\tau = R_L C$ that controls the discharge of the capacitor C. Consequently, an input signal gets differentiated (Sections 3.2.1 and Section 3.6.4.2). Figure 3.18 shows an example how loss of charge in the clamping capacitor during the presence of the signal affects the shape of the output signal.

The amount of charge that gets lost can be seen in the drooping top of the signal. Consequently, there is undershoot that preserves the original amplitude of the negative step function. As the capacitor was charged with the positive signal, the same amount of charge must be withdrawn by the negative step signal with the applicable time constant. The zero-line (baseline) is approached exponentially (Section 3.2.1), i.e., in theory it is never reached. As a consequence, the next signal starts out below the baseline, ending up even lower than the first signal. This *pile-up* continues shifting the reference line lower and lower. A diode as used in the clamp circuit speeds up the discharge as shown in Figure 3.19. As the diode has rather high impedance at *low* forward voltage (Section 2.2.7.2) a small forward bias of the diode is an essential improvement.

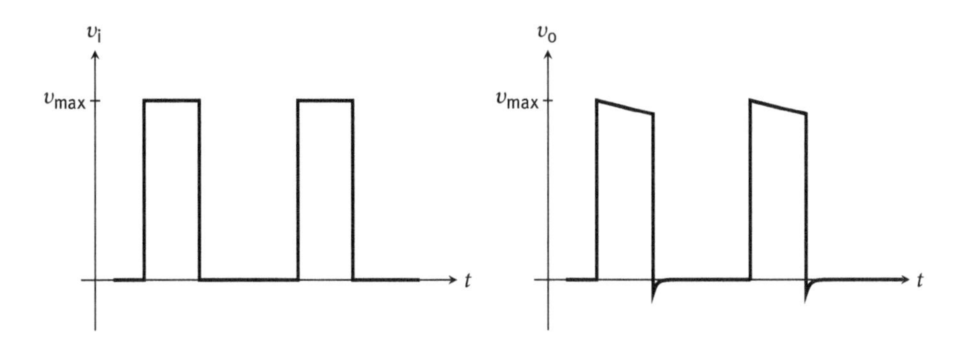

Fig. 3.18. Loss of charge in the clamping capacitor distorts output signal.

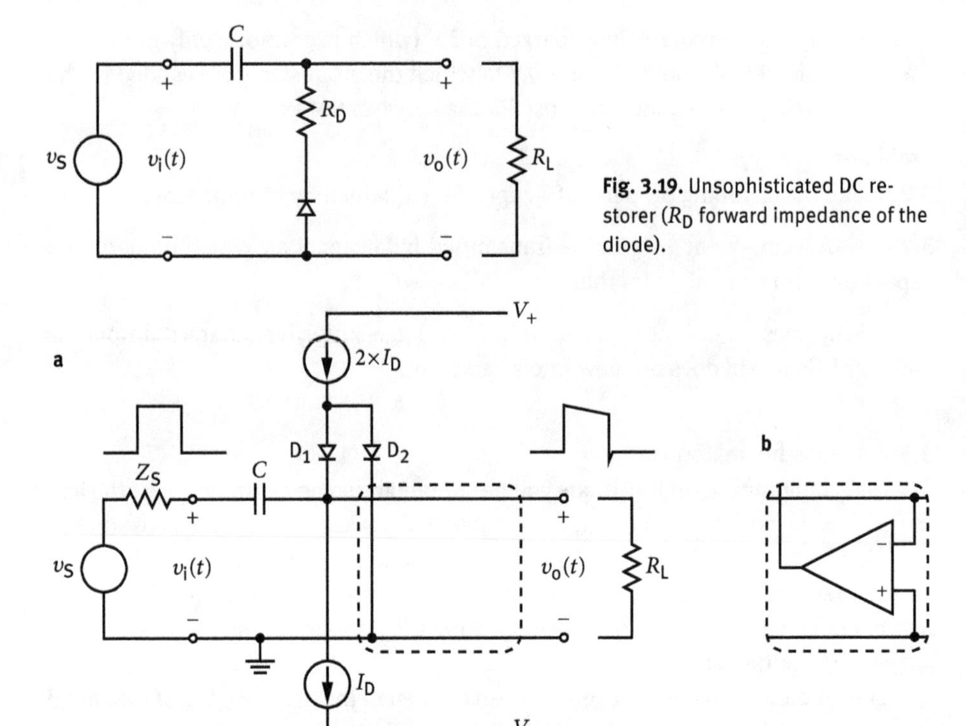

Fig. 3.19. Unsophisticated DC restorer (R_D forward impedance of the diode).

Fig. 3.20. (a) DC restorer with biased diodes. (b) Reducing the impedance of the diodes by the Miller effect.

In Figure 3.20a a matched pair of diodes D_1 and D_2 make the output voltage zero if the same current I_D flows through either diode. This is achieved by having the current source at the top deliver twice the current (e.g., 0.2 mA) of the lower one (e.g., 0.1 mA). When the output goes positive D_1 gets reversed biased with practically no conductance so that there is no shunt to the load. When the output goes negative D_2 gets cut-off, the current through D_1 increases, which speeds up the discharging of the capacitor. Thus, the discharging time constant is essentially reduced. The insert in Figure 3.20b shows how the diode impedance can be further reduced by inclusion into the feedback loop of an inverting operating amplifier (Section 2.5.2.3).

Another method to avoid undershoots is pole-zero cancelation (see Section 3.4.3.1).

Problems

3.25. What is the advantage of the current source in Figure 3.20a over an appropriately chosen resistor?

3.26. Why is the step-transition of a step pulse transmitted by a capacitance without attenuation?

3.3.2.4 Rectifying (DC power supplies)

The first step in converting bipolar signals (AC) into unipolar signals (DC) is to suppress the unwanted polarity. Obviously, diodes being current check valves (Section 2.1.6) can do this job. Figure 3.21 demonstrates how *half-wave rectification* works.

In this case, the positive half of the AC wave supplied as input voltage v_i by a transformer is passed, while the negative half is blocked. The output voltage v_o across the load resistor R_L is a unidirectional pulsating direct current. Some part of the input voltage will drop across the diode (according to the characteristic of the diode, Section 2.1.6) reducing the output voltage v_o and causing power dissipation in the diode. In Section 2.5.2.3 we learnt how the voltage drop across the diode can be minimized by feedback action.

Figure 3.22 demonstrates how *full-wave rectification works.* Full-wave rectification provides both polarities of the input wave form as direct current (voltage of one polarity). Four diodes are arranged in a bridge configuration (called *Graetz bridge*). This configuration assures that (positive) voltage supplied by the transformer is directed to the upper lead of the load resistor whereas (negative) current that is sunk in the transformer stems from the lower lead of the resistor so that there is no negative current flow through the load resistor, the voltage is unipolar.

Although the voltage across the load resistor is unipolar (DC), it is pulsating between zero and the peak value. Feeding this pulsating signal into a *storage* or *smoothing* capacitor with a sufficiently long (dis)charging time constant, provides a flat capacitor voltage with a (small) AC ripple. This ripple reflects the finite discharging time. In Figure 3.23 the circuit diagram of a voltage power supply based on full-wave rectification with a smoothing capacitor is shown.

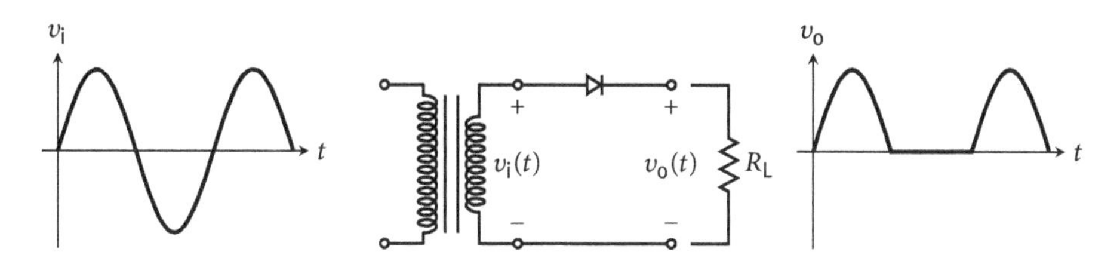

Fig. 3.21. Half-wave rectification demonstrated with a sinusoidal signal provided by a transformer.

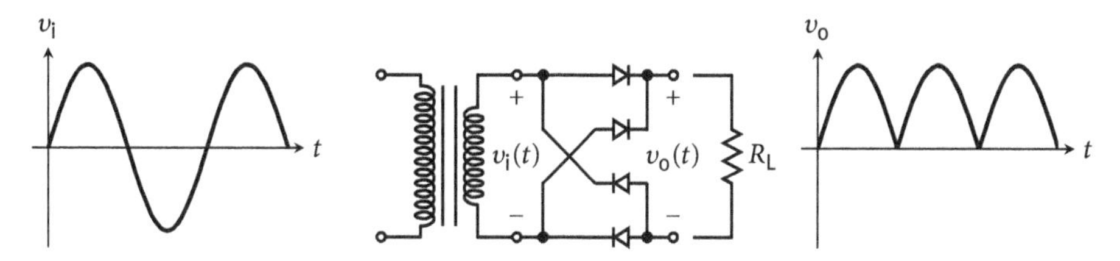

Fig. 3.22. Full-wave rectification demonstrated with a sinusoidal signal provided by a transformer.

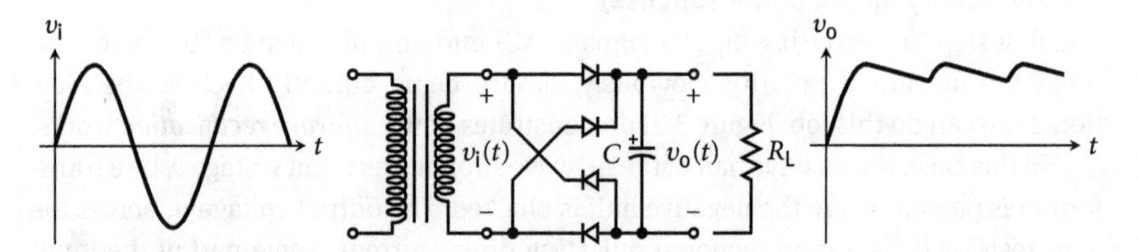

Fig. 3.23. Principle of a voltage power supply based on full-wave rectification with a smoothing capacitor C: (a) input voltage; (b) circuit; and (c) output voltage v_o with ripple.

The correct size of the smoothing capacitor depends on the load (the minimum load resistor) so that a long discharging time constant keeps the amplitude of the ripple at an acceptable level. However, the conductance of the capacitor is proportional to its size so that the peak current in the transformer secondary and the diodes will be enhanced correspondingly. In practice, it will be determined by the output impedance of the transformer. Using an active voltage regulator circuit (operation amplifier, Section 2.5.2.2) in cascade to the reservoir capacitor, allows an essential reduction of the capacity value with improved ripple performance.

Problem

3.27. When trying to understand how a circuit involving a diode works, is it better to deal with the diode current or voltage?

3.3.2.5 Diode voltage multiplying

A series configuration of two half-wave rectifiers (using the half-waves of opposite polarity) can be used to make the output voltage across the two capacitors in series twice the peak input voltage (Figure 3.24). This circuit does not only provide an output voltage of nearly double the peak AC input voltage, but the tap in the middle makes it possible to use this circuit for a split rail supply, i.e., to have two power supplies of opposite polarity and of one-half of the full value.

Another circuit of a voltage doubler (Figure 3.25) has the advantage that it can be cascaded to provide voltage multiplication. The negative half-cycle of the input voltage v_i, charges the capacitor C_1 through diode D_2 to the peak voltage v_{peak} of v_i. During the positive half-cycle, the input voltage is in series with the stored voltage v_{peak} across C_1. Thus, the capacitor C_2 gets charged through diode D_1 to $2v_{peak}$. On the next negative half-cycle, C_1 gets fully charged again.

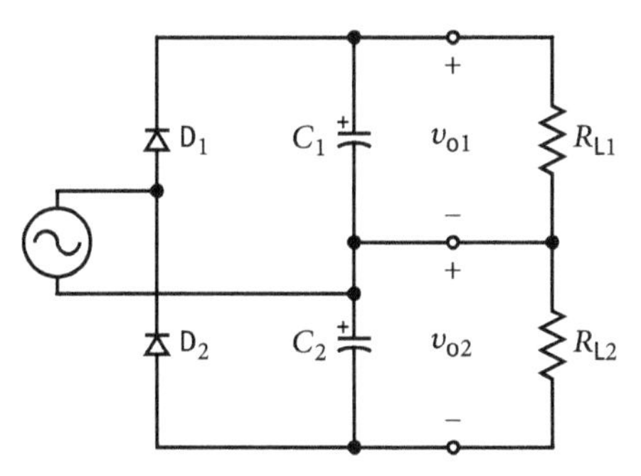

Fig. 3.24. Diode voltage doubler with a tap provision.

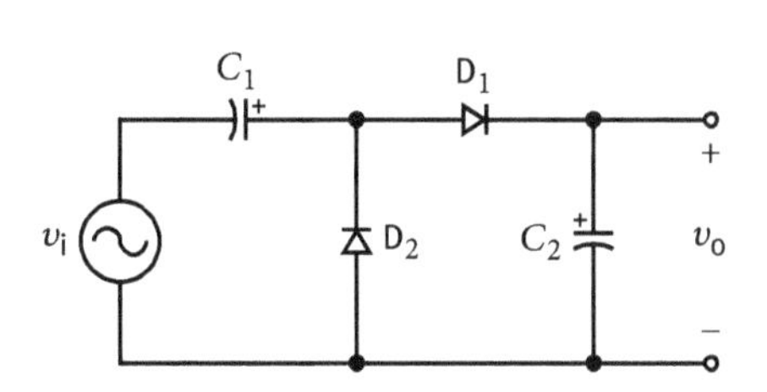

Fig. 3.25. Diode voltage doubler of cascade type ($v_O = 2 \times v_{ipeak}$).

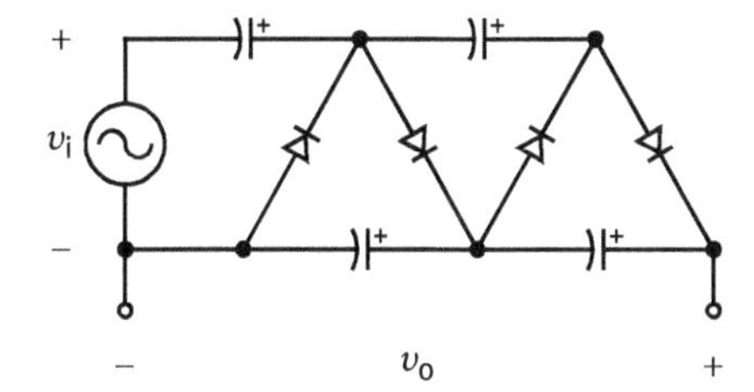

Fig. 3.26. Principle of a Cockcroft–Walton voltage quadrupler (cascaded voltage doubler).

This circuit can also be explained statically with the help of Section 3.3.2.2 (clamp circuits) and Section 3.3.2.4 (half-wave rectifiers). C_1 and D_2 form a clamp circuit that lifts the signal zero-line by v_{peak}. D_1 and C_2 constitute a half-wave rectifier which rectifies the clamped input signal.

Figure 3.26 shows the principle of a voltage quadrupler. Voltage multipliers obeying this principle are also called *Cockcroft–Walton circuits*. These circuits are capable of producing an output voltage that is tens of times the peak AC input voltage at a quite limited output current rating.

Problem

?

3.28. Why can voltage multiplying not be explained without considering the current through the diodes?

3.3.3 Current step applied to an inductor

In Figure 3.27, a conductor G is parallel to an inductor L. A simple changeover switch connects the ideal constant current source i_S alternatively to ground (*off* position) or to the inductor and conductor (*on* position).

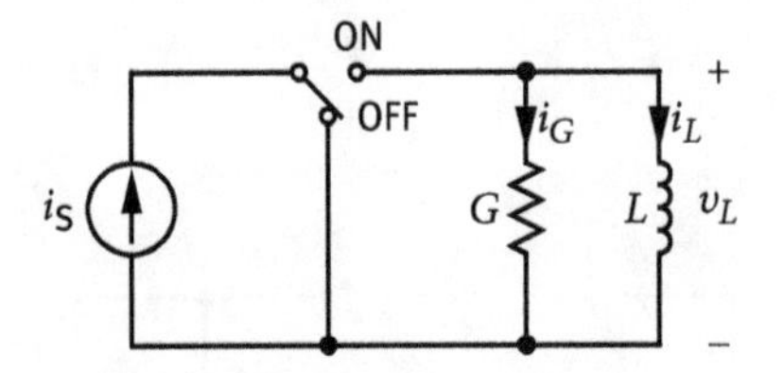

Fig. 3.27. Simulating a current step signal to charge an inductor.

The behavior of this circuit is dual to that in Figure 3.10. Consequently, we just present the relevant equations without further comment

$$\begin{aligned} i_S &= i_G(t) + i_L(t) \\ i_L(t) &= i_S - G \times v(t) \\ v(t=0) &= \frac{i_G(t=0)}{G} = \frac{i_S}{G} \\ v(t) &= \frac{i_S - i_L(t)}{G} \,. \end{aligned}$$

From (3.62)

$$v_L(t) = L \times \frac{di_L(t)}{dt} = \frac{i_S - i_L(t)}{G}$$

and with $i_L(t = 0) = 0$ we get

$$i_L(t) = i_S \times (1 - e^{-\frac{t}{GL}}) = i_S \times (1 - e^{-\frac{t}{\tau}}) \tag{3.71}$$

with τ the *time constant* $\tau = G \times L$. Using the relation $i_G(t) = i_S - i_L(t)$ we get

$$i_G(t) = i_S \times e^{-\frac{t}{\tau}} \,. \tag{3.72}$$

The current in the inductor as a response to a step signal from a real current source is again some kind of step signal, however, with a finite rise time. This rise time t_{rs} is again

$$t_{rs} = (\ln 0.9 - \ln 0.1) \times \tau = \ln 9 \times \tau \approx 2.2\tau \,. \tag{3.73}$$

3.4 Dynamic response of passive two-ports (passive filters)

The frequency dependence of a two-port comprises both the amplitude and the phase shift response of all four two-port parameters. They must be given, e.g., by using the complex notation. The presentation of input and output impedance does not differ from the presentation used for the impedance of a one-port (Section 3.2). In addition, there is the (forward) transfer function and the reverse transfer function. The smallness of the latter justifies the usual practice of not considering it. Of the four (forward) transfer parameters (Section 2.2.2), we choose the complex voltage transfer function $G_v(j\omega)$

$$G_v(j\omega) = \mathrm{Re}[G_v(j\omega)] + j \times \mathrm{Im}[G_v(j\omega)] = g_v(\omega) \times e^{j\varphi(\omega)} , \qquad (3.74)$$

with the magnitude $g_v(\omega)$

$$g_v(\omega) = |G_v(j\omega)| = \sqrt{\mathrm{Re}[G_v(j\omega)]^2 + \mathrm{Im}[G_v(j\omega)]^2} \qquad (3.75)$$

and the phase shift

$$\varphi(\omega) = \arctan \frac{\mathrm{Im}[G_v(j\omega)]}{\mathrm{Re}[G_v(j\omega)]} . \qquad (3.76)$$

Thus, the transfer function of two-ports containing at least one reactive element exhibits not only a frequency response of the amplitude transfer but also of the phase shift. In addition to (or instead of) this steady-state behavior also the transient behavior (Section 3.3) may be investigated. If not all input frequencies are transferred equally, such a two-port is called a filter.

3.4.1 Basic filter configurations

A filter (section) is a two-port with a frequency-dependent response function, i.e., the transfer parameter changes with frequency. A simple filter consists of just two elements of a different nature put into a two-port. If they are in series, they form a (complex) voltage divider, when in parallel a (complex) current divider. As we know, the forward quantities of a two-port depend on the load impedance, the reverse quantities on the source impedance. To avoid such loading conditions we presuppose that the input signal is delivered by an ideal source and the load will either be an open-circuit (for the voltage version) or a short-circuit (for the current version). Contrary to what is generally done, we will start out with the complete set of simple filters, both voltage *and* current filters.

Problem ?

3.29. A complex voltage divider consisting of two impedances Z_1 and Z_2 is fed by an ideal voltage source of angular frequency ω and amplitude of 1 V. It is loaded by Z_L. The complex impedances are as follows: $Z_1 = (20 + 30j)\,\Omega$, $Z_2 = (50 + 10j)\,\Omega$, $Z_L = (30 + 20j)\,\Omega$. Give the absolute value of each of the three currents.

3.4.1.1 Voltage filters

Figure 3.28 shows the six arrangements of pairs of different linear elements that have frequency f dependent voltage transfer. The filters (a) and (b) are called low-pass filters, because they hinder the passage of high frequency components of signals. The filters (c) and (d) are called high-pass filters because they hinder the passage of low

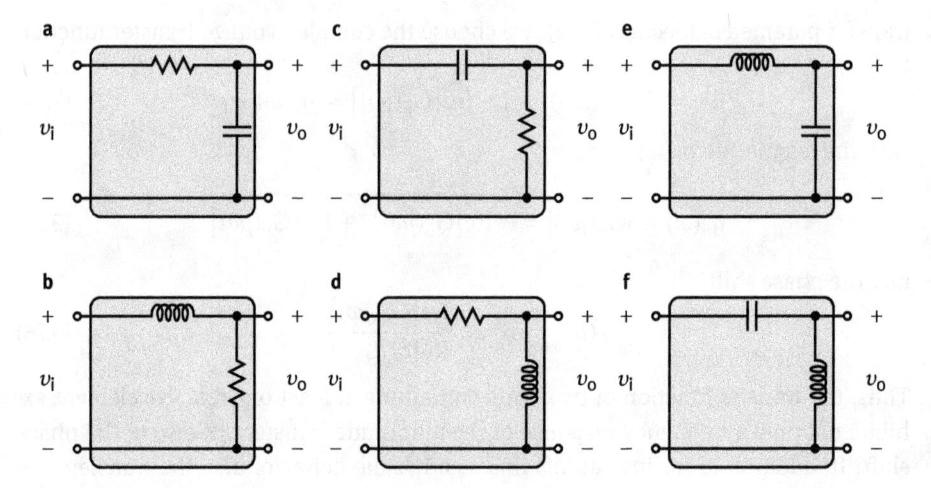

Fig. 3.28. The six basic simple frequency dependent voltage dividers.

frequency components of signals. And (e) and (f) are resonance filters, because their components form a *series resonant circuit* that is tuned to its resonance frequency.

Even if a pair of filters has the same frequency response function, they are not identical in all their properties. This can best be seen, e.g., by comparing the output impedances Z_o at very low and at very high frequencies.

Problems

3.30. Determine the input impedance Z_i of each of the 6 voltage filters from Figure 3.28 for $f = 0$.

3.31. Determine the input impedance Z_i of each of the six voltage filters from Figure 3.28 for $1/f = 0$.

3.32. Determine the output impedance Z_o of each of the six voltage filters from Figure 3.28 for $f = 0$. (As shown, with an open-circuit at the input.)

3.33. Determine the output impedance Z_o of each of the six voltage filters from Figure 3.28 for $1/f = 0$. (As shown, with an open-circuit at the input.)

3.4.1.2 Current filters

Applying duality to Figure 3.28 yields Figure 3.29. There are two current low-pass filters, two current high-pass filters, and two *parallel resonant circuits*.

Each filter type has a frequency f response of the current gain analogous to that of the voltage counterpart. Again the filter pairs differ in the value of their impedances.

As current filters are dual to voltage filters it suffices to deal with just one kind. Thus, we restrict ourselves in the following to voltage filters.

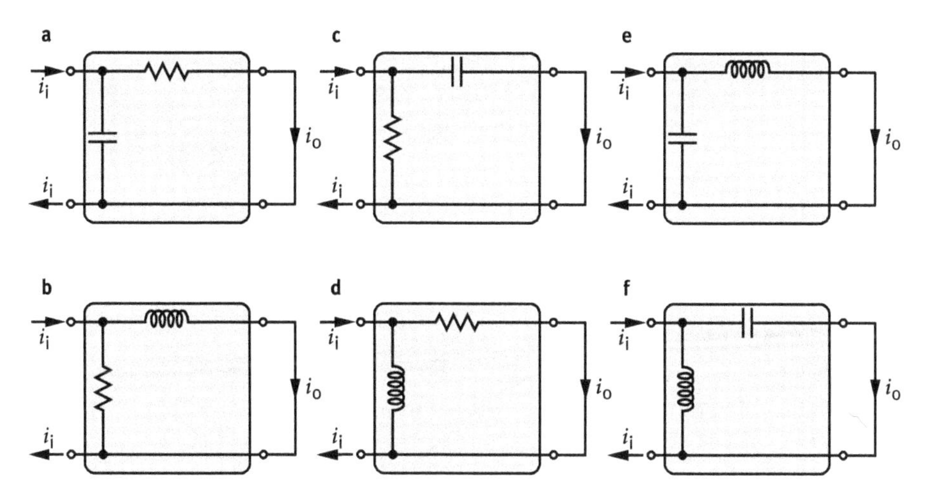

Fig. 3.29. The six basic simple frequency dependent current dividers.

Problems

?

3.34. Determine the input impedance Z_i of each of the six current filters from Figure 3.29 for $f = 0$.

3.35. Determine the input impedance Z_i of each of the six current filters from Figure 3.29 for $1/f = 0$.

3.36. Determine the output impedance Z_o of each of the six current filters from Figure 3.29 for $f = 0$. (As shown, with an open-circuit at the input.)

3.37. Determine the output impedance Z_o of each of the six current filters from Figure 3.29 for $1/f = 0$. (As shown, with an open-circuit at the input.)

3.4.2 Low-pass filters

The shape of the phase and amplitude responses is the same for all four low-pass filters except that one-half deals with voltage and the other half with current. Following the general trend we pick the voltage RC low-pass filter as example. Such a filter is shown in Figure 3.28a.

The product of resistance R and capacitance C gives the time constant τ of the filter. The turnover frequency f_u (in Hz), is determined by the time constant:

$$f_u = \frac{1}{2\pi\tau} = \frac{1}{2\pi RC} \tag{3.77}$$

or better by using the angular frequency ω_u (in units of radians/s, or simply s^{-1}):

$$\omega_u = \frac{1}{\tau} = \frac{1}{RC}\,. \tag{3.78}$$

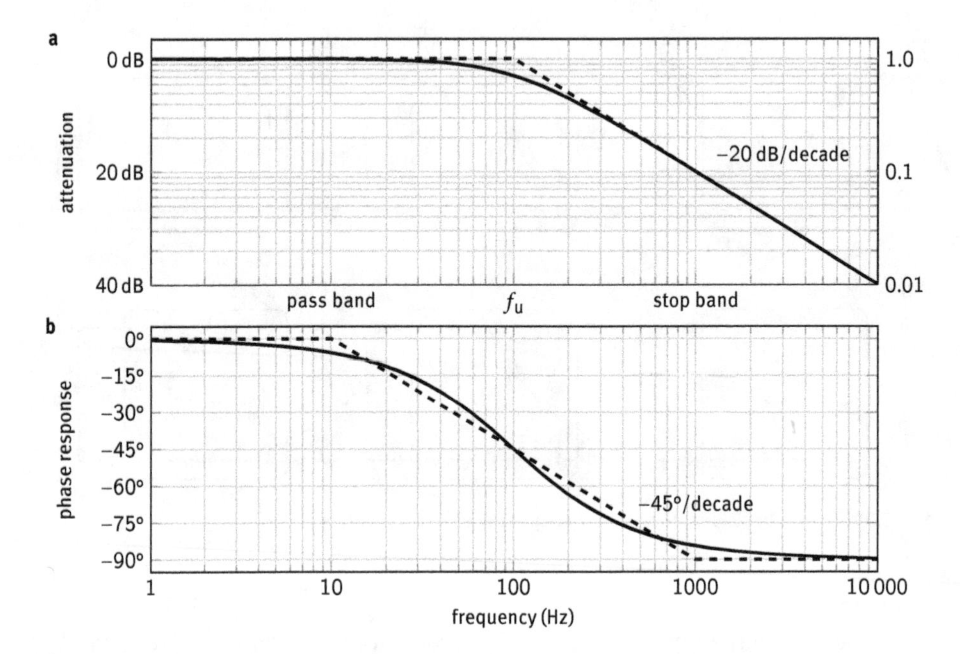

Fig. 3.30. (a) Frequency dependence of the amplitude response (Bode plot) of a simple low-pass filter together with the straight-line approximation. (b) Frequency dependence of the phase response (Bode plot) of a simple low-pass filter together with the straight-line approximation.

From Figure 3.30a it can be seen that the frequency response of the amplitude transfer is smooth. It can be linearly approximated by two straight lines intersecting at the (upper) *corner frequency* ω_u (*cut-off* or *break frequency*). The frequencies below the corner frequency are in the pass-band, those that are above are in the stop-band. The width of the pass-band is called *bandwidth BW*. The rate of frequency roll-off in the stop-band has a slope −20 dB per decade that equals −6 dB per octave. At the corner frequency, the actual amplitude response is lower by 3.01 dB, i.e., a (voltage or current) signal at the output having the corner frequency will be attenuated to $1/\sqrt{2} \approx 0.7071$ of its input value. This means that the input power will be halved (hence −3 dB), with the transmission of power through the circuit declining further with increased frequency. As discussed in Section 3.2.1 the low-pass filter has integrating properties for frequencies in the stop-band.

The phase response curve can also be approximated by (three) straight lines. At the corner frequency the curve and the linear approximation intersect at −45°. In the linear approximation, the phase shift is zero at $\omega = 0.1\omega_u$, and −90° at $\omega = 10\omega_u$, i.e., the slope is −45° per decade.

Such *maximally flat magnitude filters* have as flat a frequency response as possible in the pass-band. They are called Butterworth filters. Its order n is given by the steepness of the slope of the frequency response in the stop-band: it is $n \times 20$ dB per decade. Thus, the simple RC filter is a first-order Butterworth filter.

Using the complex notation the calculation of the transfer function is straightforward. In the voltage filter, the two elements form a voltage divider, in the current filter a current divider. From the divider equation

$$G_v(\mathrm{j}\omega) = \frac{v_\mathrm{o}(\mathrm{j}\omega)}{v_\mathrm{i}(\mathrm{j}\omega)} = \frac{Z_2}{Z_1 + Z_2}\,, \tag{3.79}$$

with $Z_1 = R$ and $Z_2 = 1/\mathrm{j}\omega C$ it becomes

$$G_v(\mathrm{j}\omega) = \frac{1}{1 + \mathrm{j}\omega\tau} = \frac{1}{\sqrt{1 + (\mathrm{j}\omega\tau)^2}} \times \mathrm{e}^{-\mathrm{j}\arctan\omega\tau}\,. \tag{3.80}$$

The first factor is the amplitude term $|G_v|$ (shown in Figure 3.30a), and the phase term (shown in Figure 3.30b) is $\varphi(\omega) = -\arctan\omega\tau$. A fast check can be made for the values with $\omega\tau = 1$. There, the amplitude term is $|G_v| = 1/\sqrt{2} = 0.701 = -3$ dB and the phase term is $\varphi = -45°$ as shown in the plots. Considering the property of the capacitor (it must get charged before a voltage occurs), it is obvious that there is a *lag* between input and output voltage. Consequently, the phase shifts in Figure 3.30b are negative.

Problems

?

3.38. A low-pass filter has a time constant τ.
(a) Give the corner frequency f_u of the frequency response function.
(b) Give the value of the amplitude transfer function g_v at $10 f_\mathrm{u}$.
(c) Give the value of the phase angle φ at $10 f_\mathrm{u}$, both exact and using the straight line approximation.

3.39. The voltage transfer of a simple low-pass filter at ω_x is $g_v(\omega_\mathrm{x}) = 0.1$.
(a) Give the exact and the asymptotic value of $g_v(0.1\omega_\mathrm{x})$.
(b) Express the bandwidth *BW* of this filter by ω_x.

3.40. The corner frequency f_u of a low-pass filter is $f_\mathrm{u} = 0.35$ MHz. Determine the rise time of the output signal when
(a) an ideal step function is applied to the input, and when
(b) a signal of the shape of the output signal of (a) is applied to the input.

3.41. Into the input of a simple low-pass filter consisting of a resistor $R = 10$ kΩ and a capacitor $C = 2$ nF a rectangular signal, 10 V high and 10 µs long, is fed.
(a) What is the maximum output voltage v_omax?
(b) What is the rise time of the output signal?
(c) What is the fall time of the output signal?

3.42. A step function is applied to the input of the circuit of Figure 3.31. How long does it take for the output voltage to reach 6.32 V?

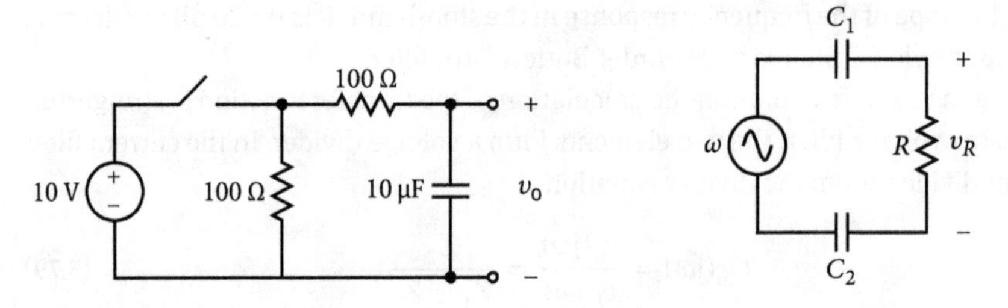

Fig. 3.31. Circuit of Problem 3.42.

Fig. 3.32. Circuit of Problem 3.43.

3.43. Determine in Figure 3.32 the angular frequency at which the voltage across the resistor is twice that of the voltage across each of the capacitors.

3.44. An ideal voltage-dependent current source is loaded with a resistor R_L shunted by a capacitor C_L. Show that the product of voltage gain $g_v(f = 0)$ times bandwidth $BW = \frac{1}{2\pi\tau}$ is independent of R_L.

3.45. Calculate the (complex) current transfer $g_i(j\omega) = \frac{i_L(j\omega)}{i_i(j\omega)}$ of a current low-pass filter consisting of a conductance G shunted by an inductor L.
(a) Present the frequency response of the amplitude transfer.
(b) Present the frequency response of the phase shift.
(c) Give the (amplitude) bandwidth.

3.46. The capacitor of an RC voltage low-pass filter consists of C_1 in series to C_2. To spare a fruitless calculational effort assume $C_1 = C_2 = C$.
(a) Think it over what the (complex) voltage across C_2 is compared to the voltage across the combination of C_1 and C_2?
(b) What is the time constant of this low-pass filter?
(c) Modify the equations of the simple RC low-pass filter to get the appropriate equations for this special case.

3.4.3 High-pass filters

A simple RC *high-pass filter* is shown in Figure 3.28c. It passes high-frequency components of signals but attenuates frequency components below the *turnover frequency* f_l as shown in Figure 3.33a. The product of resistance R and capacitance C gives the time constant τ of the filter. It is inversely proportional to the turnover frequency

$$f_l = \frac{1}{2\pi\tau} = \frac{1}{2\pi RC} \tag{3.81}$$

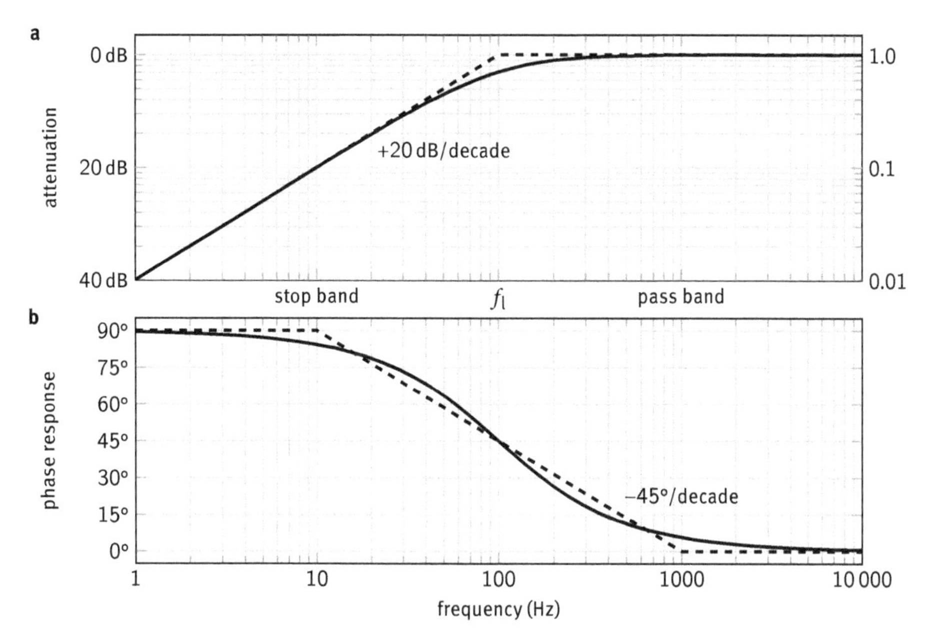

Fig. 3.33. (a) The frequency dependence of the amplitude response (Bode plot) of a simple high-pass filter together with its straight-line approximation. (b) The frequency dependence of the phase response (Bode plot) of a simple high-pass filter together with its straight-line approximation.

where f_l is in Hz, τ is in s, R is in Ω, and C is in F. It is better using the angular frequency ω_l (in units of s^{-1}):

$$\omega_l = \frac{1}{\tau} = \frac{1}{RC} . \tag{3.82}$$

At this (lower) corner frequency the amplitude response curve is lower by 3.01 dB, i.e., a signal at the output having the corner frequency will be attenuated to $1/\sqrt{2} \approx 0.7071$ of its input value which means that the input power is halved (down by −3 dB).

The frequency response curve of the amplitude is smooth. It can be linearly approximated by two straight lines intersecting at the corner frequency f_l (Figure 3.33a). The frequencies above the corner frequency are in the pass-band, those that are below are in the stop-band. The curve has a slope of +20 dB per decade that equals +6 dB per octave in the stop-band. As discussed in Section 3.2.1 the high-pass filter has *differentiating* properties for frequencies in the stop-band.

Also the phase response curve can be approximated by (three) straight lines (Figure 3.33b). At the corner frequency f_l the curve and the linear approximation intersect at 45°. In the linear approximation, the phase shift is 90° at $\omega = 0.1\omega_l$, and zero at $\omega = 10\omega_l$, i.e., the slope is 45° per decade. This simple RC filter also is a first-order Butterworth filter.

Using the complex notation the calculation of the transfer function is straightforward. In the voltage filter, the two elements form a voltage divider, in the current filter, a current divider. From the divider equation

$$G_v(\mathrm{j}\omega) = \frac{v_\mathrm{o}(\mathrm{j}\omega)}{v_\mathrm{i}(\mathrm{j}\omega)} = \frac{Z_2}{Z_1 + Z_2}, \tag{3.83}$$

with $Z_1 = 1/\mathrm{j}\omega C$ and $Z_2 = R$ it becomes

$$G_v(\mathrm{j}\omega) = \frac{1}{1 + \frac{1}{\mathrm{j}\omega\tau}} = \frac{\omega\tau}{\sqrt{1 + (\omega\tau)^2}} \times \mathrm{e}^{\mathrm{j}\arctan(\frac{1}{\omega\tau})}. \tag{3.84}$$

The first factor is the amplitude term $|G_v(\mathrm{j}\omega)|$ (Figure 3.33a), and $\arctan(1/\omega\tau)$ is the phase term $\varphi(\omega)$ (shown in Figure 3.33b). A fast check can be made for the values with $\omega\tau = 1$. There, the amplitude term is $g_v = |G_v| = 1/\sqrt{2} \approx 0.7071 = -3$ dB and the phase term is $\varphi = 45°$ as shown in the plots.

It is quite obvious that for those filters in which Z_2 is dissipative (a resistor, or a conductor, respectively) the *phase response* of the transfer function is identical with that of the input impedance (or admittance, respectively) because the output voltage (or current, respectively) is proportional to the input current (or voltage, respectively). This is also the reason for the 90° phase shift at lower frequencies. It just reflects the phase shift between voltage and current through the capacitor.

Problems

3.47. Design the dual counterpart of a voltage high-pass RC-filter.

3.48. The input impedance Z_i of a simple voltage high-pass filter with a time constant of $\tau = 10$ µs is $Z_\mathrm{i} = 14.14$ kΩ at the angular frequency $\omega_\mathrm{x} = 10^5\ \mathrm{s}^{-1}$.

(a) Give the two circuits (including the correct values of the components) that fulfill above requirements.

(b) Give at ω_x, for both circuits, the phase shift between the output voltage v_o and the current i_o through the element situated parallel to the output.

(c) Give at ω_x, for both circuits, the phase shift between the input voltage v_i and the input current i_i.

3.49. The inductance of a voltage RL high pass consists of L_1 in series to L_2. To spare fruitless calculational pains assume $L_1 = L_2$.

(a) Think it over what the (complex) voltage across L_2 is compared to the voltage across the combination of L_1 and L_2?

(b) What is the time constant of this high-pass filter? (Hint: the time constant dual to RC is GL.)

(c) Modify the equations of the simple RC high-pass filter to get the appropriate equations for this special case.

3.50. The input of a high-pass-voltage filter consisting of a capacitor of 2 µF and a resistor of 300 Ω is connected to a voltage divider (600 Ω and 300 Ω) that divides the voltage of a 9-V AC source having an angular frequency of $\omega_\mathrm{S} = 5 \times 10^3\ \mathrm{s}^{-1}$. Determine v_i and v_o of the filter.

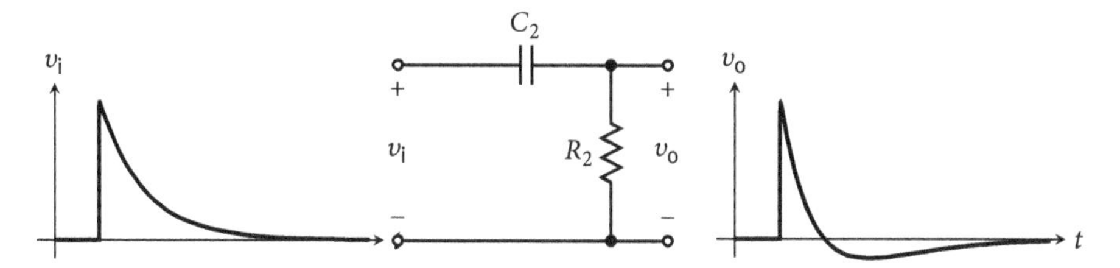

Fig. 3.34a. High-pass filter produces bipolar signal (undershoot).

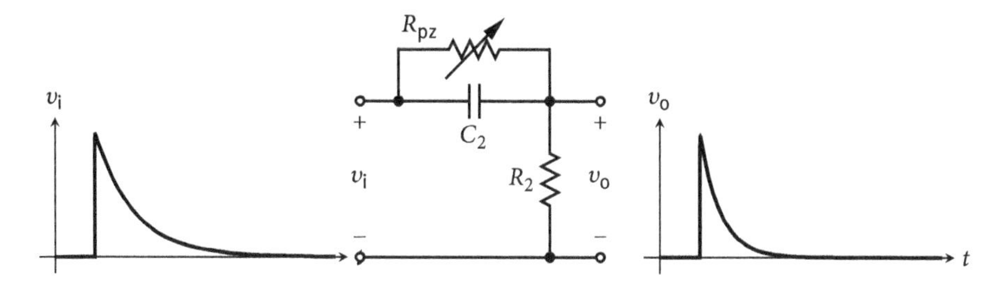

Fig. 3.34b. A resistor shunting the capacitor of the filter restores unipolarity (removes undershoot).

3.4.3.1 Pole-zero cancelation

When a unipolar signal that is exponentially decaying with a time constant τ_S to the baseline, passes a high-pass ($\tau_2 = R_2C_2$) filter (Figure 3.34a) the answer is a bipolar signal (with no DC contribution which would not pass the capacitor) where the positive and the negative lobe has the same area (each contains the same amount of charge). The ratio of the peak value v_u of the undershoot to that of the signal v_S equals the ratio of the time constant τ_2 of the high-pass filter to the decay time constant τ_S of the input signal. As this undershoot is only decaying slowly, the next signal will ride on the undershoot of a previous one reducing the peak value of that signal.

In Figure 3.34b baseline restoration is achieved by the variable resistor R_{pz} shunting the capacitor C_2. It must be adjusted to cancel the undershoot. The result is an output pulse decaying to baseline. This procedure is called *pole-zero cancellation* because it uses a zero to cancel a pole in the mathematical presentation of the transfer function. As an exact adjustment of the variable resistor is essential for good results, the variable resistor used to adjust the pole-zero cancellation must be easily accessible for the user. Some of the more sophisticated circuits simplify this task with automatic pole-zero cancellation.

As the input signal with a decay constant τ_S can be realized by differentiating a step signal by means of high-pass filter with time constant τ_S the pole-zero cancellation effectively reduces a double differentiation of a step signal to a single one.

Problems

3.51. What is the purpose of pole-zero-cancellation?

3.52. How must the time constants of the two filter sections producing a double-differentiated signal be chosen so that double differentiation has the least effect on the baseline?

3.4.4 Band-pass filters

By combining a high-pass filter with a low-pass filter, a *band-pass filter* is formed. A band-pass filter passes frequencies within a certain frequency range constituting the *bandwidth BW* and attenuates frequencies outside that range, in the two stop-bands. At low frequencies, it behaves as a high-pass filter, at high frequencies, as a low-pass filter. The phase response and the amplitude response is just the combination of the response functions of these two filters. Consequently, there are two corner frequencies ω_l and ω_u with two time constants τ_l and τ_u. Figure 3.35 shows a typical frequency response, this time also in linear presentation.

3.4.5 (Voltage) band-stop filters

Whereas a voltage band-pass filter can be produced by cascading a high-pass filter with a low-pass filter, a *band-stop* filter which is also called *band-elimination*, *band-reject*, or *notch filter* can be made by placing a high-pass parallel to a low-pass filter. Such a filter passes all frequencies above and below some frequency range called the *notch.*

3.4.5.1 Twin-T filter

As the name suggests, this filter comprises of two T-shaped filter sections. They consist of resistors and capacitors and are symmetric between input and output (Figure 3.36).

Sizing the components in the ratios as shown in the figure gives a sharp notch in the frequency response of the amplitude at the *notch frequency* $\omega_\mathrm{n} = \frac{1}{RC}$. There the attenuation is maximal. From the complex transfer function (with the abbreviation $\tau = RC$)

$$g_v(\mathrm{j}\omega) = \frac{-\omega^2\tau^2 + 1 - \mathrm{j} \times (\omega^3\tau^3 - \omega\tau)}{-5\omega^2\tau^2 + 1 - \mathrm{j} \times (\omega^3\tau^3 - 5\omega\tau)} \tag{3.85}$$

the amplitude for $\omega = \frac{1}{\tau}$ is found to be zero, and those for $\omega = 0$ and $\frac{1}{\omega} = 0$ are one. As the phase shift is quite important we give it explicitly

$$\varphi(\omega) = \arctan \frac{-\omega^5\tau^5 + 6\omega^3\tau^3 - 5\omega\tau}{\omega^6\tau^6 - \omega^4\tau^4 + \omega^2\tau^2 + 1} . \tag{3.86}$$

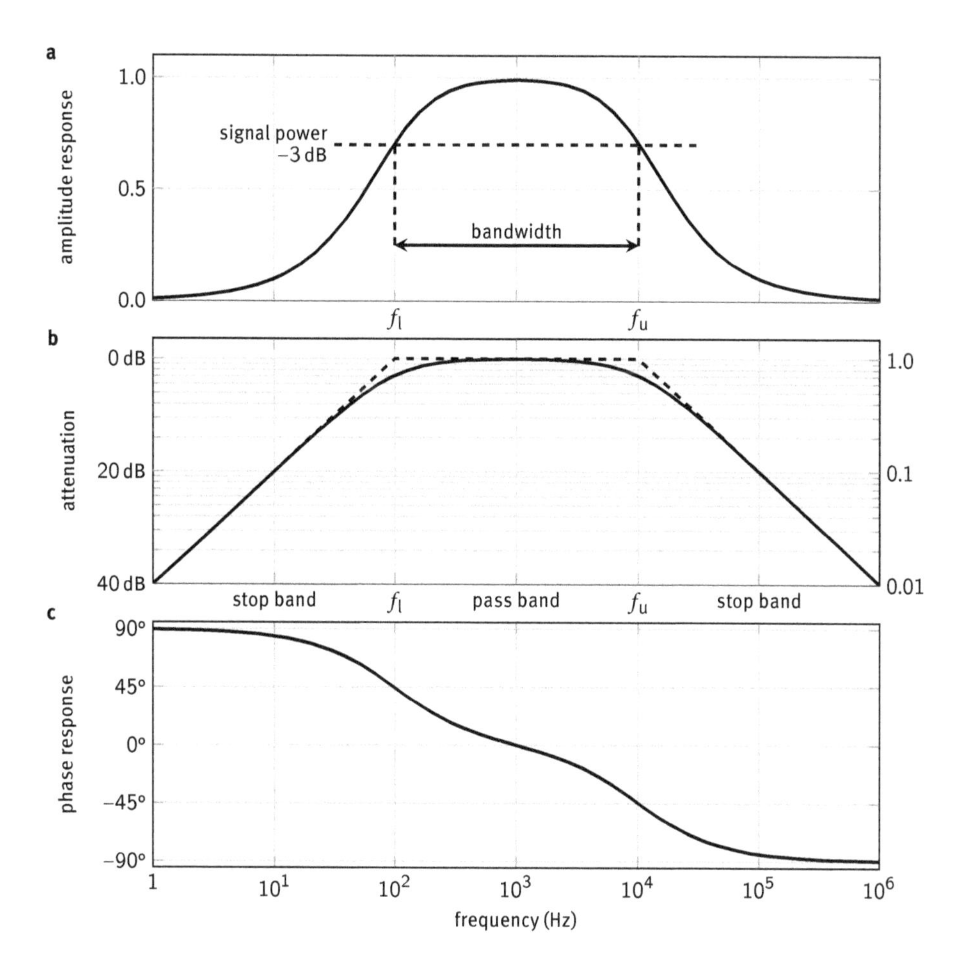

Fig. 3.35. Typical frequency response of the transfer function of a band-pass filter with lower −3 dB corner frequency f_l and upper −3 dB corner frequency f_u.

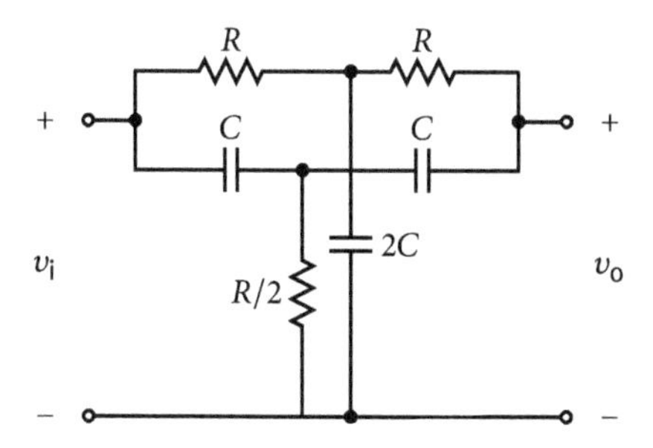

Fig. 3.36. Principle of a twin-T band-stop filter.

At ω_n the phase shift φ is zero, at low frequencies the equation for the phase shift φ degenerates to

$$\varphi(\omega) = \arctan(-5\omega\tau)$$

so that the phase shift is zero at $\omega = 0$ (as expected due to the transmission through the two resistors), and at high frequencies the equation for the phase shift degenerates to

$$\varphi(\omega) = \arctan\left(-\frac{1}{\omega\tau}\right)$$

so that the phase shift is zero at $1/\omega = 0$ (as expected due to the transmission through the two capacitors forming a short-circuit).

3.4.6 Resonant filters

Up to now we have omitted voltage and current division by the pure imaginary one-ports inductor L and capacitor C. If they are combined, they form an LC circuit, also called a *resonant*, *tank*, or *tuned circuit*. It is helpful to study first this pure form to gain a good understanding before including unavoidable stray properties present in the actual components resulting in so called RLC circuits.

LC circuits are not only used as filters (selecting a signal at a specific frequency from a complicated signal) but more so in harmonic oscillators (Section 4.2) for generating sinusoidal signals at a particular frequency. If both reactive elements in a filter are of the same nature, one gets frequency independent division because the frequency dependences drop out.

Problems

3.53. Calculate the transfer function of a voltage divider consisting of two inductors.

3.54. Calculate the transfer function of a voltage divider consisting of two capacitors.

3.55. Calculate the transfer function of a current divider consisting of two inductors.

3.56. Calculate the transfer function of a current divider consisting of two capacitors.

3.4.6.1 Series resonant circuit

Figure 3.28e and Figure 3.28f show the series connection of an inductor L with a capacitor C forming a series resonant circuit. Its (input) impedance is given by

$$Z(\mathrm{j}\omega) = \mathrm{j}\omega L + \frac{1}{\mathrm{j}\omega C} = \mathrm{j}\left(\omega L - \frac{1}{\omega C}\right) . \tag{3.87}$$

Under the condition

$$\left(\omega L - \frac{1}{\omega C}\right) = 0 , \quad \text{i.e., } \omega L = \frac{1}{\omega C} , \text{ i.e., } \omega = \frac{1}{\sqrt{LC}} \tag{3.88}$$

the imaginary part of the impedance becomes zero, and the impedance is real. Current and voltage are in phase, and *the current is maximal*. The frequency at which the phase shift vanishes (at which the impedance becomes real) is called *resonant frequency* f_r or resonant angular frequency ω_r.

Only at ω_r the total impedance Z will be zero and otherwise nonzero. Below ω_r the circuit is capacitive, above ω_r the circuit is inductive. Therefore, the series LC circuit, when connected in series with a resistive load, will act as a narrow voltage band-pass filter having zero output impedance at the resonant frequency of the LC circuit. Whereas for the ideal series resonant circuit the position of the minimum (input) impedance is at the resonant frequency it moves to higher frequencies with increasing (resistive) loss in the components of the resonant circuit.

Problem

3.57. An AC voltage source of 200 V feeds a series RLC resonant circuit. At 50 kHz the current is 5 mA and has a phase shift of 0° to the voltage. At 100 kHz the current is 3 mA. Identify the three elements R, L, and C.

3.4.6.2 Parallel resonant circuit

The parallel resonant circuit is dual to the series resonant circuit, i.e., not the voltage is divided but the current (see Figure 3.29e and Figure 3.29f).

Example 3.5 (Comparison of the two types of resonant circuits made of real components). A real inductor has a rather pronounced series resistance represented by R_S, and a real capacitor has a rather insignificant parallel conductance represented by G_p. Including these stray parameters into our resonant circuits we get the circuits of Figure 3.37. As these circuits are strictly dual to each other, it suffices to investigate just one of them. Observe, in particular, that R_S is in series to the serial resonant circuit and G_p in parallel to the parallel resonant circuit. Being real themselves they do no

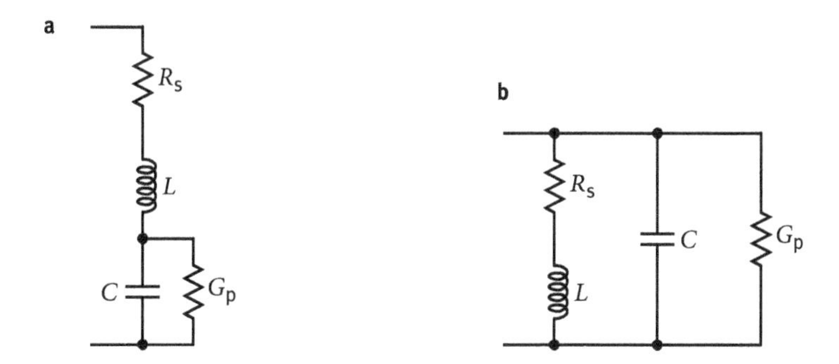

Fig. 3.37. Equivalent circuits of the two types of resonant circuits made of real components. (a) series resonant circuit and (b) parallel resonant circuit.

affect the condition for resonance. So we conclude that the resonant frequency of the series resonant circuit is independent of R_s and the dual version independent of G_p.

The impedance of the series resonant circuit is given by

$$Z(j\omega) = R_s + j\omega L + \frac{1}{G_p + j\omega C} = R_s(1 + j\omega\tau_1) + \frac{1}{G_p(1 + j\omega\tau_2)} \tag{3.89}$$

with $\tau_1 = \frac{L}{R_s}$ and $\tau_2 = \frac{C}{G_p}$. Separating the imaginary part of $Z(j\omega)$ we get

$$\mathrm{Im}[Z(j\omega)] = R_s\omega\tau_1 - \frac{\omega\tau_2}{G_p(1 + \omega^2\tau_2^2)} \tag{3.90}$$

and making it zero we get for the resonant frequency of the *serial* circuit

$$\omega_s = \sqrt{\frac{1}{LC} - \left(\frac{G_p}{C}\right)^2} = \sqrt{\frac{1}{LC} - \frac{1}{\tau_2^2}}. \tag{3.91}$$

The dual version for the resonant frequency of the *parallel* circuit is

$$\omega_p = \sqrt{\frac{1}{LC} - \left(\frac{R_s}{L}\right)^2} = \sqrt{\frac{1}{LC} - \frac{1}{\tau_1^2}}. \tag{3.92}$$

The series resonant circuit is with regard to the resonant frequency the better choice. As τ_2 is generally much larger than τ_1, the resonant frequency depends less on stray properties. This is important, as their temperature dependence makes the resonant frequency shift with temperature.

The selectivity of a resonant filter depends on the quality factor Q of the circuit. For the serial arrangement, it is the ratio of $\mathrm{Im}[j\omega L]$ to the total impedance of the circuit at ω_r:

$$Q = \frac{\omega_r L}{Z(\omega_r)} = \frac{\omega_r L}{L\left(\frac{1}{\tau_1} + \frac{1}{\tau_2}\right)} = \frac{\omega_r}{\frac{1}{\tau_1} + \frac{1}{\tau_2}} = \frac{1}{\frac{1}{\omega_r}\left(\frac{1}{\tau_1} + \frac{1}{\tau_2}\right)}. \tag{3.93}$$

The inverse is easier to remember:

$$\frac{1}{Q} = \frac{1}{\omega_r} \times \left(\frac{1}{\tau_1} + \frac{1}{\tau_2}\right). \tag{3.94}$$

As τ_1 is dual to τ_2 and vice versa, this equation applies for both resonant circuits.

As the loss tangent (Section 3.2.1) of high quality capacitors and consequently G_p can be very small, the resonant frequency of serial resonant circuits is that of the ideal circuit (3.88) and the quality factor Q_s becomes

$$Q_s = \sqrt{\frac{1}{LC}}\frac{L}{R_s} = \sqrt{\frac{L}{C}}\frac{1}{R_s}. \tag{3.95}$$

Obviously, it is essential that the series resistor of the inductor is small. There are practical limits to the increase of L. An increase in L will significantly increase R_s, too, reducing the net benefit. A decrease in C is limited by the existence of stray capacitances that should not become too eminent.

The quality factor Q reflects the bandwidth of the resonant filter. For $Q > 10$, the bandwidth BW divided by the resonant frequency $f_r = \omega_r/2\pi$ is very well given by $1/Q$

$$BW \approx \left(\frac{1}{\tau_1} + \frac{1}{\tau_2}\right) \times \frac{1}{2\pi}.$$

As the quality factor of RLC resonant circuits is of the order of 10^2, the condition $Q > 10$ would usually be fulfilled. For a series resonant circuit, the value of BW degenerates to $BW \approx \frac{1}{2\pi\tau_1}$.

Problems

?

3.58. A voltage-dependent current source ($g_m = 2$ mA/V, $Z_o = 0.5$ MΩ) feeds a parallel resonant circuit ($C = 200$ pF, $L = 0.12$ mH, $R_s = 10$ Ω). Determine
(a) the resonant frequency f_r,
(b) the admittance of the circuit at the resonant frequency,
(c) the voltage gain g_v at the resonant frequency, and
(d) the quality factor Q of the resonant circuit.

3.59. The capacitor C of a parallel resonant circuit is $C = 318$ pF, the series resistor R_s of the inductor is $R_s = 10$ Ω and its resonant frequency $f_r = 1$ MHz.
(a) Determine the quality factor Q.
(b) Which conductance G_p of the capacitor C gives the same Q as R_s of the inductor?

3.60. An AC current source of 100 mA feeds a parallel configuration of a resistance R, a capacitance C, and an inductance L. At $\omega = 5 \times 10^4$ s^{-1} is the phase shift between current and voltage 0° and the voltage is 2.5 V, at $\omega = 1 \times 10^5$ s^{-1} is the voltage 2.0 V. Find the values of the three elements.

3.61. Calculate the voltage transfer function of a resonant filter section consisting of a resistor R in series to an ideal parallel resonant circuit.
(a) Give the complex transfer function.
(b) Give the frequency dependence of the amplitude.
(c) Give the frequency dependence of the phase shift.
Solve by thinking:
(d) At which frequency is the phase shift 0° between v_o and v_i?
(e) What value has the input impedance Z_i at this frequency?
(f) What value has the output impedance Z_o at this frequency?

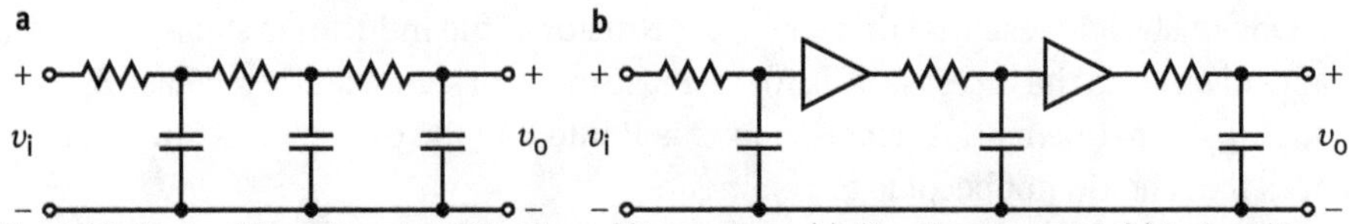

Fig. 3.38. Chain of three low-pass RC voltage filter sections (a) directly connected, (b) isolated by means of *buffer amplifiers* with a voltage gain $g_v = 1$.

3.4.7 Cascading of filter sections

Because of lack of isolation between input and output loading of a filter section must always be considered. Consequently, as in the case shown in Figure 3.38a for low-pass filter sections, all of the circuit must be analyzed as a whole, not section by section. In Figure 3.38b isolating amplifiers (voltage followers) isolate each section from the next one so that the overall response is the cumulative response of the single sections.

Problem

3.62. Are the input variables of a two-port dependent on the output variables?

3.4.8 General considerations concerning filters

An undistorted transfer of a signal containing more than one frequency component does not only require that the amplitude of all components is transferred correctly but also that the phase shift changes linearly with frequency. The latter is achieved by *linear phase* filters. All frequency components of a signal have equal delay times in such filters resulting in a distortionless signal transfer, i.e., such filters provide *constant group delay*. The group delay of a filter with *nonlinear phase* varies with frequency, resulting in *phase distortion*. The *phase deviation* $\epsilon(\omega)$ from linear phase is explained in Figure 3.39. Its equation is

$$\epsilon(\omega) = \varphi(\omega) - \omega \times \left.\frac{d\varphi(\omega)}{d\omega}\right|_{\omega=0} . \tag{3.96}$$

All practical simple filters exhibit deviations both from the ideal linear phase and a flat amplitude characteristic resulting in distortions of signals containing higher fre-

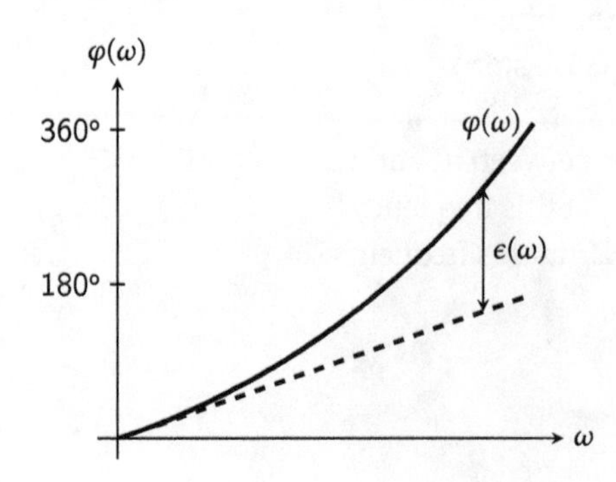

Fig. 3.39. Deviation of an actual phase response from a linear phase response.

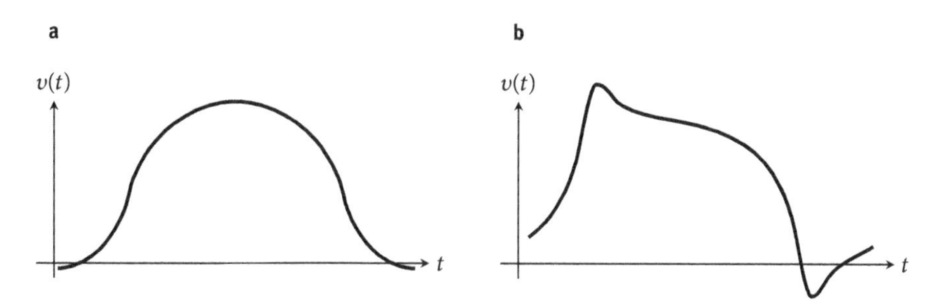

Fig. 3.40. Distortion of a single rectangular signal transmitted through a two-port; (a) insufficient amplitude bandwidth, and (b) insufficient phase bandwidth.

quencies. As the number of filter sections increases, the effective bandwidth is narrowed on cascading. The phase distortion at the corner frequency of amplitude transfer increases so much that the frequency response of the phase shift starts to dominate the transient behavior. Thus, the term *phase bandwidth* was introduced defining the frequency range where the absolute value of the difference in phase shift from a linear dependence is less than 45° ($\pi/4$). Depending on the application, another definition of the phase bandwidth is preferred. It is the width of the frequency range over which the phase-vs.-frequency characteristic deviates from linearity by less than 0.5 (about 28.7°, Figure 3.39).

Figure 3.40 gives an impression of how a rectangular signal gets distorted due to an imperfection in the amplitude transfer, and in the phase transfer.

Problems

?

3.63. Calculate the phase deviation of a low-pass filter from a linear phase at the corner frequency.

3.64. A simple voltage low-pass filter consists of a resistor $R = 1\,\text{k}\Omega$, and a capacitor $C = 10\,\text{nF}$.
(a) Give the (amplitude) bandwidth (in Hz).
(b) Give the phase bandwidth (in Hz).

3.4.8.1 Voltage step response of simple RC filters

A voltage source delivering positive (ideal) rectangular signals is loaded by a resistor in series with a capacitor. Figure 3.41 compares the source voltage signal (a) with the voltage across R (the shape after differentiation, i.e., the output signal of a high-pass filter; see Section 3.2.1) and (c) the voltage across C (the shape after integration, i.e., the output signal of a low-pass filter; see Section 3.2.1) for various lengths t_l of the signal (expressed in fractions of the time constant $\tau = R \times C$. Observe that in the case (b) the area of the positive signal equals that of the negative signal. This is so because the voltage across R is proportional to the current i that charges and discharges the

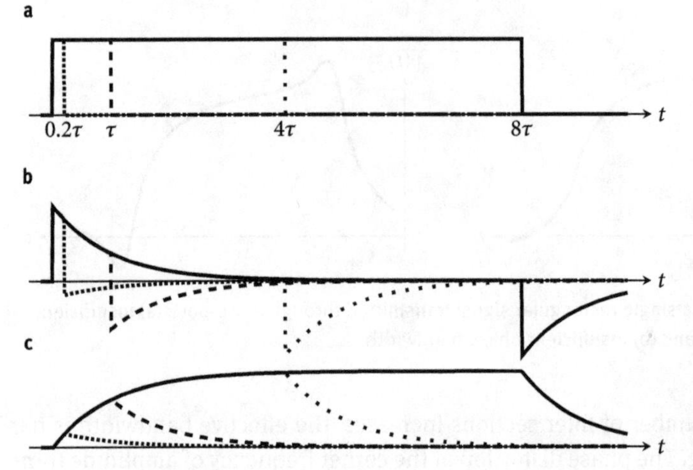

Fig. 3.41. Response of an RC combination to rectangular voltage signals of various lengths: (a) rectangular voltage input signal, (b) voltage signal across R (high-pass filter), and (c) voltage signal across C (low-pass filter).

Table 3.1. Maximum fraction of the peak voltage across the capacitor in dependence of the signal length t_l.

Fractional signal length t_l/τ	Fractional amplitude v/v_{max}
0.2	0.185
1.0	0.632
4.0	0.982
8.0	0.9997

capacitor. This fact can also be interpreted, that across R there is no voltage contribution with the frequency component $\omega = 0$, i.e., no DC component. It does not come as a surprise, that the capacitor has conductance zero at this frequency.

It is worthwhile to remember that in Figure 3.41c short signals do not attain the full height of the step. Table 3.1 lists the maximum height for several signal lengths.

Problems

3.65. A simple voltage low-pass filter has a time constant of 10 μs. Its input impedance at $\omega = 0$ is 10 kΩ.

(a) Identify the (two) linear components of the filter.

(b) What is the maximum output voltage if a single positive 10 V rectangular signal of length $t_l = 10$ μs is fed into the input?

(c) Sketch the linear approximation of the frequency response of the amplitude transfer and of the phase shift

(d) How much does the amplitude of the voltage across the inductor drop within the first 10 μs if a single positive 10 V step signal is fed into the input?

3.66. An RC low-pass filter consists of a resistor $R = 1\,\mathrm{k\Omega}$ and two capacitors of $C = 2\,\mathrm{nF}$ each, in series. The output voltage v_o is the voltage across one of the capacitors.

(a) What is the maximum amplitude and the rise time of the output signal if a positive step signal of 10 V is fed into the input?

(b) What is the maximum amplitude and the phase shift of the output signal if a sinusoidal signal of 10 V_{rms} with an angular frequency of $1 \times 10^6\,\mathrm{s}^{-1}$ is fed into the input?

3.4.8.2 Current step response of simple RLC filters (shunt compensation)

The output voltage across the resistive load R_L of amplifiers with a dependent current source at the output (like common-emitter or base circuits of bipolar junction transistors) is formed by a current low-pass RC filter (R_L and C_o are in parallel). See Figure 3.42 for this configuration. The effective capacitance C_o is often a stray capacitance of several pF. For a case with $R_L = 1\,\mathrm{k\Omega}$ and $C = 20\,\mathrm{pF}$, we get a bandwidth BW of only 8 MHz and a rise time of $t_{rs} = 44\,\mathrm{ns}$. Quite a moderate performance.

To compensate for the signal loss by the capacitance, an inductor can be placed in series to the resistor. Thus, the impedance in this branch is increased at frequencies at which the capacitance has low impedance. This is sketched in Figure 3.43.

The circuit of Figure 3.43 forms a parallel resonant circuit. Its resonant frequency is (Section 3.4.6.2)

$$\omega_p = \sqrt{\frac{1}{LC} - \left(\frac{R_L}{L}\right)^2} = \sqrt{\frac{1}{L}} \times \sqrt{\frac{1}{C} - \frac{R_L^2}{L}}. \tag{3.97}$$

For $1/C = R_L^2/L$, i.e., $L/R_L = C/G_L$, i.e., $\tau_1 = \tau_2$ the frequency becomes zero, i.e., there is no oscillation. This condition is called critical damping. The circuit is overdamped if L/CR^2 is less than one, otherwise under-damped. Table 3.2 lists several specific choices of L together with the resulting performance of the circuit. Figure 3.44 shows the improvement in the rise time of a step function through shunt compensation.

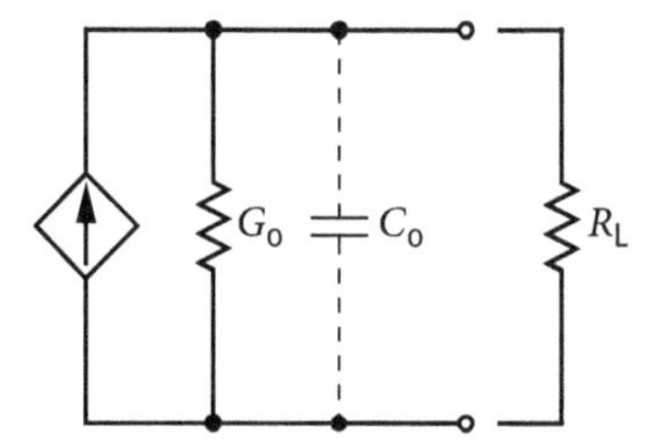

Fig. 3.42. Output condition at a dependent current source with load resistor. The stray capacitance is shown dashed.

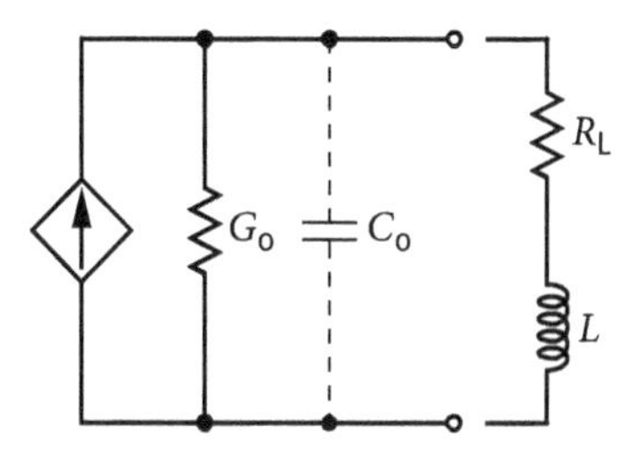

Fig. 3.43. The load impedance is increased by adding an inductor in series to the load resistor. The stray capacitance is shown dashed.

Table 3.2. Listing of L/CR^2-values and the resulting performance of the RLC circuit.

L/CR^2	**Performance**
0	$t_{rs} = 2.2RC$, no shunt compensation
0.25	$t_{rs} = 1.65RC$, no overshoot (Section 3.3)
1	one overshoot, critical damping
$1 + \sqrt{2}$	Maximally flat frequency response of amplitude transfer
$\gg 1$	oscillation at the natural frequency

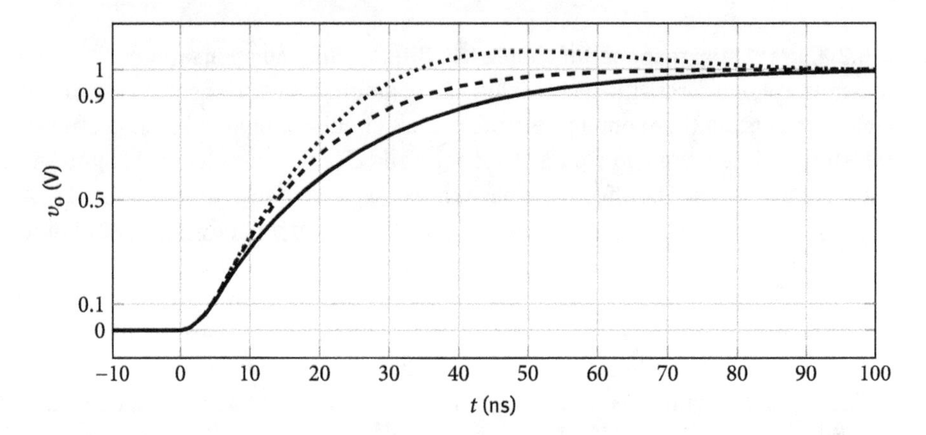

Fig. 3.44. Improvement of the rise time of a step signal by shunt compensation.

3.5 Interfacing (cascading)

In Section 2.2.2 we learnt that electronic two-ports do not isolate the input from the output which means, e.g., that the two forward parameters, input impedance and (forward) transfer parameter are functions of the amount of load at the output and the reverse parameters are functions of the amount of "load" at the input, i.e., of the source impedance. To minimize this dependence, it is necessary to feed current inputs from current sources and voltage inputs from voltage sources. Likewise, high impedance loads should be fed by a dependent voltage source at the amplifier output, whereas high admittance loads call for a dependent current source at the output of the amplifier. This method is called *impedance bridging* which means, in the case of voltage signals, that the load impedance is much higher than the source impedance. This way, a maximum of voltage is transferred at less power consumption by reduced currents. Whereas this impedance mismatch strongly supports isolation of the source from the amplifier and of the amplifier from the load, sometimes the need of maximum power transfer overrides these considerations.

In Section 2.2.4.1, we learnt that maximum power is transferred from one element to another when in each element the same power is dissipated, e.g., if in a voltage divider the divided voltage is one-half of the total voltage. If this division is done by resistors, both impedances will have the same value ($R_1 = R_2$), Thus, the term *impedance matching*. Applying power matching to complex impedances the impedance matching must be generalized to

$$Z_1 = Z_2^\star \tag{3.98}$$

where * indicates the complex conjugate. For purely real impedances, this degenerates to $R_1 = R_2$, as we already know.

The load impedance must be the complex conjugate of the source impedance to achieve optimal power transfer between them.

When conditioning signals, power matching is rarely required. Thus, we concentrate on impedance bridging, i.e., the maximum voltage transfer from element to element. Usually, the maximum current transfer is of no concern. Applying duality just would supply the correct answers.

3.5.1 Interfacing of single stages

It would take nearly an encyclopedia to list and fully describe all possible combinations of the three types of single stages of the three-terminal components belonging to the three semiconductor technology families (in both polarities)
- bipolar junction transistor (BJT),
- junction-gate field-effect transistor (JFET), or
- metal-oxide-semiconductor field-effect transistor (MOSFET).

In view of the myriads of integrated amplifiers available, the importance of this section does not lie in designing but just in understanding such two-stage combinations. Connecting single stages involves two steps
- setting the (quiescent) operating points, and
- optimizing the small-signal behavior.

Biasing individual stages depends strongly on the technology family, which can make mixing components of different families trying. Although the correct dimensioning with regard to the operating point is of high importance when designing a circuit, one can more or less disregard these complications when one just tries to understand the dynamic behavior of a circuit. In that case, the choice of the technology family is less important, too, because the vital difference between these components lies in the biasing of the devices.

Therefore, we settle for bipolar junction transistors (mixing both polarities). However, even if the common-emitter circuit is plainly the basic circuit, we will be using the

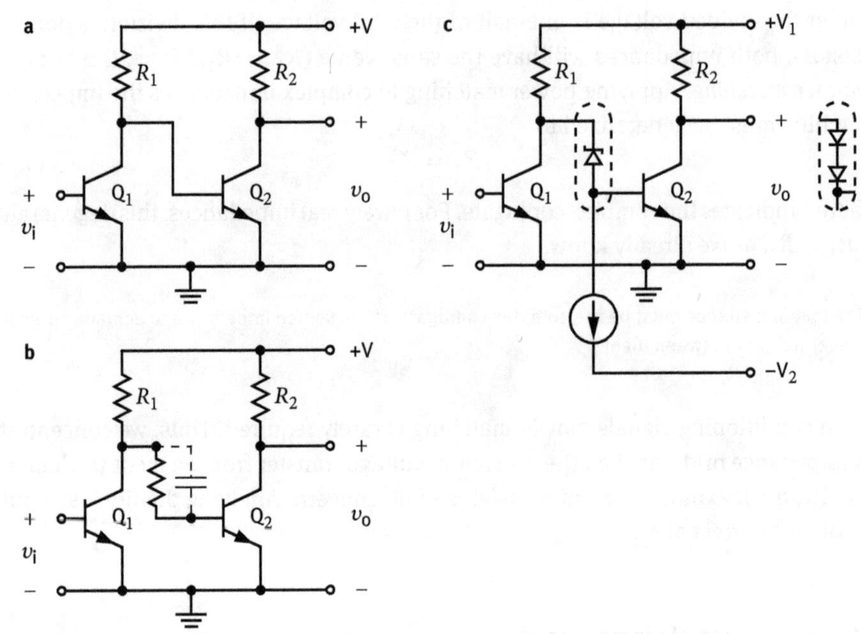

Fig. 3.45. Effect of the interface between two amplifier stages on the signal transmission and the bias setting: (a) direct coupling (b) resistive coupling (with and without) shunting capacitor, and (c) nonlinear coupling using a current source with a Zener diode or two forward biased diodes.

other two "basic" circuits as well, as this is more convenient than considering them as negatively fed back common-emitter circuits. By introducing the specific properties of the other semiconductor families ignored here, it is not too difficult to come to a grip with any of the other stages not covered.

Choosing the bipolar junction transistor has the advantage that the other types have, quite often, superior small-signal properties, so that for them the result can only be better. So we are considering the following basic stages

- common-emitter circuit, providing the highest power gain and moderate input and output impedance,
- common-base circuit (common-emitter circuit with maximum parallel–series feedback) with current gain $g_i \approx 1$, being a typical current amplifier with low input and high output impedance,
- common-collector circuit (common-emitter circuit with maximum series–parallel feedback) with voltage gain $g_v \approx 1$, being a typical voltage amplifier with high input and low output impedance.

Connecting (cascading) two stages requires more than just optimizing (small-signal) transfer. A direct connection between two stages affects both the output operating point of the first stage and the input operating point of the second stage (Figure 3.45). A *capacitor* between the stages would effectively separate them as far as operating

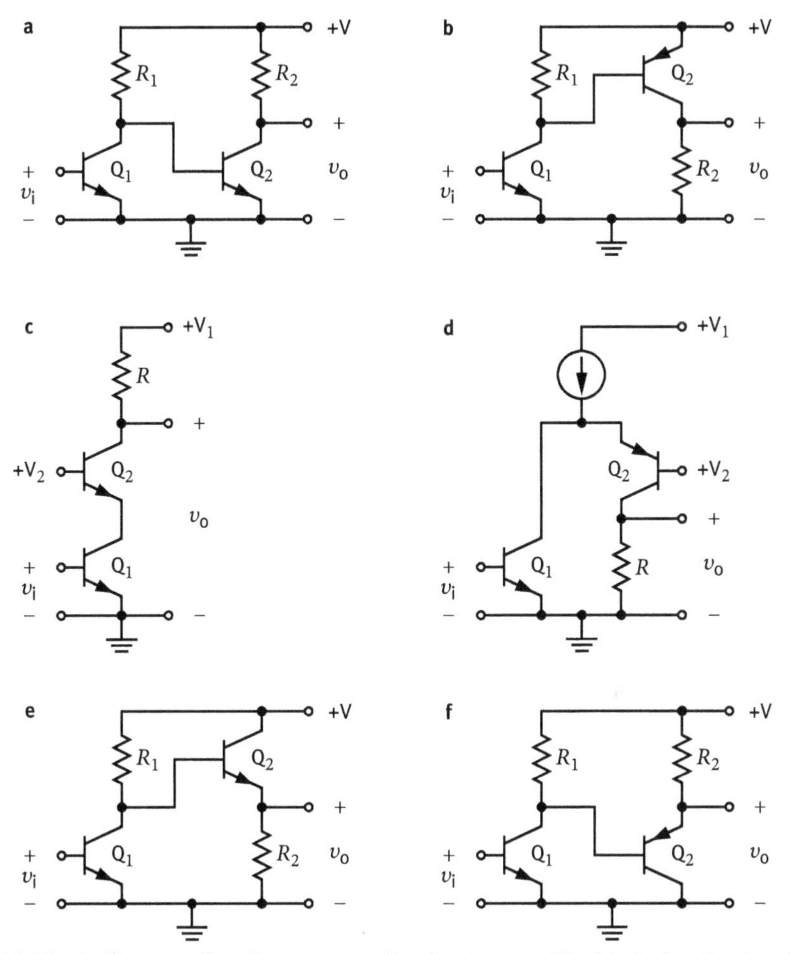

Fig. 3.46. Simple direct coupling of a common-emitter input stage with a bipolar junction transistor in the output stage.

points are concerned. However, a high-pass filter is introduced into the signal path hindering the passage of low frequencies, which will be detrimental in many applications. In particular,

capacitive coupling prevents stabilization of operating points by global feedback.

Combining two stages it will hardly be necessary to isolate them from each other (i.e., impedance mismatch, Section 3.4). Therefore, making the output impedance of the first stage equal to the input impedance of the second stage will only be desirable for providing ample power gain. Most power is delivered by common-emitter circuits. Thus, we will start with common-emitter circuits as input stage. In Figure 3.46, the six possible combinations of two directly coupled stages are shown.

Both combinations of two common-emitter circuits ((a) and (b)) are useful. However, (a) limits the output operating point voltage of Q_1 to V_{BE} of Q_2 which is a severe limitation.

The circuit (c) is important enough to have a name: *cascode* circuit. It is a cascade of a common-emitter circuit with a common base circuit. It overcomes the disadvantage of common-emitter circuits that the Miller effect (Section 2.4.3.2) increases dynamically the input capacitance, the capacitance C_{BC} between the base and the collector by multiplication with a high factor according to $C_{BC} \times (1 - g_v^*)$ limiting the bandwidth. For a cascode circuit the voltage gain of the first stage g_{v1} is

$$-g_{v1} = \frac{g_{i1} \times R_L}{Z_{i1}} \approx \frac{g_{i1} \times Z_{i2}}{Z_{i1}} \approx \frac{g_{i1}}{Z_{i1}} \times \frac{Z_{i1}}{g_{i1} + 1} \approx 1 \tag{3.99}$$

so that the dynamic value of C_{BC} is just a factor of 2 larger than the static value. (To arrive at this simple equation advantage is taken of the relation between the input impedance of the common base circuit – for small loads – and the input impedance of a common-emitter circuit – again for small loads – and assuming identical parameters for both transistors.)

The circuits (e) and (f) have a common-collector circuit as a second stage. Consequently, low output impedance is achieved.

In Figure 3.47 the six possible combinations of two directly coupled stages are shown with a common-collector circuit as input stage. The resistor R_1 in (a) may be omitted so that Q_1 works directly into Q_2. Such an arrangement is called *Darlington* pair. They exist even as a component (Darlington transistor). Their advantage is high current gain, and high input impedance because the input impedance of another common-collector circuit acts as load resistance of the first stage. However, the upper limit for the value of the input impedance is the share of the base-collector portion that stays the same, quite independent of the load.

Whereas (c) and (f) are quite impractical with regard to biasing, a symmetric version of (e) has gained high importance under the name *long-tailed pair* (Section 2.3.4.1) as input configuration of most operational amplifiers. In this case, the common-collector circuit feeds a common base circuit. The "long tail" is a constant current source that does not load the first stage. When analyzing the asymmetric long-tailed pair as a cascade of a common-collector circuit with a common base circuit, some simplifications are helpful. If the source of the input signal has low impedance, the output impedance of the voltage follower is about the internal serial emitter resistance r_e. The same is true for the input impedance of the common base circuit with a low impedance bias voltage at the base.

In both cases the internal emitter resistance r_e (in Ω) is given by $r_e = 25\,\mathrm{mV}/I_E$, i.e., in both cases it is dependent on the operating current I_E (which is given in mA to yield r_e in ohms).

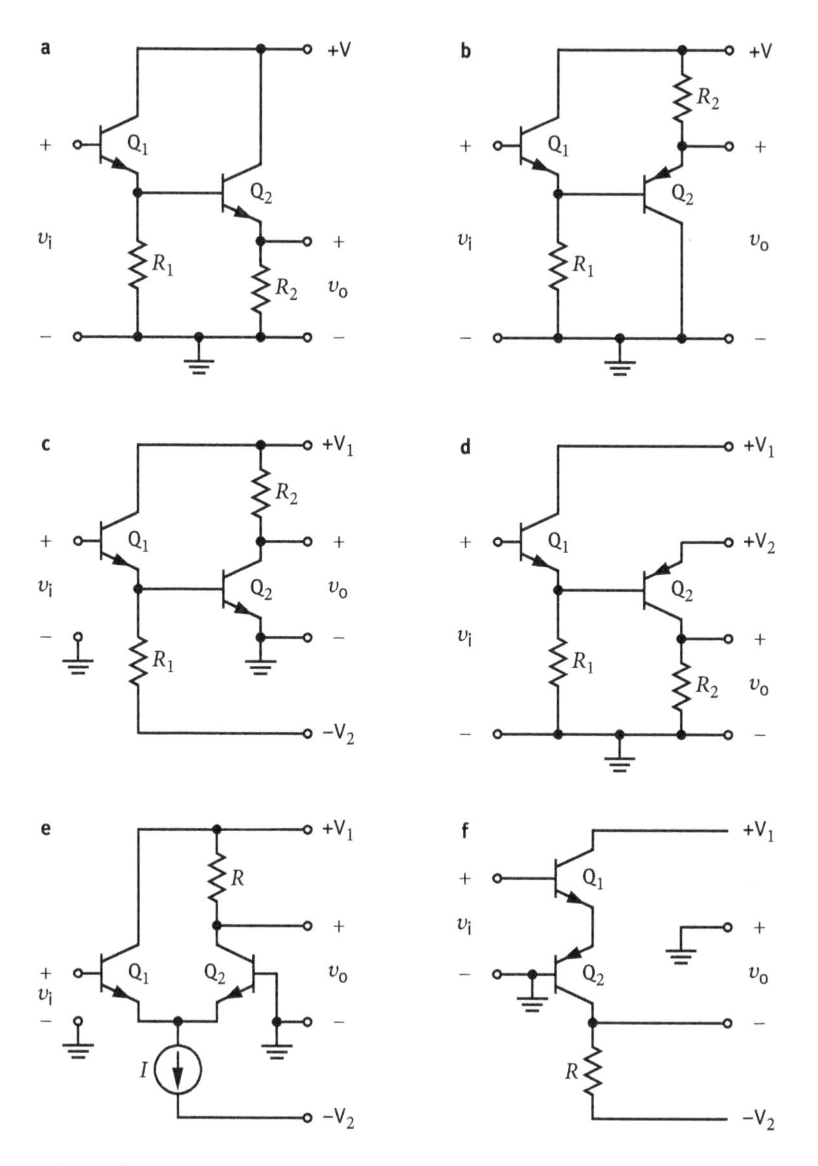

Fig. 3.47. Simple direct coupling of a common collector input stage with a bipolar junction transistor in the output stage.

Usually, a long-tailed pair is symmetrically biased so that the input impedance of the second stage equals the output impedance of the first stage. Thus, the input voltage of the second stage will be one-half of the input voltage considering that the emitter follower has an unloaded gain of one. Together with the transadmittance of the second stage

$$g_{m2} = \frac{I_C}{25\,\text{mV}}$$

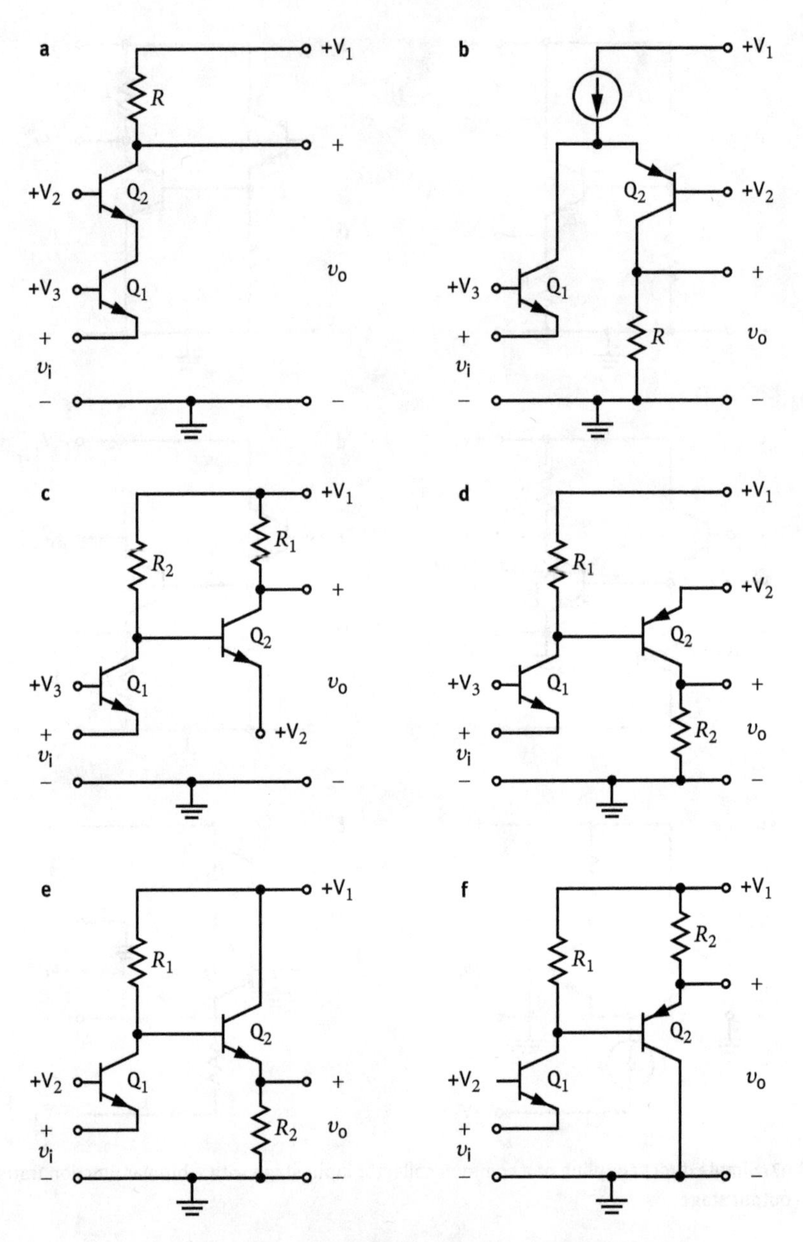

Fig. 3.48. Simple two-stage amplifiers with a common base circuit as the input stage.

the total transadmittance is obtained (in units of mA/mV) as

$$g_{\mathrm{m}} = 0.5 \times \frac{I_{\mathrm{C}}}{25\,\mathrm{mV}}\,.$$

Figure 3.48 summarizes circuits with two directly coupled stages having a common base circuit as input stage. The need of additional supply voltages for the bias is somewhat awkward. As today's circuit philosophy is mainly concentrating on high input impedance devices (i.e., voltage amplifiers), these circuits have little to no importance unless one wants to switch from voltage to current electronics.

Problems

?

3.67. Is it necessary to match the output impedance of the first stage to the input impedance of the second stage when cascading two amplifier stages?

3.68. When combining two stages of the same kind is it possible to get negative voltage gain?

3.69. Which cascade of two basic transistor circuits gives maximum voltage gain?

3.5.2 Interfacing of subsystems

Interfacing subsystems is insofar different as an "interface" circuit may be applied, i.e., a circuit that bridges both the impedances and the variables of the circuits' operating points. The simplest interface is two pieces of conductor (wire). Connecting logical circuits within the same logical family can be done this way. Subsystems from logical families are designed to allow direct connections among each other allowing both for impedance bridging and operating point compatibility. (Some require an additional *pull-up resistor*). If low frequencies need not be transferred, capacitive coupling might be appropriate. In cases with negligible difference in the operating voltage of the two subsystems impedance bridging is very well achieved by an interface made of a voltage (or emitter or source) follower.

An interface can have additional tasks aside from impedance and operating point bridging. It may be used for

- signal conditioning (amplitude, shape),
- signal inversion,
- signal delay,
- signal restoration,
- signal synchronization,
- switching, and

whatever comes to the mind of the circuit designer. This might require quite some elaborate circuitry.

Problem

?

3.70. What is the simplest method of (passive) interfacing?

3.5.3 Interfacing of systems (transmission lines)

When electronic devices are interconnected in a system a new important aspect appears. The dimensions of such systems will (by far) exceed the centimeter-range. Considering that an electric signal that travels with the speed of light covers about 30 cm in 1 ns makes clear that the space dependence of Maxwell's equations must not be disregarded when high frequencies (i.e., small time intervals) are involved.

For our purpose, we have to understand how an electromagnetic wave travels along a connecting conductor. Such conductors form a transmission line. A *transmission* line is a pair of parallel conductors showing specific electric characteristics due to distributed reactances along its length. It is designed to transmit electric signals (alternating current) with frequencies so high that their wave nature must be taken into account. An infinitely long line exhibits (input) impedance which is purely resistive called *characteristic* (or *natural*) *impedance* Z_0. It is totally different from the resistance of the conductors themselves or from the leakage conductance of the dielectric insulation between the two conductors. As a property of the *ideal* transmission line, the characteristic impedance is entirely a function of the capacitance and inductance distributed homogeneously along the line's length. The general expression for the characteristic impedance Z_0 of a *real* transmission line is, based on the transmission line model (Figure 3.49b)

$$Z_0 = \sqrt{\frac{R' + j\omega L'}{G' + j\omega C'}} \tag{3.100}$$

where
R' is the specific resistance in Ω/m,
L' is the specific inductance in H/m,
G' is the specific conductance in S/m,
C' is the specific capacitance in F/m.

For an ideal (lossless) transmission line, R' and G' are both zero, so after cancelling $j\omega$ the equation for the characteristic impedance reduces to:

$$Z_0 = \sqrt{\frac{L'}{C'}} \tag{3.101}$$

Thus, Z_0 has no imaginary part, it is purely resistive and it is frequency-independent.

It is enlightening to model a transmission line by means of cascaded (infinitesimal) lumped sections. Such a section is shown in Figure 3.49 together with a slightly simplified circuit diagram, that combines the specific resistivities of both conductors. (The specific resistivity of the inner conductor will be different from the outer, alone for geometric reasons.)

The model section (Figure 3.49b) includes the specific resistance of both conductors, the specific inductance of both conductors which are coupled by their magnetic

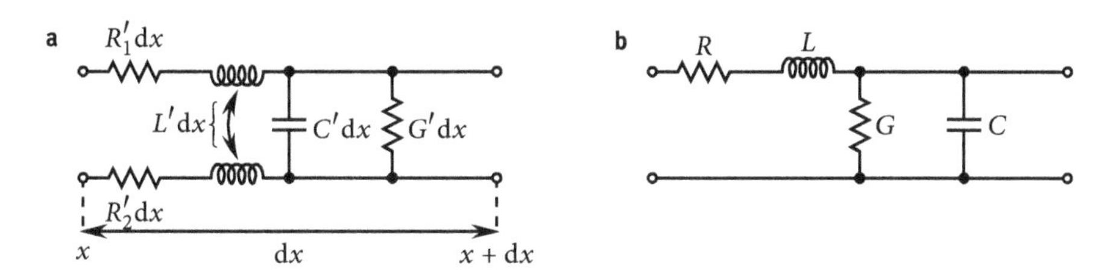

Fig. 3.49. (a) Infinitesimal model section of a transmission line. (b) Simplified circuit of a transmission line model.

field. Both the resistance and the inductance are subject to the skin effect that increases the resistance and reduces the inductance with frequency. Besides, both the specific capacitance between the conductors and the specific conductance through the insulator are included.

The transfer of high frequency signals is affected by the parasitic series inductances and parallel capacitances of conductors even if their length is just a few centimeters. Thus, even a short transmission line must be terminated with a resistor R equal to Z_0 simulating an infinitely long line (and making the input impedance purely resistive, namely Z_0). If a transmission line is not terminated with its characteristic impedance, then its input impedance is complex and not resistive. Consequently, the additional phase shift may be the cause of frequency instability (Section 3.6.3.2) in one or both systems to be connected resulting in oscillations. Note that *conducting loops* should be avoided because a changing magnetic field may induce noise in the system.

For low enough frequency transmission (when the wavelength of the highest frequency component is longer than the transmission line length), the termination resistor may be omitted because in this case the wave character of the signal can be disregarded.

3.5.3.1 Signal transmission by coaxial cables

The coaxial cable is the most important type of homogeneous transmission lines. It has

- the same structure and material composition over all its length,
- a circular cross section with,
- one or several (twisted) wires as center conductor,
- a hose of (woven) wires as outer conductor which serves, at the same time, as shielding, and
- the space between the two conductors is filled with insulating material.

The electromagnetic wave propagates in the space between the center conductor and the outer shielding, independently of the grounding conditions of the cable. Usually, the outer conductor is close to ground (earth) making the set up asymmetric with regard to ground which is, however, without consequence for the signal transport.

The outer conductor, the shield of a coaxial cable, should either be connected to ground or have low impedance toward ground so that its shielding property is most effective.

Long coaxial cables are predestined to form a *ground loop*. A ground loop is a conductor that connects two points that are supposedly at equal potential (usually earth) but, actually, are at different potentials being the cause of current flow in the conductor. These ground loops constitute a source of noise in the system. The usual remedy is the so-called *single-point ground*, where all elements of an electrical system are connected at *one* point (which is grounded). For coaxial cables, this means that the shield is connected *only* at *one* end. By breaking the ground loop, the unwanted current is prevented from flowing. As the shield may act as an antenna, the pick-up of radio frequency is furthered by this practice. *Galvanic isolation* is another way to prevent current flow in the shielding between sections of a system as no conduction occurs. Care must be taken that galvanic isolation does not limit the bandwidth of the signal path.

Efficient grounding is an art that takes a lot of experience. One proper *single-point grounding* is definitely better than any number of arbitrary groundings.

Disregarding nonideal properties the following phenomena can be perceived when an electromagnetic signal travels through a coaxial cable.

(a) The electromagnetic wave propagates in the transverse electric and magnetic mode which means that both the electric and magnetic fields are perpendicular to the direction of propagation (the magnetic field is circumferential and, the electric field is radial, Figure 3.50). In the ideal case, the velocity v of wave propagation is independent of the signal frequency. It is usually expressed by a fraction of the velocity c of light in vacuum $v = c/\sqrt{k}$, with k the relative permittivity (also called *dielectric constant*) of the insulating material between the conductors. The velocity v in coaxial cables is typically between 66% and 80% of the speed of

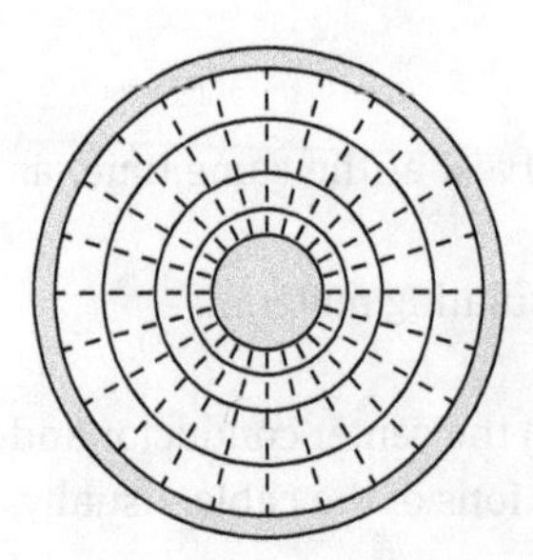

Fig. 3.50. Field lines in the cross section of a coaxial cable (electric field lines are shown dashed, magnetic field solid).

light. As the *envelope* of the wave propagates through the cable with this velocity, it is a *group velocity* (and not a phase velocity). Figure 3.50 shows an example of field lines in the cross section of a coaxial cable.

(b) As the conductors are parallel, the surfaces of constant propagation delay (or of constant phase shift) are planes perpendicular to the conductors.

(c) In the planes of constant propagation delay (or constant phase shift) both fields are constant in time. For that reason, the two conductors are oppositely charged.

(d) An infinitely long cable is purely resistive without frequency dependence. Low loss insulation materials like polytetra-fluoroethylene (PTFE), polyethylene, or polystyrol, with a *dielectric constant* $\epsilon_r \approx 2.3$ are used to produce coaxial cables with a *characteristic impedance* of 50 Ω. Such 50-Ω cables have low signal loss, and the low impedance allows the handling of higher frequencies. Compared to a 100-Ω cable the corner frequency of the unavoidable low-pass filter (Section 3.4.2, formed by the termination resistor at the output of the cable and the input capacitance of the receiving device) is a factor of 2 higher. On the other hand, twice the current must be supplied for the same voltage signal which can be demanding on the signal source.

When transmitting data through a cable, the main interest is to compare what goes in at one end with what comes out at the other. Aside from *propagation delay* and *characteristic impedance* the following additional properties must be considered, e.g.,

Signal reflection: a coaxial cable does not only unavoidably delay, but it also reflects a signal at its end. Obviously, a reflection from the end of an infinitely long cable has no effect because it would never return. Terminating a cable with its characteristic impedance simulates an infinitely long cable eliminating reflections, i.e., there is no discontinuity and the signal transmission is undisturbed. There are two extreme cases,

- that of a short-circuited end, and
- that of an open-circuited end.

If there is a short-circuit at the end of the cable, one might wonder whether there is any voltage signal at all at the cable input. There is (for a while) because how should the input "know" of the short-circuit? The wave must travel to the end and back to the input to provide this information, i.e., for twice the delay time t_d of the cable the input impedance Z_i of the cable is resistive with Z_0. After that time $Z_i = 0$.

What happens at the end of the cable, at the short-circuited output? Obviously, the output voltage v_o stays zero all the time. To simplify matters, let us consider the transmission of a step voltage signal. A step voltage of 2 V amplitude supplied through a source impedance $R_S = Z_0$ feeds a cable with a characteristic impedance of Z_0. Then a voltage step of 1 V will enter the input of the cable and travel to the end of the cable. After twice the travel time t_d the input impedance of the cable will be zero and consequently the voltage at the input zero, as well.

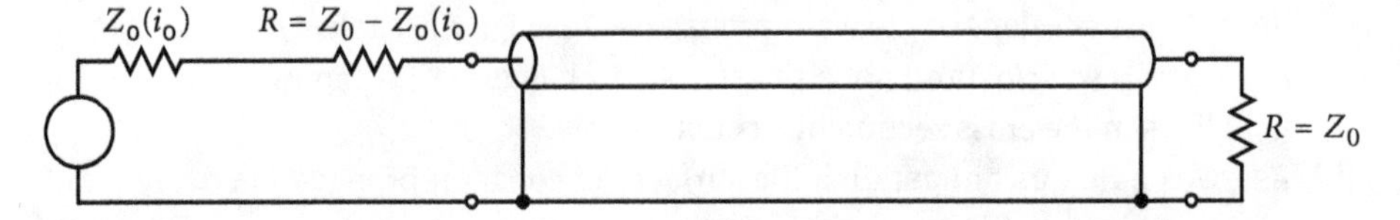

Fig. 3.51. Correct termination at the sending end (and the receiving end) of a coaxial cable with a characteristic impedance of Z_0.

Although the voltage source produced a step, the input voltage at the cable will be a rectangular voltage signal of length $2 \times t_d$, and the output voltage signal will stay zero due to the short-circuit. How does continuity allow the latter? Until the signal reaches the end of the cable, the cable must behave as if it were infinitely long, i.e., the voltage step of 1 V must be transmitted to the short-circuit. As the voltage is zero at the short-circuit, a voltage step of −1 V must be generated travelling in the other direction so that the sum is zero. This is called *reflection* of the *voltage* signal. On its way back, the −1 V-voltage wipes the +1 V-voltage step out, i.e., more and more of the cable is short-circuited until the reflection reaches the input making the input voltage zero. By then, all of the cable has the property of a short-circuit.

For an open-circuit at the output, the situation is dual. In this case, we must consider the current step signal of $1\,\text{V}/Z_0$ that enters the cable. No current can flow out of the output of the cable, i.e., the current is deleted by a current flowing in the opposite direction. This *reflected current* wipes out the primary current, i.e., no current flows which is equivalent with zero conductance. When this condition of zero conductance has reached the input, the (non-existent) voltage division with an open-circuit gives the full source voltage of 2 V at the input.

If the source impedance depends on the output current (like the output impedance of an emitter follower, Section 3.5.1), an output stage with low output impedance (e.g., a voltage follower) in series with a resistor could be used to provide an amplitude-independent impedance of Z_0 (Figure 3.51), thus, avoiding reflections due to impedance mismatch.

Any practicable coaxial cable will have connectors on either side. It must be stressed that said connectors must have the same characteristic impedance as the cable itself to avoid reflections from them and/or from the mating connectors mounted on the devices. In addition, discontinuities in the geometric dimensions (diameters of the inner and the outer conductor) must be avoided for achieving best results. For an undisturbed transmission of high frequencies, it is best to avoid connectors at all and/or to avoid cascading different types of cables, even if they have the same characteristic impedance.

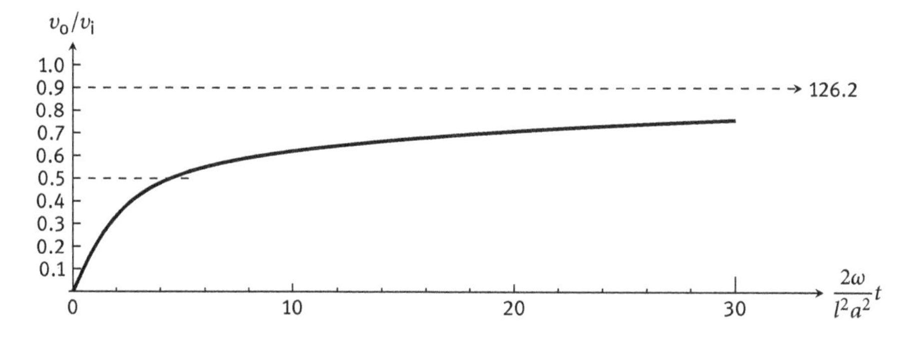

Fig. 3.52. Example of an output voltage signal of a coaxial cable as a response to a voltage step input signal.

Properties which are not found in ideal coaxial cables:

Signal attenuation (damping): When a signal travels through the cable the stray properties of the cable (resistivity of the conductors, dielectric loss in the insulator) reduce the amplitude of the signal. In addition, there is a radiative loss because of imperfect shielding. The *attenuation* increases strongly with signal frequency. Consequently, the shape of the output response to an input voltage step will depend on the cable length. This makes the usual definition of the signal rise time useless. (In the example of Figure 3.52 it takes 30 times longer to reach 90% than 50% of the maximum amplitude.) Using the bandwidth BW of the pass-band of a coaxial cable (which, of course, depends on the cable length) the rise time t_{rs} can be described as $t_{rs} = 2.2/(2\pi BW)$. Figure 3.52 displays the distortion of a voltage step function because of the frequency dependent transmission through the cable. Each cable type has its own specific attenuation, which is given at *specific frequencies* in dB/m.

Deficient shielding: The electromagnetic wave travels practically entirely in the inner volume of the cable. To this end, good *shielding* of the cable by the outer conductor is required. Cable shields do not only bar internal signals from getting out but also bar external signals (noise) from getting in. To be more effective, some cables have a double shield. Typically, double shielding adds about 20 dB of shielding at about 1 GHz.

Cable capacitance: As mentioned in Section 3.5.3, the input impedance of a coaxial cable that is not (or wrongly) terminated at the output is complex. A termination with a resistor $R < Z_0$ results in *inductive* behavior, with $R > Z_0$ in *capacitive* behavior. For $R \gg Z_0$ (e.g., without termination at all), the input impedance is just the capacity of the cable. Typical specific capacitances of coaxial cable are around or less than 100 pF/m.

When connecting dozens of subsystems with coaxial cables to form a system, it is prudent to use only proven, and tested cables from one's own stock to avoid needless cable failures.

Problems

3.71. Switching on of a DC voltage source of 2 V with $R_S = 50\,\Omega$ would give, in the ideal case, a step signal of 2 V amplitude at its output. Such a source is connected to a 50 Ω cable.

(a) What is the voltage at the input when the cable is infinitely long?
(b) What is the voltage at the input when the cable is correctly terminated?
(c) If a 10 m long 50-Ω cable, in which the signal travels with 2/3rd of the speed of light, is short-circuited at the end, how long (with respect to the time of switching) does it take for the input signal of 1 V to become 0 V? (Take for the speed of light 0.30 m/ns.)
(d) Give for the case (c) the time dependence of the voltage amplitude of the signal at the input and the output.
(e) Give for the case (c) the time dependence of the amplitude of the current signal at the input and the output.

3.72. A sinusoidal signal of 50 MHz is transmitted through a 100 m long coaxial cable, having at this frequency a specific attenuation of 0.12 dB/m. How much of the voltage amplitude gets lost?

3.5.3.2 Impedance matching of cables

The characteristic impedance of coaxial cables can be tailored within limits to meet specifications, by choosing the right material for the insulator and the correct ratio of the diameters of the outer to the inner conductor. Thus, it can be matched, e.g., to the characteristic impedance of antennae. Typical characteristic impedances of coaxial cables for analog and digital signal transmission are 93 Ω, 75 Ω, and 50 Ω. The usual selection of 50 Ω is a compromise between power-handling capability and attenuation. Besides, as mentioned above, the input capacitance of an instrument shunting the terminating resistor of 50 Ω has the least effect on the bandwidth.

Nevertheless, using coaxial cables with 93 Ω, which have the lowest specific capacitance, can be quite beneficial, in particular when the signal source is limited in its current output or when short cables may be used without termination (which improves the signal-to-noise ratio in analog applications).

When the characteristic impedance of a coaxial cable does not match the input impedance of the connected device the impedances must be matched to avoid reflections. If the cable impedance is too high, a series resistor supplying the impedance difference must be inserted in series to the input. For best performance, not a lumped resistor but a resistor with a coaxial structure should be used. If the input impedance is too high, it must be shunted by a resistor so that the combined impedance matches that of the cable.

Example 3.6 (Signal splitting).
1. Trivial case
   If two loads are located very close to each other each of them should have twice the charateristic impedance of the cable. In most cases, the reflections due to the connection between the two loads would be acceptable.
2. Professional signal splitting
   A signal splitter is a three-port with three equivalent ports that can be used either as inputs or outputs. It contains three resistors of equal value connected star like to the ports. The value of each resistor is one-third of the characteristic impedance of the cable. Of course, the value of the load impedances must equal the characteristic impedance. Besides, the signal gets attenuated which is obvious as the input current gets split into two equal output currents.

Problems

?

3.73. (a) Dimension a signal splitter for a characteristic impedance of 50 Ω.
(b) Give the attenuation caused by this signal splitter.

3.74. A 50 Ω cable is used to transmit signals to equipment with 93 Ω input impedance. Match the impedances.

3.6 Dynamic properties of active two-ports

The dynamic response of active two-ports is the result of the frequency and time response of the components inside the two-port taking the loading condition into account.

3.6.1 Signal power transmitted by a two-port

When we include the frequency dependence of the transfer function of a (linear) two-port into power considerations, we must know the (power) gain at each frequency. Up to now, we have come across the equation for the power p_{L} dissipated in a load L to be at any moment t

$$p_{\mathrm{L}}(t) = v_{\mathrm{L}}(t) \times i_{\mathrm{L}}(t) . \tag{3.102}$$

Transferring this equation to a sinusoidal steady-state signal with the angular frequency ω we get

$$p_{\mathrm{L}}(t) = \hat{v} \sin(\omega t + \varphi_v) \times \hat{\imath} \sin(\omega t + \varphi_i) . \tag{3.103}$$

The addition theorem for cosines gives us the following two equations:

$$\cos(\alpha - \beta) = \cos\alpha \cos\beta + \sin\alpha \sin\beta \tag{3.104}$$

$$\cos(\alpha + \beta) = \cos\alpha \cos\beta - \sin\alpha \sin\beta \tag{3.105}$$

By subtracting of (3.105) from (3.104) we obtain

$$2\sin\alpha\sin\beta = \cos(\alpha-\beta) - \cos(\alpha+\beta)\,. \tag{3.106}$$

With $\alpha = \omega t + \varphi_v$ and $\beta = \omega t + \varphi_i$ we get

$$\begin{aligned} p_\mathrm{L}(t) &= \hat{v}\times\hat{i}\times\frac{1}{2}\times\left[\cos(\varphi_v-\varphi_i) - \cos(2\omega t+\varphi_v+\varphi_i)\right] \\ &= \frac{\hat{v}}{\sqrt{2}}\times\frac{\hat{i}}{\sqrt{2}}\times\left[\cos\varphi - \cos(2\omega t+2\varphi_v-\varphi)\right] \end{aligned} \tag{3.107}$$

with the phase shift φ between voltage and current and the *root-mean-square values* of voltage v_rms and current i_rms:

$$\varphi = \varphi_v - \varphi_i \tag{3.108}$$

$$v_\mathrm{rms} = \frac{v_\mathrm{max}}{\sqrt{2}} \tag{3.109}$$

$$i_\mathrm{rms} = \frac{i_\mathrm{max}}{\sqrt{2}} \tag{3.110}$$

To get the mean power, one has to integrate over one cycle. The mean of any $\cos(\omega t)$ over one cycle is zero because the positive lobe is a mirror image of the negative lobe. Thus, the second term inside the brackets vanishes and we get for the power averaged over one cycle

$$P = v_\mathrm{rms}\times i_\mathrm{rms}\times\cos\varphi\,. \tag{3.111}$$

P is called *real power* and is measured in watts (W). The factor $\cos\varphi$ is called *power factor*, and $|S| = v_\mathrm{rms}\times i_\mathrm{rms}$ is called *apparent power* and is given in V A (volt-ampere). If the load is purely resistive, the two electrical variables have identical waveform, the phase shift φ is zero and $\cos\varphi = 1$. Real power is needed to provide energy. If the load is purely reactive, then the voltage and current are out of phase by 90 degree and $\cos\varphi$ is zero. This power is called *reactive power* Q. It is zero when averaged over one cycle. Power factors can be called *leading* or *lagging* depending on the sign of the current phase angle with respect to voltage. Thus, the lag of an inductive load may be (partly) compensated for by adding a capacitive load. Figure 3.53 summarizes in a vector diagram how the various kinds of power are related.

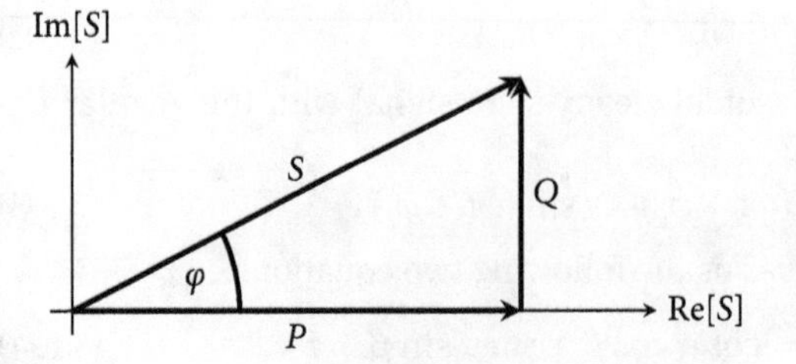

Fig. 3.53. Complex power S is the vector sum of the real P and the reactive power Q. The apparent power is the amplitude of the complex power.

Real power is along the real axis. As reactive power does not transfer energy, it is represented along the imaginary axis. This can also be expressed by using complex numbers

$$S = P + \mathrm{j}Q\,. \tag{3.112}$$

Problems

?

3.75. What is the apparent power if the phase angle between voltage and current is 60° and the real power is 500 W?

3.76. A sine wave signal with amplitude of 10 V and a frequency of $f = 1$ kHz is dissipated in a resistor of 1 kΩ.
(a) What is the dissipated power?
(b) What is the dissipated power if it is a bipolar square wave signal (±10 V, duty cycle 50%) ?
(c) What is the dissipated power if it is a unipolar square wave signal (10 V, duty cycle 50%)?
(d) If in the case (c) the (average) dissipated power is just somewhat less than the power rating of the resistor, is this still true for $f = 1$ MHz or $f = 1$ Hz?

3.77. A capacitor of 1 nF has at 5 MHz a quality factor $Q = 500$.
(a) Which series resistor R would bring forth such a factor?
(b) Which shunting conductance G would effect such a factor?

3.6.2 Dynamic properties of operational amplifiers

Reactive elements cause frequency dependence in all circuit parameters. The majority of operational amplifiers are voltage-dependent voltage sources. Such an amplifier does not show any frequency dependence if it is ideal. It has infinite open-loop voltage gain, infinite input impedances (zero input currents), infinite bandwidth, infinitely small slew rate, infinite bandwidth, and, zero output impedance. Modern integrated operational amplifiers approximate several of these ideal parameters closely (see also Section 2.3.5.2). However, the following imperfections must be taken into account when confronted with real components.

Finite bandwidth all amplifiers have a finite bandwidth and consequently a finite signal rise time.

Gain-bandwidth product Due to the finite bandwidth the voltage gain at DC does not apply to higher frequencies. The gain of a typical operational amplifier is approximately inversely proportional to frequency, i.e., its gain-bandwidth product is constant. This low-pass characteristic of an all purpose amplifier is introduced deliberately because it stabilizes the circuit by introducing one dominant time constant. Typical low cost amplifiers have a gain bandwidth product of a few MHz. Present limits of gain-bandwidth products lie far beyond 1 GHz.

Slewing is present if the rise time of a signal increases with the height of said signal. The slew rate is usually specified in V/μs. Slewing is usually caused by charging internal capacitances of the amplifier with a limited (constant) current (Section 3.2.1), in particular those capacitances used to effectuate its frequency compensation.

Input capacitance most important for higher frequency operation because it reduces the input impedance.

Noise all electronic components are subject to noise. These components will have 10^0 to 10^2 nV/$\sqrt{\text{Hz}}$ noise performance.

Not all of these imperfections must be considered all of the time. Consequently, they are dealt with whenever appropriate.

3.6.3 Dynamic feedback in linear amplifiers

The general relations derived for static feedback in Section 2.4 hold to a first order (see, e.g., Example 3.7) also in the dynamic case if the variables are either made dependent on the time t or the angular frequency ω. This is in particular true for the general feedback equation

$$A_\mathrm{F}(t) = \frac{A(t)}{1 - A(t) \times B(t)} \tag{3.113}$$

or using the complex notation to include frequency-dependent phase shift

$$A_\mathrm{F}(\mathrm{j}\omega) = \frac{A(\mathrm{j}\omega)}{1 - A(\mathrm{j}\omega) \times B(\mathrm{j}\omega)} \tag{3.114}$$

3.6.3.1 Frequency-independent negative feedback

It is worthwhile to ponder on the limitations to the approximation which we found for negative feedback with $|B| \gg |1/A|$

$$A_\mathrm{F}(\mathrm{j}\omega) \approx -\frac{1}{B(\mathrm{j}\omega)} \,. \tag{3.115}$$

For above inequality to hold, it is necessary that both variables are real, i.e., that the phase shift can be disregarded. This is easily achieved for B if it is constructed of resistors, because the (ideal) transfer function of such a resistive network has no upper corner frequency in the frequency response and is frequency independent. Thus, we have $B(\mathrm{j}\omega) = b(\omega) = b$. However, $A(\mathrm{j}\omega)$ is subject to phase shift. Assuming that its frequency response is that of a simple low-pass filter with a dominating time constant τ then up to $\omega_{\max} = 0.1 \times 1/\tau$ the phase shift may be disregarded so that up to this frequency $A(\mathrm{j}\omega) = a(\omega)$ can be assumed. With $|b(\omega_{\max})| \gg |1/a(\omega_{\max})|$ the frequency response of the amplitude of the transfer function of the feedback system becomes

$$a_F(\omega) = -\frac{1}{b}$$

for angular frequencies up to ω_{max}. Up to ω_{max} the circuit with feedback has the inverse property of the feedback circuit B. As B is built from (linear) resistors with frequency independent properties, then also the circuit with feedback will have a transfer function with a flat frequency response.

As the case of frequency independent negative feedback is of utmost importance, we will investigate this situation some more. With

$$A_F(j\omega) = \frac{A(j\omega)}{1 + |A(j\omega)| \times b} \tag{3.116}$$

and

$$A(j\omega) = A(\omega = 0) \times \frac{1}{1 + j\omega\tau} \tag{3.117}$$

we get

$$\begin{aligned} A_F(j\omega) &= \frac{\dfrac{A(\omega = 0)}{1 + j\omega\tau}}{1 + b \times \dfrac{A(\omega = 0)}{1 + j\omega\tau}} \\ &= \frac{A(\omega = 0)}{1 + j\omega\tau + b \times A(\omega = 0)} \\ &= \frac{A(\omega = 0)}{1 + b \times A(\omega = 0) + j\omega\tau} \\ &= \frac{A(\omega = 0)}{1 + b \times A(\omega = 0)} \times \frac{1}{1 + j\omega\dfrac{\tau}{1 + b \times A(\omega = 0)}} \,. \end{aligned} \tag{3.118}$$

Again, the complex transfer function with negative feedback has the shape of a simple low-pass filter with an upper corner frequency $\omega_{uF} = \frac{1+b\times A(\omega=0)}{\tau}$ which is increased by the (forward) return difference (1+ the closed-loop gain at low energies). The time constant τ_F is likewise decreased by that factor.

In Figure 3.54, the frequency dependence of the amplitude term $a(\omega)$ of the amplifier is compared with that of the circuit with feedback $a_F(\omega)$ that is reduced at low frequency by the factor $1 + b \times A(\omega = 0)$.

The amplitude term is decreased by the factor $1 + b \times A(\omega = 0)$ and the frequency ω_u of the upper turnover point is increased by the same factor to ω_{uF}, i.e., the product of the gain at lower frequencies with ω_u is a constant, i.e. it does not depend on the closed-loop gain. This effect is called the *constant gain-bandwidth product*

$$A \times \omega_u = A_F \times \omega_{uF} \tag{3.119}$$

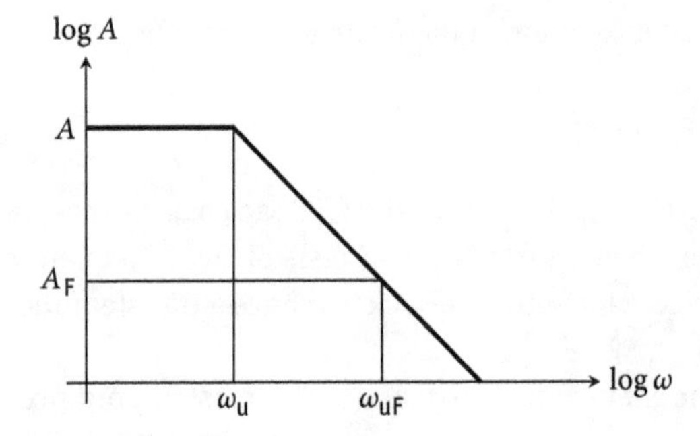

Fig. 3.54. Schematic of the frequency dependence of the amplitude term of an amplifier without feedback, and of the circuit with feedback.

Taking the actual frequency responses, it would be obvious that above about $10\omega_{\mathrm{uF}}$ the two frequency responses coincide, i.e., that above that frequency feedback has practically no effect on the transfer function. As also the phase shift of a low-pass filter does not change much at all beyond $10\omega_{\mathrm{uF}}$, it can be inferred that the frequency responses of the complex transfer functions are very close (nearly identical) for the two cases.

If the high-frequency response is identical, then the short-time response must be identical, too. This means that at the beginning of a signal there is no effect of feedback on said signal, i.e., a very short signal does not "feel" the effect of feedback. This can easily be visualized. The static concept of feedback presupposes that input and output are in equilibrium. However, even if the input signal rises instantly, the time constant of the amplifier makes the output signal rise with the rise time, i.e. it takes the output signal some time which is longer than the input signal rise time to reach the full value. The static feedback representation presumes that the full output signal is fed back. Consequently, the feedback action is bound to be smaller, if the output signal has not reached its full value. At the beginning of the input signal the output signal has not developed at all; therefore, the feedback action increases from zero at the very beginning of the input signal to its full values when the output signal without feedback would have reached its full value. Of course, these deliberations do not consider the propagation delay through the circuit that plays only a role for very short (fast) signals.

Example 3.7 (Analytical proof that (negative) feedback has no impact on very short signals). An amplifier without feedback be characterized by its gain A and one time constant τ. Applying feedback, we get $A_{\mathrm{F}} = A/(1 + AB)$ and $\tau_{\mathrm{F}} = \tau/(1 + AB)$. We apply a step signal to the input. We must compare the amplitude of the output signals at a time t which is a fraction f of τ_{F}, i.e., $t = f \times \tau_{\mathrm{F}} = f \times \tau/(1 + AB)$. For the bare amplifier the response $a(t)$ is

$$a(t) = A \times \left(1 - \mathrm{e}^{-t/\tau}\right) = A \times \left(1 - \mathrm{e}^{-f/(1+AB)}\right)$$
$$= \frac{A \times f}{1 + AB} \times \left(1 - \frac{f}{2(1 + AB)} + \frac{f^2}{6(1 + AB)^2} - \cdots\right)$$

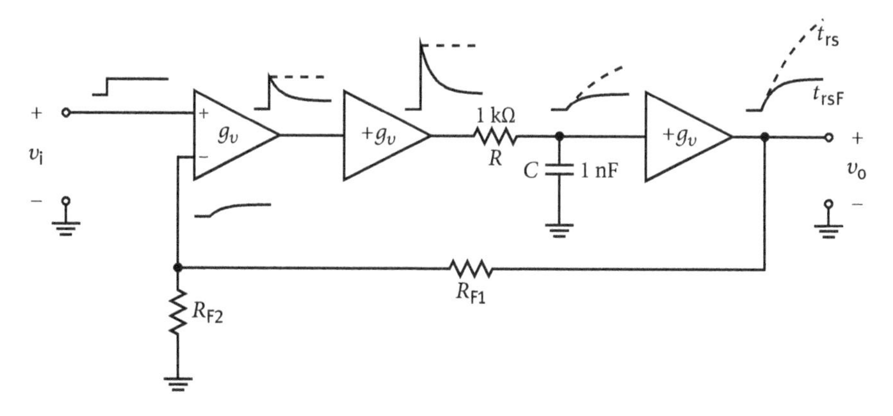

Fig. 3.55. Signal shapes in a three-stage amplifier with and without (dashed) series–parallel feedback.

and for the system with feedback

$$a_F(t) = \frac{A}{1+AB} \times \left(1 - e^{-t/\tau_F}\right) = \frac{A}{1+AB} \times \left(1 - e^{-f}\right).$$

Without feedback, we have $AB = 0$ making the second part of both equations identical, as it must be. For $f = 0$ we get the trivial answer that the output signal is zero in both cases. Let us take the ratio to find how much influence feedback has on short signals. The most difference we expect for huge closed-loop gains so that we set $1/AB = 0$. Then the ratio is

$$\frac{a_F(t)}{a(t)} = \frac{1 - e^{-f}}{f}. \tag{3.120}$$

For $f = 0.1$ this ratio is nearly 1, namely 0.9516, i.e., the response is practically unchanged, verifying our verbal reasoning that for a short time after the beginning of a signal, feedback has not had time to build up.

Figure 3.55 sketches how the voltage signals in a three stage amplifier with series–parallel feedback must look to account for the reduced rise time of the output signal in the feedback situation.

In Figure 3.55 the signal shapes after each stage of a three-stage amplifier are shown to explain how by feedback the effect of a low-pass filter after the second stage (representing the limited bandwidth of the amplifier) is reduced. All three stages provide positive voltage gain; the first is a differential amplifier with its inverting input used for negative feedback. Keep in mind that each of the amplifiers amplifies according to its intrinsic gain which is independent of any feedback. Thus, the reduction of the output signal by feedback action is achieved by reducing the effective input voltage of stage one by feeding the output signal to its second terminal which then is the cause of reduced input voltages at the two other stages, too. The effect of the low-pass filter after stage two requires an overshoot at the input of this filter to have at the output of stage three a signal that resembles the input signal of stage one. This overshoot is the result of the difference between the input signal and the attenuated output signal fed back to the inverting input.

Example 3.8 (Slewing due to an excessive signal amplitude). The relevant data of the circuit of Figure 3.55 are

- open-loop gain per stage $g_{vk} = 10$
- closed-loop gain $g_{vF} = 1 + R_{F1}/R_{F2} = 10$
- time constant $\tau = R \times C = 1\,\mu s$
- rise time without feedback $t_{rs} = 2.2\,\mu s$
- rise time with feedback $t_{rsF} = 0.022\,\mu s$

As long as the signals with overshoot are transmitted undisturbed the output signal has a rise time of 0.022 µs. By increasing the amplitude of the input signal, all internal signals will be increased accordingly. However, at a given amplitude the top of the overshoot at the output of amplifier stage 2 will leave the dynamic range of the amplifier, i.e., the upper part of the corrective signal will not be amplified any more. Consequently the rise time of the output signal will increase. Raising the input signal further will result in the loss of more and more of the overshoot increasing the rise time further. When the flat portion of the signal is at the edge of the dynamic range of stage two, no corrective overshoot will be transmitted any more, and the output signal will have the rise time of the amplifier without feedback, namely 2.2 µs. An increase of rise time with input signal amplitude is called slewing (Section 3.2.1).

3.6.3.2 Frequency instability

In equilibrium, all equations derived in Section 2.4.3 remain valid even after the time or the frequency dependence is included. As the circuits have small dimensions (on the order of cm or even much less when integrated) the finite propagation speed of signals is disregarded, i.e., the space dependence in Maxwell's equations is switched off. However, when impedances Z are involved, these are complex impedances, e.g., the input impedance will typically consist of the input resistance shunted by the input capacitance which for mechanical reasons in macroscopic circuits will always be at least on the order of 1 pF (this parasitic capacitance is called stray capacitance, Section 3.2.1.1).

The main issue, however, is the fact that lumped capacitors or internal capacities in circuit components or stray capacities between conductors of the circuit can, together with resistors of the circuit act as low-pass filters providing phase shifts up to −90° (at very high frequencies). For simplicity let us assume that three such filters dominate the frequency response of the closed-loop gain $A(j\omega) \times B(j\omega)$. Let us assume that at ω_0 the phase shifts of the three filters add up to −180° (e.g., −60°, −60°, −60°). What is the consequence? Shifting a sinusoidal signal by −180° is equivalent to inverting it, i.e., changing the sign of the signal. Thus, the equation for negative feedback

$$A_F(j\omega) = \frac{A(j\omega)}{1 - A(j\omega) \times B(j\omega)} = \frac{A(j\omega)}{1 - [-|A(j\omega) \times B(j\omega)|]} \tag{3.121}$$

becomes that for positive feedback at a given frequency ω_0

$$A_F(j\omega_0) = \frac{A(j\omega_0)}{1 - A(j\omega_0) \times B(j\omega_0)} = \frac{A(j\omega_0)}{1 - [+|A(j\omega_0) \times B(j\omega_0)|]} . \tag{3.122}$$

Furthermore, the closed-loop gain is a real number at ω_0 because at a phase shift of $-180°$ the imaginary part is zero. If the three dominating time constants lie in the active circuit A (e.g., if B is built from resistors and is, therefore, real by nature), then above equation degenerates to

$$a_F(\omega_0) = \frac{a(\omega_0)}{1 - a(\omega_0) \times b} . \tag{3.123}$$

Actually, at ω_0 the total phase shift of the *closed-loop gain* is 0°. Negative feedback provides a minus sign corresponding to a 180° phase shift and the three low-pass filters shift the phase by $-180°$. What happens now if the closed-loop gain becomes +1 (or even larger)? Mathematically one expects that a_F becomes infinite. Obviously, in the real world such a thing cannot happen. The reason should be obvious. As is usual, we used small-signal parameters, i.e., all transfer functions are supposed to be linear. In particular, the variable A decreases with amplitude, i.e., the transfer function is not linear, at both ends of the dynamic range it becomes even zero. Thus, even if the closed-loop gain is +1 (or even larger) there will be an amplitude at which it becomes smaller than one, establishing a stable (temporary) operating point. In our problem, this means that the feedback arrangement will oscillate with the angular frequency ω_0 with an amplitude at which the closed-loop gain is minutely smaller than one. With an amplifier, this is called frequency instability which, obviously, must be avoided by all means. To this end, there should be a phase margin of at least 50° at that frequency at which the closed-loop gain gets +1 and a gain margin of at least −10 dB at a phase shift of 180°. In narrow band applications, these margins may be lowered to 30° and −6 dB.

The *Nyquist criteria* give the ultimate answer on the stability of a feedback circuit. To this end the *Nyquist plot*, a parametric plot of the closed-loop transfer function, is used for assessing the stability of a system with feedback. The real part of the transfer function is plotted on the abscissa, the imaginary part on the ordinate using the frequency as parameter. Or with polar coordinates, the amplitude of the transfer function is displayed as the radial coordinate, and its phase as the angular coordinate. To

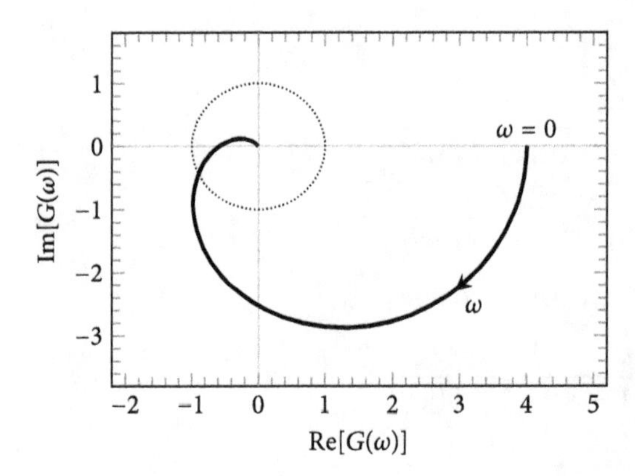

Fig. 3.56. Nyquist plot of an amplifier fulfilling the first Nyquist criterium.

measure the frequency dependence of the closed-loop transfer function the loop must be opened at some point. As identical loading conditions as in the closed-loop situation must be reestablished, it is wise to open the loop at a point were loading is not serious.

The Nyquist criteria differentiate between two cases.

- Instability of the first kind: Is the amplifier stable without feedback, so is the circuit with feedback also stable if the critical point $(-1, 0)$ in the complex number plane lies outside the Nyquist plot.
- Instability of the second kind: If the bare amplifier (without feedback) is unstable, then there must be one counter clock-wise encirclement of the point $(-1, 0)$ by the Nyquist plot for each pole (i.e., a zero in the denominator) of $A(j\omega)$ to make the amplifier with feedback stable.

What can be done if a system with negative feedback proves to be unstable according to the first criterion? We need to tailor the closed-loop transfer function accordingly. Obviously, there are two remedies. One has to change either the phase shift or the amplitude (or both).

- The obvious choice is to attenuate the open-loop gain of the amplifier for all frequencies so much that the closed-loop gain curve does not contain the point $(-1, 0)$ any more as shown in Figure 3.57c.
- Alternatively, a phase lag circuit may be inserted. Such a circuit is shown in Figure 3.58 together with its (asymptotic) Bode plots. To find the correct value of component R_1 it is necessary to account for the output impedance of the circuit (the impedance of the Thevenin source) at the position where the phase lag circuit is inserted. In Figure 3.59 its effect on the Nyquist plot is shown.
- Finally, a phase lead circuit (Figure 3.60) may be introduced into the loop reducing the phase shift so that a phase shift of $-180°$ will occur at higher frequencies where the amplitude is less than 1. A reduction of the amplitude at lower frequencies is unavoidable. It may be compensated by increasing the gain by the same amount without endangering the stability, as shown in Figure 3.61 for both the Bode and the Nyquist plot.

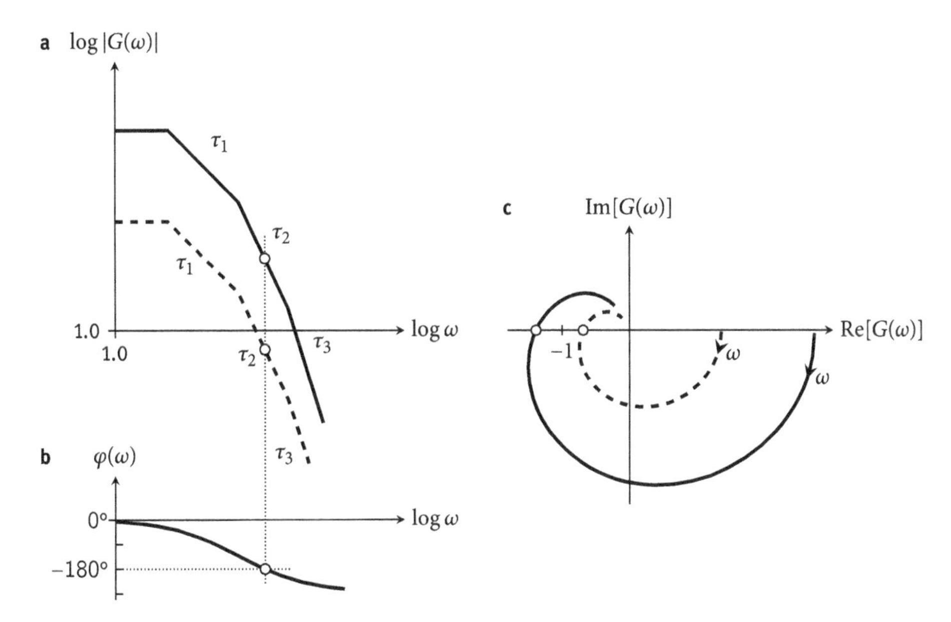

Fig. 3.57. Frequency independent attenuation: (a) Bode plot of the frequency dependence of the closed-loop gain (with three time constants). (b) Ditto of the phase shift. (c) Effect of reducing the closed-loop gain on the Nyquist plot.

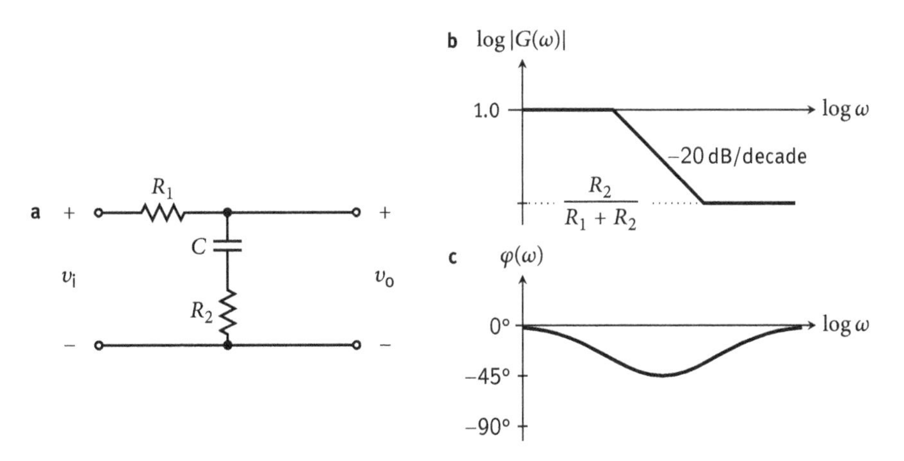

Fig. 3.58. Phase lag circuit (a), and Bode plots for amplitude (b) and phase shift (c).

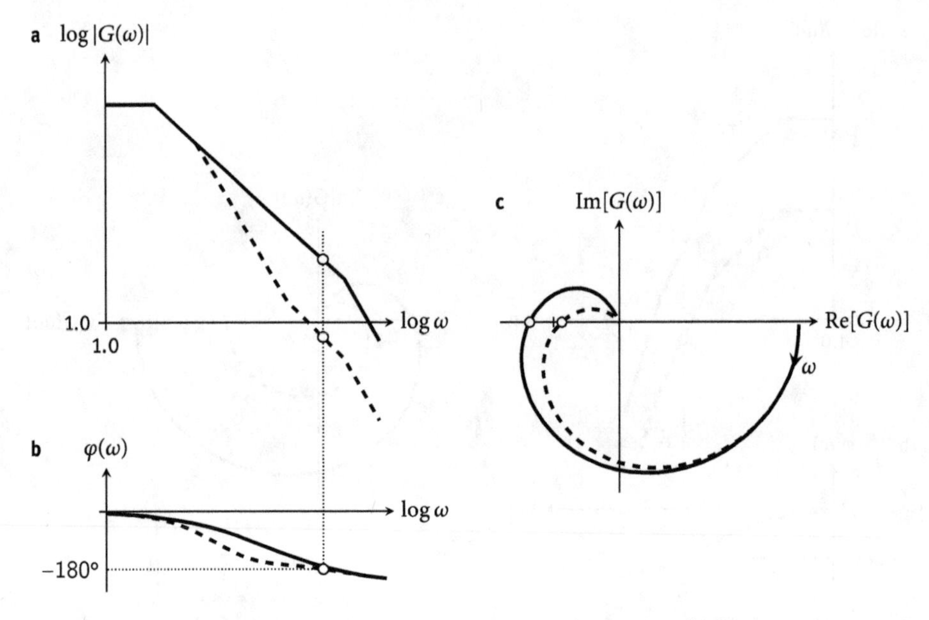

Fig. 3.59. Effect of a phase lag circuit on the closed-loop transfer function: (a) frequency dependence of the amplitude transfer function without and with phase lag circuit, (b) ditto for the phase shift dependence, and (c) Nyquist diagram without and with phase lag circuit.

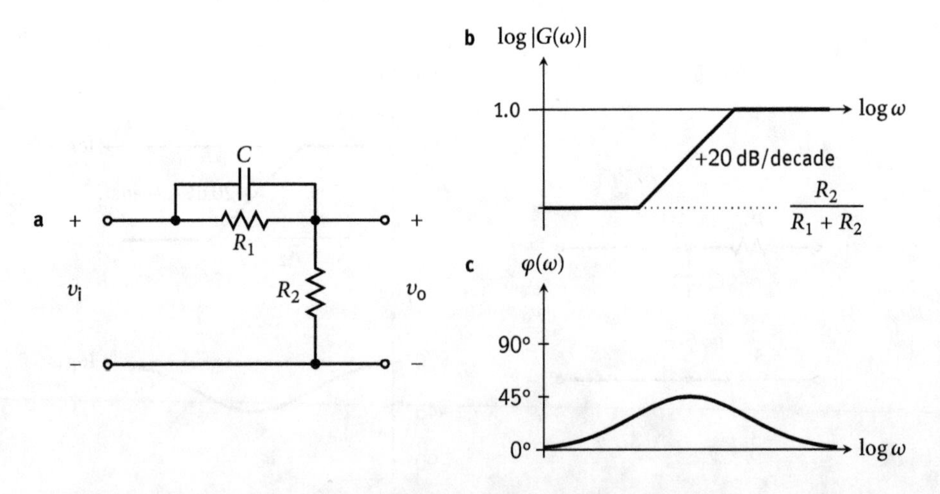

Fig. 3.60. Phase lead circuit: (a) circuitry, (b) Bode plot of the frequency dependence of the amplitude transfer function, and (c) ditto, but for the phase shift.

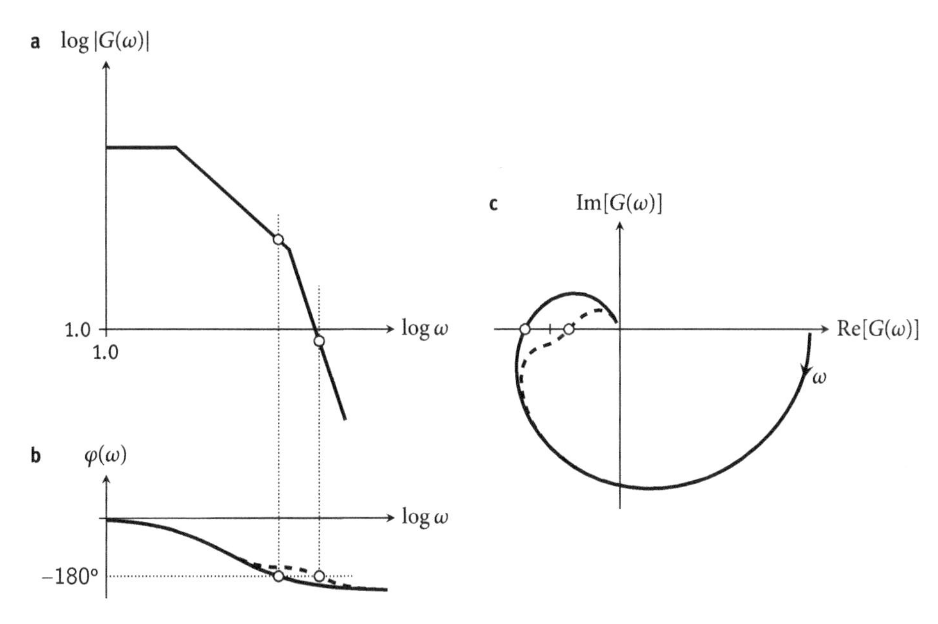

Fig. 3.61. Effect of a phase lead circuit with compensating amplification on closed-loop transfer function: (a) Bode plot of the frequency dependence of the amplitude, (b) ditto for the phase shift, (c) Nyquist plot, combining (a) and (b).

Voltage (emitter) followers have the largest closed-loop gain (B close to one). Thus, they are prone to instabilities of the first kind. A remedy for emitter followers is the insertion of a small resistor in series to the collector.

Problems

?

3.78. Give the dimension of the four *characteristic* transfer functions of the active element A, and of the circuit with feedback A_F. Namely, v_o/v_i, i_o/i_i, v_o/i_i, and i_o/v_i.

3.79. Calculate the complex voltage transfer function $G_v(j\omega)$ of the phase-lag circuit in Figure 3.58.

3.80. Calculate the complex voltage transfer function $G_v(j\omega)$ of the phase-lead circuit in Figure 3.60.

3.81. The frequency response of an amplifier is that of a simple low-pass filter with an upper corner frequency of $f_u = 3.501$ MHz and an open-loop voltage gain $g_v(0\text{ Hz}) = 10$. Increase the gain to 100 without the help of additional active elements.

(a) How can this be done? Give details.
(b) What is the rise time of a step function processed by this new amplifier?
(c) Rectangular signals with a frequency of 50 kHz and a duty cycle of 0.5 (= 50%) shall be amplified by this amplifier. Will the shape of the resulting output signal resemble that of the input signal sufficiently?

3.82. The frequency response of an amplifier is given as

$$g_v(j\omega) = \left(-\frac{10}{1 + j\omega \times 10^{-5}}\right)^3 .$$

A negative series–parallel feedback is applied with $B(j\omega) = 1/150$. Applying Nyquist's stability criterion find the gain margin in %. (There are three identical cascaded stages; investigate just one at the frequency where the phase shift is 60°.)

3.83. The frequency response of an amplifier is given as

$$g_v(j\omega) = \left(-\frac{50}{1 + j\omega \times 10^{-4}}\right)^3 .$$

A frequency-independent negative series–parallel feedback is applied. Applying Nyquist's stability criterion find the maximum allowed value of the feedback factor B.

3.84. Feedback with a return difference of $(1 + AB)$ is applied to an amplifier with one dominant time constant τ in its frequency response. Compare the rise time of the response to a rectangular input signal of length $T < \frac{\tau}{2.2 \times (1+AB)}$ with (t_{rsF}) vs. without (t_{rs}) feedback.

3.6.3.3 Frequency-dependent negative feedback (frequency compensation)

The upper corner frequency ω_u of an operational amplifier defines a time constant $\tau = 1/\omega_u$. There are two additional time constants to be considered due to

- the input capacitance C_i shunting the input resistor, and
- the output capacitance C_o shunting the output resistor.

Figure 3.62 shows how the output of an operational amplifier with series–parallel feedback is loaded by a resistive feedback network. The (ideal) voltage controlled voltage source is loaded by the output resistance R_o feeding the feedback resistor R_F and the complex impedance Z_1 at the input (R_i shunted by R_1 and C_i). R_i can be disregarded if $R_i \gg R_1$. The output capacitance C_o shunts R_o. This is a case where switching to a current source is helpful. Replacing the real voltage source by a real current source (Norton's theorem) shows at once that C_o shunts R_o. Thus, the feedback network introduces two (additional) time constants, one at the input $\tau_i = R_1 \times C_i$, and one at the output $\tau_o = R_o \times C_o$. The closed-loop voltage gain g_{vF} for low frequencies (at which capacitances can be disregarded) is given by (disregarding R_i)

$$g_{vF}(\omega = 0) \approx g_v \times \frac{R_1 + R_F + R_o}{R_1} = g_v \times \left(1 + \frac{R_F}{R_1} + \frac{R_o}{R_1}\right) . \tag{3.124}$$

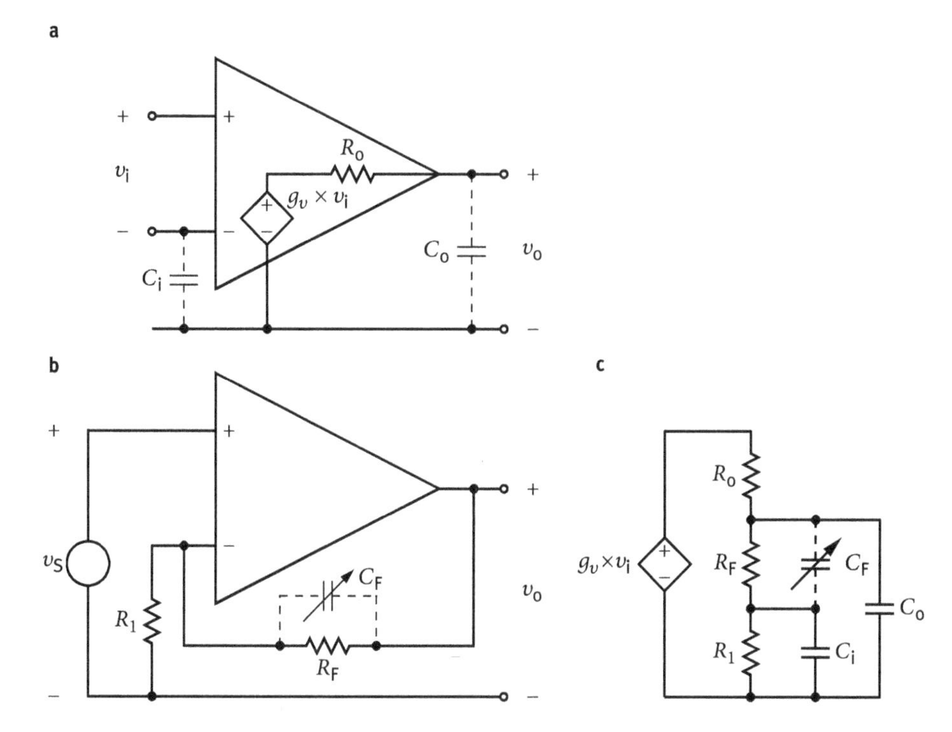

Fig. 3.62. Effective capacitive loading of the voltage controlled voltage source of an operational amplifier: (a) amplifier with stray capacitances, (b) series–parallel feedback with R_F, R_1 and C_F, (c) feedback network, redrawn.

The parallel configuration of a resistance R_x with a capacitance C_x gives a complex impedance Z_x of

$$Z_x(j\omega) = \frac{R_x}{1 + j\omega\tau_x} \tag{3.125}$$

with $\tau_x = R_x \times C_x$ so that we get for the complex closed-loop gain

$$g_{vF}(j\omega) \approx g_v(j\omega) \times \left(1 + \frac{R_F \times (1 + j\omega\tau_1)}{R_1} + \frac{R_o}{R_1} \times \frac{1 + j\omega(\tau_1 - \tau_o) + \omega^2\tau_1\tau_o}{1 + \omega^2\tau_o^2}\right). \tag{3.126}$$

With R_o and τ_o small we get

$$g_{vF}(j\omega) \approx g_v(j\omega) \times \left(1 + \frac{R_F \times (1 + j\omega\tau_1)}{R_1}\right). \tag{3.127}$$

Thus, the *resistive* feedback circuit introduces frequency dependence. By adding a *speed-up capacitor* C_F in parallel to R_F, we can make this frequency dependence vanish if $\tau_F = R_F \times C_F$ has the same value as τ_1

$$g_{vF}(j\omega) \approx g_v(j\omega) \times \left(1 + \frac{R_F \times (1 + j\omega\tau_1)}{R_1 \times (1 + j\omega\tau_F)}\right) = g_v(j\omega) \times \frac{R_1 + R_F}{R_1}. \tag{3.128}$$

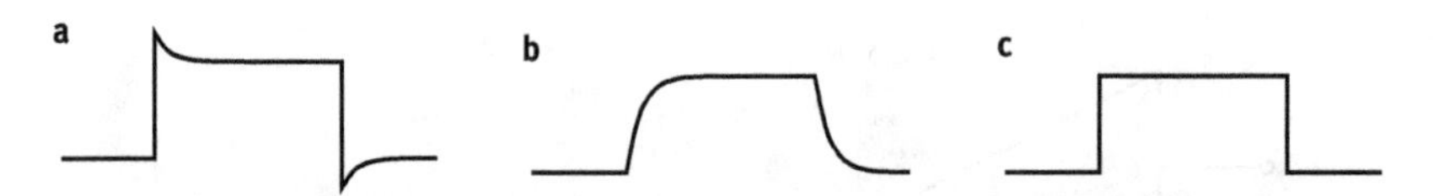

Fig. 3.63. Adjustment of the variable feedback capacitor to achieve frequency compensation: (a) C_F too small (b) C_F too large (c) C_F correct.

This method of compensating the action of a (current) low-pass filter by means of a complex voltage divider with identical time constants in both impedances is called *frequency compensation*. Observe that *no* current flows between the connection of the resistors and that of the capacitors!

Before discussing other important applications of this method, we have to show how the equality of the two time constants is best obtained. The most efficient method is to use as C_F a variable capacitor and adjust it that way that the required result (frequency independence) is achieved. It would be tedious to apply lots and lots of signals of various frequencies to check the flat frequency response.

In a one-step process, one can get the desired result. A step signal contains all frequencies. Therefore, if a step signal applied to the input gives an undistorted step signal at the output all frequencies are transferred correctly. As step signals are impractical due to their singularity, one uses rectangular signals. Figure 3.63 compares output signals for values of C_F which are too small, too large and correct.

The interpretation of Figure 3.63 is straightforward. If $C_F = 0$ or is too small, R_F feeds a low-pass filter, i.e., higher frequencies are not fed back to the same extent as lower frequencies. Consequently, their amplitude is less reduced there is a surplus of high frequencies at the output. If C_F is too large, the opposite effect is seen.

Example 3.9 (Attenuating probes). Taking measurements in a circuit (e.g., with a probe) loads this circuit (see Section 2.1.4). This load is not purely resistive but will, at least, contain a capacitive component. So there will be a frequency dependent correction factor which by itself is no real problem.

However, the input capacitance could be such that a feedback circuit that by itself is stable does not pass the Nyquist criteria when this capacitance is added. (The additional phase shift caused by the probe might suffice to make the total phase shift $-180°$ at a frequency at which the closed-loop gain is not sufficiently small.) In this case, the amplifier will oscillate, and no meaningful measurements are possible with this probe. In that case, an *attenuating probe* must be used. At the expense of signal height the capacitive loading of the circuit is reduced by the same amount as the signal height. Typical values of the input impedance of a measuring device (e.g., an oscilloscope) are $R_i = 1\,\text{M}\Omega$ shunted by $C_i = 20\,\text{pF}$. The connecting cable may add about 80 pF so that the capacitance at the tip of the measuring probe would be about 100 pF.

Figure 3.64 shows how the voltage division of a 10 : 1 attenuating probe is accomplished. Obviously, for an input resistance $R_i = 1\,\text{M}\Omega$ a resistor R_1 of $9\,\text{M}\Omega$ is needed

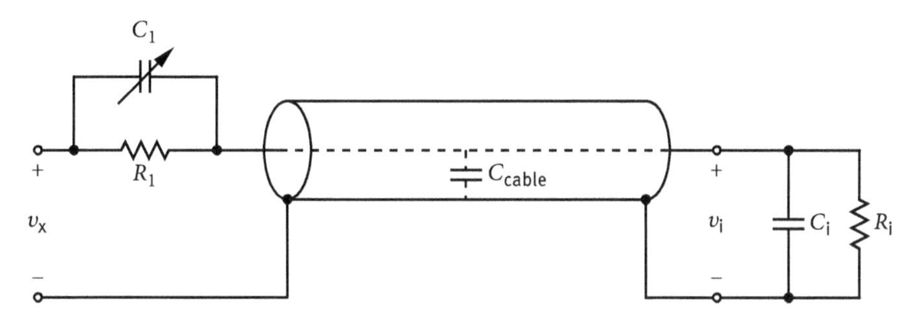

Fig. 3.64. Circuit diagram of an attenuating probe.

to attenuate lower frequencies of amplitude v_i by a factor of 10 to v_o

$$v_o = v_i \times \frac{R_i}{R_i + R_1} = \frac{v_i}{1 + \frac{R_1}{R_i}}. \tag{3.129}$$

Making $\tau_1 = \tau_i$ gives for the value of the shunting capacitor $C_1 = R_i \times C_i/R_1$. Again a variable capacitor would be so adjusted that rectangular signals (Figure 3.63) would be transmitted correctly. However, this is only a side effect. The main application is the ten times reduced capacitive load. With $\tau_i = R_i \times C_i$ and $\tau_1 = R_1 \times C_1$ the impedance Z_p of the probe is given by

$$z_p = \frac{R_i}{1 + j\omega\tau_i} + \frac{R_1}{1 + j\omega\tau_1} = \frac{R_i + R_1}{1 + j\omega\tau_p} \tag{3.130}$$

because $\tau_i = \tau_1 = \tau_p$. The last term describes the parallel configuration of a resistance $R_i + R_1$ with the total capacitance C_p of the probe in action. From $\tau_p = \tau_i$ follows that the capacity of the probe (in action) is

$$C_p = C_i \times \frac{R_i}{(R_i + R_1)} = \frac{C_i}{1 + \frac{R_1}{R_i}} = \frac{C_i}{v_i/v_o}, \tag{3.131}$$

i.e., the capacitive loading of the circuit by the measurement using an attenuating probe is reduced by the attenuation factor.

The behavior of the probe in use is just that of a tenfold increased resistance shunted by a tenfold reduced capacitance. As the main contribution to the capacitive loading is the connecting cable, an *active probe* puts an (pre-)amplifier on the tip of the probe. The capacitive loading by the probe is reduced to a few pF which can be further reduced by about a factor of 3 by introducing attenuation also for the active probe. Of course, the mechanical minimum capacity of about 1 pF will always be present.

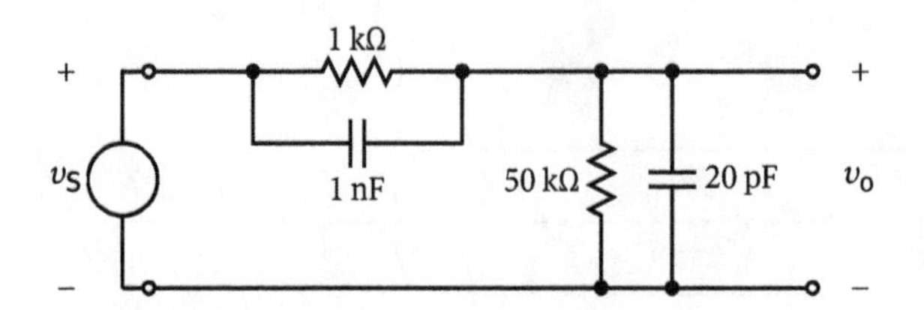

Fig. 3.65. Circuit of the probe from Problem 3.86.

Problems

3.85. Attenuating probe.

(a) Give the values (resistance, capacity) of attenuating elements in an attenuating probe. It should attenuate by a factor of 5. The impedance of the oscilloscope is 1 MΩ shunted by 30 pF, and the 1 m long cable has a specific capacity of 100 pF/m.

(b) If this probe is in use with the oscilloscope, how does it burden the circuit (which resistance shunted by which capacitance)?

(c) In which cases is the use of attenuating probes the only deliverance?

3.86. Attenuating probe.

(a) Give the (theoretical) lower and upper corner frequencies of the amplitude response of the circuit in Figure 3.65.

(b) How large is the input capacitance?

3.6.4 Dynamic behavior of operation amplifiers (active filters)

For inverting operation amplifiers, we have found that the characteristic transfer quality is the transimpedance which in the ideal approximation is

$$r_{\mathrm{mF}}(\mathrm{j}\omega) = -Z_{\mathrm{F}}(\mathrm{j}\omega)\,. \tag{3.132}$$

Considering the finite voltage gain g_v we get with $r_{\mathrm{mF}}(\mathrm{j}\omega) = g_v(\mathrm{j}\omega) \times Z_{\mathrm{iF}}(\mathrm{j}\omega)$, and with $Z_{\mathrm{iF}}(\mathrm{j}\omega) = \frac{Z_{\mathrm{F}}(\mathrm{j}\omega)}{1+g_v(\mathrm{j}\omega)}$ (Miller effect, Section 2.4.3.2)

$$r_{\mathrm{mF}}(\mathrm{j}\omega) = -\left|g_v(\mathrm{j}\omega)\right| \times \frac{Z_{\mathrm{F}}(\mathrm{j}\omega)}{1+\left|g_v(\mathrm{j}\omega)\right|} = -Z_{\mathrm{F}}(\mathrm{j}\omega) \times \frac{1}{1+\frac{1}{\left|g_v(\mathrm{j}\omega)\right|}}\,. \tag{3.133}$$

The last part of this equation – with a high enough voltage gain of the operational amplifier (throughout the frequency range of interest!) – hardly differs from the ideal answer.

The factorization in the first part gives rise to the following interpretation: an amplifier with parallel–parallel feedback has the same transfer function as a voltage amplifier with a gain of $-|g_v(\mathrm{j}\omega)|$ cascaded with an impedance that has the same value as the dynamic impedance of the feedback impedance (or vice versa). These two arrangements are compared in Figure 3.66.

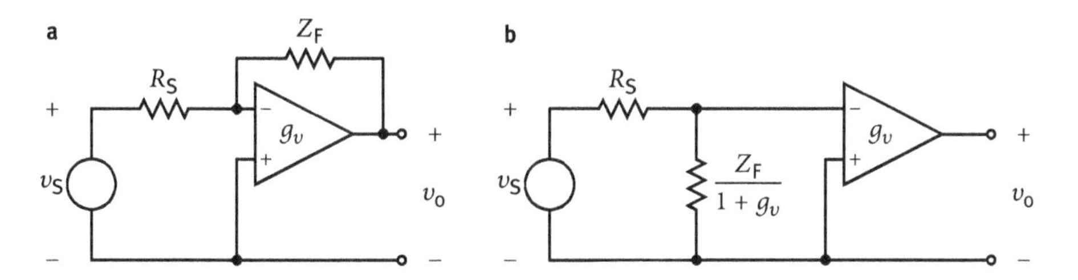

Fig. 3.66. (a) An inverting operation amplifier with voltage source, and (b) a cascade of a voltage divider and an amplifier without feedback that has the same transfer function.

Following the old-fashioned habit to stick to voltage transfer we compare in both cases a source-to-output voltage gain g_{vS}. For the output voltage v_o of the operation amplifier we get

$$v_o(j\omega) = i_i(j\omega) \times r_{mF}(j\omega) = \frac{v_S(j\omega)}{Z_S(j\omega) + Z_{iF}(j\omega)} \times r_{mF}(j\omega) \tag{3.134}$$

and as voltage gain

$$g_{vS} = \frac{v_o(j\omega)}{v_S(j\omega)} = -\left|g_v(j\omega)\right| \times \frac{\dfrac{Z_F(j\omega)}{1+\left|g_v(j\omega)\right|}}{Z_S(j\omega) + \dfrac{Z_F(j\omega)}{1+\left|g_v(j\omega)\right|}} . \tag{3.135}$$

The answer for the operation amplifier is identical with the answer of cascading a complex voltage divider consisting of the complex source impedance and the dynamically reduced complex feedback impedance cascaded with an amplifier having a voltage gain $-|g_v(j\omega)|$.

Sometimes it is more convenient to use the following equation:

$$g_{vS} = -\left|g_v(j\omega)\right| \times \frac{1}{1 + \dfrac{Z_S(j\omega)}{Z_F(j\omega)} \times \left(1+\left|g_v(j\omega)\right|\right)} . \tag{3.136}$$

As the small-signal transfer functions are identical what is then the difference between the operation amplifier arrangement and the cascade of a complex voltage divider with an amplifier of the same voltage gain, i.e., what is the difference between an active filter and a passive filter with amplification? There are three advantages of active filters that can be singled out.

- Z_F can be larger by a factor of $(1+|g_v(\omega=0)|)$ in the case of an active filter giving the same frequency dependence. If this impedance is a capacitor, this fact can be a huge advantage.
- Only the amplifier in the active filter case has feedback (by means of Z_F) contributing to linearity and stability of the system.

- The post-amplifier must have an upper corner frequency that is about ten times higher than the upper corner frequency of the filter so that the phase shift introduced by it is negligible, i.e., its bandwidth is ten times wider than for an active filter. This would result, e.g., in additional (white) noise.

3.6.4.1 Active low-pass filter

Figure 3.67 shows a simple active low-pass filter using an operational amplifier. In this case $Z_S(j\omega) = R_S$ and $Z_F(j\omega) = -j/\omega C_F$. Then, according to (3.135) the voltage transfer from the source to the output is given by

$$\begin{aligned}\frac{v_o}{v_S} &= -\left|g_v(j\omega)\right| \times \frac{1}{1 + j\omega C_F \times (1 + \left|g_v(j\omega)\right|) \times R_S} \\ &= -\left|g_v(j\omega)\right| \times \frac{1}{1 + j\omega\tau_F}\end{aligned} \tag{3.137}$$

with τ_F the effective time constant taking into account the dynamic increase of C by the factor $(1 + |g_v(j\omega)|)$ due to the Miller effect (Section 2.4.3.2). This means that an active low-pass filter has the combined frequency response of the operational amplifier with an upper corner frequency ω_{uA} and that of a low-pass filter with a corner frequency $\omega_{uF} = 1/\tau_F$. If the frequency response of the amplifier is adequate ($\omega_{uA} > 10\omega_{uF}$), then its frequency response has only little impact. The frequency response is that of a low-pass filter with τ_F multiplied by the low frequency voltage gain (the open-loop gain) of the amplifier $-g_v(\omega = 0)$. With $\omega_{uA} > 100\omega_{uF}$ the situation is even better because at $0.01\omega_{uA}$ the phase shift is less than 0.6° off the ideal value of −90°.

In the ideal case $(1/g_v(j\omega) = 0)$ above equation degenerates to

$$\frac{v_o}{v_S} = -j\frac{1}{\omega R_S C_F}\,. \tag{3.138}$$

The voltage gain is inversely proportional to ω for all ω values, i.e., its decrease with frequency is 20 dB per decade. Such an ideal filter has *integrating* properties in the full frequency range (Section 3.2.1).

The frequency response of the transfer function above the corner frequency ω_{uF} is based on the open-loop gain of the amplifier. Therefore, it is not stabilized by feedback

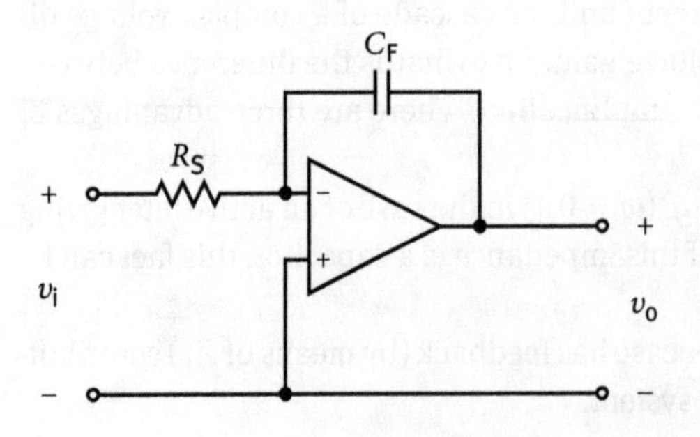

Fig. 3.67. Active low-pass filter.

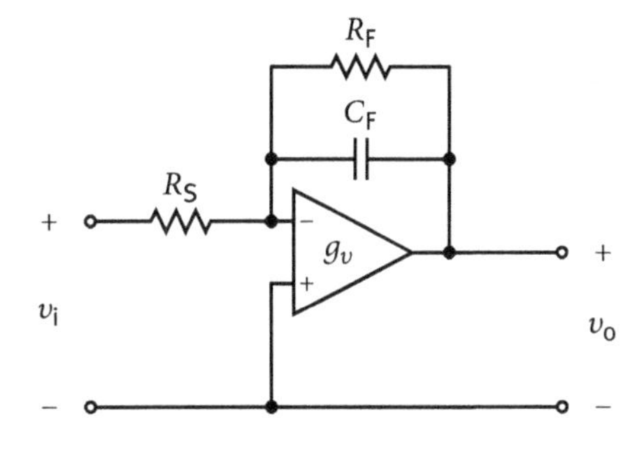

Fig. 3.68. Active low-pass filter with feedback resistor R_F.

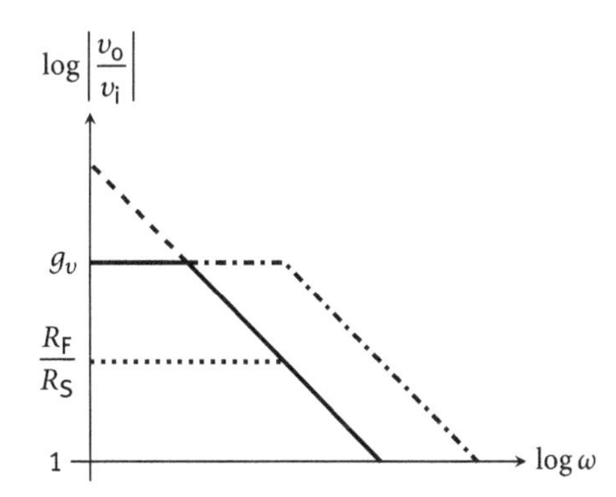

Fig. 3.69. Bode plots of several cases of low-pass filters: bare amplifier, feedback with basic R_SC_F low-pass, feedback with added resistor R_F, and for the ideal low-pass.

action. Thus, the position of the quiescent operating point will reflect all changes in the circuit (e.g., the input offset voltage) to the full extent. This can be remedied by shunting the capacitor by a feedback resistor R_F as shown in Figure 3.68. This resistor stabilizes the transimpedance and consequently also the source-voltage gain. The equation using finite g_v-values

$$\frac{v_O}{v_S} = -\left|g_v(j\omega)\right| \times \frac{1}{1 + \left(j\omega C_F + \frac{1}{R_F}\right) \times \left(1 + \left|g_v(j\omega)\right|\right) \times R_S} \tag{3.139}$$

degenerates in the ideal case $(1/g_v(j\omega) = 0)$ to

$$\frac{v_O}{v_S} = -\frac{R_F}{R_S} \times \frac{1}{1 + j\omega R_F C_F} \,. \tag{3.140}$$

At lower frequencies, the second term is irrelevant, so that the voltage gain is given by the ratio of two resistors. For frequencies beyond the corner frequency $\omega_{uF} = 1/R_F C$ the roll-off with 20 dB per decade reflects this low-pass filter property. Figure 3.69 compares the Bode plots of the operational amplifier, of the ideal active low-pass filter, of the simple low-pass filter and of a low-pass filter with an additional feedback resistor.

An important application of an inverting amplifier with parallel–parallel feedback by means of a capacitor is the *charge sensitive amplifier* as shown in Figure 3.70. It is used in connection with detectors that respond to radiation of all kind by producing free charge, e.g., photo-electrons. As the negative parallel–parallel feedback effectively forces all the input current i_i to flow into the capacity C_F, the output voltage v_o does not differ much from the voltage v_C across C_F (the inverting input of the amplifier is at virtual ground)

$$v_o \approx v_C = \frac{Q}{C_F} = \frac{1}{C_F} \times \int i_C \, dt \,. \tag{3.141}$$

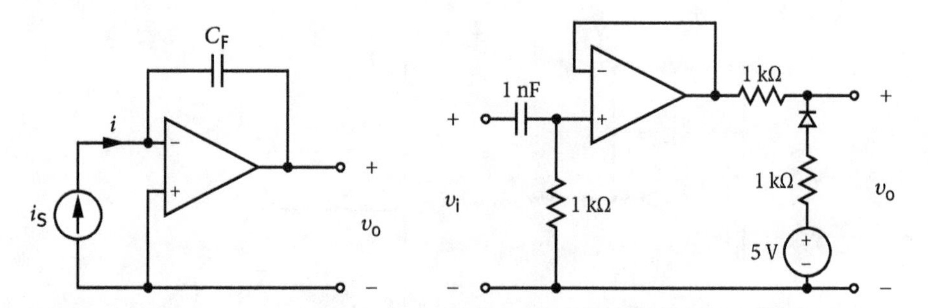

Fig. 3.70. Principle of a charge sensitive amplifier.

Fig. 3.71. Circuit of Problem 3.87.

Although the principle is simple, much know-how is involved in the design of a practical charge-sensitive amplifier, in particular when a low-noise instrument is needed.

If i_C is constant (e.g., provided by an invariable current source), one gets

$$v_C(t) = \frac{i_C}{C_F} t \,. \tag{3.142}$$

The formation of the ensuing voltage ramp is discussed in Section 3.2.1. This relation is used in all circuits for the conversion of voltage amplitude to a time interval and vice versa (Section 4.5).

Problem

3.87. A single rectangular pulse (5 V high, 0.1 ms long) is fed into the input of the circuit of Figure 3.71. The amplifier has a dynamic range of ±10 V, the diode is assumed to be ideal.

(a) What is the maximum v_{omax} and what the minimum voltage v_{omin} at the output?

(b) What is the quiescent output operating voltage?

3.6.4.2 Active high-pass filter

Figure 3.72 shows a simple active high-pass filter using an operational amplifier. In this case $Z_F(j\omega) = -R_F$ and $Z_S(j\omega) = \frac{1}{j\omega C_S}$. Then, according to (3.135) the voltage

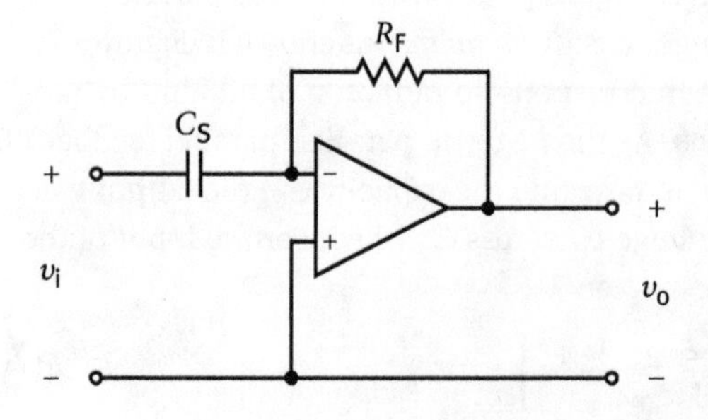

Fig. 3.72. Principle of an active high-pass filter.

transfer from the source to the output is given by

$$\frac{v_o}{v_S} = -\left|g_v(j\omega)\right| \times \frac{1}{1 + \frac{1}{j\omega C_S} \times \left(1 + \left|g_v(j\omega)\right|\right) \times \frac{1}{R_F}}$$
$$= -\left|g_v(j\omega)\right| \times \frac{1}{1 + \frac{1}{j\omega\tau_F}} \tag{3.143}$$

with $\tau_F = C_S \times \frac{R_F}{1+|g_v(\omega=0)|}$ the effective time constant τ_F takes into account the dynamic decrease of R_F due to the Miller effect (Section 2.4.3.2).

With an ideal operational amplifier (with $1/g_v(\omega) = 0$), the voltage transfer from source to output is given by

$$\frac{v_o}{v_S} = -j\omega R_F C_S \tag{3.144}$$

which means that it is proportional to the angular frequency ω for all ω values, i.e., its increase with frequency is 20 dB per decade in the full frequency range. Such an ideal filter has differentiating properties (Section 3.2.1). Thus, the output voltage is proportional to the derivative of the voltage across the capacitor. For sinusoidal signals, the minus sign, which stems from the negative gain of the operational amplifier, is tantamount to a phase difference of 180° between the output and the input signal.

A real active high-pass filter has the combined frequency response of the operational amplifier with an upper corner frequency ω_u and that of a high-pass filter with a lower corner frequency $\omega_l = 1/\tau_F$. If the frequency response of the amplifier is adequate ($\omega_u \gg 10\omega_l$), then its frequency response has no impact, and the total lower frequency response is that of a high-pass filter with τ_F, multiplied by the low frequency voltage gain of the amplifier $-|g_v(\omega = 0)|$ up to about $0.1\omega_u$. Actually, the active high-pass filter is a band-pass filter with a bandwidth $BW = \omega_u - \omega_l$. Thus, to perform properly up to higher frequencies an active high-pass filter needs an operational amplifier with a particularly high upper corner frequency ω_u, i.e., a wide bandwidth.

In Figure 3.73b, the frequency response of the amplitude transfer is shown for several active high-pass filter arrangements. If, as shown in Figure 3.73b, ω_u is not high enough, the frequency dependence of the filter does not intersect the flat portion of the frequency response of the amplifier, but the negative slope of −20 dB/decade, i.e., at the corner frequency there is a change in the slope by 40 dB/decade corresponding to two phase shifts of 90° each, indicating the danger of instability, because of positive feedback (Section 3.6.3.2). In addition, for angular frequencies above ω_l there is no practical benefit of negative feedback action. To counteract this situation a resistor R_S can be introduced (Figure 3.73a) resulting in a time constant $\tau_S = R_S \times C_S$. Under ideal conditions ($1/g_v(\omega = 0) = 0$) we get

$$\frac{v_o}{v_S} = -\frac{R_F}{R_S} \times \frac{1}{1 + \frac{1}{j\omega\tau_S}} \,. \tag{3.145}$$

Above the angular frequency $\omega_{IF} = 1/\tau_S$, the gain is $-R_F/R_S$, independent of ω until the curve hits the roll-off of the amplifier curve.

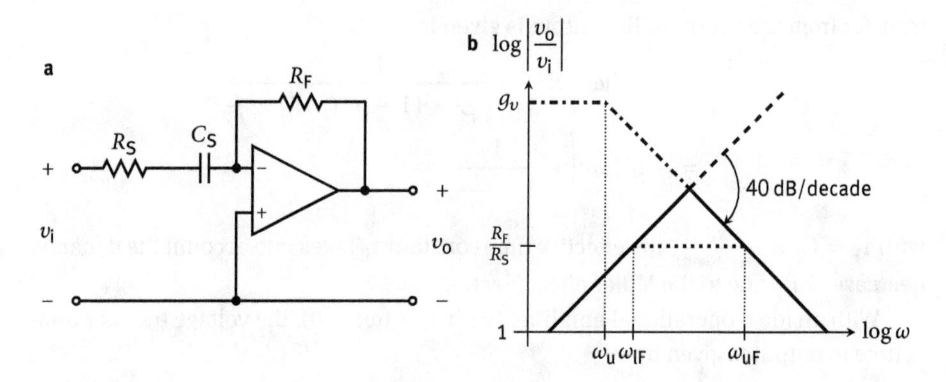

Fig. 3.73. (a) Circuit diagram of an active high-pass filter with added source resistor R_S. (b) Frequency response of the amplitude transfer of the amplifier and several active high-pass filter arrangements.

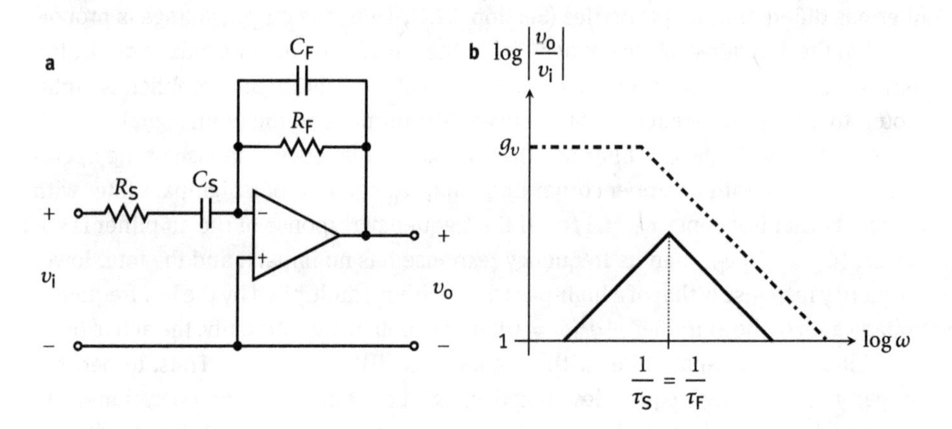

Fig. 3.74. (a) Combined active high-pass low-pass filter (b) Frequency response of the amplitude transfer.

By shunting R_F with a capacitor C_F so that $\tau_F = R_F \times C_F = \tau_S$ a frequency response as shown in Figure 3.74b is obtained. It is a narrow band band-pass filter that below the corner frequency acts as a high-pass filter and above as a low-pass filter. In this case, the gain is determined by feedback over all the frequency range, different from the analogous frequency dependence pictured in Figure 3.73.

Problems

3.88. How does feedback change the product of gain $A(\omega = 0)$ times bandwidth BW?

3.89. Name three advantages of active filters over passive filters (with appropriate amplification).

3.6.5 Dynamic positive feedback (gyrators)

A gyrator is a linear two-port that interrelates the current into one port of a two-port with the voltage on the other port and vice versa. The output variables (across the load) are connected to the input variables by

$$v_L = -R \times i_i \,, \quad \text{and} \tag{3.146}$$

$$i_L = -G \times v_i \tag{3.147}$$

where R is a conversion factor with the dimension of a resistance and $G = 1/R$. Thus, a gyrator is *selfdual*, i.e., it is at the same time its dual counterpart, just like a resistor and a switch. It inverts the current–voltage characteristic of an electronic component or circuit. In the case of linear elements, the inversion of the characteristic results in the inversion of the impedance. It does not only make a capacitor behave as an inductor, but also a parallel LC circuit behave as a series CL circuit which is another feature of duality. It is primarily used in replacing inductors which are bulky, expensive and are in their electric properties much farther away from an ideal inductance than a capacitor from an ideal capacitance.

In Figure 3.75, the realization of a gyrator by means of two negative impedance converters (NICs, Section 2.5.4.1) is shown. The analysis of this circuit is best done in three steps.

1. Let us call the impedance at the inverting input of NIC_1 Z_A. Then the input impedance of the circuit is $Z_i = -Z_A$.
2. Let us call the impedance at the inverting input of NIC_2 $Z_B = -R$.
3. Let us introduce Z_C which is Z_B shunted by $(R + Z_L)$

   $Z_C = -(R + Z_L) \times \dfrac{R}{Z_L}$.

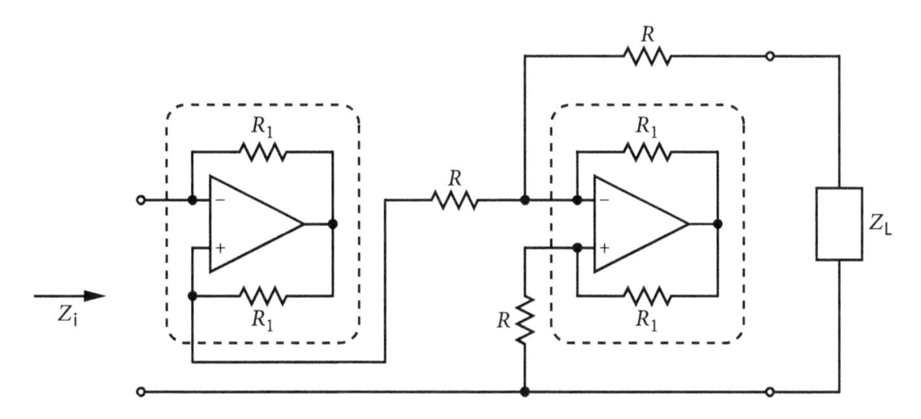

Fig. 3.75. Example of a gyrator based on two NICs.

By inspection, it is clear that Z_A is the resistor $R + Z_C$ and consequently

$$Z_i = -Z_A = \frac{R^2}{Z_L}\,. \tag{3.148}$$

In case that $Z_L = 1/j\omega C_L$ the input impedance Z_i behaves as an inductance $L_i = j\omega R^2 C_L$. This inductance is not only based on a capacitor, but its value can easily be adjusted by proper choice of the resistors R.

Problem

3.90. If the load of a gyrator is a real inductor with series resistance and parallel capacitance, what stray properties of the capacitance produced by this gyrator are expected?

4 Time and frequency (oscillators)

Time t always means time difference. Its unit is the second (s). Time is an analog variable just as its inverse, the frequency f. Frequency is measured in units of hertz (Hz). As the trigonometric functions require that the angle is given in radians, we use the angular frequency $\omega = 2\pi f$. Its unit is (s^{-1}). As in everyday life, *time* has several faces also in electronics. There is

- the *absolute time* which is actually relative to *General Mean Time* (GMT),
- the time difference within a signal (rise time, fall time, signal length, etc.)
- the time difference between signals (*synchronization*), and
- the minimum time interval that defines *simultaneity*.

Obviously, this boils down to two kinds of time differences (intervals), those within a signal and those between two (or more) signals.

The question of time within a steady-state signal can be reduced to timing within one *cycle* (*period*). For sinusoidal signals often the phase within this cycle is used instead of the fraction of the cycle time. Thus, phase is another analog variable of a signal.

When comparing the timing of two or more signals the question of simultaneity is reduced to a question of *coincidence*. Two signals are said to be coincident if they occur within a finite *resolution time* which is subject to the speed of the electronics involved. Obviously, if signals are coincident, it does not mean that the events generating these signals are simultaneous. However, *simultaneity* itself can never be definitely assured because, for practical purposes, a time difference of zero does not exist (experimentally).

Observe that a time difference is an analog quantity.

Timing within one signal has been covered at several places. It is connected with the following terms: rise time, fall time, propagation delay, signal length, phase shift, etc.

The establishment of time marks representing the instant of an event requires *triggers* which are based on a relaxation oscillator as we will see in this chapter.

Time (intervals) can be measured (relatively) to the highest precision of all physical quantities as frequencies can be measured precisely by counting. The precision with which time can be measured is better than 10^{-16} when using an ytterbium clock. This number is so small that only a comparison can help. The (relative) uncertainty of the age of the universe would be less than half a minute.

As time is a conjugate variable of the angular frequency ω (which is the basis for the Fourier transform, Section 3.1.1) time and frequency belong together. Thus, as a first step we will investigate how repetitive signals are produced by so-called frequency generators. Even if harmonic oscillations supplying the base for Fourier analysis are most important, we will start with rectangular oscillations because they are less sophisticated.

Problems

4.1. Is the mains frequency of 60 Hz (or 50 Hz) an analog or a digital quantity?

4.2. Name the analog variables that can be related to signals.

4.3. Has an isolated digital quantity (an integer) an uncertainty?

4.1 Degenerated amplifier output at two levels (hysteresis)

In Section 2.5.4, we discussed stable positive feedback with a closed-loop gain of < 1. Figure 4.1 shows an operational amplifier with positive series–parallel feedback. The characteristic transfer parameter for such an operation amplifier is the voltage gain g_{vF}. From our general discussions on feedback, we get its relation to the voltage gain of the amplifier g_{vA} as

$$g_{vF} = \frac{g_{vA}}{1 - AB} = \frac{g_{vA}}{1 - g_{vA} \times \frac{R_1}{R_1 + R_F}} . \tag{4.1}$$

Figure 2.47a shows the linearized voltage transfer function for $R_1 = 0$ (no feedback, $AB = 0$, dotted line), for $R_1 < \frac{R_F}{|g_{vA}|-1}$ (for $AB < 1$, dashed line), for $R_1 = \frac{R_F}{|g_{vA}|-1}$ (for $AB = 1$, full line), and for one example with $AB > 1$ (dashed–dotted curve). In the last case there is hysteresis: When the input voltage v_i reaches the upper threshold voltage $v_{thr,u}$ the output voltage v_o jumps from the low output value v_{oL} to its high value v_{oH}. When lowering the input voltage below the lower threshold voltage $v_{thr,l}$ which is lower than $v_{thr,u}$ the output voltage jumps back to v_{oL}. Thus, we obtain a transfer characteristic with a *hysteresis*. Usually the output voltage range is close to the voltage range of the power supply so that $v_{oL} \approx -V_-$, and $v_{oH} \approx +V_+$. Then the width of the hysteresis $v_{thr,u} - v_{thr,l}$ is approximately the following fraction f of the

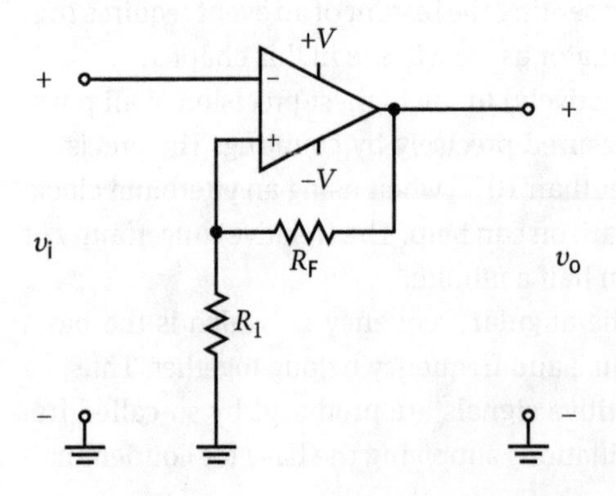

Fig. 4.1. Operational amplifier with positive series–parallel feedback.

output voltage range:

$$f = \frac{1 + AB}{|g_{vA}|} = \frac{R_1}{R_1 + R_F} - \frac{1}{g_{vA}} . \tag{4.2}$$

By means of the resistor R_1 the closed-loop gain and, therefore, the width of the hysteresis can be adjusted.

Amplifiers that, with the help of positive feedback, have only two output states (low L and high H) are called relaxation oscillators. The time duration of the transition between the states can be very small and is disregarded. Comparing the transfer functions of Figure 2.47a leads to the following conclusions:

- Without positive feedback, there exists an input range in which the output is somewhere between L and H. Such a gap can be very small if the gain is high (comparator), but it is not tolerable when only two states are allowed.
- The limiting case with a positive feedback having a closed-loop gain of 1 gives theoretically the wanted performance but is utterly impractical. Firstly, it is impossible to set the closed-loop gain exactly to 1 (and to keep it at this value). As soon as it is below 1, stable positive feedback sets in (Section 2.4.2). Therefore, a gain margin is necessary to ensure a starting value of the closed-loop gain > 1. Secondly, an input signal that would have exactly the voltage of the threshold would confuse the circuit because its fluctuations (noise) would make the output jump erratically between H and L. To avoid this erratic behavior, the *slew rate* of the input signal passing the threshold must be fast enough to ensure a clean output transition.
- Thus, only a hysteresis can provide a circuit having an output with exactly two well-defined states. The possibility to adjust the width of the hysteresis (by means of the closed-loop gain) is an important factor: the width must be wide enough to ensure that input noise does not get confused with a signal and, on the other hand, small enough so that the difference between the two thresholds does not disturb.

A hysteresis in the transfer function is the key for two well-defined output levels.

A circuit of an amplifier with stable positive feedback is called *comparator* (Section 2.3.4.2), if the transfer function has a hysteresis it is called *Schmitt trigger* (Section 4.1.2.1). Thus, when using three-terminal amplifier stages two stages (either two common-emitter circuits or a common base and a common-collector circuit) are necessary for a realization of the *noninverting voltage amplifier* needed for positive feedback.

Problem

4.4. What is the difference between a comparator and a Schmitt trigger? ?

4.1.1 Introduction to relaxation oscillators

A relaxation oscillator is a circuit that returns to equilibrium conditions after being disturbed. After each perturbation, the system relaxes until the next disturbance occurs.

An (operational) amplifier with (unstable) positive feedback is just such a system. Any effective disturbance will result in a jump from one output state (H or L) to the opposite (L or H). As listed in Table 4.1, we have to consider two times three cases, depending on the quiescent operating point and the frequency dependence of the feedback circuit.

There are two distinctly different situations regarding the quiescent operating point.

- If it is so situated on the transfer characteristic, that the gain is insufficient to make the closed-loop gain > 1, e.g., on the L or H branch of the transfer curve, we have the situation of a class C amplifier (Section 2.3.4). It takes an input signal of the correct polarity and sufficient amplitude to bring such an amplifier into the amplifier range.
- If there is no unique quiescent operating point, this operating point is either not stable or it may lie on either branch of the transfer curve (L or H).

Then we have to distinguish between three cases of frequency dependence of the closed-loop transfer function. Cases (a) and (b) are relaxation oscillators whereas case (c) are harmonic oscillators.

(a) The closed-loop gain is larger than 1 already for $\omega = 0$ (bistable circuit and Schmitt trigger).

(b) The closed-loop gain is larger than 1 for angular frequencies $\omega > \omega_0$. (Astable and monostable multivibrator)

(c) The closed-loop gain is larger than 1 just at one angular frequency $\omega_0 \neq 0$ (free running or gated harmonic oscillator).

Quite obviously group (a) cannot be realized with a transformer in the feedback loop as transformers do not transfer signals with $\omega = 0$. Circuits of group (b) that apply pos-

Table 4.1. List of positive feedback oscillator circuits.

Quiescent operating point	ω values for which $AB \geq 1$	Circuit type
Unique and stable	$\omega \geq 0$	Schmitt trigger
	$\omega = \omega_0$	Gated oscillator
	$\omega \geq \omega_0$	Monostable
Not unique or not stable	$\omega \geq 0$	Bistable
	$\omega = \omega_0$	Free oscillator
	$\omega \geq \omega_0$	Astable

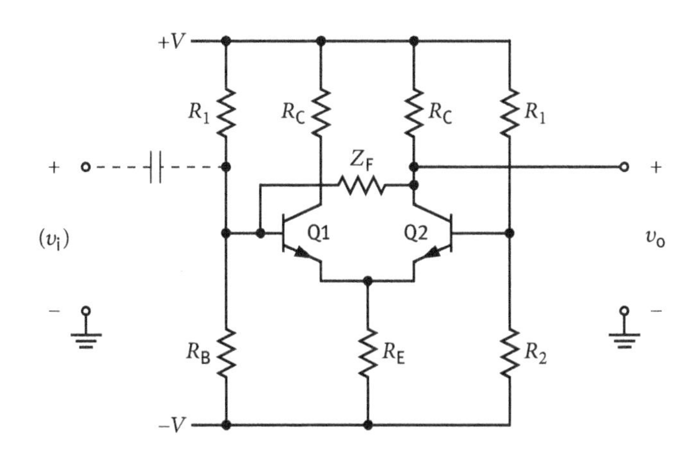

Fig. 4.2. All six types of relaxation oscillators can be realized with a long-tailed pair circuit.

itive feedback using transformers are called *blocking oscillators*. Figure 4.2 presents a long-tailed pair circuit that with appropriate circuitry can operate in any of these two times three modes. By appropriate biasing (choice of R_B), the two stages are either symmetrically (current) biased, or one of them works as C class amplifier. Making the feedback impedance Z_F either resistive, or capacitive, or in the form of a series resonant circuit (Section 3.4.5.1) the threefold variety is achieved.

Problem

4.5. What kind of (positive) feedback is active in the circuit of Figure 4.2?

4.1.2 Gated (asymmetric) relaxation oscillators

When an amplifier is of class C, a gating signal is required to take the operating point into the amplification region. Depending on the position of the quiescent operating point (L or H) a positive or negative signal is required. Such a signal may be very short, just the amplitude (the amount of charge) must be sufficient.

Problem

4.6. Does the gating signal of a class C amplifier have positive or negative polarity?

4.1.2.1 Schmitt trigger

A Schmitt trigger is a comparator (Section 2.3.4.2) with hysteresis. Consequently it is often called comparator (also in data sheets). As discussed, the addition of hysteresis to a comparator response becomes necessary in a noisy environment to prevent the circuit from toggling between L and H (when the input signal stays at the switching threshold for a while). Noise will not cause a Schmitt trigger to switch states unless it is larger than the width of the hysteresis. A major problem can arise if L and H are not symmetrical around zero, resulting in a hysteresis that is not symmetrical around the

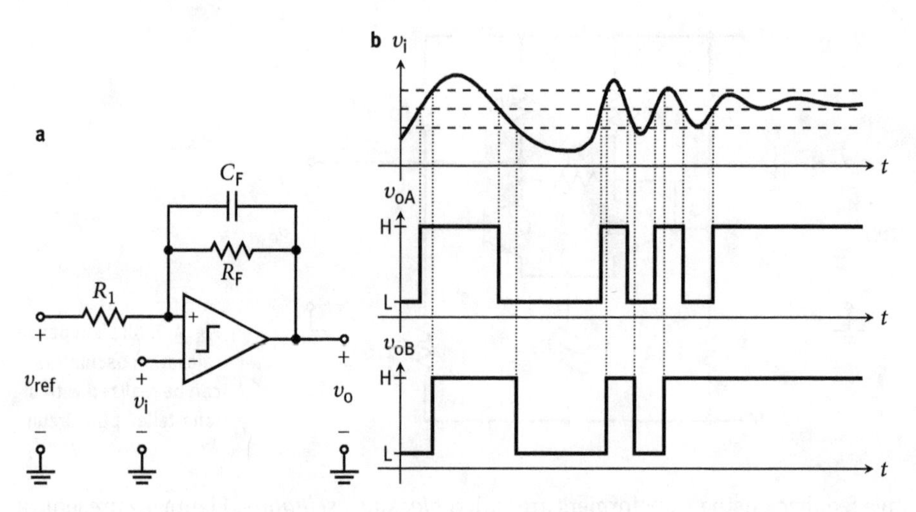

Fig. 4.3. (a) Basic circuit of a Schmitt trigger (C_F is just for frequency compensation) (b) Time dependence of some input voltage v_i and the resulting output voltages of a comparator v_{oA} (i.e., with zero-width hysteresis) and a Schmitt trigger v_{oB} (with a hysteresis).

nominal threshold voltage v_{ithr}. (See discussion in Section 4.1 regarding the width of the hysteresis.)

Figure 4.3a shows the circuit of a Schmitt trigger using an integrated comparator as an active element. The input signal is applied at the inverting input, the purely resistive positive feedback goes to the noninverting input. The quiescent voltage at the noninverting input (the common mode voltage) stems from the reference voltage v_{ref} which establishes the threshold voltage. A speed-up capacitor C_F across the feedback resistor R_F speeds up the operation (reduces the rise-time) by increasing the closed-loop gain at higher frequencies. It compensates for the effect of the input impedance of the amplifier (Section 3.6.3.3).

In Figure 4.3b the response of the Schmitt trigger to some input voltage is sketched. The middle line is for a dummy zero-width hysteresis, the other two lines indicate the width of the hysteresis. This figure is self-explaining.

Clearly, the signal input voltage that may be applied to a comparator (Schmitt trigger) must be within limits so that the circuit is not destroyed. Usually, a dynamic input range from rail-to-rail (i.e., within the supply voltage(s)) can be expected. The same is true for the reference input voltage.

The primary use of a comparator (and a Schmitt trigger) is as *discriminator* which decides, whether an (analog) input signal is larger than a given value (the threshold voltage) or not. This decision is clearly a binary decision resulting in an output signal of H or L. In this respect, the Schmitt trigger is a link between analog and digital (binary) electronics.

An alternative use is as *trigger* (Section 4.4.1). In this case, the emphasis lies on the other analog property of a signal, its time of occurrence.

An important application, in particular in high speed computing, is the *recovery* of clock timing signals. High speed signals transmitted even over a rather short distance become deformed due to stray reactances. Application of a high speed Schmitt trigger recovers the waveform while maintaining a minimum of delay. Nowadays, such fast circuits have a resolution time on the order of 10^{-1} ns.

Problem

4.7. Which property of the circuit determines the size of the hysteresis?

4.1.2.2 Monostable multivibrator (one-shot)

The name *multivibrator* was originally applied to the astable, free-running version of the circuit. Nowadays, there are three types of multivibrator circuits: astable, monostable, and bistable ones, depending on the number of quiescent operating points: none, one, or two.

The monostable multivibrator has one stable state (L or H) which is determined by the quiescent operating point, with the other state (H or L) transient. An input signal causes the circuit to move the operating point into the amplifier region so that the positive feedback becomes active and pushes the operating point into the unstable state. From the unstable state, the circuit will relax into the stable state after a preselected time. Thus, the term relaxation oscillator. This circuit is also known as a *one-shot* multivibrator.

The traditional circuit of a monostable multivibrator is shown in Figure 4.4. It uses two npn-junction transistors. In the quiescent state Q_1 is saturated, its collector is in the L state. Q_2 is cut-off, its collector is in the H state. Applying a negative signal to the base of Q_1 gives a positive signal at the collector of this transistor which opens Q_2 lowering the collector voltage of Q_2. This signal is fed through C_1 to the base of Q_1 rein-

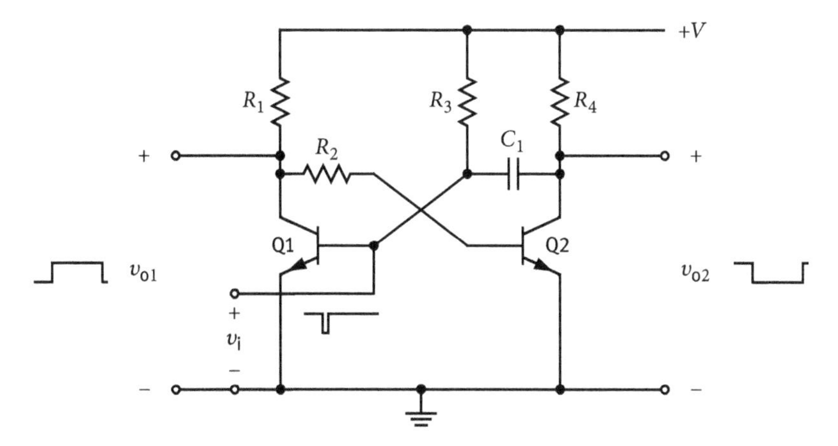

Fig. 4.4. Traditional circuit of a monostable multivibrator with two npn transistors.

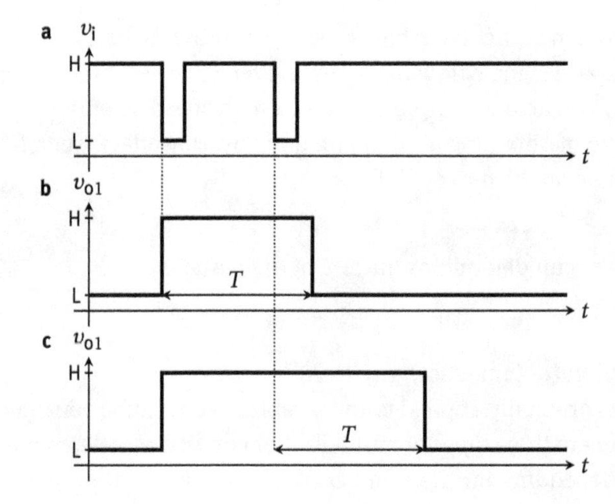

Fig. 4.5. Resolution time of a standard monostable multivibrator (a) input signal (b) standard output signal, and (c) output signal when second pulse restarts the charging process.

forcing the input signal. If the returned signal is larger than the input signal, we have the case of an unstable positive feedback (closed-loop gain > 1). The output signal of Q_1 switches to H terminating the positive feedback action due to zero gain (at $i_{C1} \approx 0$). However, C_1 got charged during the signal transmission to the difference between H and L (see Section 3.3.2) so that with the collector of Q_2 in the L state, v_{BE} of Q_1 is negative, keeping Q_1 in the H state. Only after C_1 has sufficiently discharged through R_3, the operating point of Q_1 will be back in the amplifier region. Therefore, feedback action is again possible, this time for a positive signal at the base of Q_1, bringing the collector down to L and that of Q_2 up to H, i.e., the quiescent condition is reestablished. Thus, the length of the transient state is determined by the time constant R_3C_1 (disregarding the small output impedance of Q_2 which is saturated).

In Figure 4.5 is shown what happens if during the discharge time of C_1 a new negative input signal occurs. Obviously, the discharging process of the capacitor is terminated, it becomes charged again, and the discharging starts anew. As can be seen from Figure 4.5c, the output signal v_o has not its standard length (shown in Figure 4.5b), which is determined by the time constant. The time difference between the two pulses is added to the standard length. Thus, the *resolution time* of standard monostable multivibrators is about the length of one standard output signal.

In Figure 4.2 the long-tailed pair version of a monostable multivibrator is shown with $Z_F = C_F$.

In Figure 4.6 a monostable multivibrator based on an operational amplifier is shown. At the input there is a high pass that provides a short negative spike to trigger the one shot.

To avoid the loss of signals during the resolution time of a standard monostable multivibrator one can make the circuit retriggerable by forcing the discharge of the timing capacitor each time an input signal arrives, e.g., by means of an electronic switch shunting the capacitor. Thus, in a *retriggerable* monostable, an additional input pulse received during the resolution time initiates a new signal of standard length.

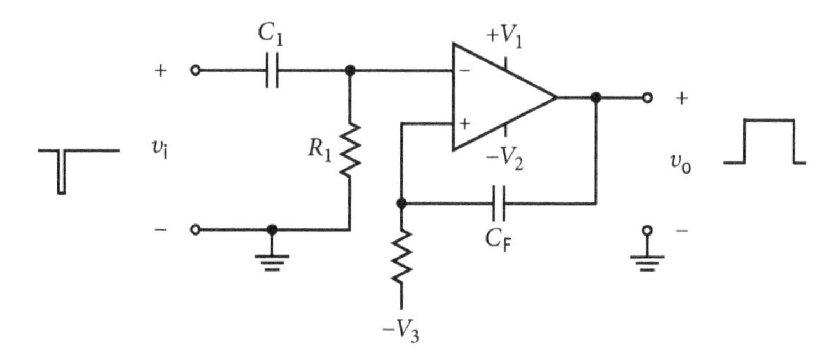

Fig. 4.6. Monostable multivibrator based on an operational amplifier.

One important application of monostable multivibrators is the *debouncing* of switches (Section 3.3.1). As mechanical oscillations (of small parts) have a frequency on the order of 10^{-1} kHz, the time constant of such a debouncing one-shot is of the order of 10^{1} ms. More elegant is the use of an SR flip–flop (Section 5.3.2.1).

Problem

4.8. Is there bouncing in mercury switches?

4.1.3 Symmetric relaxation circuits

Symmetric relaxation circuits do not have a preferred quiescent operating point condition. It is L or H or neither.

4.1.3.1 Astable multivibrator

An astable multivibrator has no stable operating point, in either state. It switches repeatedly from the L state to the H state and back. As the circuit operates in the amplifier region it does not require an input signal to take it there.

Figure 4.7 shows the traditional astable multivibrator circuit with two common-emitter stages. This circuit is strictly symmetric. Whenever the collector of Q_1 is H that of Q_2 is L, and vice versa.

The capacitor C_1 will be charged with the time constant R_3C_1 which determines the duration of the L state (at output 1). The duration of the H state (at output 1) depends on the time constant R_2C_2 responsible for charging C_2. A duty cycle different from 50% can easily be achieved by using different time constants. From the approximate duration t_1 of state L and t_2 of state H we get the total period T of the oscillation as

$$T = t_1 + t_2 = \ln 2 \times R_3C_1 + \ln 2 \times R_2C_2 \,, \tag{4.3}$$

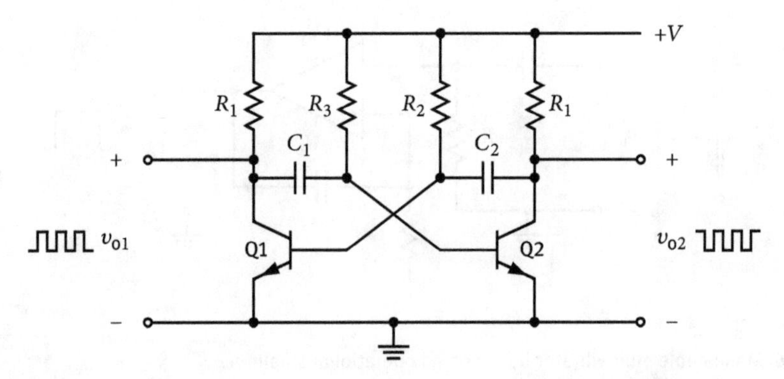

Fig. 4.7. Traditional circuit of a symmetric astable multivibrator using npn transistors.

and the repetition frequency as

$$f = \frac{1}{T} = \frac{1}{\ln 2 \times (R_2C_1 + R_3C_2)} \approx \frac{1}{0.693 \times (R_2C_1 + R_3C_2)} . \tag{4.4}$$

For a 50% duty cycle both time constants τ must be the same $\tau = RC$ so that the frequency becomes

$$f = \frac{1}{T} = \frac{1}{\ln 2 \times 2RC} \approx \frac{0.721}{RC} . \tag{4.5}$$

In Figure 4.8, the circuit diagram of a symmetric astable multivibrator based on an operational amplifier is given. Aside from the obvious use as clock generators, astable multivibrators easily synchronize with some reference frequency. A free-running astable multivibrator accurately locks to a reference frequency which may be two to 10 times higher than its basic frequency, thus, dividing this reference frequency.

In Figure 4.2, the long-tailed pair version of an astable multivibrator is shown, when a capacitor C_F as feedback impedance Z_F is used and the current biasing of the two stages is symmetric.

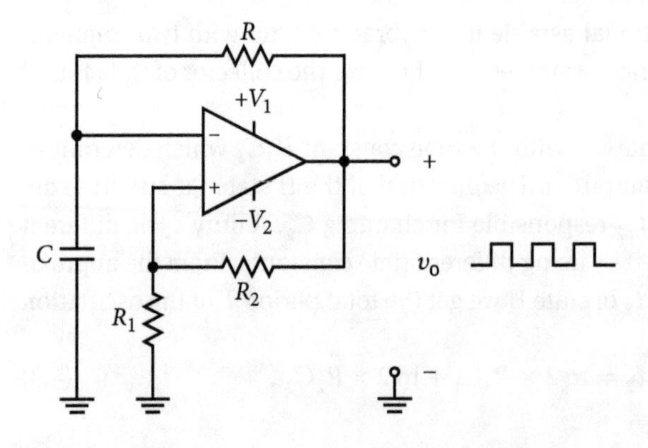

Fig. 4.8. Circuit diagram of a symmetric astable multivibrator based on an operational amplifier.

Problems

?

4.9. Give the duty factor of the output signal delivered by an astable multivibrator.

4.10. Under what condition does the duty factor of an astable multivibrator equal 0.5?

4.1.3.2 Bistable multivibrator

A bistable multivibrator is a circuit that can stay indefinitely long in either state. By an external signal the state of the circuit can be flipped, i.e., reversed. Therefore, this circuit is also called *flip–flop*. By substituting the capacitors of an astable multivibrator by resistors one arrives at a bistable multivibrator (see Figure 4.9). Consequently the positive feedback works already for $\omega = 0$. This basic circuit needs two inputs: a (positive) signal at the set input S (at the base of Q_1) makes the output v_{o1} H, one at the reset input R (at the base of Q_2) makes the output v_{o1} L.

R_{F2} and R_{F1} may be shunted by speed-up capacitors, increasing the closed-loop gain at higher frequencies, thus speeding up the flipping. This *frequency compensation* of the voltage divider compensates for the slowing down effect of the input capacitance of the transistors (see Section 3.6.3.3) decreasing the transition time (rise time of the transitions).

Example 4.1 (Elementary analysis of the transition in a flip–flop from L to H). To flip the bistable multivibrator of Figure 4.9 into the X = H state a positive signal must be applied to the S(et) input. The starting condition (X = L state) is that Q_2 is saturated and Q_1 in the off state. Both active elements are class C amplifiers, i.e., not in the amplifier region. The open-loop gain is zero (very small) and therefore, the closed-loop gain as well. The loop for a *set* signal goes from the base of Q_1 to its collector then via R_{F2} to the base of Q_2. Then from its collector via R_{F1} back to the base of Q_1. Taking X as the output (at the collector of Q_2), we have a positive parallel–parallel feedback with R_{F1} as feedback element. As mentioned, speed-up capacitors (C_{F1} and C_{F2}) shunting the feedback resistors compensate for the effect of the input capacitances of the transistors by frequency compensation (Section 3.6.3.3).

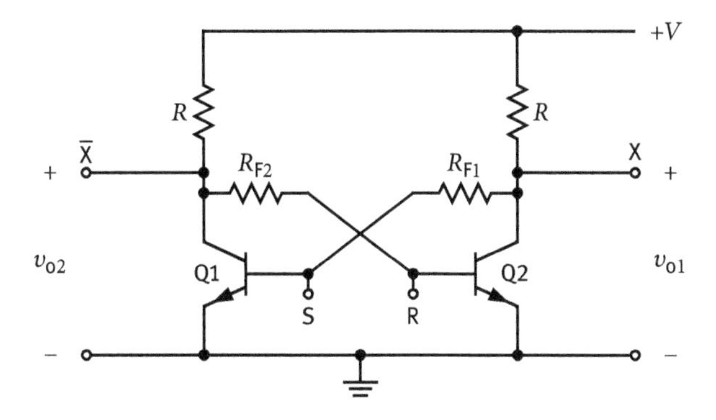

Fig. 4.9. Circuit of an SR bistable multivibrator using npn transistors.

A positive base current of sufficient amplitude (e.g., from the H state of another circuit) at the S input takes Q_1 into the amplifier region producing an amplified negative voltage signal at its collector which gets to the base of Q_2. By the negative signal at its base Q_2 gets into the amplifier region and produces a positive signal at its collector. Via the feedback resistor R_{F1}, this positive signal increases the original input signal by parallel feedback so that Q_1 is driven into saturation. This makes the input voltage v_{BE2} of Q_2 so small that no collector current flows and the output is in the H state. As in the beginning, both transistors are now class C amplifiers with negligible open-loop gain, there is no feedback any more. The speed-up capacitors do not only make the transition time shorter by increasing the amplitude of fed back high frequencies but they must get discharged before the next transition occurs (see Section 3.3.2), introducing a minimum resolution time. Therefore, their values should not be larger than frequency compensation (Section 3.6.3.3) requires.

In Figure 4.2 the long-tailed pair version of a bistable multivibrator is shown, when a resistor R_F as feedback impedance Z_F is used and the current biasing of the two stages is symmetric.

Flip–flops are elementary circuits in binary electronics and will be covered in Chapter 5.

Speed-up capacitors in the feedback branch decrease the rise time of the output signals due to frequency compensation.

?

Problems

4.11. Identify the loop for a reset signal in the circuit of Figure 4.9.

4.12. Which input variable, current or voltage, does the positive feedback in Figure 4.9 increase?

4.1.4 Relaxation oscillators involving transformers

As an open-loop gain not much greater than 1 suffices to yield a closed-loop gain ≥ 1, a single-stage amplifier (with positive gain) is adequate. In the case of a common-emitter circuit, the voltage gain is negative necessitating an inverting feedback circuit to provide positive feedback. The only passive component that can provide signal inversion is a *transformer*. Also in the case of a common-collector circuit a transformer is necessary to boost the open-loop voltage gain beyond 1. However, a transformer cannot transfer direct current, i.e., frequency components with $\omega = 0$. Thus, neither an equivalent to a bistable multivibrator nor a Schmitt trigger can be realized by such circuits. Traditionally, relaxation oscillators having a transformer in the feedback circuit are called *blocking oscillators*. The minimal configuration of a free-running blocking oscillator requires only a few discrete electronic components namely a three terminal element as amplifier, one or a few resistors for biasing it, and a transformer. The name stems from the blocking action at the three-terminal element (it is driven into the cut-off region, resulting in an H signal). A blocking oscillator is another binary circuit because it switches between H and L. The transformer is the vital component of a blocking oscillator. A pulse transformer with excellent high-frequency transmission is required to create rectangular pulses with fast rise and fall times.

Blocking oscillators used to be important circuits in TV monitors employing step-up high voltage transformers.

Transformers do not transmit DC components of signals.

Problem

4.13. Why do transformers with an iron core not qualify for use in relaxation oscillators?

4.1.4.1 Monostable (triggered) blocking oscillator

In Figure 4.10, the circuit of a monostable version of a blocking oscillator is shown. R_E ensures that without a signal the transistor Q is in the cut-off state due to the $v_B \approx 0$ biasing by R_1. A positive spike transmitted through C_1 takes Q into the amplifier region. The reverse biased diode D_2 which shunts the input impedance of the transistor is in this moment irrelevant because of its high impedance. The voltage drop, due to the increased collector current in Q and the high impedance of the primary windings of the pulse transformer for a fast signal (for high frequencies), is transformed by the transformer into a positive voltage signal in the secondary windings. This signal opens

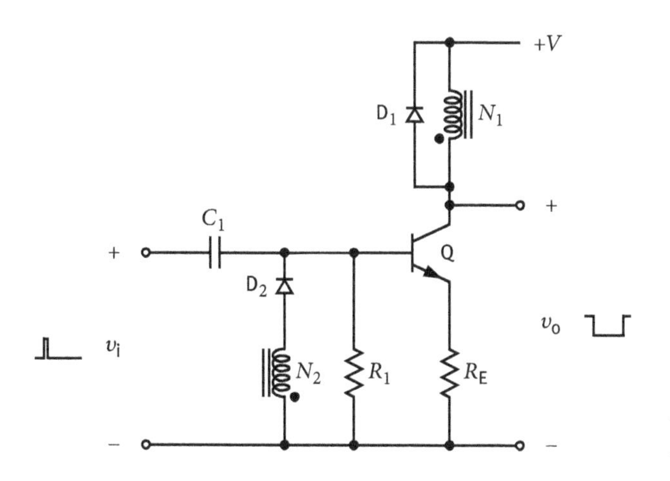

Fig. 4.10. Monostable blocking oscillator using a common emitter bipolar junction transistor.

D_2 and positive feedback reinforces the input signal so that Q gets fully opened resulting in the L state of the output. (The transformer's "winding sense" with regard to the magnetic flux is indicated by the heavy dot, i.e., with reference to this dot, voltage changes go the same direction.) As the inductance discharges with the time constant L/R the primary (and the secondary) current decreases until the L state at the output cannot be maintained anymore and the output voltage moves toward H. This positive signal is transformed to a negative signal in the secondary windings reverse biasing the diode and making $v_{BE} \leq 0$. Thus, Q is cut-off again resulting in an H output voltage. D_1 protects Q against a too high collector voltage (Section 2.2.7.2). It connects the output to V_+ if the output voltage rises beyond V_+ due to a voltage spike produced by the inductance of the transformer as a result of a fast current change. The pulse length T, i.e., the duration of the L state, is approximately given by

$$T \approx \ln\left(1 + \frac{n \times R}{r_e + R_E}\right) \times \frac{L}{R} \tag{4.6}$$

with L the inductance of the transformer, n the ratio of the number of secondary windings over that of the primary ones, R the resistance over which L discharges, i.e., the sum of the parasitic resistance of the windings, the saturation resistance of Q, the internal emitter resistance r_e, and the external emitter resistance R_E.

Problems

4.14. Why is there no bistable blocking oscillator?

4.15. Why is there no blocking oscillator with the function of a Schmitt trigger?

4.1.4.2 Astable blocking oscillator

Figure 4.11 shows the simplified circuit diagram of an astable version of a blocking oscillator. Except for the biasing it is very similar to that of the monostable version. Q is biased by R such that its operating point is in the amplifier region. C is a decoupling capacitor that ensures that the operating point is not influenced by the low

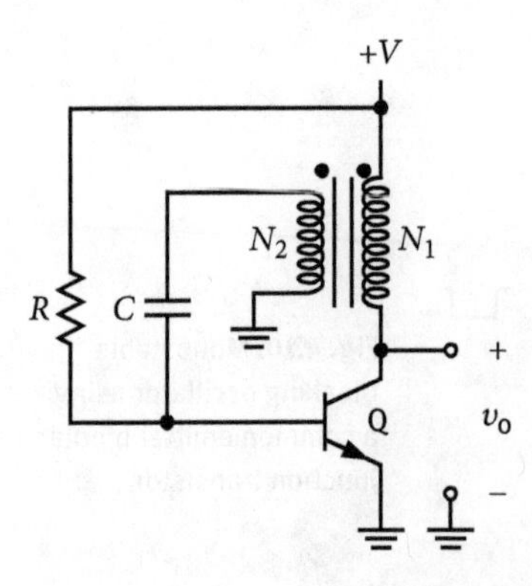

Fig. 4.11. Circuit diagram showing the essentials of an astable blocking oscillator based on an npn transistor.

resistance of the secondary windings of the transformer. Any disturbance (noise) with a frequency component for which the closed-loop gain is ≥ 1 pushes the operating point from the amplifier region either in the saturation region L (if this disturbance has a positive lobe) or into the cut-off region H (if the lobe is negative). From there on the discharging and charging of the inductance makes the output flip from H to L and vice versa, i.e., in this regard the circuit behaves like a symmetric astable multivibrator.

Problem

4.16. Why are blocking oscillators predestined to deliver high voltage output signals?

4.1.5 Switched (astable) delay-line oscillator

The start/stop oscillator in Figure 4.12 uses feedback in a completely different way. The start signal is recirculated from the output of a comparator (Section 2.3.4.2) or a binary OR gate (Section 5.3.1.3) via a delay line (Section 3.5.3.1) to the second input generating the next signal with a time distance determined by the propagation delay in the circuit and the value of the delay in the delay line. A fine adjustment of the period length can be achieved by the low-pass filter consisting of the variable resistor and the capacitor (determining the rise time of the recirculated signal and, therefore, the trigger moment of the comparator). Such an arrangement ensures that the pulse distance is constant between all signals independent of their position in the pulse train. As only delay lines for rather short delay times are practical, such oscillators are particularly well suited for high-frequency oscillators up to 100 MHz and beyond.

Problem

4.17. What is the advantage of delay-line oscillators when a synchronous gated train of equidistant pulses is required?

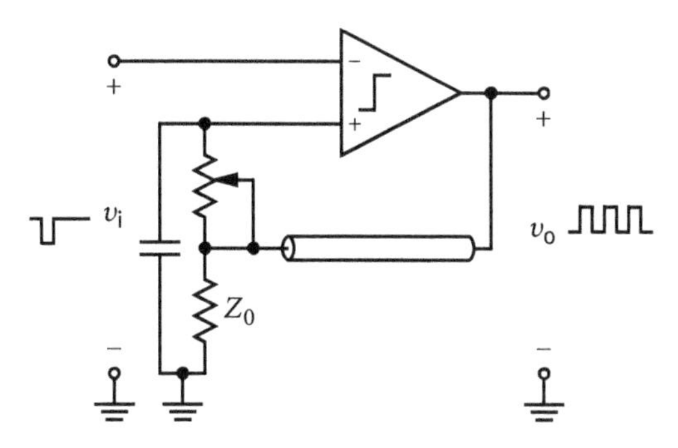

Fig. 4.12. Delay line oscillator using a comparator with fine adjustment of the period.

4.2 Harmonic feedback oscillators

If the output of an oscillator is a (single frequency) sine wave, it is a harmonic oscillator. Reversing Nyquist's stability criterion (Section 3.6.3.2), one arrives at the condition under which a feedback system will oscillate at a distinct frequency ω_0. It will oscillate when the amplitude of the complex closed-loop transfer function is one at this frequency (magnitude criterion), and the phase shift is 0° or a multiple of 360° (phase criterion). As the inversion of a sine wave by an amplifier corresponds to a phase shift of 180° this means that for inverting amplifiers the phase shift of the positive feedback circuit must be 180°. To get a stable oscillation *both* criteria must be fulfilled

- the frequency ω_0 of the oscillation is that frequency at which the total phase shift of the closed-loop gain is zero or a multiple of 360° (frequency condition), and
- (to have a stable oscillation at the frequency ω_0) the amplitude of the closed-loop transfer function at this frequency must be exactly 1 (stability condition).

Table 4.2. Overview over types of (passive) feedback circuitry depending on the polarity of the amplification used for harmonic oscillators.

Sign of amplification	Type of feedback circuitry
Positive	Bridge with differential amplifier
Negative	Cascaded filter sections
Either	Transformer coupling
	LC circuit with split inductance
	LC circuit with split capacitance

4.2.1 Oscillators using three-terminal devices

As the closed-loop gain requirement is very moderate (≥ 1), single-stage amplifiers suffice for the realization of basic harmonic feedback oscillators. However, if signal inversion is detrimental, two stages are necessary. Even if operational amplifiers are readily available (and, maybe, even cheaper than a basic three-terminal component), we do not favor this alternative. Three-terminal components are in most cases the better choice as an active element in *high-frequency oscillators* because of their higher unity-gain frequency.

Problem

4.18. Are integrated operational amplifiers always the best choice when designing an oscillator?

4.2.1.1 Oscillators of Wien-bridge type

These oscillators are based on noninverting amplifiers. For practical reasons, we choose voltage amplifiers and combinations of resistors R with capacitors C as voltage divider elements providing the phase shift. The task is to find (simple) complex voltage dividers that either have *selectively* the phase shift 0° exactly at one frequency (different from $\omega = 0$) or, if the phase shift is zero at more than one frequency selecting one of these frequencies by a maximum in the magnitude of the transfer function. Taking two (complex) impedances Z_1 and Z_2 of the form $Z_k = \mathrm{Re}[Z_k] + \mathrm{j}\mathrm{Im}[Z_k] = R_k + \mathrm{j}X_k$ to form a voltage divider from output to input one gets

$$\frac{v_\mathrm{i}}{v_\mathrm{o}} = \frac{Z_2}{Z_1 + Z_2} = \frac{1}{1 + \dfrac{R_1 + \mathrm{j}X_1}{R_2 + \mathrm{j}X_2}} = \frac{1}{1 + \dfrac{R_1R_2 + X_1X_2 + \mathrm{j} \times (X_1R_2 - X_2R_1)}{R_2^2 + X_2^2}} . \tag{4.7}$$

For the phase shift to be zero the ratio in the denominator must be real, i.e., $X_1R_2 = X_2R_1$ must be true. The trivial solution $X_1 = X_2 = 0$ must be disregarded because then the voltage division by purely resistive components provides zero phase shift for all frequencies and in addition a frequency-independent amplitude. The other trivial solution with a purely reactive voltage division we will investigate later. Assuming that $X_2 \neq 0$ (and $R_2 \neq 0$ because otherwise also $R_1 = 0$ which we excluded) we get

$$\frac{\mathrm{Im}[Z_1]}{\mathrm{Im}[Z_2]} = \frac{\mathrm{Re}[Z_1]}{\mathrm{Re}[Z_2]} . \tag{4.8}$$

As only passive components are used, both real parts are positive and consequently the imaginary parts must be of the same nature. Restricting ourselves to capacitors each Z_k may either be a resistor in series to a capacitor

$$Z_k = R_k \times \left(1 - \frac{\mathrm{j}}{\omega\tau_k}\right) , \tag{4.9}$$

or in parallel

$$Z_k = \frac{R_k}{1 + \omega^2\tau_k^2} \times (1 - \mathrm{j}\omega\tau_k) \tag{4.10}$$

with $\tau_k = R_kC_k$.

If both impedances are either a parallel configuration or a serial one, the voltage division is independent of frequency and not selective. However, mixing the configurations does the trick

$$\frac{\mathrm{Im}[Z_2]}{\mathrm{Im}[Z_1]} = \frac{-\omega \times \tau_2 \times \mathrm{Re}[Z_2]}{-\dfrac{1}{\omega \times C_1}} = \frac{\mathrm{Re}[Z_2]}{\mathrm{Re}[Z_1]} = \frac{\mathrm{Re}[Z_2]}{R_1} . \tag{4.11}$$

It yields a unique frequency ω_0 at which the transfer function is real

$$\omega_0^2 = \frac{1}{\tau_1\tau_2} . \tag{4.12}$$

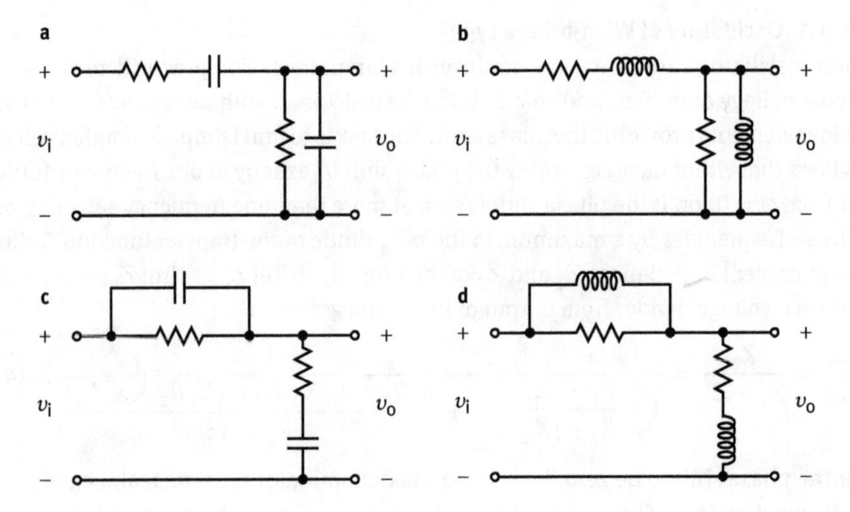

Fig. 4.13. The four candidates for voltage feedback circuitry providing at ω_0 a phase shift of 0°.

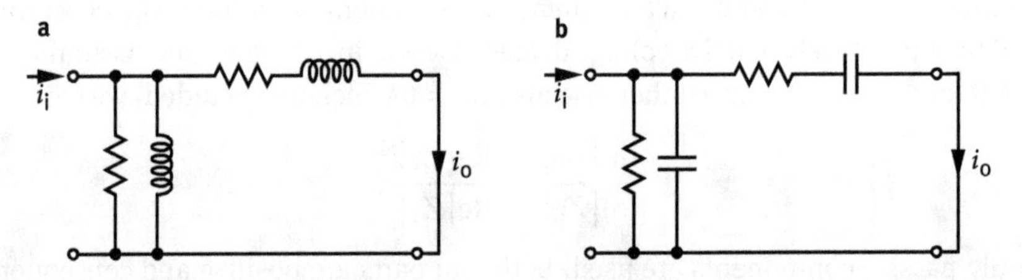

Fig. 4.14. Current dividers providing only at one frequency, at $\omega_0 = 1/\sqrt{\tau_1 \times \tau_2}$ a transfer function with zero phase shift.

This final relation also applies when inductors rather than capacitors are used as reactive components. Figure 4.13 summarizes the four elementary voltage dividers that give for ω_0 zero phase shift. However, as can easily be seen, version (c) and (d) also provide zero phase shift for $\omega = 0$ (with maximum transmission!), so that only (a) and (b) fulfill both conditions. In Figure 4.14, the dual counterparts of Figure 4.13a and b are shown.

From the stability condition with $AB = 1$ at ω_0 we get

$$A(\omega_0) = \frac{1}{B(\omega_0)} = \frac{\mathrm{Re}[Z_1] + \mathrm{Re}[Z_2]}{\mathrm{Re}[Z_2]} = 1 + \frac{\mathrm{Re}[Z_1]}{\mathrm{Re}[Z_2]} \tag{4.13}$$

so that with

$$\mathrm{Re}[Z_1] = R_1 ,$$

$$\mathrm{Re}[Z_2] = \frac{R_2}{1 + \omega_0^2 \tau_2^2} , \quad \text{and}$$

$$\omega_0^2 = \frac{1}{\tau_1 \tau_2}$$

we get

$$A(\omega_0) = 1 + \frac{R_1}{R_2} + \frac{C_2}{C_1} . \tag{4.14}$$

How must R_2 be chosen if R_1 and ω_0 is given so that $A(\omega_0)$ is minimal? The answer to this question is that τ_1 must equal τ_2, making

$$\omega_0 = \frac{1}{\tau} \quad \text{and}$$

$$A(\omega_0) = 1 + \frac{2 \times R_1}{R_2} .$$

For $R_2 = R_1$ the closed-loop gain becomes $A(\omega_0) = 3$.

In Figure 4.15, the frequency dependence of the amplitude

$$a(\omega)b(\omega) = \frac{3}{\sqrt{7 + \omega^2\tau^2 + \frac{1}{\omega^2\tau^2}}} \tag{4.15}$$

and of the phase shift

$$\varphi(\omega) = \arctan \frac{\frac{1}{\omega\tau} - \omega\tau}{3} \tag{4.16}$$

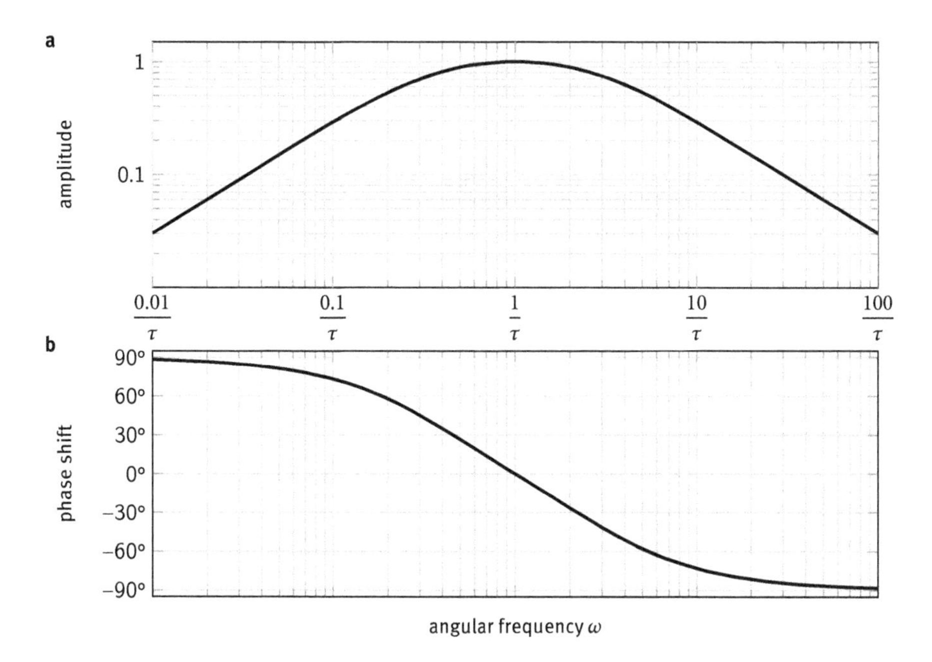

Fig. 4.15. Frequency dependence of the closed-loop transfer function of a Wien-bridge type oscillator: (a) amplitude (b) phase shift.

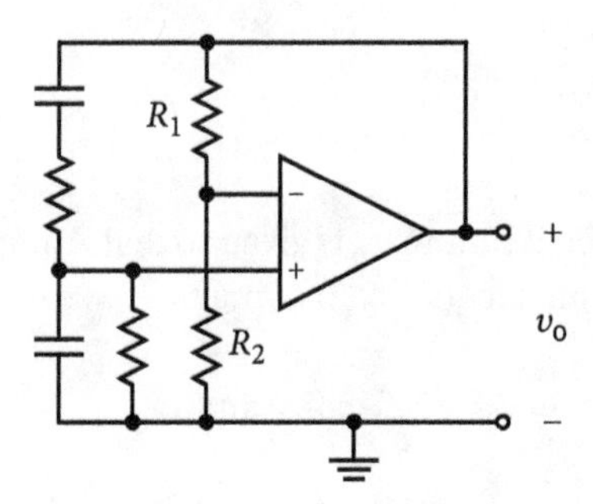

Fig. 4.16. Schematic of a Wien's bridge oscillator using a differential amplifier.

of the closed-loop gain is shown. As can be seen, the amplitude peaks at ω_0 so that all other frequencies have less amplification.

Of the two times two possible feedback configurations, only one became popular, namely the voltage division using capacitors. These oscillators are called Wien's bridge oscillators. Why this name includes bridge will be clear at once. For convenience, such oscillators are built with equal pairs of resistors and capacitors. By mechanically coupling variable capacitors, it is then possible to continuously vary the frequency ω_0.

Figure 4.16 gives a schematic circuit diagram of a Wien's bridge oscillator. Both the negative and the positive feedback circuits are voltage dividers with the divided voltages fed into a differential amplifier. Thus, the name bridge. The branch of the positive feedback provides the selectivity for ω_0 fulfilling the frequency condition. The negative feedback circuit is there to fulfill the stability condition $AB = 1$.

How is it possible to fully fulfill the stability condition? For a linear system (in the mathematical sense) this would not be possible. Even if the system was absolutely stable, the "integer" value 1 could never be exactly dimensioned. Fortunately, the voltage transfer function is only approximately linear and outside the amplifier range even strongly nonlinear. So the slope (the differential voltage gain) changes with the instantaneous operating point. It becomes even zero for very small and/or very large excursions. Thus, at some output level the closed-loop gain will reach the value 1 if it was larger than that in the quiescent operating point. The oscillator will oscillate with exactly that output amplitude.

As we have seen in Section 2.4.1.3, negative feedback improves linearity so that we encounter a rather linear transfer function, except for the two nonlinearities beyond the linear range (where feedback ceases). Without further provisions the instantaneous operating point, where the stability condition would be fulfilled, would lie outside the linear range. This means that a strongly nonlinear part of the transfer function is involved which would distort the signal. A soft approach to the stability conditions would strongly reduce this distortion. This can be achieved either by using a resistor R_1 with positive temperature coefficient or choosing for R_2, the second resistor of the voltage divider, a resistor with negative temperature coefficient. With increased output amplitude, the increased current heats the resistors of the voltage divider. Their values are changed in such a way that the loop gain of the positive feedback (that for small-signals is just above 1) will be reduced dynamically to 1.

In a more recent method R_1 is replaced by a dynamic impedance. An n-channel JFET is used as a voltage-controlled resistance. Raising the gate voltage of the JFET which operates in the ohmic region increases the drain–source resistance. When there is no output signal, the gate voltage is zero and the drain–source resistance is at the minimum. Under this condition, the closed-loop gain is greater than 1, starting an oscillation that is accompanied by an output signal. Using the rectified output signal to bias the JFET increases the drain–source resistance of the JFET to such a value that the amplitude of the output voltage becomes stable (because the closed-loop gain has reached the value 1).

Originally, a filament bulb (which has a resistance with a pronounced positive temperature coefficient) was used for R_1 so that one can see the light of this bulb inside the housing of traditional Wien's bridge oscillators. Of course, the temperature of the filament inside the lamp depends on the equilibrium between electric power dissipated in it and radiation power emitted from it. Clearly, at so low frequencies of the oscillation that the filament cools down within one period of the oscillation the stability of the output amplitude is jeopardized. To have a stable equilibrium (i.e., a stable resistance) also for low but not too low frequencies, the temperature should not be too high, i.e., the current through the lamp should be rather small. Under this kind of operating condition, the life of the lamp should be almost infinite. With these precautions, an amplitude stability of 0.1% can easily be achieved even with a simple two-stage amplifier. Using an operational amplifier that linearizes the transfer function very well a distortion of the sine wave of the order of 10^{-5} can be achieved. The ultimate performance limit of this rather simple bridge configuration lies in the limited common-mode rejection of the differential amplifier.

Example 4.2 (Wien's bridge oscillator is an undamped resonant circuit). As shown in Figure 4.17, we calculate the impedance at the noninverting input of the amplifier. It has a voltage gain of 3 so that at ω_0 the closed-loop gain becomes 1, as required for a stable oscillation. The input impedance of the operational amplifier is so high that it can be disregarded, and its output impedance is so low that it can be disregarded, as well. Then the input current i_i can be expressed by

$$i_i = \frac{v_i - v_o}{R_s + \frac{1}{j\omega C_s}} = \frac{v_i \times (1 - g_v)}{R_s + \frac{1}{j\omega C_s}} \tag{4.17}$$

and the input admittance Y_i (with $\tau_s = R_s C_s$) by

$$Y_i = \frac{i_i}{v_i} = \frac{(1 - g_v) \times \left(R_s - \frac{1}{j\omega C_s}\right)}{R_s^2 + \frac{1}{\omega^2 C_s^2}} = \frac{g_v - 1}{1 + \frac{1}{\omega^2 \tau_s^2}} \times \left(\frac{1}{-R_s} + \frac{1}{j\omega C_s R_s^2}\right). \tag{4.18}$$

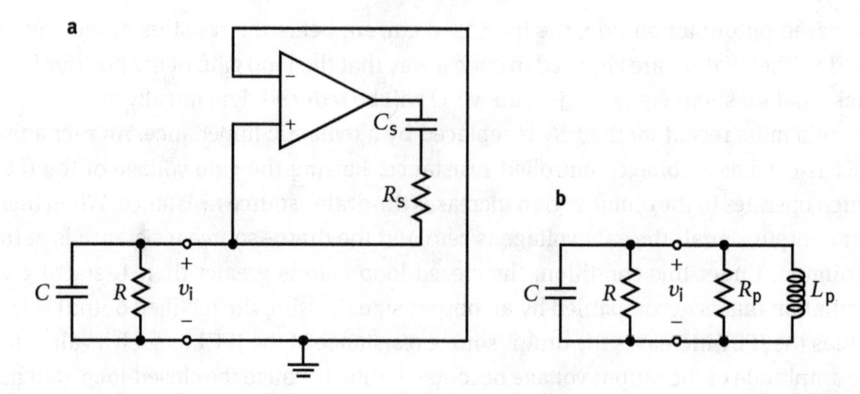

Fig. 4.17. Frequency-selective positive feedback (a) schematic circuit diagram (b) reduced circuit diagram.

The admittance is that of a resistance R_p proportional to $-R_s$ shunted by an inductance L_p proportional to $R_s^2 C_s$. This circuit acts as a gyrator (Section 3.6.5). The capacitance C_s behaves like an inductor. As shown in Figure 4.17b C, R, R_p, and L_p form a parallel resonant LC circuit (Section 3.4.6.2). For $R = |R_p|$

$$R = \frac{R \times \left(1 + \frac{1}{\omega^2 \tau^2}\right)}{g_v - 1} \tag{4.19}$$

this resonant circuit is purely reactive, it is undamped. From this equation we get

$$g_v = 2 + \frac{1}{\omega^2 \tau^2}. \tag{4.20}$$

The resonant frequency ω_r of an undamped resonant circuit is given by

$$\omega_r = \frac{1}{\sqrt{LC}} = \sqrt{\frac{g_v - 1}{\left(1 + \frac{1}{\omega^2 \tau^2}\right) \times C^2 R^2}} = \frac{1}{\tau} = \omega_0, \tag{4.21}$$

and the gain at ω_r becomes with $\omega_r = 1/\tau$

$$g_v(\omega_r) = 3. \tag{4.22}$$

Not surprisingly, the answers are identical with those of the general analysis. This result supports the nomenclature which names amplifiers and elements with negative impedance *active* elements.

Example 4.3 (Output impedance of a Wien's bridge oscillator at ω_0). For small-signals, i.e., under the assumption that linearity applies, it is clear from Figure 4.16 that the total output impedance $Z_{oF}(\omega_0)$ of the oscillator has three components, which lie in parallel. Using this information and all the other diligently will allow to arrive at the *exact* answer to this rather complicated task without much mathematical effort. We know that

- the total output impedance consists of three individual impedances in parallel. Thus, it will be wise to deal with its inverse value, the admittance, rather than with the impedance;
- the voltage gain g_v of the amplifier at ω_0 is exactly 3, i.e., the reverse voltage gain is 1/3;
- $\omega_0 = \frac{1}{\tau}$ and $\tau = RC$; and
- the dynamic value of an impedance Z that connects the output with the input of a voltage amplifier is $\frac{Z}{1-\frac{1}{g_v}}$ (Section 2.4.3.2).

Thus, the impedance Z_{F-} of the negative feedback circuit is given by

$$Z_{F-} = \frac{R_1}{1 - \dfrac{1}{g_v}} = \frac{R_1}{2/3}, \tag{4.23}$$

that of Z_{F+}, the impedance of the positive feedback circuit, by

$$Z_{F+} = \frac{R_S - \dfrac{j}{\omega_0 C_S}}{1 - \dfrac{1}{g_v}} = \frac{R_S \times \left(1 - \dfrac{j}{\omega_0 \tau}\right)}{1 - \dfrac{1}{g_v}} = \frac{R_S \times (1 - j)}{2/3}. \tag{4.24}$$

The capacitive contribution is expected because the voltage division with the complex admittance at the noninverting input which has a capacitive component, too, must result in phase shift zero at ω_0.

The total impedance Z_{oF} is obtained by shunting the output impedance of the amplifier Z_{oA} with Z_{F-} and Z_{F+}. Switching to the admittance Y_{oF}, we get the moderately simple relation

$$Y_{oF} = Y_{oA} + \frac{2/3}{R_1} + \frac{2/3}{R_S \times (1 - j)} = Y_{oA} + \frac{2/3}{R_1} + \frac{1/3}{R_S} \times (1 + j) \tag{4.25}$$

As it must be, the capacitive component of Z_{F+} appears unchanged in the equation of the total admittance

$$\mathrm{Im}\left[\frac{1}{Z_{F+}}\right] = \frac{1}{3R_S}. \tag{4.26}$$

Problems

?

4.19. With a complex voltage divider made from elements of the same nature in the feedback circuit the closed-loop gain becomes real not only at the frequency ω_0 but also at $\omega = 0$. How would such a circuit influence the operating point?

4.20. Which is the most popular feedback configuration and why?

4.21. Designing a Wien's bridge oscillator based on current division:
(a) What kind of amplifier would one use?
(b) What type of (positive and negative) feedback would one apply?

4.22. Is the oscillator frequency of a Wien's bridge oscillator primarily selected by the amplitude or by the phase shift?

4.2.1.2 LC oscillators

Returning to the simple frequency selective voltage dividers, we found in the previous section that for the phase shift to be zero $X_1R_2 = X_2R_1$ must be true. This condition is also met with purely reactive voltage division. If the reactances are of the same kind, the phase shift of the divided signal is zero for all frequencies. As the amplitude transfer is frequency independent so that no frequency gets selected, these cases must be disregarded.

For reactances having different sign Figure 4.18 applies. There the stability condition which requires voltage division to provide a closed-loop gain of 1, was anticipated.

The voltage divider of Figure 4.18 provides the positive feedback with

$$B(j\omega) = \frac{jX_2}{j\left(\omega L_1 - \frac{1}{\omega C_1} + X_2\right)} = \frac{1}{1 - \frac{1}{X_2} \times \left(\frac{1}{\omega C_1} - \omega L_1\right)} \tag{4.27}$$

Again we get a transfer function that is real for all frequencies. However, the amplitude is frequency dependent. Let us now investigate the two possibilities for X_2

1. $\frac{1}{X_2} = -\omega C_2$, and
2. $\frac{1}{X_2} = \frac{1}{\omega L_2}$.

In the first case, we get with the series configuration C of C_1 and C_2

$$C = C_1 \times \frac{C_2}{C_1 + C_2} \tag{4.28}$$

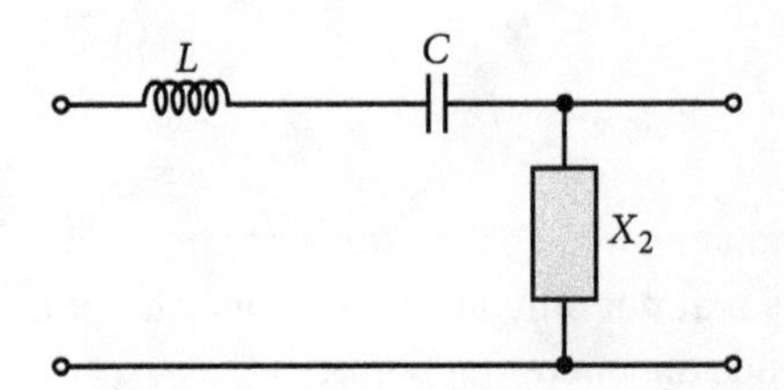

Fig. 4.18. Reactive voltage divider with inductance, capacitance, and a general reactive element.

and with

$$\omega_r = \sqrt{\frac{1}{L_1 C}} \tag{4.29}$$

the denominator in (4.27) becomes zero, i.e., $A(j\omega_r)$ may even be zero to provide the necessary closed-loop gain.

In the second case, we introduce

$$L = L_1 + L_2 \tag{4.30}$$

and with

$$\omega_r = \sqrt{\frac{1}{LC_1}} \tag{4.31}$$

the denominator of $B(j\omega)$ becomes zero, and again $A(j\omega_r)$ may even be zero to provide the required closed-loop gain of 1 for a stable (undamped) oscillation.

Of course, we get the dual answer for current division. So we have arrived at the ideal series resonant circuit, and the ideal parallel resonant circuit with capacitive and inductive gain adjustments.

In two respects, these answers are unsatisfactory:

1. if an ideal resonant circuit could be realized it would not need amplification to oscillate without damping, and
2. the losses in the circuit (stray resistance and conductance, the emission of electromagnetic radiation) need not be replenished by an active device.

Consequently, we must deal with more realistic conditions.

The stray properties of the reactive components and both the output impedance Z_{oA} and the input impedance Z_{iA} of the amplifier can be taken into account with the circuit of Figure 4.19a. In Figure 4.19b, this circuit is converted into a simple voltage divider.

To spare space we drop $(j\omega)$ in the following calculations, i.e., tacitly all impedances are assumed to be complex. Then we get the (complex) transfer function of the feedback network as follows, after replacing the circuit across Z_1 using Thevenin's theorem

$$B = \frac{Z_1}{Z_{oA} + Z_1} \times \frac{Z_3}{Z_2 + Z_3 + \frac{Z_{oA} \times Z_1}{Z_{oA} + Z_1}} = \frac{Z_3}{Z_{oA} + \frac{Z_2 + Z_3}{Z_1} \times (Z_{oA} + Z_1)} . \tag{4.32}$$

The first factor takes care of the voltage reduction v_{Th}/v_o. The complex transfer function of the simple complex voltage divider is given by

$$B = \frac{Z_3}{Z_3 + Z_4} \tag{4.33}$$

so that by comparison one gets

$$Z_4 = Z_2 + \frac{Z_{oA}}{Z_1} \times (Z_1 + Z_2 + Z_3) \tag{4.34}$$

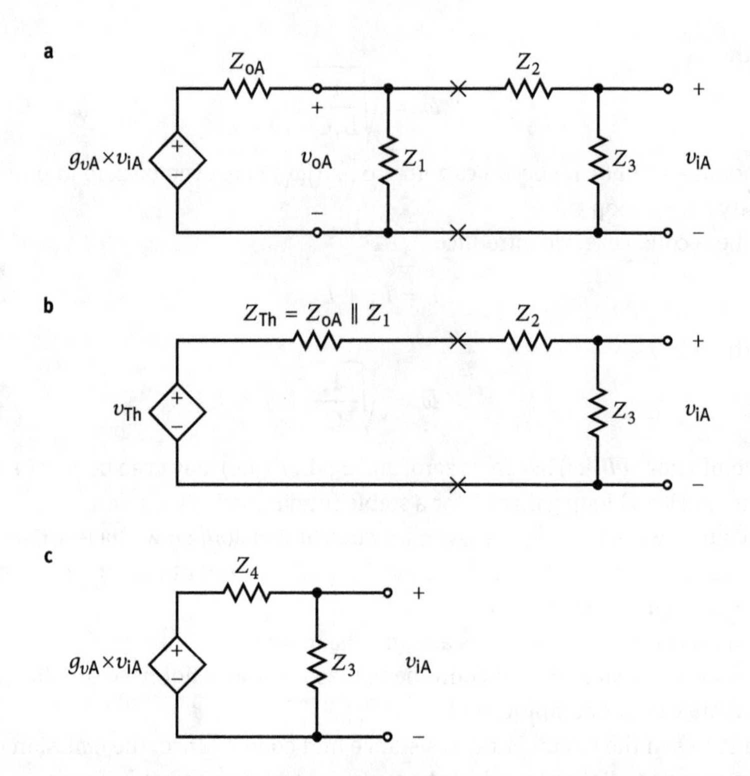

Fig. 4.19. (a) Circuit diagram of a realistic LC frequency-selective positive feedback circuit. (b) Simplified circuit by means of Thevenin's theorem. (c) Equivalent simple complex voltage divider.

We are investigating purely reactive voltage division, so that

$$\mathrm{Re}[Z_3] = \mathrm{Re}[Z_4] = 0\,. \tag{4.35}$$

Thus, the stray resistance or conductance and the input resistance of the amplifier must be disregarded. $\mathrm{Re}[Z_4] = 0$ means

$$\mathrm{Re}[Z_4] = R_2 + R_{\mathrm{oA}} \times \left(1 + \frac{R_1 R_2}{R_1^2 + X_1^2} + \frac{X_1 \times (X_2 + X_3)}{R_1^2 + X_1^2}\right) = 0\,. \tag{4.36}$$

This gives the general frequency equation

$$-X_1 \times (X_2 + X_3) = \frac{R_2 + R_{\mathrm{oA}}}{R_{\mathrm{oA}}} \times (R_1^2 + X_1^2) + R_1 R_2\,. \tag{4.37}$$

The right-hand side of this equation is positive, consequently not all of the X_k can be of the same kind.

If R_1 and R_2 can be disregarded, above equation degenerates to

$$X_1 + X_2 + X_3 = 0\,. \tag{4.38}$$

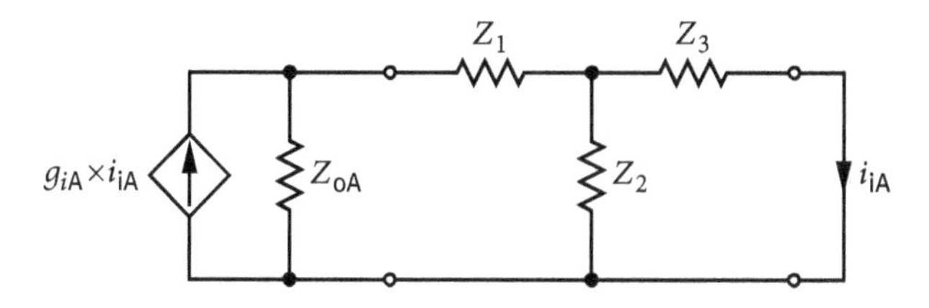

Fig. 4.20. Positive frequency dependent feedback at a current amplifier.

Above conditions make sure that the phase shift is zero. In addition, the closed-loop gain must be 1 so that the voltage gain g_v of the amplifier must be (for a reactive voltage divider)

$$g_v = \frac{1}{B} = \frac{X_3 + X_4}{X_3} = \frac{1}{X_3} \times \left(X_3 + X_2 + R_{oA} \times \frac{R_1 \times (X_2 + X_3) - R_2 X_1}{R_1^2 + X_1^2} \right) . \quad (4.39)$$

Insertion of the general frequency relation yields the stability condition

$$g_v = -\frac{1}{X_1 X_3} \times \left[R_2 \times (R_1 + R_{oA}) + \frac{R_{oA} + R_2}{R_{oA}} \times (R_1^2 + R_1 R_{oA} + X_1^2) \right] . \quad (4.40)$$

For $g_v > 0$ we need $X_1 X_3 < 0$, i.e., these reactances are of a different nature they must be dual to each other.

With $\mathrm{Re}[Z_1] = \mathrm{Re}[Z_2] = 0$ this equation degenerates to

$$g_v = -\frac{X_1}{X_3} = \frac{X_2 + X_3}{X_3} = 1 + \frac{X_2}{X_3} . \quad (4.41)$$

From above equation, it is clear that X_2 and X_3 are of the same nature as for $B < 1$ it is necessary that $g_v > 1$. Applying duality leads to Figure 4.20. The dual counterpart to (4.38) is

$$X_1^{-1} + X_2^{-1} + X_3^{-1} = 0 \quad (4.42)$$

and to (4.41)

$$g_i = -\frac{X_1^{-1}}{X_3^{-1}} = \frac{X_2^{-1} + X_3^{-1}}{X_3^{-1}} = 1 + \frac{X_2^{-1}}{X_3^{-1}} . \quad (4.43)$$

In Figure 4.21, real electronic components are used to accomplish the *voltage* di-

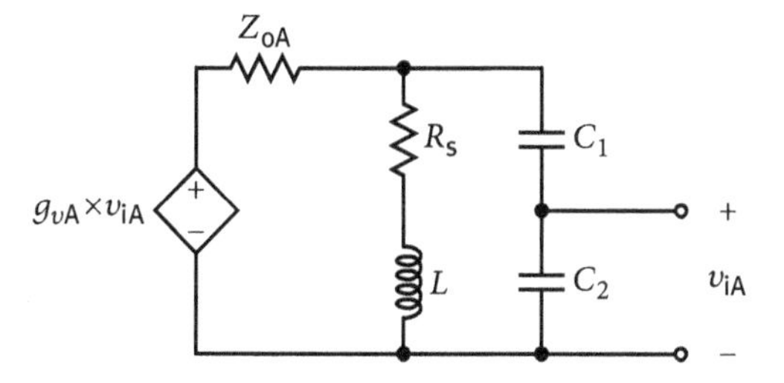

Fig. 4.21. Parallel resonant circuit.

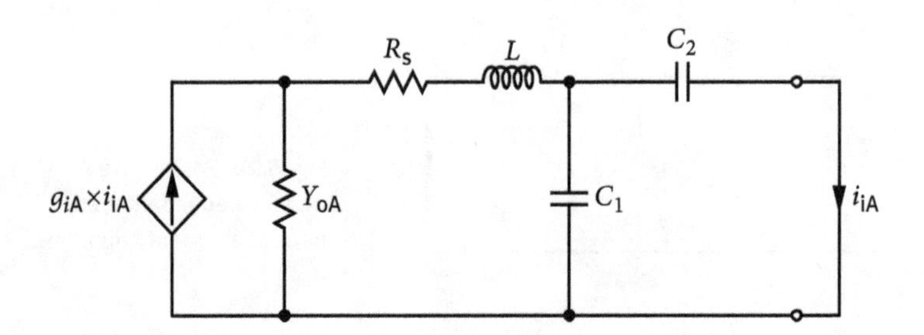

Fig. 4.22. Series resonant circuit.

vision in the feedback circuit. Z_1 is realized by an inductor characterized by its inductance L and a parasitic series resistance R_s. Both Z_2 and Z_3 are high quality capacitors with negligible stray properties. From the frequency equation we get

$$R_s^2 + \omega^2 L^2 = -\omega L \times \left(-\frac{1}{\omega C_1} - \frac{1}{\omega C_2}\right) \tag{4.44}$$

from which with

$$\frac{1}{C} = \frac{1}{C_1} + \frac{1}{C_2} \tag{4.45}$$

one arrives at the well-known relation for the resonant frequency of a parallel resonant circuit

$$\omega = \sqrt{\frac{1}{LC} - \left(\frac{R_s}{L}\right)^2}\,. \tag{4.46}$$

From the stability equation we get

$$g_v = \frac{C_2}{C} \times \left(1 + \frac{R_{OA} R_s C}{L}\right) = \frac{C_2}{C} \times \left(1 + \frac{\tau_2}{\tau_1}\right) \tag{4.47}$$

with

$$\tau_1 = \frac{L}{R_s} \quad \text{and} \tag{4.48a}$$

$$\tau_2 = R_{OA} C. \tag{4.48b}$$

In Figure 4.22 real electronic components are used to accomplish the current division in the feedback circuit. Y_1 is the admittance of an inductor (characterized by its inductance L and a parasitic series resistance R_s). Both Y_2 and Y_3 are the admittances of high quality capacitors with negligible stray properties. From the frequency equation we get

$$\frac{1}{R_s^2 + \omega^2 L^2} = \frac{\omega L}{R_s^2 + \omega^2 L^2} \times \omega (C_1 + C_2) \tag{4.49}$$

from which with $C = C_1 + C_2$ one arrives at the well-known relation for the resonant frequency of a series resonant circuit

$$\omega = \sqrt{\frac{1}{LC}}\,. \tag{4.50}$$

Table 4.3. Some numerical values for the dependence of the impedance bandwidth BW on the quality factor Q.

$\frac{1}{Q}$	$BW \times \frac{2\pi}{\omega_r}$
0.001	0.001
0.010	0.010
0.100	0.098
0.300	0.278
0.400	0.360
0.500	0.438

From the stability equation we get

$$g_i = \frac{C}{C_2} \times \left(1 + \frac{R_S}{R_{oA}}\right) = \frac{R_{oA} + R_S}{R_{oA}} \times \frac{C_1 + C_2}{C_2} . \tag{4.51}$$

The second part of the equation illustrates that the current gain must compensate for the current division by R_S (i.e., the impedance of the series resonance circuit at resonance) and by C_2.

Note that the resonant frequency of the series resonant circuit does not depend on R_S or any other resistance in series to it. Such resistances do not change the precondition of a resonant behavior: the transfer function of the feedback circuit (of the closed-loop gain) must be real at the resonant frequency. Even if the resonant frequency ω_r of a series resonant circuit is independent of the series resistance of the inductor its impedance is not. In the ideal case, it is zero, with a finite value of R_S it is identical with R_S. Applying duality it is clear that the admittance of an ideal parallel resonant circuit at the resonant frequency is zero.

To answer the question how the impedance (admittance) of a resonant circuit changes with frequency we must return to the quality factor Q of a resonant circuit (Section 3.4.6.2). The -3 dB bandwidth BW of the frequency dependence of the impedance (admittance) value is given by

$$\left(1 - \frac{2\pi BW}{2\omega_r}\right)^4 - \left(1 - \frac{2\pi BW}{2\omega_r}\right) \times \left(2 + Q^{-2}\right) + 1 = 0 \tag{4.52}$$

with the approximate solution

$$BW \approx \frac{\omega_r}{2\pi} \times \frac{1}{Q} . \tag{4.53}$$

Table 4.3 gives an impression how good this approximation is. For Q values larger than 10 (which is a low value for actual oscillators) the approximation is better than 2%.

Different from relaxation oscillators, which oscillate between two levels L and H with no information contained in the actual amplitude, harmonic oscillators are expected to oscillate with *stable amplitude* which requires that there is a stable closed-loop gain of 1. As discussed before (Section 4.2.1.1), the unavoidable nonlinearity of the closed-loop (voltage) transfer function is the key to make the closed-loop gain 1 by having it slightly higher than 1 in the quiescent operating point so that, with increasing amplitude, it is self-adjusting to 1. Obviously, the nonlinearity can occur in the active device or in the feedback circuit. A very good solution was demonstrated in Section 4.2.1.1 where the amplitude dependence of the gain was achieved dynamically in a negative feedback loop. One advantage of such an arrangement is that negative feedback provides a stable operation of the amplifier.

Aside from stable amplitude also *stable frequency* may be required. In the case of a series resonant circuit this requires stable values of the inductance and the capacitance, for a parallel resonant circuit also changes in the parasitic series resistance R_s must be avoided. The main reason for a frequency change lies in *temperature dependence.*

There are (at least) two contributions to temperature dependence both of the inductor and the capacitor. In both cases, the geometry changes because of thermal expansion; in addition there is a material-dependent component. The linear expansion of the wire of a wire-wound inductor contributes about 2×10^{-5}/°C (at 25 °C) to the temperature coefficient. By burning silver wire into a ceramic substrate, this expansion can be minimized. They have a temperature coefficient as low as 10^{-5}/°C. The *internal* flux in the inductor rises with temperature as the skin depth decreases with increasing resistance of the wire. This effect contributes about as much or even more to the temperature coefficient so that it adds up to roughly $+5 \times 10^{-5}$/°C.

Aside from inductors with air cores (without core dependent contribution), there are inductors with ceramic (ferrite) cores, or with silicon steel cores. The relative permeability (i.e., relative to air) of useful magnetic core materials ranges from 10 to 10 000, or, more practical in the range of 100 to 1000. The inductance can be raised by this factor using an appropriate core. The use of a ceramic core increases the temperature coefficient by typically 9×10^{-5}/°C. To comply with the need of miniaturization there exist thin film chip inductors.

The temperature coefficient of capacitors is mainly a matter of the kind of dielectric that is used. Ceramic capacitors with graded negative temperature coefficients and with high stability and low losses are available for resonant circuit application. By matching the negative coefficient with the positive coefficient of the inductor, the temperature dependence of the resonant frequency can be minimized. As ceramic capacitors are quite noninductive compared to the other classes of capacitors, they are well suited for high-frequency work. An alternative is a capacitor with an insulator of polystyrene that has a typical temperature coefficient of -10^{-4}/°C. Such a coefficient is needed to yield stable tuned circuits compensating the positive 10^{-4}/°C coefficient of the inductor. It should be obvious that also the amplifier's properties influence the stability of the oscillation frequency. At higher frequencies changes in the phase shift of the amplifier's transfer function cannot be disregarded.

A superior capacitor uses *mica* for insulation. The parasitic inductance and conductance are particularly small, as well as the temperature coefficient of about

$+10^{-5}/°C$. It does not compensate for an inductor's temperature coefficient but can be very useful in special applications requiring high stability.

Problem

4.23. Show that the impedance of a series resonant circuit is zero at the resonant frequency.

4.2.1.3 Types of LC oscillators

Table 4.4 gives an overview over all possible simple LC feedback oscillators.

Table 4.4. Overview over all simple LC feedback oscillators.

Dimension of AB	Resonance type	Divider type	Name
Voltage ratio	Parallel	Capacitor	Colpitts
		Inductor	Hartley
		Transformer	Armstrong
Current ratio	Serial	Capacitor	No name
		Inductor	No name
		Transformer	No name

Colpitts oscillator

The frequency of a Colpitts oscillator (Figure 4.23a) is determined by a parallel resonant circuit consisting of an inductor L and two capacitors C_1 and C_2 in series performing the voltage division. Thus, the oscillation frequency is approximately (disregarding the impedances of the active component) given by

$$\omega_r \approx \sqrt{\frac{1}{L} \times \left(\frac{1}{C_1} + \frac{1}{C_2}\right)} \tag{4.54}$$

Clapp oscillator

To maintain the stability condition (closed-loop gain equals 1), when varying the resonant frequency, both capacitors of a Colpitts circuit must be varied simultaneously so that the amount of voltage division stays constant. This approach is not viable. Decoupling the fulfilment of the phase criterion from that of the amplitude criterion is done in the Clapp oscillator by introducing an additional capacitor C_0 (Figure 4.24). C_1 and C_2 perform the voltage division, whereas C_0 controls almost entirely the resonant frequency if it has a small enough value

$$\omega_r \approx \sqrt{\frac{1}{L} \times \left(\frac{1}{C_0} + \frac{1}{C_1} + \frac{1}{C_2}\right)} \tag{4.55}$$

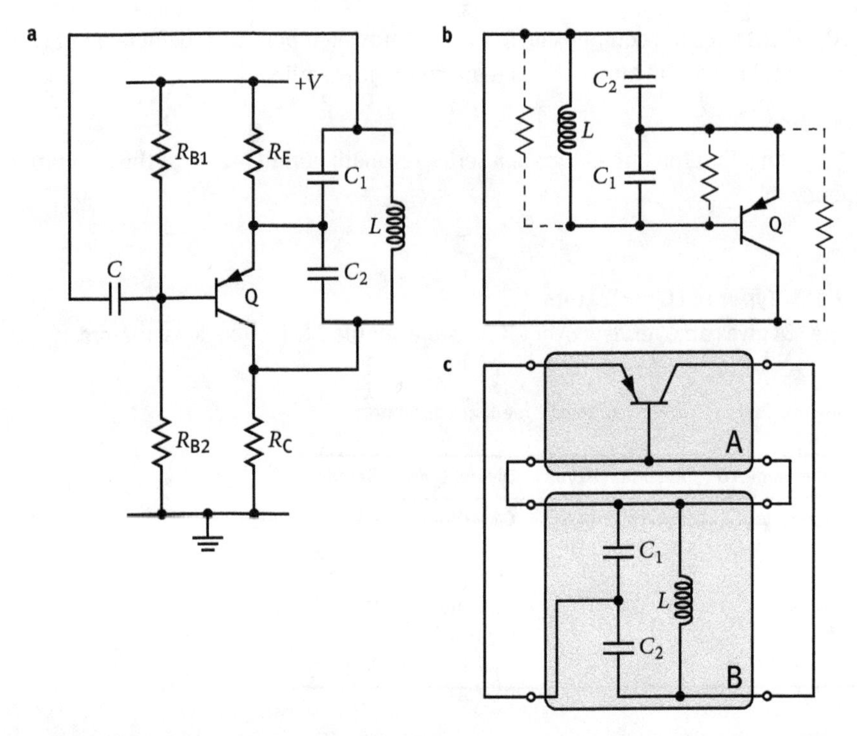

Fig. 4.23. Schematic of a Colpitts oscillator with a PNP transistor as active component: (a) circuit, (b) equivalent circuit for small-signals, and (c) redrawn as feedback configuration with two-ports A and B (without resistors in A).

Thus, tuning of C_0 does not affect the closed-loop gain but acts predominantly on the resonant frequency ω_r. Therefore, a Clapp circuit is often a better choice than a Colpitts circuit when a variable frequency oscillator must be built.

Hartley oscillator

The frequency of a Hartley oscillator (Figure 4.25) is determined by a parallel resonant circuit consisting of a capacitor C and two inductors L_1 and L_2 in series performing the voltage division. Considering the (magnetic) coupling factor k between the two coils the total effective inductance L_0 that determines the frequency of the oscillation is

$$L_0 = L_1 + L_2 + k\sqrt{L_1 L_2}\,. \tag{4.56}$$

Thus, the oscillation frequency is approximately (disregarding the impedances of the active component) given by

$$\omega_r \approx \sqrt{\frac{1}{L_0 C}}\,. \tag{4.57}$$

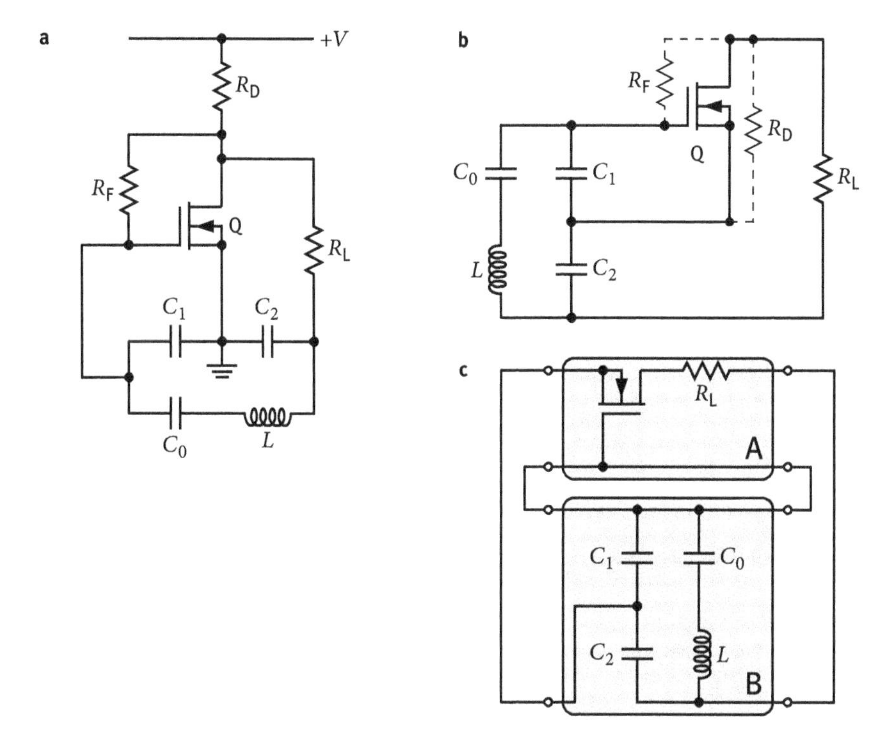

Fig. 4.24. Clapp oscillator based on an n-channel MOSFET as the active component: (a) circuit (b) small-signal equivalent circuit (c) feedback configuration with two-ports.

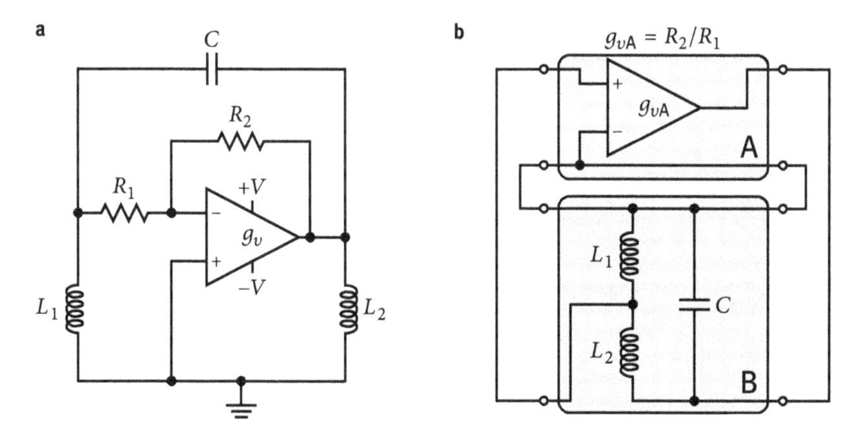

Fig. 4.25. Principle of a Hartley oscillator based on an operational amplifier: (a) circuit (b) feedback configuration with two-ports.

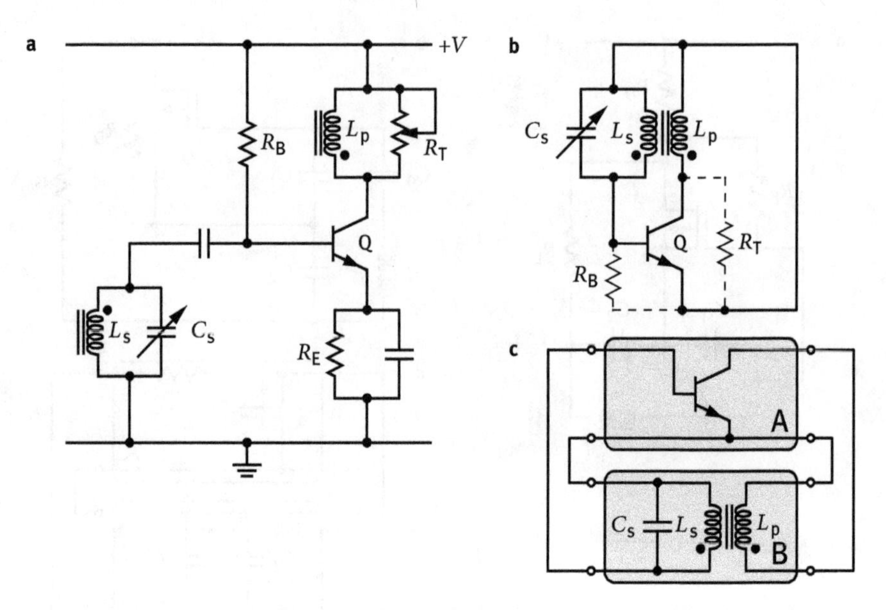

Fig. 4.26. Schematic of an Armstrong oscillator with an NPN transistor as an active component: (a) circuit, (b) small-signal equivalent circuit, and (c) feedback configuration with two-ports.

Armstrong oscillator

This type of LC feedback oscillator uses a transformer in the feedback circuit, as shown in Figure 4.26. As a transformer is symmetric with regard to input and output, there are two options for placing the resonant circuit. The original Armstrong oscillator had the resonant circuit at the input of the active component (a valve) with the transformer providing the voltage division from output to input. The effective inductance is primarily the inductance L_s of the secondary windings of the transformer so that the resonant frequency is obtained as

$$\omega_r = \sqrt{\frac{1}{L_s C}} . \tag{4.58}$$

The variant with the resonant circuit at the output of the active element is called *Meissner oscillator*. In this case, the effective inductance is primarily the inductance L_p of the primary windings of the transformer so that the resonant frequency is obtained as

$$\omega_r = \sqrt{\frac{1}{L_p C}} . \tag{4.59}$$

Due to their relative clumsiness and cost, transformers are usually avoided. For that reason, Armstrong oscillators are not popular.

For analogous reasons, feedback oscillators with current division have found so little interest that there are not even names for them. The reason is that of convenience:

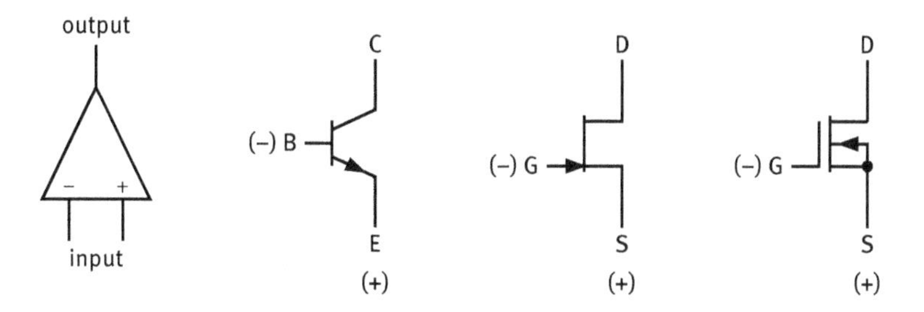

Fig. 4.27. Symbol of a general active three-terminal component vs. equivalent symbols of actual components.

Table 4.5. Combinations of reactances applied to active three-terminal components in dependence of the voltage gain to yield stable oscillation by positive feedback.

Voltage gain g_v	X_1 vs. X_3	X_2 vs. X_3
$g_v < 0$	$\mathrm{sign}X_1 = \mathrm{sign}X_3$	$\mathrm{sign}X_2 \neq \mathrm{sign}X_3$
$0 < g_v < 1$	$\mathrm{sign}X_1 \neq \mathrm{sign}X_3$	$\mathrm{sign}X_2 \neq \mathrm{sign}X_3$
$g_v > 1$	$\mathrm{sign}X_1 \neq \mathrm{sign}X_3$	$\mathrm{sign}X_2 = \mathrm{sign}X_3$

Amplifiers are still optimized to resemble ideal voltage amplifiers so that designing a feedback circuit with current division would be awkward.

Example 4.4 (Degeneration of the three basic circuits of three-terminal components to a single circuit when used in an LC feedback oscillator circuit). Three-terminal active components have a natural differential input (in the case of the bipolar transistor between the base as inverting and the emitter as noninverting input) and a natural output terminal (in the case of a bipolar transistor the collector). Consequently we use the triangular amplifier symbol to cover all technology families of active three-terminal components (Figure 4.27).

Now let us investigate how reactances in three-terminal oscillator circuits must be arranged to yield an oscillation with stable amplitude. For simplicity let us assume that the resistive contributions to the complex impedances are so small that they can be disregarded. Then we have $X_1+X_2+X_3 = 0$ as frequency condition, and from (4.41) $g_v = -\frac{X_1}{X_3}$ as the condition for stable oscillation. Combining these two equations we get an additional way to express g_v as

$$g_v = \frac{X_2}{X_3} + 1. \tag{4.60}$$

Table 4.5 lists the combinations of reactances for the three basic three terminal oscillator circuits providing stable oscillation by positive feedback.

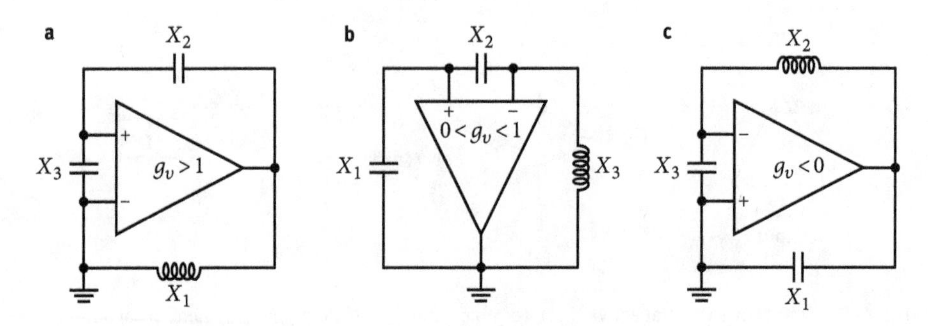

Fig. 4.28. Disregarding the position of the grounding point which has no impact on the oscillation, there is no difference between the three LC feedback oscillators in the three basic circuit configurations of active three-terminal components.

Figure 4.28 displays the three cases listed in Table 4.5 using purely reactive components for an LC oscillator.

As demonstrated in the example, no specific basic circuit configuration of the active three-terminal component can be identified in feedback LC oscillators. The reason is quite obvious: there is no input (necessary) and consequently none of the terminals are common to both input and output. Of course, this is not true if a transformer is used in the feedback circuit. A transformer is a two-port forcing the active element to have an input and an output. Consequently, for oscillators with transformer feedback there exists the trinity of the basic circuits of three-terminal amplifiers.

? **Problem**

4.24. Why, except in cases with transformer coupling, it does not make any difference which basic circuit is chosen to explain a three-terminal oscillator?

4.2.1.4 Quartz oscillators

If the bandwidth of the impedance (or admittance) of an oscillator is small, then a small frequency change is accompanied by a large phase shift change. Therefore, a high quality factor Q guarantees that jitter in the phase shift does not adversely affect the resonant frequency. On the other hand, a high quality factor Q makes it difficult for a harmonic feedback oscillator to gain the equilibrium, i.e., it lasts long until the amplitude of the oscillation has reached its final (stable) value. To attain an oscillator with excellent frequency stability, one needs a resonator with a high quality factor. Such elements are readily available by way of quartz crystals. They provide quality factors up to 10^6. This is several orders of magnitude more than a simple *LC* oscillator can achieve.

Example 4.5 (Relation between quality factor (width of the resonance curve) and the rise time of the oscillation amplitude). The quality factor is given by

$$\frac{1}{Q} = \frac{1}{\omega_r} \times \frac{1}{\tau_1}, \quad \text{so that} \tag{4.61}$$

$$\tau_1 = \frac{Q}{\omega_r} \tag{4.62}$$

with $\tau_1 = L/R$, and disregarding τ_2 which depends on the tiny conductance of the capacitor. The first time constant governs the rise of the oscillation amplitude after starting the positive feedback action.

In Section 4.2.1.2, we have shown that

$$\frac{1}{Q} \approx \frac{2\pi BW}{\omega_r} \tag{4.63}$$

so that we get

$$\tau_1 = \frac{1}{2\pi BW}. \tag{4.64}$$

Some actual numbers measured at a feedback oscillator

$$\begin{aligned} \tau_1 &= 30\ \text{s, and} \\ f &= 7.0\ \text{kHz}\ , \text{ i.e.}\ , \\ \omega &= 44 \times 10^3\ \text{s}^{-1} \quad \text{then} \\ Q &= 30 \times 44 \times 10^3 \approx 1 \times 10^6\ . \end{aligned}$$

Such a high quality factor indicates that this oscillator is a quartz oscillator. It takes this oscillator a very long time after start-up to build up the stable output signal.

When a crystal of quartz is deformed it gets electrically charged. On the other hand, if one applies a voltage across a properly cut quartz it gets deformed. This property is known as piezoelectricity. A solid plate of quartz has various mechanical resonances. Depending on the cut of the plate, the plate will be stimulated to oscillate at one of these resonances. Oscillating at a mechanical resonance frequency is energetically most favorable so that electrical stimulation at the resonance frequency will require the least power, i.e., the production of the piezoelectric charge (i.e., the gain) has its maximum at this frequency. For electronic purposes, the response of a properly cut and mounted quartz crystal can be described by a resonance circuit as shown in Figure 4.29b.

Actually, this circuit diagram can be interpreted in two ways:

- The quartz may be used in place of a series resonant circuit. In this case, C_1 is not part of the resonant circuit but just a capacitor shunting the series resonant circuit. The resonant frequency is

$$\omega_{rs} = \sqrt{\frac{1}{LC_2}} \tag{4.65}$$

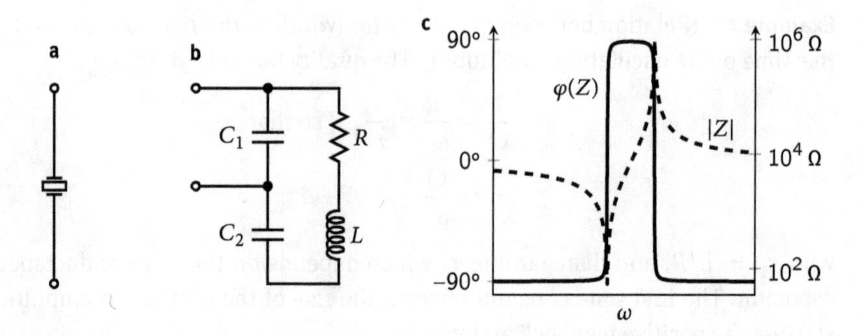

Fig. 4.29. Quartz oscillator: (a) symbol of an electronic quartz, (b) equivalent electrical circuit, and (c) amplitude and phase term of the impedance.

and the quality factor

$$Q = \omega_{\mathrm{rs}} \times \frac{L}{R}. \tag{4.66}$$

– The quartz may be used in place of a parallel resonant circuit. In this case, we have a series connection of two capacitors having a capacity C of

$$C = \frac{C_1 \times C_2}{C_1 + C_2}. \tag{4.67}$$

The resonant frequency of a parallel resonant circuit depends on the series resistor R so that the resonant frequency is

$$\omega_{\mathrm{rp}} = \sqrt{\left(\frac{1}{LC} - \frac{R^2}{L^2}\right)} \tag{4.68}$$

and the quality factor

$$Q = \omega_{\mathrm{rp}} \times \frac{C}{G_{\mathrm{p}}} \text{ with} \tag{4.69}$$

$$G_{\mathrm{p}} = \frac{RC}{L}. \tag{4.70}$$

Taking the ratio of the resonance frequencies one gets

$$\frac{\omega_{\mathrm{rp}}^2}{\omega_{\mathrm{rs}}^2} \approx \frac{C_2 + C_1}{C_1} > 1 \tag{4.71}$$

which means that in Figure 4.29c the resonance at the higher frequency is that of the parallel resonant circuit. This figure does not only show the frequency dependence of the impedance but also that of the phase shift. At frequencies between the two resonance frequencies the phase shift becomes 90°, i.e., at these frequencies the quartz crystal is equivalent to an inductor. However, this is a high quality

Table 4.6. Orders of magnitude of the sizes of the elements describing the electrical performance of a resonating quartz crystal.

Element	Size and unit
Inductor L	10^1 H
Resistor R	10^1 Ω
Capacitor C_1	10^1 pF
Capacitor C_2	10^{-2} pF

inductance which cannot be realized by a coil. To get some feeling for the size of the elements that are the electronic equivalent of a resonating quartz crystal Table 4.6 gives an idea on the orders of magnitude involved.

The frequency stability of a quartz oscillator is jeopardized by aging and by temperature effects. A typical aging effect is 10^{-7} per month. The resonant frequency of a properly cut quartz crystal will not change by more than $\pm 2 \times 10^{-6}$ in the temperature range from 0 °C to 50 °C. A temperature-dependent voltage control of the oscillator frequency can vary the frequency by typically several parts per million (ppm) only (because the high Q factor of the crystals does not allow larger changes) which is sufficient to reduce the temperature dependence by about a factor of 10.

Placing such a frequency-controlled oscillator into a temperature-controlled oven at a constant temperature (a higher than ambient temperature, e.g., around 80 °C) improves the stability of the oscillator frequency further by at least one order of magnitude.

Problem

?

4.25. A harmonic feedback LC oscillator is based on a class C amplifier. By means of a gating signal, the operating point of this amplifier can be moved into the amplifier region so that positive feedback with a closed-loop gain > 1 is achieved. Why is the use of a quartz crystal counter productive for such an oscillator?

4.2.1.5 Phase-shift oscillators

Taking advantage of the fact that inverting a sine wave has the same effect as shifting its phase by 180° spares that much of phase shift, when an inverting amplifier is used inside the positive feedback loop. The remaining 180° must be furnished by a feedback circuit with frequency-dependent phase shift. Concentrating on simple (two-component) filters it is obvious that (at least) three filter sections are needed to accomplish a phase shift of plus or minus 180° because the maximum phase shift of such filters is 90° which, however, cannot be obtained with a finite frequency. Figure 4.30 shows three filter sections of the same kind connected to the dependent voltage source of an inverting voltage amplifier. The output of this filter chain is connected to the input of said amplifier.

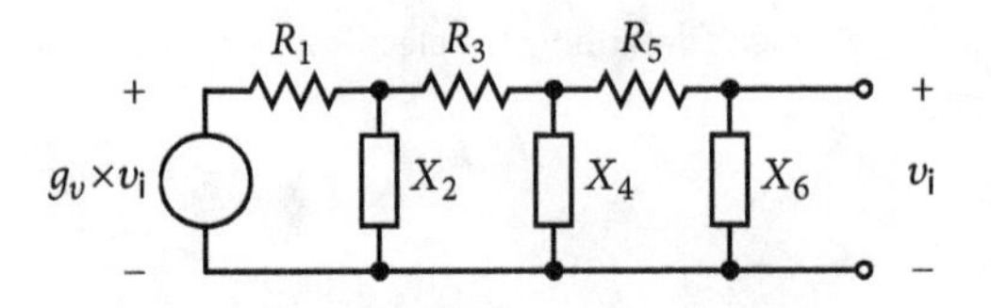

Fig. 4.30. Schematic of a feedback network of a phase shift oscillator.

Each filter section may consist of a resistor R_k and a reactance X_{k+1} with $k = 1, 3, 5$. By setting the imaginary part of the transfer function equal zero, we get the frequency equation

$$\frac{R_1 \times R_3 \times R_5}{X_2 \times X_4 \times X_6} = \frac{R_1 + R_3 + R_5}{X_6} + \frac{R_1 + R_3}{X_4} + \frac{R_1}{X_2} . \tag{4.72}$$

From setting the closed-loop gain equal 1, the required amplifier gain for stable oscillation is obtained as

$$-g_v = \frac{R_1 \times R_3 \times R_5}{X_2 \times X_4 \times X_6} \times \left(\frac{X_6 + X_4}{R_5} + \frac{X_4 + X_2}{R_3} + \frac{X_2}{R_1} \right) - 1 . \tag{4.73}$$

Now let us specialize: all filter sections shall have identical time constants, i.e.,

$$\frac{Z_1}{Z_2} = \frac{Z_3}{Z_4} = \frac{Z_5}{Z_6} , \tag{4.74}$$

or with the constants c_1 and c_2 (positive real numbers)

$$Z_3 = Z_1 \times \frac{Z_4}{Z_2} = c_1 \times Z_1 \quad \text{and}$$

$$Z_5 = Z_1 \times \frac{Z_6}{Z_2} = c_2 \times Z_1 .$$

Then we get for the frequency equation

$$\left(\frac{Z_1}{Z_2} \right)^2 = 3 + \frac{c_1^2 + c_1 + c_2}{c_1 \times c_2} \geq 3 \tag{4.75}$$

and for the stability equation

$$-g_v = \left(\frac{Z_1}{Z_2} \right)^2 \times \left[\left(\frac{Z_1}{Z_2} \right)^2 - \frac{1}{c_2} \right] - 1 \geq 8 . \tag{4.76}$$

If the filters are *low-pass* filters, then

$$\left(\frac{Z_1}{Z_2} \right)^2 = R^2 \omega^2 C^2 = \omega^2 \tau^2 = \omega^2 \times \left(\frac{L}{R} \right)^2 \tag{4.77}$$

depending on whether the *RC* or the *LR* combination is used. Thus, the frequency equation becomes

$$\omega^2 \tau^2 = 3 + \frac{c_1}{c_2} + \frac{1}{c_1} + \frac{1}{c_2} , \tag{4.78}$$

and the oscillation frequency ω_0 is

$$\omega_0 = \frac{1}{\tau} \times \sqrt{3 + \frac{c_1}{c_2} + \frac{1}{c_1} + \frac{1}{c_2}} > \frac{1}{\tau} \times \sqrt{3} \tag{4.79}$$

requiring for stable oscillation an amplifier gain g_v of

$$-g_v = (\omega\tau)^2 \times \left[(\omega\tau)^2 - \frac{1}{c_2}\right] - 1 . \tag{4.80}$$

For *high-pass* filters the inverse reaction applies

$$\left(\frac{Z_2}{Z_1}\right)^2 = R^2\omega^2C^2 = \omega^2\tau^2 = \omega^2\frac{L^2}{R^2} \tag{4.81}$$

depending on whether the CR or the RL combination is used. Thus, the frequency equation becomes

$$\frac{1}{\omega^2\tau_2} = 3 + \frac{c_1}{c_2} + \frac{1}{c_1} + \frac{1}{c_2} , \tag{4.82}$$

and the oscillation frequency ω_0 is

$$\omega_0 = \frac{1}{\tau} \times \frac{1}{\sqrt{3 + \frac{c_1}{c_2} + \frac{1}{c_1} + \frac{1}{c_2}}} < \frac{1}{\tau} \times \frac{1}{\sqrt{3}} \tag{4.83}$$

requiring for stable oscillation an amplifier gain g_v of

$$-g_v = \frac{1}{(\omega\tau)^2} \times \left[\frac{1}{(\omega\tau)^2} - \frac{1}{c_2}\right] - 1 . \tag{4.84}$$

One finding is important: the oscillation frequency is in both cases inversely proportional to the time constant, e.g., inversely proportional to the value of the capacitor.

It is convenient to choose $c_1 = c_2 = 1$. This is particularly helpful in the case of a variable frequency oscillator because three identical variable capacitors can be coupled mechanically so that any variation of them affects each capacitance by the same amount. Using high-pass filters we get

$$\omega_0 = \frac{1}{\tau} \times \frac{1}{\sqrt{3+3}} = \frac{1}{\tau} \times \frac{1}{\sqrt{6}} \tag{4.85}$$

requiring for stable oscillation an amplifier gain g_v of

$$g_v = 6 \times (6 - 1) - 1 = 29 . \tag{4.86}$$

In Figure 4.31 the frequency dependence of the amplitude of the closed-loop gain

$$|AB| = \frac{29}{\sqrt{1 + \frac{26}{\omega^2\tau^2} + \frac{13}{\omega^4\tau^4} + \frac{1}{\omega^6\tau^6}}} \tag{4.87}$$

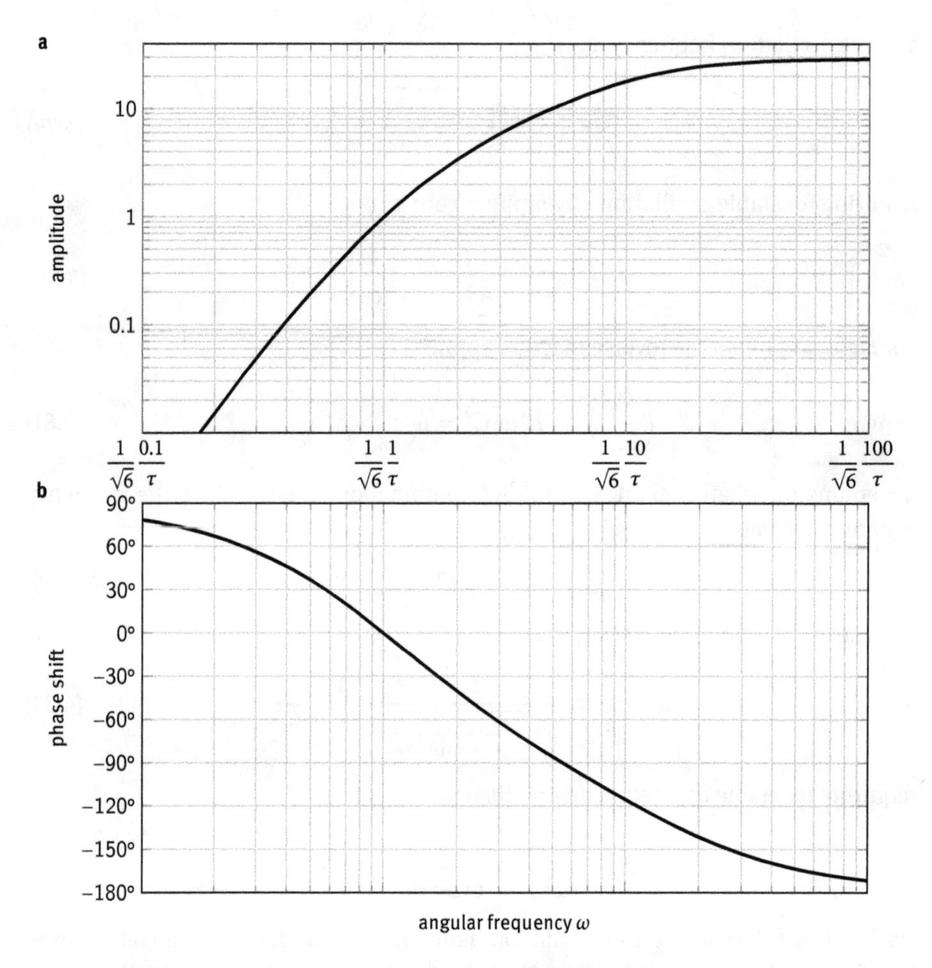

Fig. 4.31. Frequency dependence of the closed-loop gain (a) amplitude $|AB|$ (b) phase shift $\varphi(AB)$.

and of the phase shift φ of the closed-loop gain

$$\varphi = \arctan\left(\frac{1/\omega\tau - 6\omega\tau}{5 - \omega^2\tau^2}\right) = \arctan\left(\omega\tau \times \frac{\frac{1}{\omega^2\tau^2} - 6}{5 - \omega^2\tau^2}\right) \tag{4.88}$$

are displayed.

Example 4.6 (Phase shift oscillator using three equal unloaded filter sections). If the loading of each filter section by the following one may be disregarded (i.e., if buffer amplifiers, e.g., voltage followers, are used to isolate them), each filter section must provide 60° phase shift so that 180° is achieved. Taking high-pass filters the amplitude of the transfer function of one filter section is given by (Section 3.4.3)

$$|g_v| = \frac{1}{\sqrt{1 + \frac{1}{\omega^2 \tau^2}}} \tag{4.89}$$

and the phase shift

$$\varphi = \arctan \frac{1}{\omega \tau} \tag{4.90}$$

For a 60° phase shift (per section) $\frac{1}{\omega\tau}$ must be $\tan(60°) = \sqrt{3}$ so that $\omega_0 = \frac{1}{\tau\sqrt{3}}$ just as before as the limiting case. At this frequency, the attenuation per section is ½ so that a total (negative) gain of $2^3 = 8$ is required to make up for the attenuation. Again this value is the limiting value from the general deliberation. Phase shift oscillators based on current filters have not gained any practical importance. Even if the application of the duality principle makes life easy, we disregard them here, too.

Problem

?

4.26. Is the oscillator frequency of a phase shift oscillator selected by the amplitude or by the phase shift?

4.2.1.6 Twin-T oscillator

This type of oscillator is completely different from all the others. The frequency selective circuit is *not* the positive feedback circuit but the negative feedback circuit. It resembles the Wien's oscillator insofar as the two inputs of the differential amplifier form a *bridge* between the voltage dividers formed by the two feedback circuits (Figure 4.32). The signal in the capacitive branch of the Twin-T filter (Section 3.4.5.1) is leading, that in the resistive branch lagging, so that they cancel each other at the notch frequency $\omega_n = 1/RC$. As the signal transmission vanishes at the notch frequency, also the negative feedback action at this frequency vanishes, i.e., the amplifier operates at full open-loop gain. At low and high frequencies the gain of the negative feedback circuit is 1, just as in the case of a voltage follower (Section 2.5.2.1). The resistors R_1 and R_2 in the positive feedback branch of the circuit adjust the closed-loop gain at the notch frequency to the required value 1.

As $|AB| > 1$ is true only for frequencies inside the notch, this circuit produces a sine wave with extremely low distortion. Its circuit is more complex than that of the Wien-bridge oscillator, and its output amplitude can be varied more easily.

Problems

?

4.27. It was shown in Section 3.4.5.1 that a twin-T filter provides zero degree phase shift also at $\omega = 0$. Why does the oscillation not occur at $\omega = 0$?

4.28. Which feedback branch determines the oscillation frequency of a twin-T oscillator?

4.29. Is the oscillation frequency of a twin-T oscillator selected by the phase shift or by the amplitude of the closed-loop transfer function?

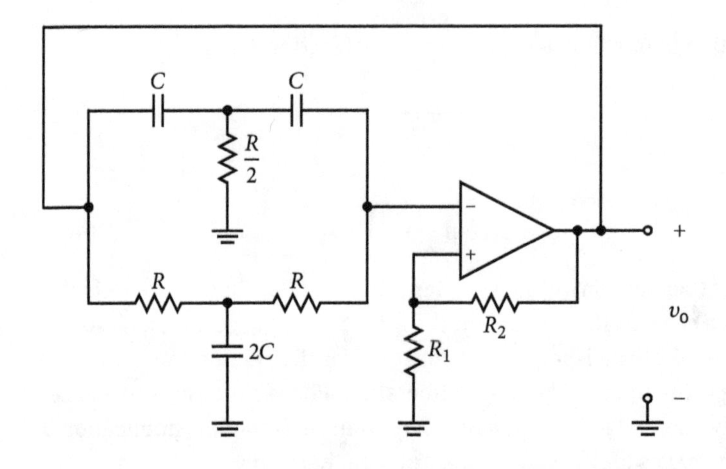

Fig. 4.32. Schematic of a twin-T bridge oscillator.

4.2.1.7 Summary on harmonic positive feedback oscillators

In this section, we want to determine how many variants (i.e., how many different *basic* circuit configurations) of harmonic feedback oscillators are possible.

All feedback oscillators have positive feedback from the output to the input of the active device. The oscillation property lies in the (complex) closed-loop transfer function. Consequently there are two conditions to be considered, the

- phase criterion, and the
- magnitude criterion.

The phase criterion delivers the frequency equation which gives the frequency (-ies) at which the phase shift of the closed-loop transfer function is zero (or a multiple of 360°). The magnitude criterion delivers the stability equation from which the gain of the active device can be determined so that the closed-loop gain (at the oscillation frequency) is exactly 1. These two criteria may be realized the following ways:

- both are realized in the positive feedback circuit (e.g., in the standard LC oscillators)
- the phase criterion is realized in the positive feedback circuit, the magnitude criterion in the amplifier by negative feedback (e.g., in Wien's bridge oscillator)
- the positive feedback circuit is frequency independent, the negative feedback selects the oscillation frequency, (e.g., twin-T oscillator)

Further, the closed-loop transfer function may have the dimension of a voltage gain or a current gain. If we proceeded with the voltage gain only, is this in the tradition of circuit design. But, actually, for each voltage oriented circuit there is a replica in the current world, i.e., there are twice as many circuits to be considered. This omission is not serious as the current versions can easily be constructed from their voltage counterpart by application of duality.

Then the phase criterion may be fulfilled either by

- resonant LC circuits or by
- nonresonant filters.

There are *two* choices for the reactance of a simple filter section. Only filters using capacitors are being used because inductors are expensive, clumsy and have undesired stray properties. Again, one-half of the possibilities remain idle.

As we have shown for one case (Section 4.2.1.1), the two methods of fulfilling the phase criterion are closely related. The amplifier with a capacitor in the positive feedback circuit is a gyrator (Section 3.6.5) behaving like a (dynamic) inductor so that the circuit can be explained as resonant circuit.

Adding up all variants we count at least 24 (considering that the phase-shift oscillator works both with high-pass and with low-pass filter sections). About one-third of them have been realized in practical circuits.

As the frequency selection is often done by varying the capacitor C on which the frequency depends, one distinction is of how this dependence looks like. There are *two* distinctly different dependences. Feedback oscillators based on RC filters have $1/C$ dependence, those based on LC resonant circuits $\frac{1}{\sqrt{C}}$ dependence.

Oscillators with RC feedback circuits

RC-feedback oscillators are best suited for frequencies up to about 1 MHz. The three types of harmonic feedback oscillators that use RC circuits to select the frequency are the Wien-bridge oscillator, the twin-T oscillator, and the phase-shift oscillator. As the Wien bridge is by far the most widely used type of harmonic RC-feedback oscillator, it was an obvious candidate for starting the discussion of harmonic oscillators.

Oscillators with LC feedback circuits

Normally, LC-feedback elements are used in oscillators that deliver higher frequencies of oscillation. Because of the frequency limitation (lower unity-gain frequency) of most operational amplifiers, *three-terminal components are in most cases the best choice as an active element in high-frequency feedback oscillators*. The common positive feedback circuits are of the Clapp, Colpitts, Hartley, Armstrong, and quartz type.

All feedback oscillators must obey both the phase and the magnitude criterion.

Problem

4.30. Why are oscillators based on current loop gain not presented? ?

4.3 One-port oscillators

As we have seen in Section 2.2.5 power gain can also be achieved by means of active one-ports, i.e., with one-ports having a characteristic where a portion of it has negative impedance (admittance). On the other hand, we have shown in Section 2.5.4.1 that negative impedances can be produced dynamically using amplifiers. Furthermore, we have seen in Section 4.2.1.1 that an amplifier with positive feedback is electrically equivalent to a circuit containing a negative impedance. Thus, it should be clear that there is a substantial equality between a one-port having negative impedance and an amplifier with positive feedback.

Therefore, positive feedback oscillators have their counterparts in oscillators using active one-ports. Figures4.33 and 4.34 juxtapose feedback oscillators to the corresponding active one-port oscillators.

In Table 4.7, the instability conditions of active one-ports are listed (Section 2.5.4.1).

As always when dealing with small-signal behavior we must not forget the biasing. The operating point must be set in the region with negative impedance, or it must be moved there by an activating gate signal. Naturally, the choice of the biasing net-

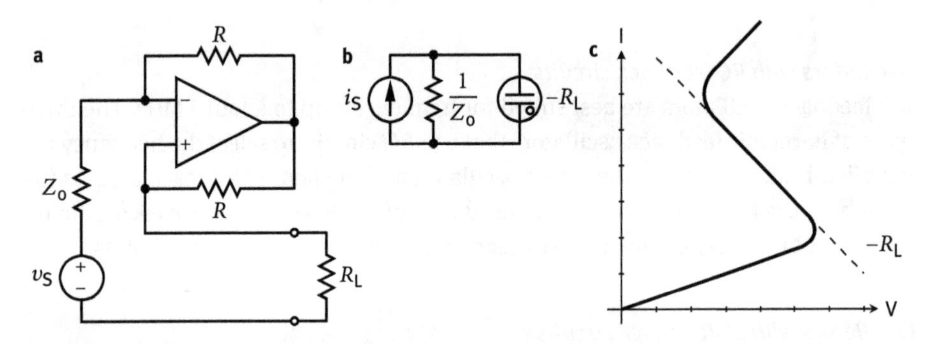

Fig. 4.33. Elements with S-shaped i-v-characteristics. (a) A two-port that produces dynamically the impedance $-R_L$ at the inverting input. (b) A one-port (neon bulb) having a (negative) impedance of $-R_L$. (c) Exemplary characteristic with a (negative) impedance of $-R_L$. To produce instability, the impedance Z_o must be $\leq |R_L|$.

Table 4.7. Instability conditions of active one-ports.

Characteristic type	Load condition	Oscillator type
S-shaped i-v-characteristic	Series resonant circuit	Harmonic
	Conductive, $> 1/R_L$	Bistable, Schmitt trigger
	Capacitive	Astable, mono stable
N-shaped i-v-characteristic	Parallel resonant circuit	Harmonic
	Resistive, $> R_S$	Bistable, Schmitt trigger
	Inductive	Astable, mono-stable

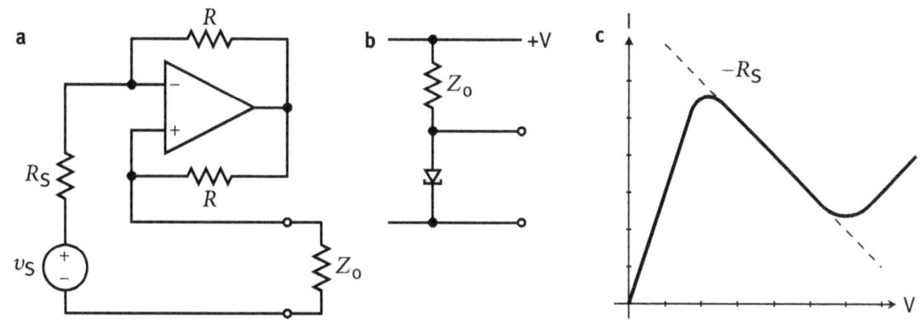

Fig. 4.34. Elements with N-shaped i-v-characteristics. (a) A two-port produces dynamically the impedance $-R_S$ at the noninverting input, (b) A one-port (tunnel diode) having a (negative) impedance of $-R_S$. (c) Exemplary characteristic with a (negative) impedance of $-R_S$. To produce instability, the impedance Z_0 must be $\geq |R_S|$.

work must be such that the stability condition of the characteristic in question is met. Remember: Elements with S-shaped i-v-characteristics are stable against open-circuit (any horizontal load line intersects the i–v-characteristic in one point, only) whereas N-shaped ones are stable against short-circuit (any vertical load line intersects the i–v-characteristic in one point, only).

Problems

4.31. For two-port oscillators with a transformer in the feedback circuit there do not exist one-port equivalents. Why is that so?

4.32. What kind of reactive element is required for a one-port with negative impedance of the N-type to form an oscillator?

4.33. What kind of reactive element is required for a one-port with negative impedance of the S-type to form an oscillator?

4.3.1 Harmonic one-port oscillators

Harmonic oscillators with one-ports are based on resonant circuits (Section 3.4.6). The instability condition is given only at the resonant frequency where the impedance (resp. admittance) is particularly small. As they are not zero as would be the case for ideal resonant circuits, they are the cause of the damping of the circuits prohibiting an oscillation with constant amplitude. The resistive portions of the resonant circuit must be counterbalanced by the negative impedance, provided, e.g., by the active one-port. Then, one gets ideal behavior of the circuit with an oscillation of constant amplitude.

The damping resistor of a series resonant circuit lies in series to L and C. Consequently, an active one-port with an N-shaped i-v-characteristic must be put in series to it so that its negative impedance matches the impedance of the resonant circuit in (absolute) magnitude. This undamps the resonant circuit which then oscillates with constant amplitude.

Again, the nonlinearity of the characteristic makes such an oscillation with stable amplitude possible. As long as the negative impedance at the operating point is larger than the impedance of the resonant circuit, the oscillator will increase its amplitude until the amplitude is reached at which the exact match of the two impedances is given. Then the oscillator will oscillate with this amplitude.

For the parallel resonant circuit the situation is dual. When an active one-port with an S-shaped i-v-characteristic shunts a parallel resonant circuit its damping conductance must be compensated for by the negative conductance of the one-port. This situation was shown in depth in Section 4.2.1.1.

Problem

4.34. Design a one-port oscillator equivalent to an Armstrong oscillator.

4.3.2 One-port relaxation oscillators

In one-port relaxation oscillators the instability condition is fulfilled for all frequencies above a minimum frequency f_{min}. Consequently, there are two groups, depending on the value of that minimum frequency f_{min}:

- $f_{min} = 0$, for bistable multivibrators and Schmitt triggers, and
- $f_{min} > 0$, for astable and monostable multivibrators.

Again, there are the following alternatives for the biasing:

- there is a unique quiescent operating point (with a high H output, or a low L output) from which an input (gate) signal moves the operating point to the portion of the characteristic with negative impedance to provoke instability, or
- there is no unique quiescent operating point.

This situation is sketched in Table 4.8.

If the operating point is in the lower branch of the positive impedance (conductance) of the characteristic, the output state is L if it is in the upper branch it is H. Thus,

Table 4.8. Overview over one-port relaxation oscillators.

Quiescent operating point	f_{min} (hertz)	Kind of oscillator
Unique	0	Schmitt trigger
	> 0	Monostable multivibrator
Not unique	0	Bistable multivibrator
	> 0	Astable multivibrator

N-shaped i-v-characteristics will result in "rectangular" voltage signals, whereas S-shaped ones produce "rectangular" current signals.

Different from the feedback relaxation oscillators the practical restriction to capacitors is not feasible because the N-shaped i-v-characteristic which results in current output signals, requires (for duality reasons) an inductor.

Problems

?

4.35. Can all four types of relaxation oscillators be built with active one-ports?

4.36. Can blocking oscillators be built with active one-ports?

4.4 Time and frequency as analog variables

When an (analog) signal occurs usually two pieces of information are of interest
- its amplitude, and
- its timing.

Additional information can sometimes be gained from its shape, e.g., its rise time.

With regard to signal height, discrimination against noise (i.e., unwanted signals) is most common. Such circuits are called discriminator with its typical representatives the Schmitt trigger (Section 4.1.2.1) and the comparator (Section 2.3.4.2). As the information lies in the discriminator threshold (a voltage level), such an action is also called *level triggering*. To get the full information on the signal height an analog-to-digital-converter (Section 6.4.2) is needed.

In this section and the following sections, we concentrate on timing. As discussed in the introduction to this chapter, absolute simultaneity cannot be determined. The crucial property of time measurements is the *resolution* (or *resolving*) *time*. This is the smallest time difference that can be captured (measured). Usually, the time of the signal is correlated in time with some event so that actually the instant of that event and not that of the signal is of interest. In an ideal case, the *instant of an event* is recorded as the step of a step signal. However, electrical step signals have finite rise time (Section 3.3.3) so that the trigger moment depends on the (absolute) discriminator level and the signal height. If the required time resolution is shorter than the rise time of the signal, the discriminator level is set so low that it discriminates against noise, and the effect of the pulse height is disregarded. In the case of binary signals which have all the same pulse height, only such *edge triggering* is needed. In the case of signals having different height (e.g., analog signals), there are special circuits available (Sections 4.4.1.2 and 4.4.1.3) that make the trigger instant quite independent of the pulse height and rise time.

Obviously, frequency being the inverse of time must be analog, too, even if it is often given as an integer variable (e.g., "60 Hz"). Originally, frequency was only assigned to a steady-state (sinusoidal) oscillation. However, it is common to call the inverse of the time length of a period (a cycle) frequency. For signals that arrive statistically, the term *signal rate* is more appropriate than frequency. The unit of a signal rate is events/s and not Hz. The usual incongruity of inverse variables at the limits of their validity exists in this case, too: the frequency zero (DC current) cannot be expressed via the time and the time interval zero not via the frequency.

Fully digital signals are digital with regard to amplitude and time.

Problems

4.37. Which electronic variable is more important time or frequency?

4.38. How many frequencies does a step function contain?

4.39. Why is it not possible to produce an ideal step-function electronically?

4.4.1 Trigger

As the same device serves two purposes, it does not surprise that the names discriminator and trigger get intermixed. Thus, we have three types of triggers, all of them called discriminators, namely
- leading-edge discriminator,
- zero-crossing discriminator, and
- constant-fraction discriminator.

The time determined by a trigger signal is not the instant at which the triggering event occurred. Relative to the instant there will be some (*propagation*) *delay* and there will be some *time dispersion*, i.e., the “delay” time will not be constant but vary (statistically) around a mean value.

Problem

4.40. What is the difference between a discriminator and a trigger?

4.4.1.1 Leading-edge trigger (time jitter)

Leading-edge triggers are, generally, based on the Schmitt trigger principle. They are optimum for binary (digital) signals, as there is no amplitude variation involved. The sole (relative) timing uncertainty is due to *time jitter*. Jitter in the timing of a signal is a momentary, noncumulative variation of the signal position on the time axis. It

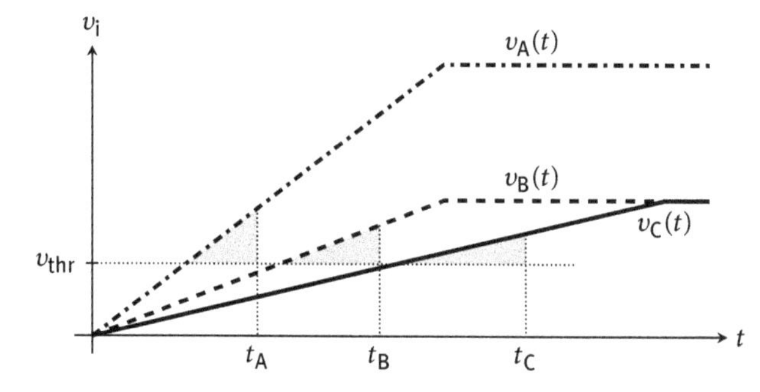

Fig. 4.35. Walk of the trigger instant at a given discriminator threshold.

is something like timing noise. As the width of the time distribution of jitter can be kept quite small ($\ll$ 1 ns), it affects only high-speed applications. The cause of jitter is mainly electronic noise of all kind. If a reduction of the amount of jitter is required in a specific application, the cause of it (e.g., thermal noise, cross talk, electromagnetic interference) must be (partially) eliminated.

Figure 4.35 illustrates how the trigger instant is moved depending in the rise time of the input signal. There is not only the "geometric" effect that it takes longer for a signal with longer rise time to cross the threshold voltage v_{thr}, but there is the additional effect that the charging of the input capacity takes longer with a slow rising signal. This effect is indicated by the shaded triangle (indicating equal areas) in the figure. The dependence of the trigger instant on the rise time (and at a constant rise time on the amplitude) is called *walk*.

If, in analog applications, walk must be essentially smaller than the rise time of the triggering signals, one of the other two triggering methods should be used.

Problems

?

4.41. What is time jitter?

4.42. Which type of trigger is prone to walk?

4.4.1.2 Zero-crossing trigger

The accumulation (summing) of voltage step signals (i.e., unshaped signals) would soon make the operating point leave the operating range of an amplifier. Therefore, it is necessary to make the voltage of the instantaneous operating point go back toward the quiescent operating point (to the baseline). This can be done by a high-pass filter, i.e., by differentiating the step signal (Sections 3.2.1 and 3.4.3). Differentiating a step signal twice (at best with the same time constant) results in a signal with an oppositely charged undershoot so that no direct current contribution to the signal remains (see also Section 3.4.3.1). Thus, in the long run, no shift of the baseline occurs. In Figure 4.36 it is shown that such double-differentiated signals cross the baseline at the same instant, independent of their amplitude. Setting the discriminator threshold to

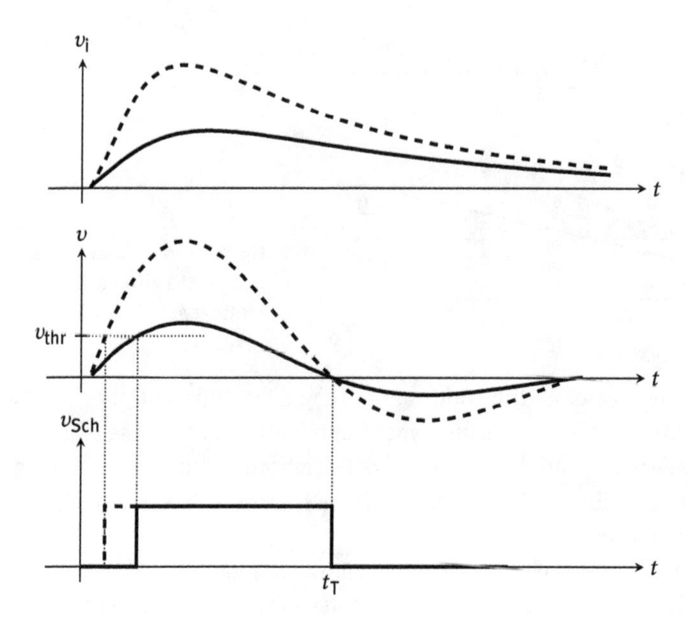

Fig. 4.36. Reduced walk with zero-crossing triggering compared to edge triggering.

zero volts catches these moments of zero-crossing. This moment occurs (much) later than the crossing of the leading edge. This delay might be disadvantageous.

The walk of the trigger instant is essentially reduced to the difference in time it takes to charge the input capacitance (in Figure 4.35 the amount of charge needed is symbolized by the shaded triangles).

Problem

4.43. What kind of signals are required for zero-crossing triggers?

4.4.1.3 Constant-fraction trigger

Zero-crossing of a bipolar signal that is especially tailored for this purpose has proved to provide the smallest walk due to both rise time and amplitude variations. Such a circuit is called constant-fraction trigger as some fraction of the input signal (determined experimentally) is used to produce a composite signal for the zero-crossing trigger.

The incoming (unipolar) signal is split in two signal branches. One delays by t_d (some fraction of the rise time), the other inverts and attenuates to some fraction f. The constant fraction signal is a composite of the inverted, attenuated signal and the delayed unattenuated signal. The proper choice of the delay time t_d minimizes walk due to rise time and amplitude variations, whereas the optimum choice of f minimizes jitter. Input signals that lie outside the dynamic range of the trigger (or of the amplifier feeding the trigger) will result in saturated pulses (having a different pulse shape). Such pulses might cause multiple output signals. The same remedy as used in "debouncing" of switches (Section 4.1.2.2) can be applied to suppress multiple output signals.

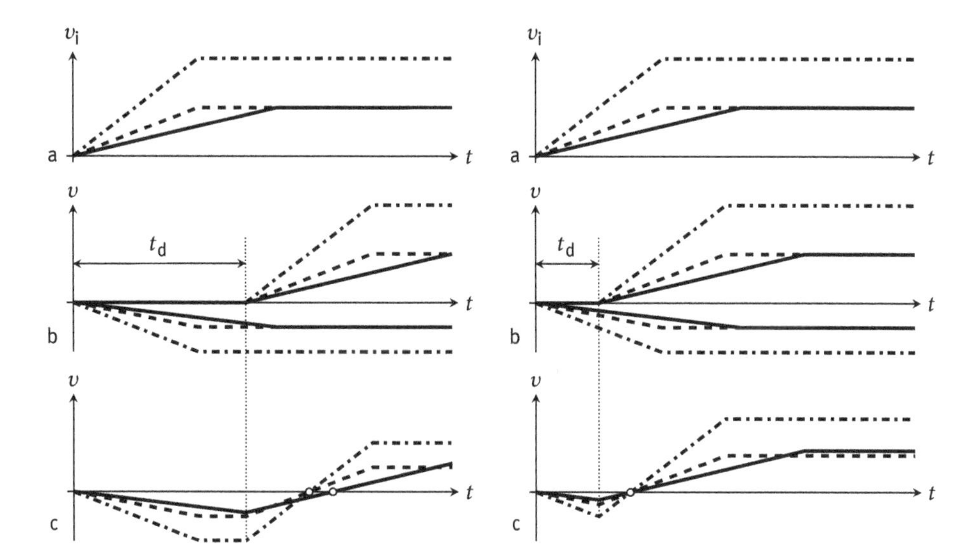

Fig. 4.37. Pulse shapes in a true-constant-fraction discriminator: (a) input signals; (b) modified input signals; (c) bipolar timing signals.

Fig. 4.38. Timing relations of the input signal applied to an amplitude-to-rise time-compensated (ARC) constant-fraction discriminator: (a) input signals; (b) modified input signals; (c) bipolar timing signals.

There are several ways to arrive at a constant fraction trigger. The true-constant-fraction (TCF) trigger (Figure 4.37) triggers effectively at the same input-signal threshold. To achieve that, t_d must be at least somewhat longer than the rise time of a typical signal. If there is a signal with longer rise time, the timing will be off as shown in Figure 4.37c.

To minimize the rise time dependence the delay t_d must be reduced to less than the minimum rise time of the input signals. Such a circuit is realized in the *amplitude-and-rise-time-compensated* (ARC) constant fraction discriminator. Its timing response is sketched in Figure 4.38.

In an ARC constant fraction discriminator the effective pulse height threshold at which the discriminator triggers is inversely proportional to the rise time.

Constant-fraction discriminators are the best choice with regard to minimal timing errors (walk, jitter) except for signals with a narrow range both in amplitude and rise time. In these cases leading edge timing might provide better timing resolution.

Example 4.7 (Unreflected use of a constant fraction discriminator). At an international experiment in Japan it seemed as if the percentage loss of recorded counts increased with decreasing primary count rate. This paradox was traced back to the constant fraction discriminator employed for standardizing the (analog) detector signals. The lower the primary count rate became, the higher the percentage of background counts became. A good deal of the background counts originated from cosmic rays having high energy and, consequently, high signal amplitude. Signals with an amplitude outside the dynamic range of the amplifier system get distorted, i.e., clipped at about the height of the power supply voltage. Therefore, signals of high energetic events have a flat top. This unexpected signal form cannot be handled properly by a constant fraction discriminator. Usually it will output multiple signals during the duration of this clipped input signal. Even if there was only one event, many signals will be recorded as lost which fully explained the observed mystery.

? **Problem**

4.44. What happens when an out-of-range signal (i.e., a signal of a wrong shape) is processed by a constant fraction discriminator?

4.4.2 Coincidence circuits (logic gates)

Logic gates belong to analog electronics (just as the other circuits with binary output we covered up to now) when used as coincidence circuits even if the name *logic* gates suggests a binary nature. The reason for this possibly amazing statement is based on the fact that this circuit is used in timing which is an analog process.

Simultaneity between two events as determined by the simultaneity of their signals can only be found out within some resolution time. Signals that occur within a given resolution time are called *coincident*. If one kind of the signals must be delayed to be coincident with those of the other kind, this is called a *delayed coincidence*.

In cases with time-correlation in the occurrence of two signals the time distances give rise to a time (interval) spectrum that can be analyzed either directly or with an analog-to-digital converter after being converted into a pulse height spectrum by means of a time-to-pulse-height converter (Section 4.5.1).

A straightforward way of determining a coincidence between output signals of two triggers is to sum the output signals (e.g., by means of a summing amplifier, Section 2.5.1.1). An output signal, that has twice the height of an output signal stemming from a single input, indicates that both signals occurred within the *resolution time* which is the sum of the length of both signals. This *overlap coincidence* gets into trouble if the trailing edge of one signal intersects the leading edge of the other. There will be a short output signal with a height that depends on the amount of time overlap. Therefore, output signals must be shaped, e.g., by a monostable multivibrator which shapes the output signals to signals of constant length and pulse-height H having short rise-time.

If the only purpose of such a circuit is to find out whether two (or more) signals at its input(s) are present at the same time, it is called an AND gate. In binary logic, truth tables are powerful means to describe the relation between inputs and outputs of logic elements (Section 5.3). Although we are dealing here with circuits having analog

Table 4.9. Truth table of a coincidence circuit (an electronic two-input positive logic AND gate).

Signal at input A	Signal at input B	Signal at output X
L	L	L
L	H	L
H	L	L
H	H	H

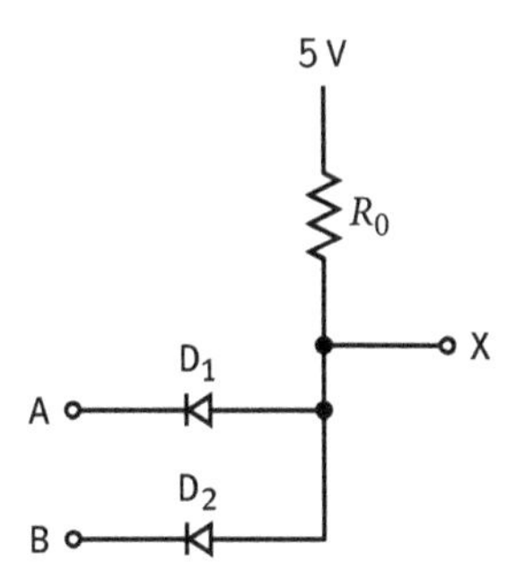

Fig. 4.39. Two-input AND gate in diode logic.

inputs and binary outputs truth tables are useful already in this context. A truth table of a two-input AND gate is shown in Table 4.9. By its nature no information on the analog property (timing) is contained in such a truth table. Obviously there will be an output signal only then when the two input signals overlap in time. There are several ways of realizing logic circuits. This will be covered in Chapter 5.

A simple realization of a two-input AND gate in *diode logic* (DL) is shown in Figure 4.39. The low level L is about 0 V, the high level H about 5 V. With one input or both inputs at L (0 V) gives an output signal L, the voltage of a forward biased diode ($\approx$ 0.7 V). If both inputs are at H (5 V), the diodes are reverse biased, and the output is connected to the 5-V supply voltage via the resistor R_0, i.e., it is at H.

Problems

?

4.45. Why is a coincidence circuit an analog circuit?

4.46. Which information on timing properties are contained in a truth table?

4.4.2.1 Bothe and Rossi circuit

In Figure 4.40 two three-terminal components are cascaded to make a coincidence circuit that is basically a Bothe circuit. Again we have a two-input AND gate, this time in transistor logic. If any of the inputs is L, no current will flow as the corresponding transistor is in the cut-off region. Only if *both* inputs are H, both transistors get saturated and the output voltage is close to the power supply voltage, i.e., high H.

Arranging transistors in parallel is another way to arrive at an AND gate. The advantage of a parallel over the cascaded arrangement, discussed above, is the simplicity of extending the circuit to more than two inputs. This is not so easily achieved with

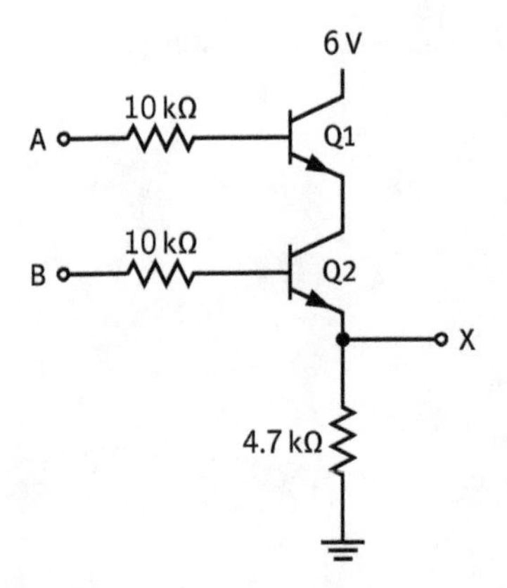

Fig. 4.40. Principle of a Bothe AND gate shown with bipolar npn junction transistors.

the Bothe circuit, because of problems with the biasing. Figure 4.41 gives an example how the Rossi version of a two-input AND gate using bipolar junction transistors can be designed. Exactly speaking it is a positive logic NAND gate, i.e., an AND gate with inversion of the output signal (see truth Table 4.10).

Problem

4.47. In which regard is the Rossi circuit dual to the Bothe circuit?

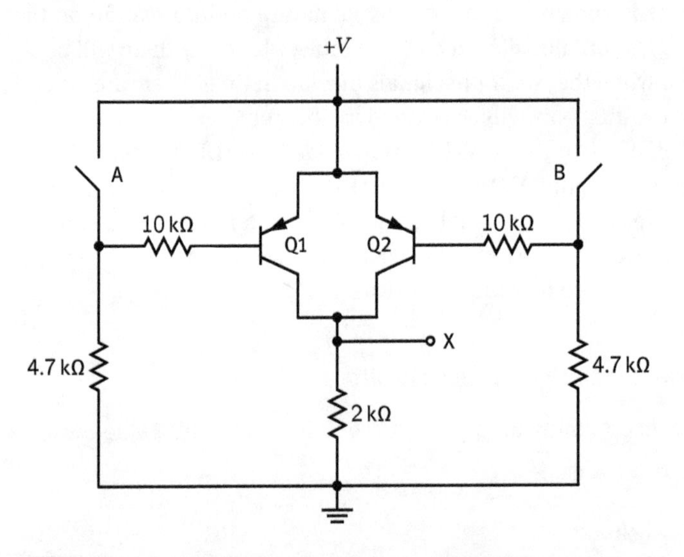

Fig. 4.41. Principle of a Rossi positive logic NAND gate shown with bipolar PNP junction transistors.

Table 4.10. Truth table of an two-input positive logic NAND gate.

Signal at input A	Signal at input B	Output signal X	Inverted X signal
L	L	H	L
L	H	H	L
H	L	H	L
H	H	L	H

4.4.2.2 CMOS technology for logic gates

The term *metal-oxide-semiconductor* describes the physical structure of the field-effect transistors in question which have a metallic gate electrode placed on an oxide insulator, which in turn is placed on semiconductor material. Although complementary metal-oxide-semiconductor (CMOS) is a technology for constructing integrated circuits which by itself is not topic of this book, circuits that became easily feasible by it, are of interest. The word *complementary* refers to the design method of using complementary and symmetrical pairs of p-type and n-type metal oxide semiconductor field effect transistors in the same circuit. By applying this method, load resistors become obsolete because they are replaced by complementary switched FETs (Figure 4.42). Thus, in both binary states (L and H) one of the cascaded FETs is in the cut-off state so that its quiescent operating current is close to zero and, therefore, the quiescent power as well. Power is only dissipated momentarily during the transition from L to H and vice versa keeping the circuit "cool".

The two-input positive logic NAND gate in Figure 4.42 can be understood as cascade of a Bothe circuit (Q_1 and Q_2) with a Rossi circuit (Q_3 and Q_4). With A and B at L, the output is H. Q_1 and Q_2 are in the cut-off region, Q_3 and Q_4 completely open (with negligible current). The state H at the output remains that way as long as at least one input signal is low. Only when both input signals are H Q_1 and Q_2 are fully open and Q_3 and Q_4 in the cut-off region resulting in an L state at the output.

A reminder: In Section 2.2.4.1 we learnt that the maximum power transfer between two elements is then when the voltage is equal across both elements. Obviously, this condition occurs in the middle of the transition from L to H and vice versa. Therefore, it is quite obvious that the power consumed by CMOS circuits depends on the switching frequency. The power-saving feature of complimentary circuits was already shown in Section 3.3.2.1 for the complimentary emitter follower.

Problem

4.48. Does the power consumption of a CMOS circuit depends on the clock frequency?

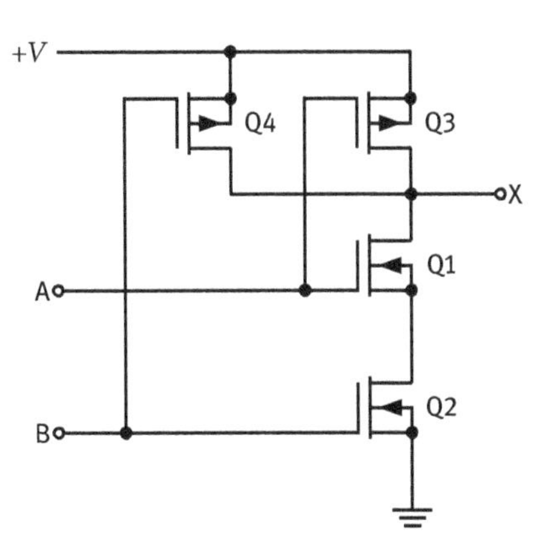

Fig. 4.42. Two-input positive logic NAND gate in CMOS technology.

Table 4.11. Truth table of a two-input anticoincidence circuit.

Signal (at input A)	Veto-signal (at input B)	Signal at output X
L	L	L
L	L or H	L
H	L	H
H	H	L

4.4.2.3 Anticoincidence circuit

Such a circuit is also called *inhibit* circuit. As the name suggests, one of the inputs provides a veto that stops the transmission of a signal at the other input(s). Table 4.11 shows the truth table of such a two-input circuit. In the case of an overlap coincidence the veto signal must cover all of the time the signal to be vetoed is present, i.e., it must cover the resolution time which will be longer than the length of the other signal.

Problem

4.49. Is an anticoincidence circuit an analog circuit, i.e., does it retain the timing information?

4.4.3 Linear gates

A linear gate resembles an anticoincidence circuit insofar as, again, the two inputs are not equivalent. However, Table 4.11 does not really apply, as the output of a linear gate is analog and not binary in amplitude. Therefore, there is one analog input, and one binary gate input and the output signal has analog amplitude. Figure 4.43 shows a circuit that transmits an analog signal only during the duration of a gate signal, the so-called *acquisition time*. The operating point of a class C amplifier is moved by

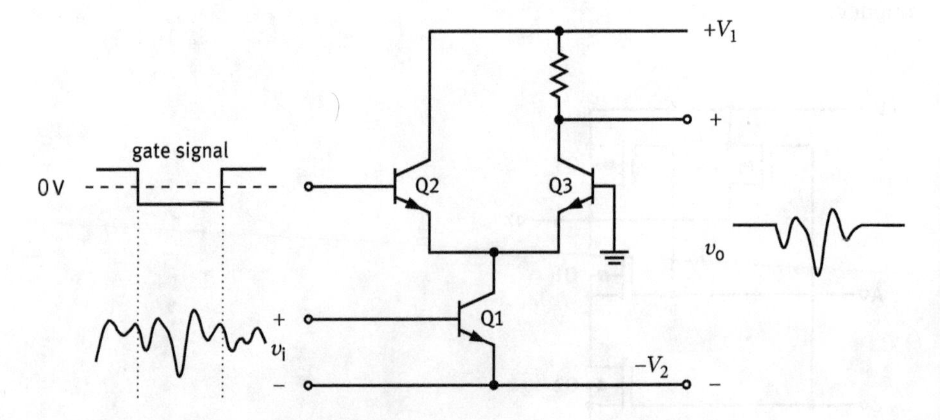

Fig. 4.43. Linear gate based on a long-tailed pair circuit.

the gate signal temporarily into the amplifier (class A) region so that the analog input signal is properly amplified as long as the gate signal lasts. As so often a variation of the long-tailed pair is used. Only when Q_2 is cut-off by means of a negative signal the analog signal at the input of Q_1 gets transmitted to the output.

Problems

?

4.50. What is the difference in the output signal of a linear gate and a logic gate?

4.51. May a linear gate be used with logic signals?

4.52. When active, the cascade of Q_1 and Q_3 in Figure 4.43 has a distinct name. What is it?

4.4.3.1 Sampling

The principle of linear gates is also applied for time digitizing of continuous analog signals. This operation is called *sampling*. In sampled signals, the analog amplitude information is maintained whereas the timing information is digitized by sampling with a frequency that is appropriate for the task. However, often it is more advantageous first to digitize the amplitude before it is sampled. Digital music (conversion of a sound wave to a sequence of discrete-time signals), digital telephone, digital cameras, digital television, and the use of digital measuring instruments are well-known fields which are based on sampling. The sampling frequency or sampling rate is as low as 16 kilosample/s for digital telephone applications and up to 40 gigasample/s in very high-speed digital oscilloscopes.

An ideal sampler generates samples equivalent to the instantaneous value of a continuous signal at the desired time instants spaced by the sampling interval. From looking at the output, it cannot be told, how the input looked between the sampling instants. If the input changed slowly compared to the sampling rate, then the value of the signal between two sample moments was somewhere between the two sampled values so that a reconstruction by means of interpolation is justified. However, for a rapidly changing input signal this procedure is not correct because for too wide sample intervals the reconstruction will fail, because of a misinterpretation within the interpolation process. This failure is called aliasing (Section 6.1).

Aliasing can be avoided by following the *Nyquist–Shannon sampling theorem*

It specifies that reconstruction of any signal is only possible when the sampling frequency is at least twice that of the highest frequency component of the signal being sampled. The *Nyquist frequency* (half the sample rate) must exceed the highest frequency component of the signal being sampled.

Other reasons for a flawed reconstruction are
- *noise* (affecting the exact amplitude definition)
- *jitter* (affecting the exact timing definition, Section 4.4.1.1), and
- *aperture effect* (loss of information).

Due to the finite time resolution, the amplitude of the sample is a time average over the resolution width, rather than that of a sampling instant.

Problem

4.53. The human ear is receptive for sound of frequencies between about 16 Hz and 16 kHz, depending on age.
(a) What is the Nyquist frequency when sampling sound that is meant for human ears?
(b) Hi-fi audio compact disks are based on a sampling rate of 44.1 ksamples/s. Is this sampling rate sufficient to obey the Nyquist–Shannon sampling theorem?

4.4.3.2 Sample and hold (peak detector)

When a (continuous) analog signal was sampled at some instant (i.e., averaged over some acquisition time), its analog value must be preserved until it has been used (e.g., until it has been processed). Its analog value is *held* (*frozen* or *locked*) at this value for a specified time length as needed for processing the signal. The obvious storage device for voltage is a capacitor. As the voltage across a capacitor is inversely proportional to its capacitance and proportional to its charge (Sections 3.2.1 and 3.3.2) a large capacitor with small leakage of the charge results in a small droop. The capacitor is invariably discharged by its leakage current and by unavoidable load currents, which makes the storage inherently volatile. It is advantageous to use dynamically enhanced capacities (Miller effect, Section 2.4.3.2) in particular by way of integration amplifiers (Section 3.6.4.1).

Analog information cannot be stored forever. The loss of voltage (voltage droop) in a sample-and-hold circuit is given by the droop rate in units of V/s.

A related principle is applied in *peak detection* (of single pulses). Figure 4.44 explains the principle. The capacitor C is charged as long as the input voltage signal (minus forward voltage of the diode) exceeds the voltage across the capacitor. If the

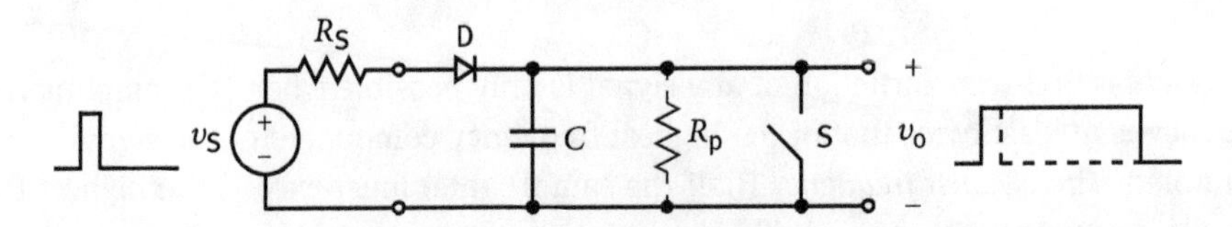

Fig. 4.44. Principle of a peak detector.

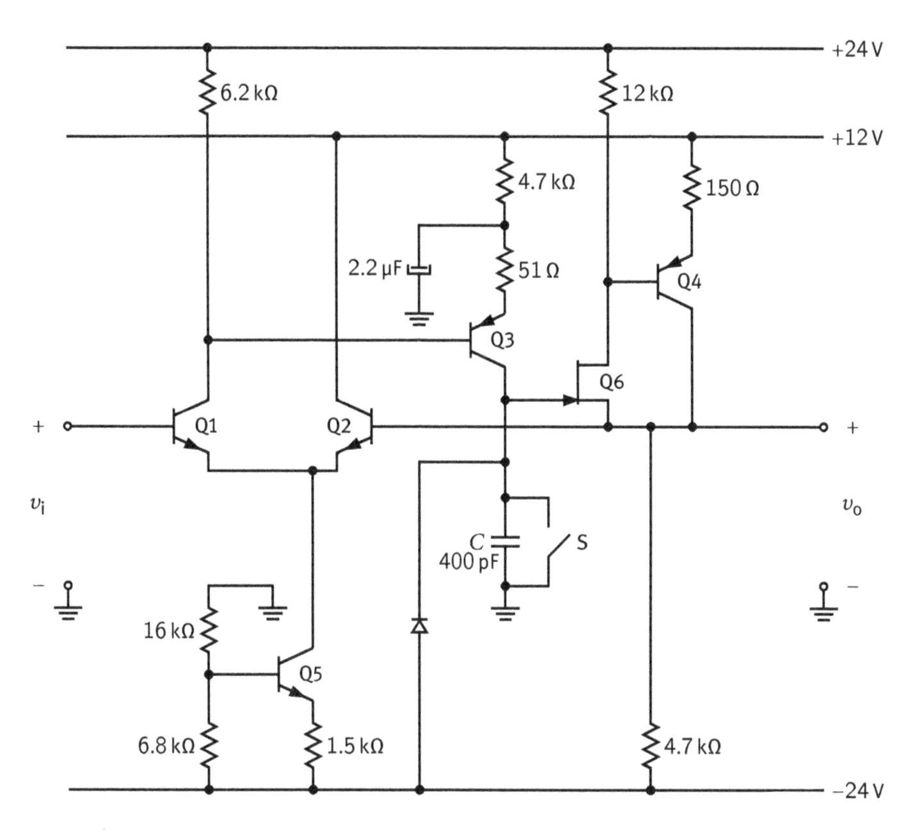

Fig. 4.45. Example of a sample-and-hold circuit with feedback.

input voltage gets smaller, the diode is reverse biased and the voltage across the capacitor stays constant (disregarding its discharge by leakage current). After the analog information is not needed any more, the capacitor is discharged by closing the switch. Such a circuit is effectively a *pulse-lengthener*. A low droop peak detector using a 1 nF polystyrene capacitor with an unbelievable low droop of 1 mV/s is reported.

Let us use Figure 4.45 to explain in more detail, how a sample-and-hold circuit works. Via Q_1 of the long-tailed pair and Q_3 the capacitor C gets positively charged if a positive input signal is applied. Q_1 and Q_2 of the long-tailed pair form a differential amplifier. By feedback through the FET and Q_4, the momentary voltage at the capacitor is fed to the input of Q_2 so that the voltage across the capacitor will rise until there is no voltage difference at the inputs of the differential amplifier. After discharging the capacitor by the switch, the circuit is ready for the next signal. (The leakage current of the diode compensates for the leakage current of Q_3).

In monolithic FET-input integrating amplifiers, sample-and-hold circuits can be found with internal capacitors of typically 10^{-10} F. A switch with very low leakage is essential. These circuits have a typical droop rate of 1 V/s and aperture times from

10^{-5} to 10^{-8} s to be able to fulfill a great variety of requirements. In moderately fast circuits, an agreement of the stored voltage with the input voltage of better than 10^{-4} is standard.

Sample-and-hold circuits are essential in analog-to-digital conversion (Chapter 6), in particular in connection with voltage comparing ADCs using the successive approximation method (Section 6.4.2.5). Another typical application would be if signals of several input channels that were sampled by a common sample clock must be handled (multiplexer).

As pointed out, the very low leakage of the charge in the capacitor is essential. Therefore, it must not be loaded to any noticeable degree which calls for a very low admittance of all components (e.g. FET gates) shunting the capacitor.

Problem

4.54. What is the loss of voltage in a hold circuit called?

4.4.4 Analog delays

The purpose of analog delays is the preservation of analog information for future use. Analog information of all kind (acoustic, optical, etc.) can be stored for long times by magnetic or other nonelectronic media. Some of this storage is nonvolatile. However, here we are disregarding nonelectronic methods.

If it is the matter of preserving the voltage value occurring at some instant, a sample-and-hold circuit (previous section) will do the job. If one must preserve the peak voltage value of a single signal, a peak detector (previous section) will do the job. An interesting way of storing (delaying) sampled analog voltage signals in the millisecond range is an analog shift register that moves from cell to cell the charge of a CCD (charge-coupled device) which is proportional to the analog values. Such circuits are available in very-large-scale integration (VLSI) technique. Analog charge packets are shifted from one cell to another by clock pulses. Depending on the clock rate and the number of cells the cell content is made available at a later time. This way a sequence of sampled analog values can be stored for later use.

Storing the full time-dependence of an analog signal for a given time interval t_d means moving it on the time axis by t_d, i.e., delaying it by that amount. For very long delays, the only practical method is recording and reproduction using nonelectronic devices. The only reasonable way of accomplishing analog delay by purely electronic means is to take advantage of the propagation delay (Section 3.5.3).

Propagation of signals in transmission lines is somewhat slower than that of light in vacuum (0.3 m/ns). Consequently, 75.0 m of a 50 Ω-coaxial cable (type RG-58) can provide a delay of 383 ns. However, as we have learnt the transmission through coaxial cables is lossy, in particular at the highest frequencies. Thus, only a restricted spectrum of frequencies can be "stored" resulting in a rise time of a voltage step signal of about 6 ns after transmission through above cable.

Coaxial cables are useful for delays measured in nanoseconds, with an upper limit of a few microseconds. Lumped delay lines consisting of LC-filter sections are rarely used, partly, because of the unpopular inductors even if they extend the useful delay range due to the smaller signal propagation velocity.

A coaxial cable, with a center conductor wound helically around an insulating (preferable ferrite) core, can provide specific delays of 20 to 3000 ns/m. Not surprisingly these delay lines have higher characteristic impedance (on the order of 1 kΩ) and have rather bad higher frequency transmission properties. For a 1 μs delay, the bandwidth is around 10^7 Hz, which restricts their use to not too fast applications.

Problem

?

4.55. Name electronic ways of storing analog amplitude information.

4.4.5 Signal shortener

In several applications, the information resting in a signal is the (maximum) pulse height, i.e., the portion of the signal after the peak is of little interest. This statement is particularly true if the original signal is a step signal.

Shortening of step signals is particularly simple because it only takes a high-pass filter with a short time constant to remove most of the signal except for the step the height of which is preserved (differentiation, Section 3.4.8.1). By means of delay lines, step signals can be shortened as well, either by reflection from a short-circuited line (Section 3.5.3.1) shortening the signal to twice the delay time of the line, or by inserting a delay line into one of the input branches of a difference amplifier (Section 2.5.3).

A circuit with the following features provides a straightforward way of shortening a pulse. By having the aperture time of a linear gate (Section 4.4.3) just so short that the peak voltage is included and the rest of the signal gets suppressed, the signal is shortened without losing the information stored in the peak value.

Problems

?

4.56. Why can it be beneficial to shorten pulses?

4.57. A single positive rectangular pulse can be shortened by means of a difference amplifier and a delay line. What does the output signal look like?

4.5 Analog conversion of signal attributes

The conversion of any of the analog signal variables (amplitude, time, frequency) into any other analog signal variable is a purely analog process even if it is done in preparation for digitizing. Therefore, we are dealing with it here without reference to digitizing which will be covered in Chapter 6. The substitution of one variable for the other is entirely natural and should always be employed when a better performance can be expected without too much additional circuitry.

Problem

4.58. Does the conversion of one analog variable into another belong to the field of analog or digital electronics?

4.5.1 Time-to-amplitude (-pulse-height) converter (TAC or TPHC)

Basically, two methods are applied to achieve time-to-amplitude conversion
- the overlap method, and
- the start-stop method.

If the time distance of signals is converted into amplitudes by the *overlap method*, there are special requirements on the quality of these (rectangular) signals:
- their lengths must not only agree with the maximum time distance to be converted, but
- they must be highly stable, as well.

If two such signals are fed into a two-input AND gate (Section 4.4.2), the length of the (rectangular) output signal is that time span which these signals overlap. The maximum overlap occurs for coincident signals, the minimum at the edge of the range given by twice the length of the input signal. By integration (with an active low-pass filter, Section 3.6.4.1), the rectangular output signals are converted into ramp signals with a height which is inversely proportional to the time distance of the two signals. Although the linearity of the conversion is far from perfect, this method has its merits when processing high counting rates because of its simplicity. Figure 4.46 shows a simple circuit of an overlap time-to-amplitude converter.

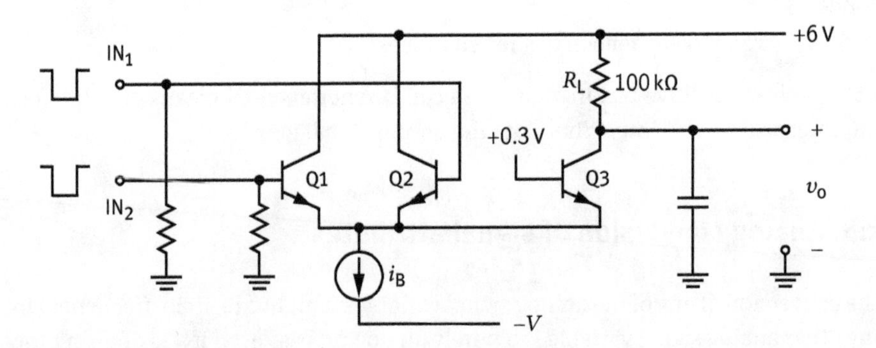

Fig. 4.46. Principle of an overlap time-to-amplitude converter.

A time-to-amplitude converter, based on the *start-stop method*, delivers an analog output signal with an amplitude that is proportional to the time difference between the leading edges of the start and stop input pulses. The start signal opens the switch across the converter capacitor so that it begins to be charged at a rate determined by a constant-current source. The current linearly charges the capacitor, producing a ramp voltage (Section 3.2.1). The leading edge of the stop signal stops the charging of the capacitor. As the charging current is constant, the voltage across the capacitor is linearly proportional to the current and the time interval, and inversely proportional to the value of the capacitance. The conversion constant (in units of V/s) can be chosen by varying either the value of the source current or that of the integrating capacitor. The integrating capacitor retains the charge it received during the conversion. The resulting voltage signal is passed through a buffer amplifier (with high input and low output impedance) to a linear gate. The linear gate transmits the voltage pulse, a rectangular signal with a height that is proportional to the time interval between the start and stops pulses, through the output amplifier to the output. At the end of the output pulse, the integrating capacitor is discharged to ground potential by closing the switch that shunts it. The circuit is ready for the next pair of start and stop signals.

Thus, the spectrum of the time intervals converted into a pulse height spectrum can indirectly be measured with standard pulse height spectroscopy (Section 6.4).

Example 4.8 (Adjusting the full-scale (FS) range of a TAC to the need at hand). According to (4.91) the FS-range of a TAC determines the range of the time intervals that can be converted. In a commercial TAC various FS-time intervals are offered by switching the size of the conversion capacitor. If now the built-in FS-ranges are, e.g., 0.3 and 0.5 μs but 0.4 μs is needed, the second range could only be used at the loss of resolution. A simple remedy is to shunt the appropriate conversion capacitor with one that has one-third of its value, effectively changing this FS-range from 0.3 to 0.4 μs. To have as little temperature dependence as possible this shunting capacitor should be of the mica type.

Problem

?

4.59. Which of the two signals of an overlap coincidence serves as start signal?

4.5.2 Amplitude-to-time converter (time-length modulator)

Conversion of a voltage amplitude to time or vice versa is based on

$$v_C = \frac{Q_C}{C} = \frac{i_S}{C} \times t \tag{4.91}$$

because of

$$Q_C = \int i_C \, dt = i_S \times t \tag{4.92}$$

with $i_C = i_S$ = constant in time.

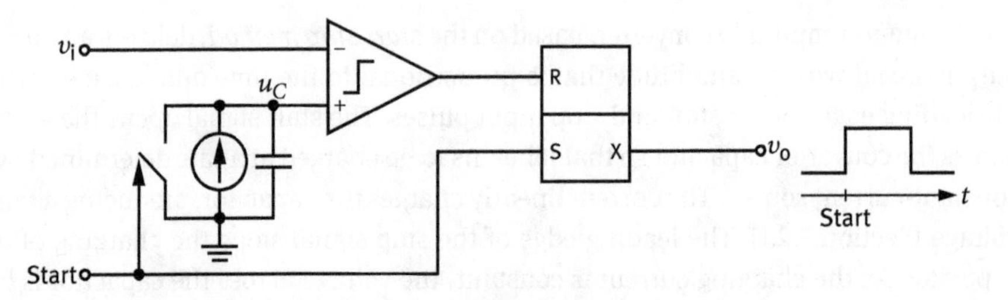

Fig. 4.47. Principle of an amplitude-to-time converter.

Thus, the conversion factor is i_S/C which can be tailored to the requirement at hand.

Problem

4.60. Is the conversion of time interval into pulse height (and vice versa) a digital process?

4.5.2.1 Direct voltage-to-time converter

Figure 4.47 shows the principle of such a converter. A comparator compares the voltage amplitude to be converted with the voltage across an integrating capacitor C that is charged by a constant current i_S. When the ramp voltage signal v_C across the capacitor is as high as the input amplitude, the output signal that was started at the beginning of the conversion process is terminated by resetting an SR-flip–flop (Sections 4.1.3.2 and 5.3.2.1). Thus, the length of the output signal is the charging time which is linearly related to the voltage amplitude of the input signal. The conversion constant C/i_S has a unit s/V. The stability of this conversion factor depends both on the stability of the current source and that of the capacitor.

Problem

4.61. Will the duty factor of the string of output signals of a voltage-to-time converter usually be higher than that of the input signals?

4.5.2.2 Compound voltage-to-time converter

The dependence of the direct voltage-to-time converter on the capacity value of the capacitor limits the stability of the conversion. In the following scheme, the capacity value drops out. The input voltage v_i is converted by a stable resistor R into the input current $i_i = v_i/R$. This current is used to charge a capacitor C for a given time length t_{ch} so that the capacitor gets charged to the voltage $v_{ch} = i_i \times t_{ch}/C$. (The input current follows the input voltage changes so that the accumulated capacitor charge is a measure of the input voltage integrated over the charging time t_{ch}.) Then the capacitor gets discharged by a constant current i_S from a current source. The discharging time is t_{dis} (i.e., the time from the instant at which the current source is activated until the voltage across the capacitor is zero). From $t_{\mathrm{dis}}/t_{\mathrm{ch}} = v_{\mathrm{i}}/(R \times i_{\mathrm{S}})$ the conversion constant $t_{\mathrm{ch}}/(i_{\mathrm{S}} \times R)$ is obtained which is independent of C and has as unit s/V. The stability of this conversion factor depends on the stability of the current source, of the resistor and of the length of the charging time. By using the same time base for t_{ch} and t_{dis}, any longer-term instability of the time base drops out.

Problem

4.62. Is it possible to determine whether a conversion of a voltage into a time interval inside a black box is done directly or by a compound device?

4.5.2.3 Time interval magnification

Cascading a time-to-amplitude converter and an amplitude-to-time converter makes it possible to stretch or compress a time length at will. The "time gain" of such a system is the product of the two conversion factors having the dimension of a time ratio. Because of the sequential nature of all signals such a system is rather limited in the count rate it can handle. Merging both circuits into one circuit results in a particularly simple solution needing just one capacitor. If it takes the time t_{dis} to discharge a capacitor with a constant current i_{dis} that was charged over a time interval t_{ch} with a current i_{ch}, then the "time gain" is inverse proportional to the current ratio

$$\frac{t_{\mathrm{dis}}}{t_{\mathrm{ch}}} = \frac{i_{\mathrm{ch}}}{i_{\mathrm{dis}}} . \tag{4.93}$$

Problem

4.63. Does the duty factor of a train of pulses increase by expansion of the pulse length?

4.5.3 Amplitude (voltage-)-to-frequency conversion (VFC)

Amplitude-to-frequency conversion is also known as *frequency modulation*. The output signal may either be a harmonic, a square wave, or a pulse frequency.

A straightforward conversion can be done with voltage-controlled oscillators (VCOs) that have a highly linear relation between applied voltage and frequency. A VCO is an electronic oscillator (Section 4.2) whose oscillation frequency is controlled by a (time-varying) voltage. Such oscillators are either harmonic or relaxation oscillators.

The frequency of harmonic oscillators (Section 4.2) has $1/\sqrt{C}$ or $1/C$ dependence. Of course, for RC oscillators there is an additional dependence on R which can also be used.

The frequency of relaxation oscillators (Section 4.2) depends mainly on the charging time of a capacitor C. Thus, there are two parameters that can be varied, the capacity and the (dis)charging current (R).

In both cases, a voltage applied to a *varactor* (=variable capacitor) diode that is part of the circuit's capacitance varies the frequency because a varactor diode has a voltage dependent capacity.

Harmonic VCOs are usually more stable than the relaxation VCOs, whereas for the latter the tuneable frequency range is usually wider. In addition, they offer the option to vary the charging rate of the capacitor by means of a voltage controlled current source which is particularly helpful at low frequencies. The most widely used harmonic VCO circuits are based on the Colpitts oscillator (Section 4.2.1.3). The conversion curve of simple VCOs will have a limited linear range only. To qualify as voltage-to-frequency converter (VFC) a highly linear response over a wide range of input voltages is substantial.

Aside from VCOs two basic VFC architectures are common, one based on a *current-steering multivibrator*, the other uses the *charge balanced method* which exists in two forms, the *asynchronous* and the clocked (*synchronous*) version. Such amplitude-to-frequency conversion (with pulse-frequency modulation) was originally applied to transmit analog signals, in particular by telemetry links (e.g., transmitted from remote sensors). Because of the analog system noise superimposed on directly transmitted analog signals, the quality of the transmitted pulse stream is superior. The analog signals are recovered with a low-pass filter after reshaping the received pulses. Even in the presence of electronic noise it is possible to recover the leading and trailing edges of the transmitted pulse stream thus rejecting the noise contribution. The accuracy of the transmission process is reduced to the accuracy with which the analog input signal can be recovered from the transmitted pulse stream.

The principle of the current-steering VCF is shown in Figure 4.48. A voltage follower (Section 2.5.2.1) with a source follower (Section 2.4.3.5) as output stage converts via R_S the voltage into current. This (drain) current i_D negatively charges a capacitor C. When the drain voltage in the form of a negative ramp drops below v_{ref}, the Schmitt trigger (Section 4.1.2.1) fires, switching the flip–flop (Section 4.1.3.2). The flip–flop steers a changeover switch reversing the terminals of the capacitor C, i.e., the charge of the capacitor reverses its sign with regard to i_D. Consequently, i_D continues charging the capacitor negatively which means that due to reversed polarity the capacitor is being discharged. The voltage across the capacitor has triangular wave form (i_C in Figure 4.48) because the change from charging to discharging occurs always at the same amplitude. The next steering of the changeover switch finds an empty capacitor. Therefore, changing the polarity is without consequence for the value of the drain voltage v_D. The voltage v_D at the input of the Schmitt trigger has the shape of a saw tooth. It can be best explained by the subtraction of the triangular signal v_C across the capacitor from the power supply voltage and paying attention to the reversal of the terminals of the capacitor, i.e., the voltage increase decreases this input voltage, ditto the voltage decrease because of the reversed capacitor polarity. Such a VFC is simple in

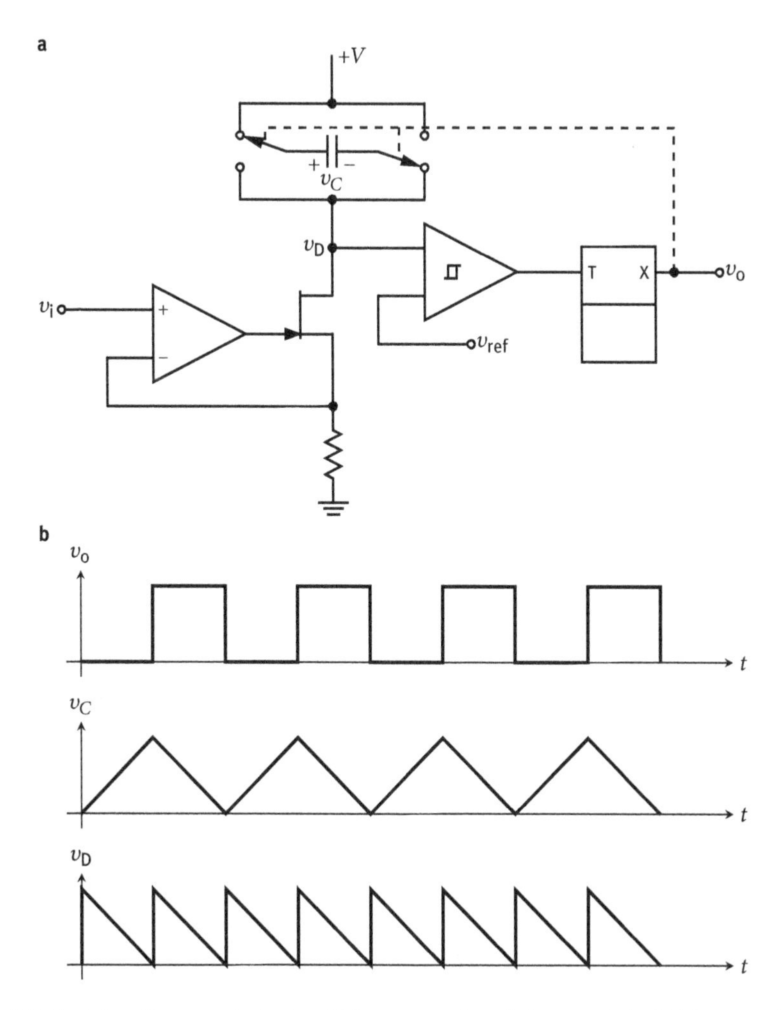

Fig. 4.48. Principle of the current-steering VFC, using a voltage follower, a Schmitt trigger, and a toggle flip–flop.

its design, it provides good accuracy, and does not need much power. Consequently, it is a favorite in telemetry applications.

The principle of a *charge balance VFC*, which is more demanding and has better performance, is shown in Figure 4.49. The conversion capacitor is part of a charge-sensitive amplifier (Section 3.6.4.1). If its output voltage v_{o1} passes the threshold value v_{ref}, a Schmitt trigger (Section 4.1.2.1) fires and a monostable multivibrator delivers an output signal v_o of a precise length. During the duration of this signal, a precise current source removes a fixed amount of charge from the conversion capacitor reducing the output voltage of the amplifier v_{Ao}. As the input current into the amplifier continues to flow, no input charge gets lost during this discharging process. Additional charge at the input raises the output voltage of the amplifier again beyond the thresh-

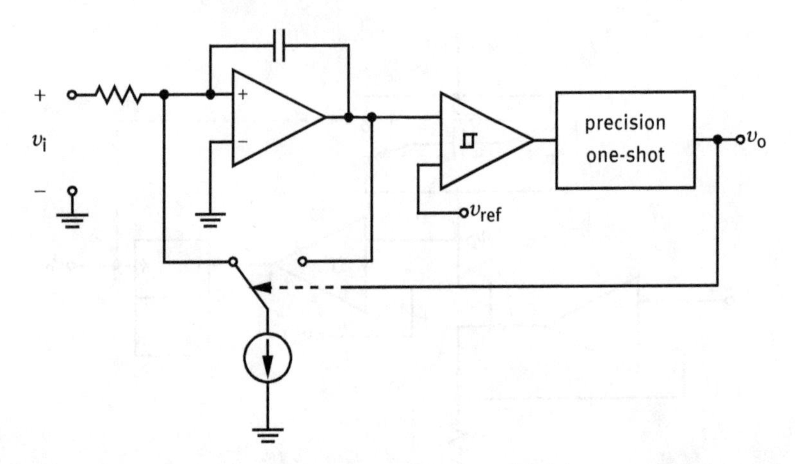

Fig. 4.49. Principle of a charge balance voltage-to-frequency converter, consisting of an integrating operation amplifier, a constant current source, a Schmitt trigger and a one-shot.

old repeating the process described above. The time difference between the responses of the Schmitt trigger (and one-shot multivibrator) depends on the magnitude of the current flowing into the input, i.e., on the input voltage v_i. The output pulse rate is proportional to the rate at which the removed charge is replenished by the input signal.

The conversion quality of such a circuit is very good, it depends on the stability both of the current source and the length of the output signal of the one-shot (its timing capacitor), and on charge losses in the conversion capacitor (not on its value and its stability).

The same principle is used in the clocked VFC except for the one-shot that is replaced by a D flip–flop that is synchronized to a clock signal. Synchronization provides easier handling of the output data transfer in particular when there are signals from several channels.

Although the circuitry differs only slightly between the asynchronous and the synchronous version there is a very distinct difference in their performance: the output frequency of the synchronous circuit is not analog any more, it has become *digitized* by their coincidence with clock signals. The output signals are digital both with regard to amplitude as with regard to time, they are of the *digital-digital* type (Chapter 5).

Voltage-to-frequency converters are used as input part of composite ADCs (Section 6.6). The output part would be a frequency digitizer to convert that frequency into a digitally coded output signal representing the input voltage. In telemetry the two parts may be widely separated, e.g., with the frequency signal transmitted wirelessly. This is an easy way to digitize signals from a remote sensor.

Problem

4.64. The quality of the charge balance VFC does depend neither on the stability nor on the capacity of the conversion capacitor. Why?

4.5.4 Frequency-to-amplitude (-voltage) conversion

Electronic instruments for frequency-to-voltage conversion have been around for a long time by way of rate meters. In radiation protection, the position of the indicator of a dose-rate meters depends on the voltage derived from the detected number of events per time unit. Converting a count rate based on statistically emitted radiation into voltage is more demanding than regular frequency conversion.

The principle has remained, sophistication has increased. Figure 4.50 shows the circuit of a diode pump together with an example for the output voltage when equally spaced rectangular input signals occur. The output signal is called stair-case signal as will be obvious with the active version of the diode pump (Figure 4.51).

The positive spike of a rectangular signal differentiated by C_1 passes through the diode D_2 and charges C_2. The following negative spike can pass D_1 but not D_2 so that the voltage across C_2 remains unchanged. At the same time the negative voltage across D_1 cannot become larger than its forward voltage so that after the negative spike has subsided, there is about the same operating point voltage behind C_1 as at the beginning, i.e. C_1 is discharged. Only that part of the next positive spike is effective that has a higher voltage than the voltage across the capacitor because D_2 does not conduct

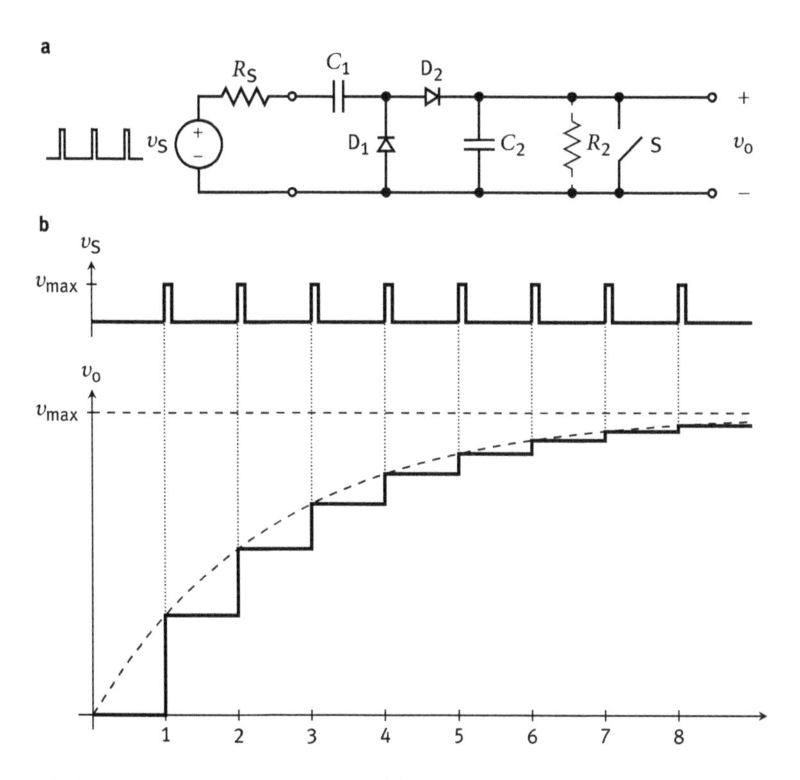

Fig. 4.50. Diode pump as a stair-case generator: (a) basic circuit; (b) input and output voltage.

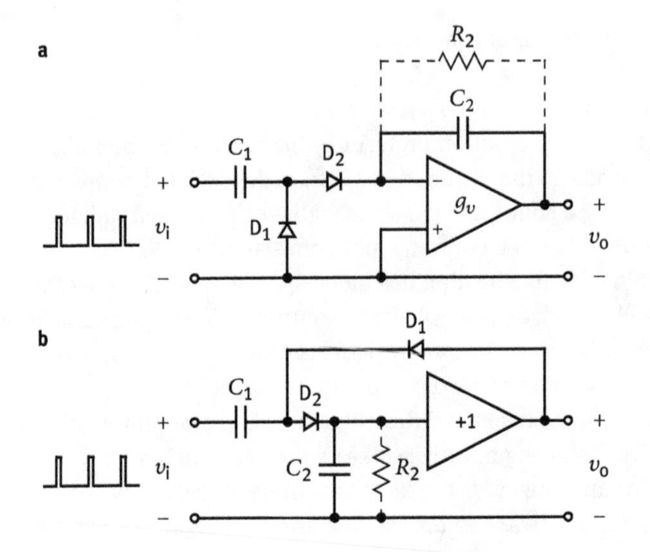

Fig. 4.51. Diode pump in active circuits: (a) making use of the negative feedback; (b) applying positive feedback.

when reverse biased. Consequently, each output step is smaller than the previous one resulting in an output signal form shown in Figure 4.50 b).

Using the basic relation $C = Q/v_C$ there is an even shorter way to explain the function of a diode pump. The positive charge per pulse of C_2 equals the positive charge of C_1; it is $Q_+ = C_1 \times v_{C1\,\mathrm{max}}$. The negative charge per pulse flows through D_1 to ground. Thus, C_2 does not get discharged but accumulates the positive charge of all input signals.

There are two remedies to improve linearity, i.e., to make the step size equal for each pulse

- markedly reducing the output voltage behind the diode D_2, or
- avoiding the reverse biasing of D_2.

Figure 4.51 shows how this can be done using feedback (Section 2.4.3.2).

In Figure 4.51a D_2 works against the virtual ground of the amplifier which hardly changes its voltage when C_2 gets charged. Thus, every input signal delivers the same amount of charge into the charge sensitive amplifier (i.e., transimpedance amplifier) that charges C_2.

In the bootstrap version of Figure 4.51b, the diode D_1 does not work against ground but against the voltage across C_2 provided by a voltage follower. Again for each input signal the diode D_2 has the identical operating conditions so that for each input signal the same amount of charge is delivered into C_2.

By reversing both diodes, a diode pump will respond to negative signals delivering a negative output. The staircase output signal can be utilized for *prescaling* the input frequency by a factor of n by making the reference voltage of a comparator equal to the voltage of n steps. Thus, the comparator triggers after n steps controlling a switch to discharge C_2.

Diode pumps are integrating circuits because their output signal is proportional to the sum of the accumulated input signals. To use such a circuit as a *frequency converter*, equilibrium between incoming and discarded charge must be reached, i.e., the charge collecting capacitor C_2 must be discharged at a constant rate, e.g., by a shunting resistor R_2. By the choice of the time constant R_2C_2 the speed with which the circuit responds to a frequency change can be adjusted.

A modern monolithic FVC reverses the charge balancing technique of the VFC (Section 4.5.3) by injecting a fixed amount of charge into a charge sensitive amplifier (Section 3.6.4.1) each time an input signal arrives. By permanently discharging the capacitor, as discussed before, a dynamic equilibrium is reached so that the output voltage reflects the input frequency. Again a short time constant is needed to follow quick frequency changes.

Problem

?

4.65. Name the two methods of linearizing the diode pump output.

5 Fully digital circuits

An analog circuit is designed so that after switching on the power supply it will assume a unique quiescent operating point. Circuits that have their quiescent (output) operating point in one of the two flat portions of the transfer function (with zero gain) are in one of two output states, L and H (Section 4.1). They can stay (temporarily) in either state. *Binary electronics* is based on the existence of two distinctly different (output) states which are usually described by their voltage values.

In this chapter, circuits that are fully binary are dealt with, i.e., both the amplitude and the timing of their output signals do not reflect corresponding properties of an analog (input) signal. Other circuits commonly thought to be digital, only because of their two-state (L and H) output were dealt with in previous sections, e.g.,

- comparators (Section 2.3.4.2),
- relaxation oscillators (Section 4.1) including triggers (Section 4.4.1) and flip–flops (Section 4.1.3.2), and
- gates (coincidence circuits) (Section 4.4.2).

When used in digital electronics, all these circuits lose their analog timing property by synchronization of the output signal with an independent master clock.

Digital circuits may be categorized as:

- logic circuits,
- storage circuits (registers, memories),
- interface circuits (level shifters, code converters, serializer, deserializer, etc.),
- transmitter and receiver circuits,
- driver circuits (hardware management), and
- processors (arithmetic and others).

For obvious reasons any circuit that is computer specific is excluded, i.e., only elementary digital circuits of moderate complexity are dealt with which are found, e.g., in traditional digital measuring instruments.

Problem

5.1. Does a fully digital signal still contain analog component information?

5.1 Basic considerations

For those not so familiar with operating points, a switch is usually used as an example of binary behavior. In Figure 5.1, a simple ON/OFF switch connected to a real voltage source with voltage v_S and source impedance R_S is shown. The so-called truth table, i.e., Table 5.1 summarizes its behavior.

Table 5.1. Truth table of an ON/OFF switch.

Input A	Output voltage		Output current	
Switch position	v_{SW}	X	i_{SW}	Y
ON	0 V	L	I_S	H
OFF	V_S	H	0 mA	L

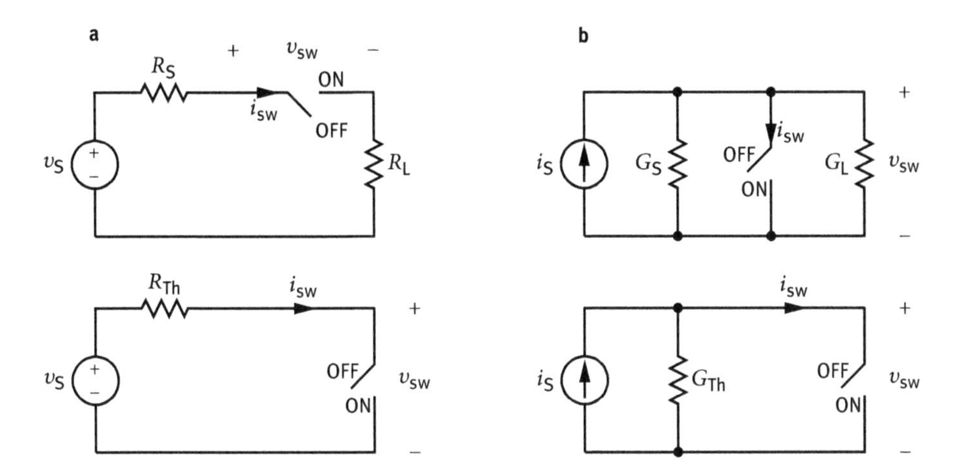

Fig. 5.1. Switching (a) a real voltage source and (b) a real current source.

A *truth table* is a listing of the values of all possible relations between the input(s), e.g., designated as A, (B, C, etc.,) and of the corresponding output(s), often called Q or X, (Y,) and also $\overline{Q}$ or X_1, (X_2, etc.).

In Figure 5.1a, the switch lies in series to the source. After converting the real voltage source to a real current source (Norton's theorem), the switch lies in parallel to the source. A switch is self-dual. If it is OFF (open-circuit, zero admittance), the voltage across it is H (and the current through it is L). If it is ON (short-circuit, zero impedance), the voltage is L (and the current H).

The two states of binary circuits, also called logical levels (either L/H or OFF /ON), can be correlated to the two binary digits 0_2 and 1_2 or the two binary logic values F(=False) and T(=True).

A logical system (in short *logic*) is called *positive* if H is assigned the value 1_2 (or T=True). If H is assigned 0_2 (or F=False) the logic is *negative*. In circuit design, it is sometimes beneficial to switch from positive logic to negative logic or vice versa.

Both types of logic are equivalent even if positive logic is preferred. In some cases, it pays to use a mixed logic when it spares circuit components. Thus, depending on the employed logic, the same binary circuit can fulfill two different logic functions. Therefore, we prefer the electronic states L and H rather than the logical states 0 and 1 in the truth tables whenever applicable.

We will refrain from digging deep into binary logic. In particular, we do not show how the number of logic elements can be minimized. We will concentrate on binary electronic components. This is the more justified as conversion of logical patterns into circuits has been done most efficiently by the suppliers of integrated digital circuits.

The logical side is only half the truth. At least as important is the behavior of the circuit in the time domain. As any process consumes time, there are, at least, two effects to be considered:

- Synchronization that is necessary to avoid spurious signals.
- Existence of busy time that signals the potentiality of signal loss.

Both effects cannot be accounted for by binary logic (truth tables). One bad example of signal misalignment is demonstrated in Section 5.3.1.3. To avoid such problems which usually will be very serious, synchronization of logic devices by applying a "clock" signal to the strobing (enable) input is unavoidable.

Already the impossibility to generate a signal with zero rise time makes clear that it is not possible to resolve two signals that occur within the rise time. Depending on the operation performed inside the device to which a signal is applied, the resolving time will be much larger than the rise time. Example 5.1 is intended to clarify the situation.

Example 5.1 (Consequence of the sequential nature of information). As the flow of time is a natural phenomenon, singular (real one-shot) events (without any "history") are difficult if not impossible to capture. The counting process which requires incrementing is definitely a process that can only be performed by allocating a *busy time* to it. Additional signals to be counted arriving during this time cannot be accepted, the counter is said to be dead. Thus, the term *dead time* is common for this phenomenon. In most cases the loss of signals to be processed is utterly undesirable. Therefore, it is necessary to avoid such losses if possible. One has to discriminate between three cases:

(a) Signals having a regular frequency or a minimum time distance between successive members of the signal train.
In these cases it suffices to use (very) high speed electronics to make the busy time shorter than the minimum time distance between the signals. If this is not possible, the number of lost signals can be counted by devising a coincidence between the busy time and the input signals yielding the number of the signals that were not processed.

(b) Statistically arriving signals.
Signals that arrive randomly in time, as, e.g., is the case with signals from nuclear counters measuring radioactive decay, cannot be processed without loss. A simple calculation will prove that point. If 100 000 events/s are measured with a busy

time of 1 μs, the device will be busy for 100 000 μs/s, i.e., 10% of the time. Therefore, 10% of randomly arriving signals will not be processed, as the probability of occurrence is the same in each time interval. Decreasing the busy time by a factor of 100 to 10 ns reduces the loss accordingly to 0.1%. As is obvious, the loss cannot be made to equal zero. As it is not possible to make the number of lost signals zero it is necessary to measure the time the device is dead, or even better, the time during which the device is receptive. This can best be done by devising an anticoincidence between the signals of the timer and the busy time establishing the so-called *live time*; the actual time the device was operative.

(c) Statistical signal arriving in regular time bursts.
In this situation, a combined solution is advisable. If the time interval between the burst is larger than the dead time there is no difference to case (b). Otherwise it is necessary to count the lost bursts according to (a). To get a grip on the signal loss within each time frame it is necessary to count (with as small dead time as feasible) all arriving signals and those signals that get processed. This gives, to a first order, the additional correction for the statistical nature of the signals.

Problems

?

5.2. Name as many self-dual electronic elements as possible.

5.3. Is positive logic "better" than negative logic?

5.1.1 Logic codes

Binary logic is the base for handling binary values, numbers with the two binary digits 0_2 and 1_2 and the logical variables F and T. Using arithmetic operations (adding, subtracting, multiplying, dividing, etc.) and logical decisions (e.g., equality, non-equalities, inversion) logical codes can be designed to tackle nearly any thinkable task. In all more complicated cases, such codes are not designed as direct binary codes but in a computer language to be used in a microprocessor (computer).

It is an obvious fact that the time element makes it necessary to have a step-by-step procedure when more than one logical operation must be accomplished. In binary electronics, this is called serial operation or *serial logic*. To keep track of the meaning of each step, we need a *protocol* that allocates significance to the position (in space and time) and to the electrical state (to the *binary digit* = *bit*) of each binary element. To avoid faults in the transmission of data, protocols contain *redundancies*, i.e., specific information on the data at hand. By checking these redundancies, faulty data may be recognized (e.g., by the *parity check* or the *cyclic redundancy check* (CRC)). Even if these topics are highly important for data transmission we disregard this part of digital circuit design as it is not basic.

A straightforward way to represent a number in a code is by weighting its places. The usual decimal system uses a weighted code based on the number 10 by assigning the (decimal) digit at the nth place the weight of 10^{n-1}, e.g., the decimal integer $3456_{10} = 3 \times 10^3 + 4 \times 10^2 + 5 \times 10^1 + 6 \times 10^0$. The straight binary presentation of numbers follows the same pattern: the binary integer $1011010_2 = 1 \times 2^6 + 0 \times 2^5 + 1 \times 2^4 + 1 \times 2^3 + 0 \times 2^2 + 1 \times 2^1 + 0 \times 2^0$ (which is in the decimal system 90_{10}).

We must not only decide on the type of code but also generate a protocol which decides

- on the maximum number N_{max} of digits that the circuit can handle. Usually, this will be a multiple of 4 (up to 128), and
- whether this number N_{max} of digits is handled serially (sequentially) or in parallel. In the first case, a single binary element is used N_{max} times to present the number in question with the protocol assigning the correct weight to each step, and in the second case, N_{max} single binary elements must be present at the same time (in parallel) so that in one step the complete number is electronically present. In the latter case, the correct weight must be assigned to each element. Of course, there could be a mixed presentation, e.g. two steps and $1/2 \times N_{max}$ binary elements.

As is easily seen, using binary notation of numbers means an inflation of digits (in the above example seven binary digits are needed for the presentation of the decimal number 90). Triples of digits are combined in the octal code, and quadruples in the hexadecimal code reducing (formally) the number of digits to be considered. However, this does not reduce the number of outputs in the circuit with the states L and H. Thus, for circuits only binary codes are of importance. Even then there is, aside from the weighted binary code, a plentitude of possibilities of representing all numbers from 0 to $2^n - 1$ in binary form. There are $2^n!$ variations possible. Of these, only a few usually based on a weighted binary code are of importance.

Some codes are hardware oriented, in particular to meet the needs of counters or encoders. In a *binary-coded-decimal* (BCD) code (which is a concession to our decimal world), each decimal place is mapped into a number (≥ 4, ≤ 10) of bits with (≥ 6) unused code patterns. We will introduce the appropriate number codes as needed when introducing the electronic circuits in question.

Weighted codes facilitate the visual output of numbers.

? **Problem**

5.4. The symbolic presentation of a number depends on the base that is used. Present the number 10 using the base 10, 9, 8, ... to 1.

5.1.2 Parallel vs. serial logic

As life is sequential any process has a serial component. (It does not make sense to build a one-shot electronic system.) Consequently, some processes as *counting* (based on repeatedly incrementing an existing number) can only be done in series. Other operations that can be done within a (short) resolution time can be viewed as being done instantly, e.g., determining a coincidence (Section 4.4.2) or the use of a parallel ADC (Section 6.4.2.1). In view of the fact that a timing component is unavoidable the notion of a parallel (=one-shot) operation is difficult to realize. Let us call a (short) one-step operation that occurs at some (naturally given) instant a *parallel* operation. *Serial* operations are repetitive and are synchronized to some clock signal.

In many cases, the same result is obtained by using one element n times (serial operation) or by using n elements once (see discussion in Section 6.4.2.1). Serial operation spares hardware, parallel operation time. There is a binary duality between serial logic and parallel logic: the number of operational steps is inverse to the number of operational elements. This duality between "time" and "space" is incomplete insofar as there is no "single step" electronics. It will always be used sequentially. Although single-step logic is not practical, a purely serial system is feasible. However, circuits have become extremely cheap so that there is no need to spare circuits. Consequently, we are refraining from dealing with purely serial digital circuits (e.g., serial memories, serial shift registers) because pure serial logical circuits have de facto vanished.

Miniaturization is beneficial for both types of logic: more circuits can be realized in the same area and the reduction of the length and the width (i.e., of the area) of intracircuit connections reduces capacity speeding up the circuit allowing higher internal clock rates.

The usual configuration in a binary system is the repeated (=serial) use of circuits which are arranged in parallel. This parallelism is expressed through the number of bits that can be handled in one step by said circuit. Thus, essentially all digital systems use parallel logic sequentially.

Usually, one encounters mixed logic: parallel logic used serially.

Problem

?

5.5. Does miniaturization conform to parallel or to serial logic?

5.2 Integrated logic technologies (miniaturization)

Integrated circuit technology is based on semiconductor cells on a substrate (a chip). These cells function either like bipolar junction transistors or like (metal-oxide-semiconductor) field-effect transistors. Most of the circuits have a silicon substrate. Some use GaAs (gallium–arsenide) or SiGe (silicon–germanium) technology. Over the decades, essential improvements mainly with regard to speed and power consumption were introduced (e.g., the Schottky transistor, low power supply voltages), allowing us to integrate larger and faster circuits. Some developments were prompted by the production process, like the NMOS n-channel metal-oxide technology and the PMOS p-channel metal-oxide technology, which were easier produced than the CMOS complementary metal-oxide technology which is far superior with regard to power consumption (Section 4.4.2.2) allowing higher cell densities. Since the introduction of the first integrated circuits the speed increased by more than a factor of 10^3 and the scale of integration by more than 10^6.

Miniaturization usually has the following benefits:

- small mass (less weight, less material cost),
- small volume (the small surface is detrimental because it limits the transfer of heat generated by the dissipated power requiring the use of power-saving technology and of enforced cooling),
- small power consumption (allowing high density of components, i.e., complex circuits),
- higher reliability (which is inversely proportional to some power of the strongly reduced number of components used in the fabrication of a digital system),
- lower cost (cheaper assembly, allowing cheap mass production), and above all
- strongly reduced internal capacitances (resulting in the increased speed of the internal clock and reduced propagation delay).

An integrated circuit, in short IC, is monolithic, i.e., it uses a single semiconductor chip as a substrate. There may be analog, digital, and mixed circuits on the same chip. The miniaturized circuits of the next chapter need a mixed architecture. In the following, we will sketch the evolution of integrated digital circuits. In the beginning, an integrated digital circuit contained transistors numbering in the tens performing logic operations (by way of logic gates). We now call this technology "small-scale integration" (SSI). The demand for increased integration resulted in "medium-scale integration" (MSI). Each chip contained hundreds of transistors. The next step was "large-scale integration" (LSI), with tens of thousands of transistor functions per chip. Already at the beginning of LSI single chip processors became reality. Further development of integration was dominated by the needs of producing chips for computers and mobile phones. Over the decades the number of transistor functions on integrated circuits doubled approximately every two years. CMOS technology (Section 4.4.2.2) made it possible to integrate millions of logic gates on one chip by way of "very large scale integration" (VLSI). An important by-product of this miniaturization is the much increased internal clock frequency (beyond 1 GHz) due to the extremely small capacitance associated with the short and thin internal connections. The term "ultra-large-scale integration" (ULSI) is used for a chip complexity of more than 1 million transistor functions. Semiconductor memory chips with a capacity of several tens of billions transistor functions have been produced. With chip integration in several layers by way of three-dimensional integrated circuits (3D-IC) faster operation due to optimized signal paths and reduced power consumption can be expected. Obviously, design and production of such circuits is not straightforward. Integrated digital circuits have become so cheap that designing and building discrete digital circuits is an expensive and, in addition, time-consuming enterprise. Consequently, it does not make sense to discuss the circuits in detail. (The basic principles have been covered in parts of Chapter 4.) However, knowledge of the types of circuits and their general behavior is indispensable. Thus, we concentrate on that.

Problem

5.6. Make sure to understand all the reasons for miniaturization.

5.2.1 Logic families

Integrated digital circuits belong to one of several logic families. Members of one family can directly act with one another, i.e., they can be interfaced (Section 3.5) by plain wires. The output and input levels (L and H), the power supply voltage and the clock rate are standardized within a family so that interoperability among them is guaranteed. Logic families usually have many members, each tailored for a specific task. They can be used as "building blocks" to create complex circuits.

Among the presently most prominent families, there are two based on bipolar NPN junction transistors (TTL and ECL), two based on field-effect transistors (CMOS and NMOS for computer chips) and one family incorporating both types of active elements (BiCMOS).

Over the years integrated circuits have become faster and more complex. After having minimized the parasitic capacitances, time constants can only be made shorter by making resistances smaller. Lowering the resistances requires lowering the power supply voltage to keep the dissipated power small. Thus, circuits operating with low supply voltages were designed. The greatest power saver is, however, the use of complementary circuits (CMOS, Section 4.4.2.2). Only this technology made VLSI possible.

The difference in technology between the families resulted in different supply voltages and different voltage levels for L and H. Interconnecting circuits of any two logic families requires special interfacing techniques, e.g., so-called pull-up resistors. Thus, CMOS circuits were developed that are completely compatible with the TTL family, even pin-compatible. In the case of TTL the introduction of Schottky technology was a breakthrough with regard to speed. Merging TTL technology with CMOS technology got hold of the best of both technologies in the BiCMOS family.

Programmable logic devices (PLD, firmware) are in competition to the fixed-wired circuits of the logic families. Different LSI-type functions such as logical gates, flip–flops and adders can be implemented on a single chip. Present field-programmable gate arrays contain tens of thousands of LSI circuits which can be operated at high speed.

As a rule the logic states high H and low L are represented by two voltage levels quite different for each logic family. Inside a circuit the logic levels are narrowly de-

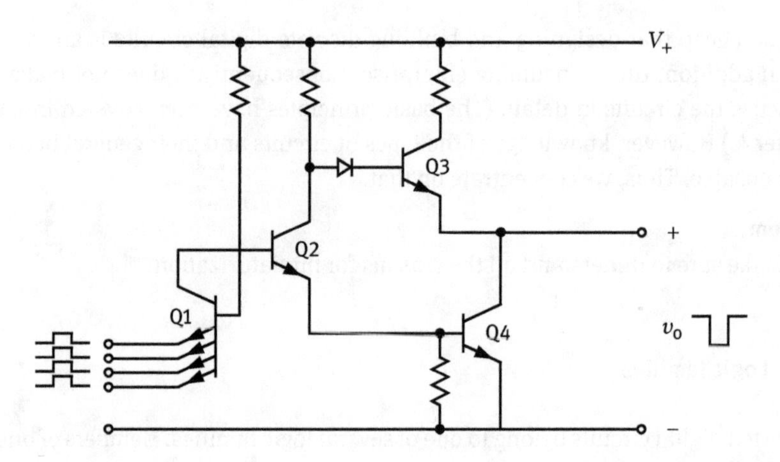

Fig. 5.2. SSI: Schematic of a 4-input positive logic NAND gate in TTL technology.

fined as, e.g., we discussed for the Schmitt trigger (Section 4.1.2.1). Aside from these *logic levels* there are the so-called *wire levels* for outside signals allowing some tolerance in the voltage levels. These tolerances are necessary to take into account the *fan-out*, i.e., the loading of the circuits. Between the L-band of allowed voltages of L-levels and the H-band of allowed voltages of H-levels there is an invalid, intermediate voltage range in which a state would be undefined, representing a faulty condition. Of course, during logic transitions the instantaneous operating points sweep through this range. If this sweep is fast enough, i.e., the rise and fall time of the input signal short, the existence of this forbidden band is not noticed. However, if by chance or on purpose the slew rate (or rise time) is too slow, oscillations between L and H may occur (see also meta-stability, Section 5.3.2).

Problem

5.7. In which regards do logic families differ from each other?

5.2.1.1 TT (Transistor–transistor) logic

Transistor–transistor logic TTL is based on bipolar NPN transistors. Figure 5.2 shows a schematic of a positive logic 4-input NAND gate in TTL technology. The input transistor has four equivalent emitters that perform the logical AND function (such an arrangement is called fourfold *fan-in*). Since a transistor of a standard TTL circuit gets saturated in the LOW state, minority-carrier storage time in the junction limits the speed of the device. The introduction of Schottky transistors overcame this speed limitation.

Problem

5.8. What are the fastest TTL circuits?

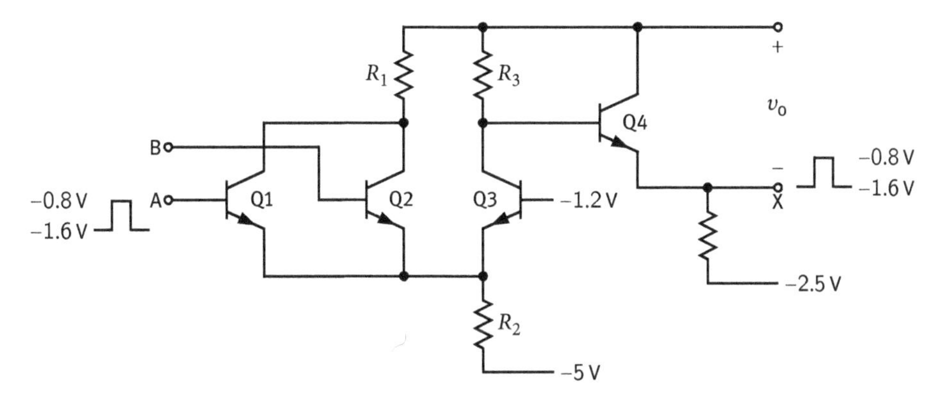

Fig. 5.3. Circuit diagram of a 2-input positive logic OR gate in ECL technology.

5.2.1.2 EC (Emitter-coupled) logic

Emitter-coupled logic ECL is based on bipolar NPN transistors. When we discussed the long-tailed pair in Section 2.3.4.1, we found that emitter-coupling is a cascade of a common-collector circuit with a common-base circuit. Both configurations have maximum feedback (Section 2.4.3.5) giving the maximum increase of the bandwidth (Section 3.6.3.1). Consequently ECL is especially apt for very high-speed applications. Figure 5.3 gives an example of an ECL gate.

Problem

5.9. Why are ECL circuits in principal faster than TTL circuits?

5.2.1.3 CMOS (Complementary metal-oxide semiconductor) logic

In Figure 4.42, a two-input NAND gate in CMOS technology is shown. Complementary arrangements of n- and p-channel field effect transistors make the quiescent power negligible as discussed in Section 4.4.2.2. Only when the state of the gate is switched, current flows from the power supply into the internal (and external) capacitances. Thus, the dissipated power of CMOS devices increases with the switching rate. However, it takes longer to charge capacitors with small currents, reducing the speed of such circuits.

Problem

5.10. Name the advantage and the disadvantage of CMOS circuits.

5.2.1.4 BiCMOS (bipolar-CMOS mixed technology) logic

The reason for the emergence of several technology families lies in the production process of integrated circuits. Fusing different digital technologies on one chip is extraordinary difficult. With the knowledge that both TTL and ECL have no pnp transistors one can conclude that it is more difficult to produce integrated circuits with pnp transistors and even more so to produce integrated circuits based on complementary bipolar junction transistors which, on the other hand, has been done with CMOS FETs for a long time already. Combining bipolar junction transistors (with their high speed and high gain) with CMOS FETs (with their high input impedance) constitutes an excellent condition for constructing fast low-power logical gates. Combination of the two technologies has existed for a long time in discrete circuits, however, without implementation in integrated circuits such circuits were not frequently used.

First, the BiCMOS technology was applied in operational amplifiers and other analog circuits (e.g., comparators, voltage regulators, DACs). High current circuits use metal-oxide-semiconductor field-effect transistors for efficient power control, whereas bipolar elements perform specialized tasks. Several integrated microprocessors are based on BiCMOS technology. BiCMOS will not replace CMOS in pure digital logical circuits because it cannot compete with the low power consumption of CMOS. It is used for special tasks as, e.g., bus transceivers. Reducing the number of chips in electronic systems by combining two chips into one increases the speed and reliability, and reduces cost and size.

5.3 Basic circuits

Serial use of practical digital circuits requires synchronization (Section 5.5.1), i.e., their answer is “scanned” by a clock; the output is coincident with the (in most cases positive) clock signal. In several digital circuits, this synchronization aspect can be disregarded simplifying the situation because there is one signal less to consider. Although only in special cases we will consider the coincidence with the clock signal, it must be kept in mind that

> practically all binary circuits will have a clock input necessary for synchronizing the output because of the intrinsic necessity of serial operation.

The clock signal “enables” the transmission of the logical result to the output. Consequently, we use the letter E for the input of the clock signal, the enable signal or the strobe signal.

There are circuits without feedback, so-called combinational circuits, for which a simple truth table relates all input values to all output values. Then there are the so-called sequential circuits, for which the current output values are a function of the values of the current inputs, past inputs, and past outputs.

Problem

5.11. Which timing information contains a fully digital signal?

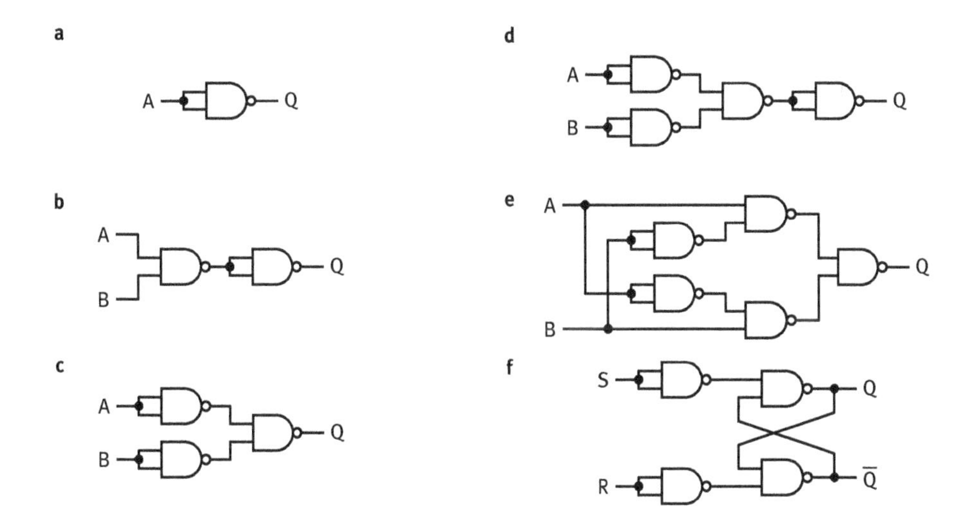

Fig. 5.4. Composition of six basic logical circuits by using two-input NAND gates: (a) NOT gate, (b) AND gate, (c) OR gate, (d) NOR gate, (e) XOR gate, (f) SR flip–flop.

5.3.1 Basic gate (combinational) circuits

It is curious that any basic logic circuit can be built from two-input NAND or NOR gates which for that reason are called *universal gates*. Figure 5.4 shows six configurations with NAND gates representing six basic logic circuits. The symbol used in this figure is that of a two-input NAND gate. At the moment, we are not going to discuss it any further.

Problem

?

5.12. Redo Figure 5.4 using two-input NOR gates.

5.3.1.1 NOT circuit

The most basic binary circuit is the NOT circuit, a logic inverter. Any inverting class C amplifier (Section 2.3.4) can act as NOT circuit if the input signal drives the output to the other binary output level. Its truth table is shown in Table 5.2. For completeness, the translation of the behavior of an inverter into positive logic NOT gate (sub p) and into negative logic NOT gate (sub n) is provided, too.

Table 5.2. Truth table of an inverter circuit (logical NOT).

A	Q	A_p	Q_p	A_n	Q_n
L	H	0	1	1	0
H	L	1	0	0	1

A—▷o—Q Fig. 5.5. Logical symbol of a NOT gate.

In a schematic a NOT gate is symbolized by the triangular symbol of an amplifier with a rather small circle at the output. Inversion either of the input signal or an output signal is indicated by placing such a small circle at the appropriate place (Figure 5.5).

The negation of a logical value is indicated by a *negation bar* (overline, overbar or overscore), e.g., $\overline{1}$ is 0. For AND and OR gates it is indicated by a leading N, i.e., NAND and NOR.

5.3.1.2 AND and NAND gate

The properties of the two-input logical AND function is defined in the truth table which is given in Table 5.3 together with its realization as two-input positive logic AND gate. The inverted output $\overline{\mathrm{Q}}$ is the answer for the NAND function.

The logical symbols for both gates are shown in Figure 5.6.

The number of functions that can be realized in a single integrated circuit is limited by the number of pins an integrated component has. In SSI, a 14-pin component named quad 2-input NAND gate has four two-input NAND gates on one chip (two pins are used for supplying power).

Problems

5.13. Generate the truth table of a two-input AND gate made of two NAND gates.

5.14. Construct a NAND circuit using four NOR gates.

5.15. Generate the truth table of a two-port NAND gate used as a NOT gate.

5.3.1.3 OR and NOR gate

The properties of the two-input logical OR function is defined in the truth table which is given in Table 5.4 together with its realization as the two-input positive logic OR gate. The inverted output $\overline{\mathrm{Q}}$ is the answer for the NOR function.

Table 5.3. Truth tables of the AND gate (Q), the NAND gate ($\overline{\mathrm{Q}}$), and for positive logic.

A	B	Q	$\overline{\mathrm{Q}}$	A	B	Q	$\overline{\mathrm{Q}}$
0	0	0	1	L	L	L	H
0	1	0	1	L	H	L	H
1	0	0	1	H	L	L	H
1	1	1	0	H	H	H	L

a A, B —D— Q   b A, B —Do— Q

Fig. 5.6. Logical symbol of a two-input AND (a) and a two-input NAND gate (b).

Table 5.4. Truth tables of the OR gate (Q), the NOR gate ($\overline{Q}$), and for positive logic.

A	B	Q	$\overline{Q}$	A	B	Q	$\overline{Q}$
0	0	0	1	L	L	L	H
0	1	1	0	L	H	H	L
1	0	1	0	H	L	H	L
1	1	1	0	H	H	H	L

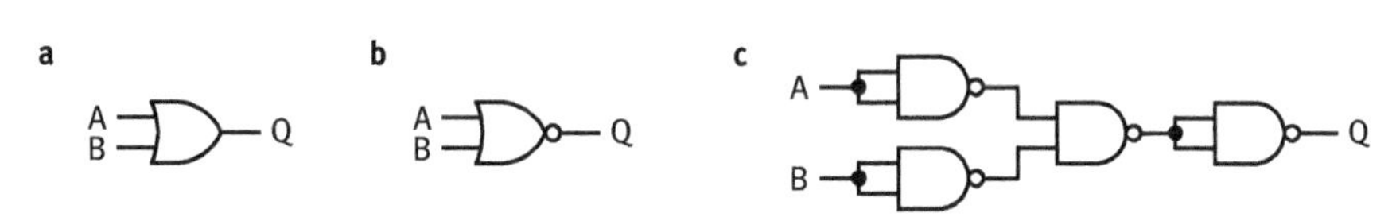

Fig. 5.7. (a) Logical symbol of an OR gate, (b) logical symbol of a NOR gate, and (c) two-input NOR function realized by four two-input NAND gates.

If one compares the truth table of the positive logic AND gate with the negative logic OR gate, it is obvious that these truth tables are identical, i.e., one circuit acts either as the AND gate (using positive logic) or as the OR gate (using negative logic). Using the letters L and H rather than 0 and 1 (as usually done) for the two output levels of binary electronic circuits is without ambiguity. If done differently, one must state the type of logic that is used. Then, the ambiguity disappears even when using the logical values 0 and 1.

In the following, we are going to follow this practice to deal with logical functions rather than with electronic circuits.

Figure 5.7 shows the logical symbol for the OR and the NOR gate, and the realization of a two-input NOR gate by four two-input NAND gates.

Let us investigate the electronic behavior of a specific NOR circuit (Figure 5.8a). In this case, the signal to the second input passes through three inverters so that its value is inverted. A truth table would show that the output value is 0 independent of the input signal. However, when looking at the voltage levels (Figure 5.8b) – assuming positive logic – we notice a short H signal at the output with a length of three propa-

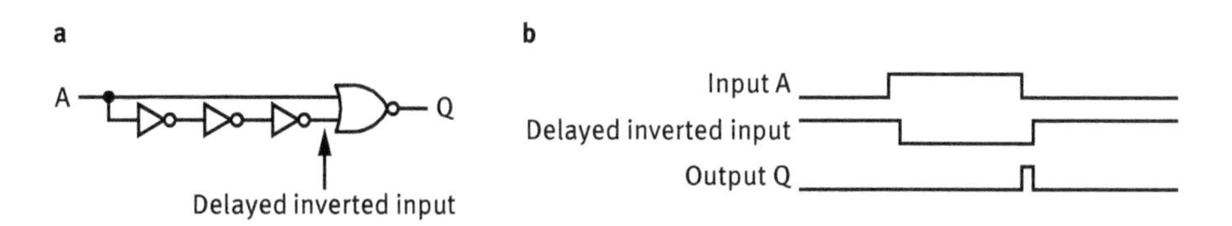

Fig. 5.8. (a) NOR circuit with three inverters in front of one input, and (b) voltages at the two inputs and at the output.

gation delays of a NOT circuit. This teaches us the difference between binary logic and digital circuitry. To avoid such "surprises" one must either make sure that the path lengths of signals do not differ too much or use a clock signal that defines the moment of the correct answer (as given by the truth table).

NAND or NOR circuits are called *universal gates* because they may be used to build any other logical circuit.

Problems

5.16. Generate the truth table of a two-input OR gate made of three NAND gates.

5.17. Generate the truth table of a two-input NOR gate, made of four NAND gates.

5.18. Construct a NOT circuit from a two-input NOR gate and give the truth table.

5.3.1.4 XOR (EOR, EXOR) gate

The XOR gate implements an exclusive decision. The output will be 1 only then when just one of the inputs is 1. It will be 0 if both inputs are the same. This behavior is summarized in Table 5.5 under the header S. It has the inequality function, i.e., the output is only then 1 if the input values are not the same. An XOR gate can be made from four NAND gates as shown in Figure 5.9b.

An important application of the XOR gate is the half-adder which consists of an XOR gate and an AND gate (Figure 5.10). As can be seen from its truth table (Table 5.5), it performs the addition of two one-bit binary numbers with the result given by two bits, the Carry bit and the Sum bit.

Problem

5.19. Generate the truth table of a two-input XOR gate made of five NAND gates.

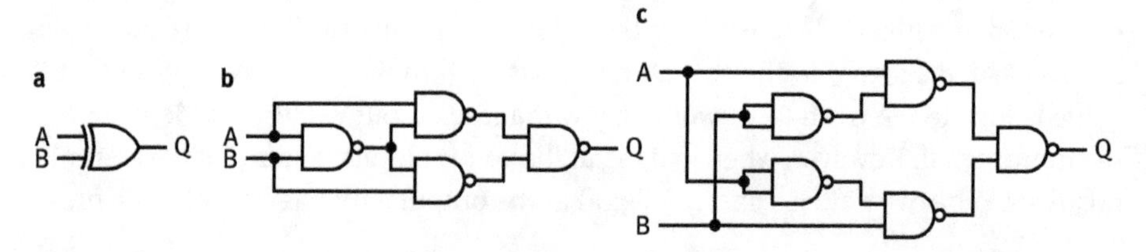

Fig. 5.9. (a) Logical symbol of an XOR gate, and the function realized by two-input NAND gates (b) and (c).

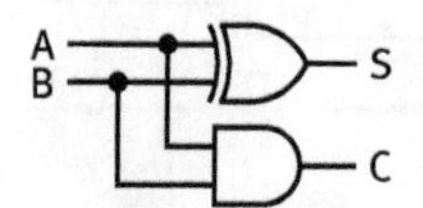

Fig. 5.10. Realization of a half-adder. S is the sum output, and C the carry output.

Table 5.5. Truth table of a half-adder.

A	B	C	S
0	0	0	0
0	1	0	1
1	0	0	1
1	1	1	0

5.3.2 Basic bistable (sequential) circuits

In Section 4.1.3.2, we came across the bistable multivibrator with the two stable quiescent operating points (voltage levels) L and H. A typical bistable electronic circuit is symmetric, i.e., it has two primary inputs and two outputs, Q and $\overline{Q}$. The output $\overline{Q}$ delivers the inverted value of Q, i.e., if one output is H, the other is L, and vice versa (see the symbol in Figure 5.11).

Without signal, the output operating point will stay in its binary state as long as the circuit is operative, i.e., as long as the necessary power for its operation is supplied. Thus, the binary information is stored allowing the use of this circuit as a *memory cell*. Such a memory is *volatile*, i.e., the information gets lost if the supply voltage is gone. It is not within the scope of this book to discuss memories in general, neither their structures nor the many ways they can be realized.

Bistable circuits are used extensively in semiconductor registers and memories. If a bistable circuit is intended for *data storage*, it is called a *latch*. A latch where the output value is the instant result of a signal at the input is called *transparent*. By means of an additional input (named *enable*), it becomes nontransparent (or *opaque*).

By means of input signals, the output level may be switched forth and back from one state to the other. If switching the circuit is the primary intention, then the name *flip–flop* for such circuits would rather be used. Flip–flops are used in counters or for synchronizing input signals fluctuating in time to some reference timing signal. The latter application requires clocked (synchronous or edge-triggered) flip–flops.

Feedback as applied in Section 4.1.3.2 to active three-terminal components may be applied to (integrated) gates, as well. As shown in Figure 5.12, a latch can be con-

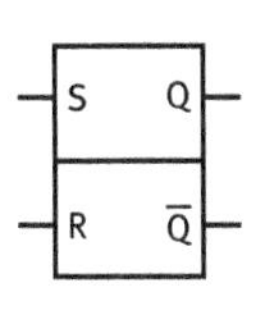

Fig. 5.11. Symbol of an SR latch. The symmetry of a bistable circuit with the two inputs S and R and the complementary outputs Q and $\overline{Q}$ is obvious.

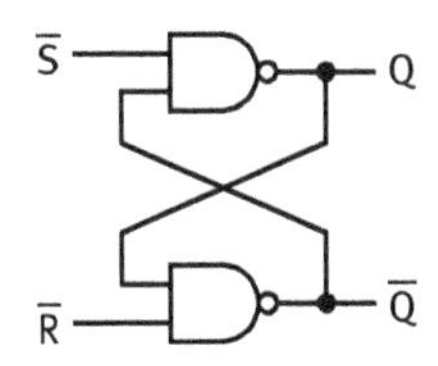

Fig. 5.12. NAND gate latch.

Table 5.6. Truth table of a NAND gate latch; Q_{n-1} previous (output) state, Q_n (output) state with input signal applied.

$\overline{R}$	Q_{n-1}	Q_n	$\overline{S}$	$\overline{Q}_{n-1}$	$\overline{Q}_n$
1	0 or 1	Q_{n-1}	1	1 or 0	$\overline{Q}_{n-1}$
0	0 or 1	1	1	1 or 0	0
1	0 or 1	0	0	1 or 0	1
0	0 or 1	Irregular	0	1 or 0	Irregular

structed by NAND gates forming the so-called *NAND-gate latch*. Its truth table is given in Table 5.6. A 1 at input R resets the output Q to 0, a 1 at input S sets the output Q to 1. If both input levels are 0, no action is taken, i.e., the state of Q remains unchanged, namely Q_{n-1}.

Note that the feedback of the NAND gates is mutual; each one is in the feedback loop of the other. Since either is inverting, the two stages connected in series provide the noninverting amplification needed for positive feedback.

Flip–flops with an additional enable input can be *clocked* (*pulsed* or *strobed*). Such devices ignore the signal at the inputs except at the transition of an enabling signal. Clocking affects the flip–flop to either change or sustain its output signal dependent on the value of the input signals at the time of the transition. A flip–flop changes its output state either on the rising edge of the clock signal or on the falling edge. The data transfer from input to output at the moment of the enabling signal (clock or otherwise) introduces two additional properties that must be considered

- propagation delay, and
- meta stability.

Propagation delay of a flip–flop is the enable-signal-to-output delay (commonly called t_{CO} or t_P in data sheets). The time for a H-to-L transition (t_{PHL}) may differ from the time for an L-to-H transition (t_{PLH}). *Meta stability* occurs when two input signals, such as the set (or reset) value and the enable value, are changing simultaneously, i.e., within the resolution time of the circuit. In that case, the output may settle unpredictably to one state or the other with an unpredictable propagation delay. Meta stability in flip–flops can be avoided by keeping the input data constant during the setup time (t_{su}), a specific period before, and the hold time (t_{hold}), a specific period after the clock pulse. These times are specified in the data sheet of the circuit, and can be extremely short in modern devices.

Logical circuitry built around a NAND gate latch is used to bring about flip–flops with dedicated properties. These are the SR (*set–reset*), the D (*data* or *delay*), the T (*toggle*), and the JK types.

Problem

5.20. Which provision makes any flip–flop to a fully digital circuit?

5.3.2.1 SR (set/reset) flip–flop

Figure 5.11 shows the symbol of an SR flip–flop without clock input, i.e., without provision for synchronization.

Table 5.7. Truth table of an SR latch; Q_{n-1} previous (output) state, Q_n (output) state with the input signal applied.

R	S	Q_{n-1}	Q_n	Function
0	0	0 or 1	Q_{n-1}	Storage
1	0	0 or 1	0	Reset
0	1	0 or 1	1	Set
1	1	0 or 1	n/a	Irregular

From the truth table (Table 5.7), we see that the set signal (value 1 at the S input) produces the state 1 (at the Q output) independent of the previous state. The reset signal (value 1 at the R input) produces the state 0 (at the Q output) independent of the previous state. Obviously it would be irregular to have the value 1 at both inputs.

Figure 5.13 shows a schematic of an SR latch. It is a NAND gate latch with inverters at the inputs.

To overcome the irregular state, one can add logic circuitry that converts the input pair (S,R) = (1,1) to one of the allowed combinations. That can be:

- an S-latch, converting (1,1) to (0,1), i.e., Set is dominant,
- an R-latch, converting (1,1) to (1,0), i.e., Reset is dominant, and
- a JK-latch with $Q_n = \overline{Q}_{n-1}$ for inputs (1,1), i.e., the (1,1) combination would toggle the output.

If, as in Figure 5.14, the two input NAND gates are activated by a common enable signal E, a synchronous SR latch (or clocked SR flip–flop) results.

The enable input signal applied at E can be called enable, read, or write, or strobe, or clock signal. When the flip–flop is enabled S and R signals can pass through to the

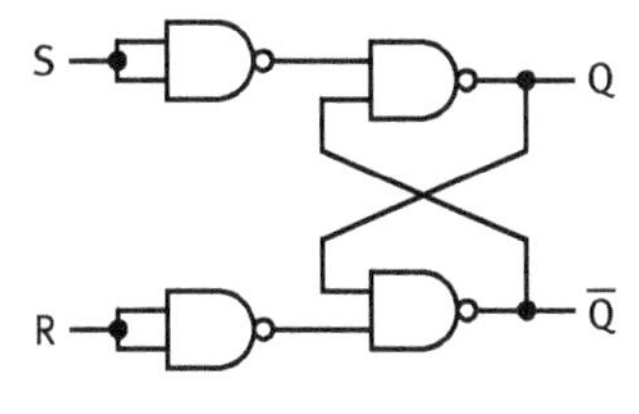

Fig. 5.13. Schematic of an SR latch made of four two-input NAND gates.

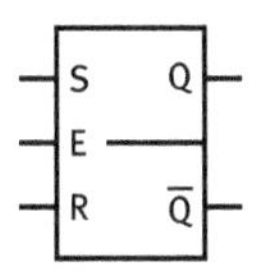

Fig. 5.14. Logical symbol of a synchronous SR flip–flop.

($Q, \overline{Q}$) outputs, i.e., the latch is transparent. When not enabled the latch is closed, and the outputs remain in that states that existed at the end of the last enabling signal.

Problem

5.21. Generate the truth table of an SR flip–flop using four two-port NAND gates.

5.3.2.2 D (data, delay) flip–flop

Figure 5.15 shows the symbol of a D flip–flop with enable input that can be used for synchronization with clock signals. The D flip–flop transfers the value of the D-input at a given instant of the clock cycle to the Q output. A change of the output value at Q is not possible at any other instants. A D flip–flop can serve as a memory cell.

Table 5.8. Truth table of an edge-triggered D flip–flop (with S = R = 0 and E = 1).

Clock	D	Q_n
Rising edge	0	0
Rising edge	1	1
Else	0 or 1	Q_{n-1}

From the truth table one can see that without enable/clock signal the level of the D input has no effect on the output. With the rising edge of the enable signal the output value becomes that of the D input. If, as in Figure 5.15, both an S and an R input is present, these inputs have the same function as in the SR flip–flop, overriding the signal at the D input.

Cascading two D memory cells as shown in Figure 5.16 makes it possible to have during some time interval (the duration of the clock signal) both the new and the previous D signal at one's disposal. Such an arrangement is called *master–slave edge-triggered D flip–flop*. The enable signal for the second stage is inverted so that at the time of the 1 to 0 edge of the enable signal a 0 to 1 edge is created which is delayed by

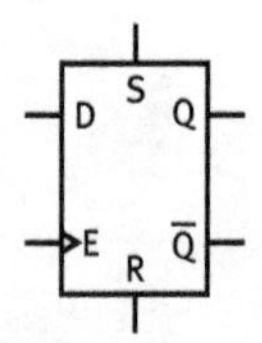

Fig. 5.15. Logical symbol of a clocked D flip–flop with enable (E), reset (R), and set (S) input.

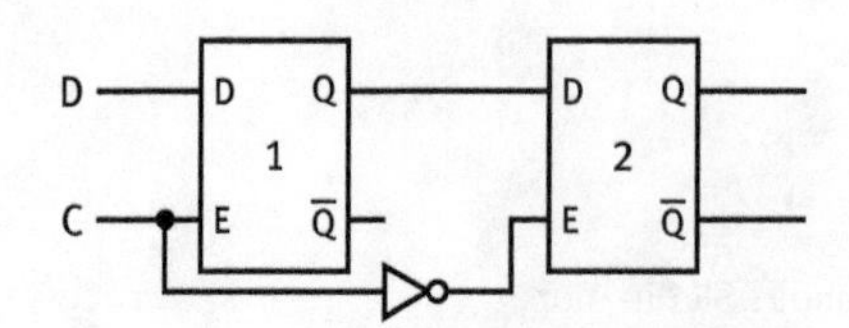

Fig. 5.16. Symbolic presentation of a master–slave D flip–flop.

Table 5.9. Truth table of a master–slave D flip–flop.

D	$Q_{2,n-1}$	Effective edge	$Q_{2,n}$
0	0 or 1	1 to 0	0
1	0 or 1	1 to 0	1

the length of the clock signal. Therefore, the output of the first stage reflects already the new input state D_n whereas that of the second stage still has the information on the previous input state D_{n-1}. Only after the 1-state of the clock signal is over (zero) both flip–flops are in the same state. Table 5.9 presents the truth table of a master–slave D flip–flop. The inversion of the clock signal for the slave latch results in a *synchronous system with a two-phase clock*, where the operation of the two latches with different clock phases prevents data transparency.

Problem

?

5.22. What is the purpose of master–slave flip–flops?

5.3.2.3 T (toggle) flip–flop

In Figure 5.17, the symbol of a strobed T flip–flop is shown.

If the T input is 1, the T flip–flop changes its state ("toggles") whenever it is strobed by the clock input. If the T input is 0, the flip–flop sustains the previous value. The basic truth table of a T flip–flop is given in Table 5.10.

The T flip–flop can be used "in the reverse." Holding T high the flip–flop will be toggled by the enable signal. If this stems from a clock with the frequency f, then the signal at the output Q will have the frequency $f/2$, it will be prescaled (divided) by two which is the basic operation of binary counters (Section 5.4.3). Obviously, no signals are applied to the T input in this application. Thus, T and E have exchanged their function: T acts as the enable input, whereas the signal input is E.

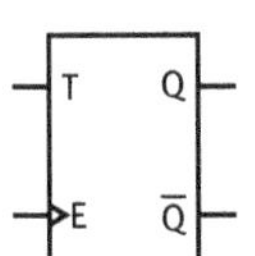

Fig. 5.17. Logical symbol of a strobed T flip–flop.

Table 5.10. Truth table of a T flip–flop when the enable is active (H).

T	Q_{n-1}	Q_n	Action
0	0	0	Hold state
0	1	1	Hold state
1	0	1	Toggle
1	1	0	Toggle

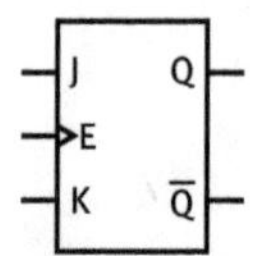

Fig. 5.18. Logical symbol of a strobed JK flip–flop (with the enable input E).

Table 5.11. Truth table of a basic JK flip–flop with the active enable input.

J	K	Q_n	Action
0	0	Q_{n-1}	Hold state
0	1	0	Reset
1	0	1	Set
1	1	$\overline{Q}_{n-1}$	Toggle

Problem

5.23. Name the main use of T-flip–flops.

5.3.2.4 JK flip–flop

In Figure 5.18, the symbol of a strobed JK flip–flop is shown.

When comparing the truth table Table 5.11 with that of the SR flip–flop (Table 5.7) we see that J and S perform the set, and K and R the reset task. However, now the S = R = 1 (i.e., the J = K = 1) condition makes the flip–flop toggle, i.e., it changes the output to the logical complement of its current value.

The JK flip–flop is the *general type flip–flop* because it can be used as an

- SR flip–flop by making S = J and R = K, a
- D flip–flop by setting K = $\overline{J}$ (the complement of J), or a
- T flip–flop by setting K = J.

The timing diagram of Figure 5.19 should support a better understanding of the performance of a JK flip–flop.

Problem

5.24. Verify that a JK flip–flop is a combination of a T flip–flop and an SR flip–flop.

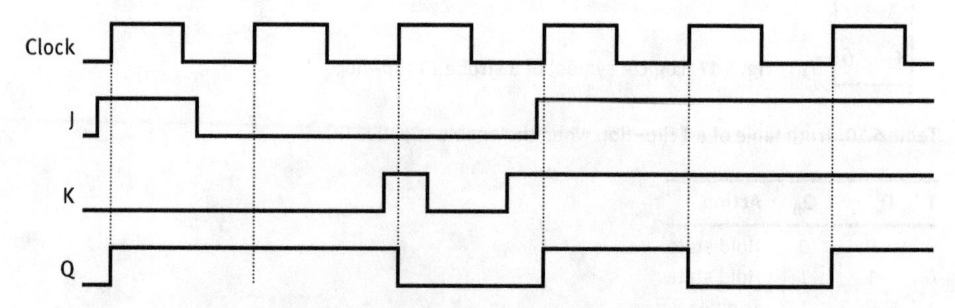

Fig. 5.19. Timing diagram of a positive-edge-triggered JK flip–flop.

5.4 Registers (involved sequential circuits)

A register is an array of binary storage elements (e.g., latches) which stores information and makes it readily available to logical circuits. To this end, information must be written into the register and be read from the register. In special applications, the information may be modified, as well. One flip–flop can store one bit of information. A group of n flip–flops forms an n-bit register. The *state* or *content* of a register is stored bitwise in the flip–flops. Obviously, the number of 0–1 combinations in a register is finite. It is convenient to embrace the bitwise data in words of n bit with typical word lengths of 2^2, 2^3, 2^4, 2^5, or 2^6 bit. For that reason, flip–flops are united to form n-bit registers storing *data words* of n bit. All flip–flops of a register are synchronously controlled by a single clock line.

If the output of each flip–flop of a register is accessible, such a register has *random access*, i.e., any bit can be read at any time.

Problem

?

5.25. By connecting the end of a delay line (coaxial cable) via a signal regenerating circuit (e.g., a Schmitt trigger) to the input, binary information (bits) can be stored in this volatile *serial register*.

(a) Which properties of the delaying medium limit the number of bits that can be stored?

(b) Can the stored information be accessed randomly?

5.4.1 Shift register

If n flip–flops are cascaded (i.e., output connected to the input of the next flip–flop) and all flip–flops are synchronously controlled by a single clock line, we have an n-bit shift register. Figure 5.20 is a schematic of a 4-bit shift register using D flip–flops.

Because of the synchronous operation the signal on the D input is captured only at the instant the flip–flop is clocked and is ignored at other times. Signals at optional additional inputs (set or reset) may act either asynchronous or synchronous with the clock.

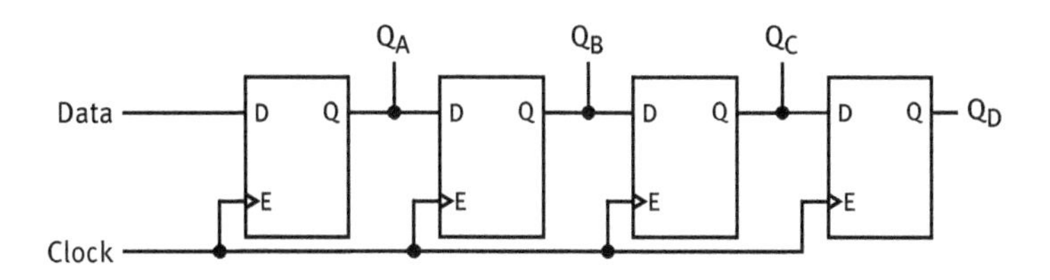

Fig. 5.20. Schematic of a 4-bit shift register using D flip–flops.

On each active transition of the clock, a shift register shifts the contents of one flip–flop to the next (to the right). The first flip–flop accepts the input data for storage. The data stored in the *n*th stage is removed. It may be used by other logic circuits. Thus, a shift register has one input. As the signals at the output of each flip–flop are accessible, it has n outputs, i.e., n bits (i.e., the register's complete content) may be read simultaneously.

From mathematics, we know that shifting the decimal point in a number by n places reduces or increases the said number by 10^n, depending on the direction of the shift. Thus, it is no surprise that shifting a weighted binary number that is stored in a shift register n times reduces or increases this number by a factor of 2^n depending on the direction of the shift. Thus, shift registers may be used for multiplying or dividing by 2^n, depending on protocol, i.e., whether the most significant bit is assigned to the first or the *n*th flip–flop.

? **Problem**

5.26. A weighted binary number stored in a shift register is shifted by one place. The *n*th flip–flop stores the most significant bit.
(a) Is the resulting binary number divided by 2 or multiplied by 2?
(b) Name two conditions necessary to yield an exact result.

5.4.1.1 Deserializer (serial-to-parallel converter)

If a sequential binary word of n bit is fed into a shift register by applying n clock pulses, the register contents is shifted n times until all of the word is stored in the shift register at the end of the shifting process. Then the n-bit word is available simultaneously in parallel from at the n outputs. Such a process is called *deserialization* or *serial-to-parallel conversion* or SIPO (*serial in–parallel out*). Of course, in the data protocol the (serial) position in time must correspond to the correct (parallel) position in the circuit. Observe that it takes n steps for one conversion. As the input information is sequential anyway, this process does not add dead time. The shift register accumulates the serial word, hence the name *accumulator*.

Problem

5.27. What is the advantage (and the disadvantage) of transmitting information in a serial manner?

5.4.1.2 Serializer (parallel-to-serial converter)

If an n-bit word is stored in an n-bit shift register, n clock signals would move the stored bit pattern one by one to the output of the *n*th flip–flop forming a sequential word of n bits at the rate of the clock frequency. Such a process is called *serialization* or *parallel-to-serial conversion* or PISO (*parallel in–serial out*). Again it takes n steps for one conversion. However, in this case dead time during the process is unavoidable.

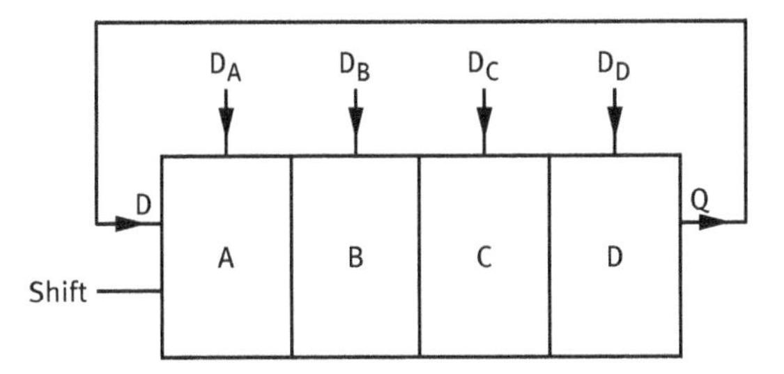

Fig. 5.21. Cyclic 4-bit shift register, configured as a ring counter, i.e., with the output fed back into the input.

5.4.2 Cyclic register

If the output of a register is fed back into its input, a cyclic register results. Thus, loss of the data in the nth bit can be avoided. After n shifts the original register content is restored. The data pattern present in the shift register recirculates as long as clock pulses are entered.

5.4.2.1 Straight (overbeck) ring counter

A cyclic n-bit shift register with a preloaded 1 in the first flip–flop can serve as ring counter. For each signal at the clock input the state 1 is shifted to the next flip–flop. After n shifts the original state is reestablished. For example, in the schematic of Figure 5.21 the data pattern will duplicate every four clock pulses. However, we must load a data pattern first, e.g., by the D input of the flip–flops (Figure 5.21). If a 1 is loaded into stage A and zeros in all other stages, this pattern is repeated for every fourth clock signal constituting a *prescaler* by four.

A more practical application is a binary decimal counter using a 10-bit shift register. The tenth signal at the clock input shifts the state 1 of the output of the tenth flip–flop to the first flip–flop. This same signal is also used to move the preloaded 1 of the next decade from the first to the second flip–flop. Thus, for each decade a 10-bit cyclic register is needed. Only when a ring counter is used for counting clock pulses it is synchronous. Of course, a straight ring counter wastes hardware: 10 flip–flops are needed to count from 0_{10} to 9_{10}; just 4 would be needed to do it with a weighted binary counter. As to any time just one flip–flop is *hot*, i.e., is in the 1 state, the code of a straight n-bit ring counter is called the *one-out-of-ten* code. Note that the frequency at the output of the tenth flip–flop is prescaled by a factor of 10 as compared to the frequency at the (shift) input.

Problems

?

5.28. How many stages (bits) does a shift register configured as a ring counter need to repeat its pattern at every fifth clock signal, i.e., to prescale by a factor of 5?

5.29. What is the disadvantage of a straight ring counter over the standard binary counter?

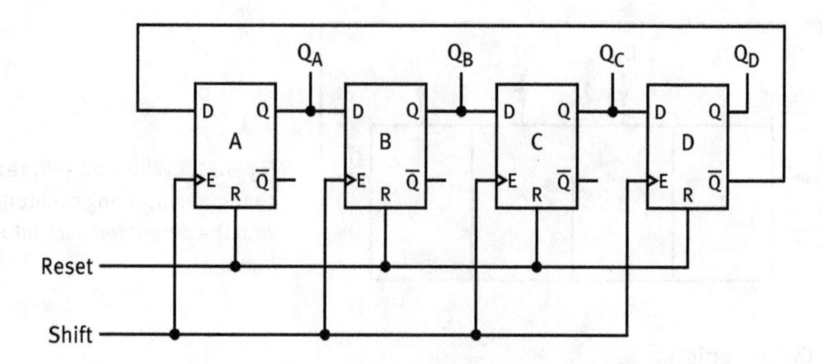

Fig. 5.22. Switch tail ring counter with a 4-bit shift register using D flip–flops.

5.30. What is the advantage of a straight ring counter over the standard binary counter?

5.31. A binary number given in the one-out-of-ten code reads: 0000000001 0000000010. What is its decimal equivalent?

5.4.2.2 Switch-tail (twisted, Johnson) ring counter

In an n-bit switch tail ring counter, the complement output of the nth flip–flop is connected to the input of the first flip–flop. If the content of the register at the beginning is zero (all flip–flops in the 0 state), the first shift would enter a 1 into the first flip–flop, the second again, and so on. When the register is full with *ones*, the next shift would enter a 0 into the first flip–flop, the second again, and so on until the content of the register is zero again. Figure 5.22 shows the schematic of a 4-bit switch-tail ring counter. Its performance is detailed in Example 5.2.

Example 5.2 (Performance of a 4-bit switch-tail ring counter). By resetting all flip–flops (by a signal to the input R), the switch-tail ring counter is put into the starting condition with a bit pattern of 0000 (Table 5.12). The first (shift) signal received at the

Table 5.12. Decimal equivalent of the output pattern of the switch tail ring counter of Figure 5.22.

Q_A	Q_B	Q_C	Q_D	
0	0	0	0	0_{10}
1	0	0	0	1_{10}
1	1	0	0	2_{10}
1	1	1	0	3_{10}
1	1	1	1	4_{10}
0	1	1	1	5_{10}
0	0	1	1	6_{10}
0	0	0	1	7_{10}

enable input shifts Q_A to B, Q_B to C, Q_C to D, and $\overline{Q}_D$ to A so that the new pattern is 1000. After four input (shift) signals the pattern is 1111, and after seven 0001. Thus eight signals are needed to restore the original pattern 0000. For an *n*-bit switch-tail ring counter it takes 2*n* shifts to restore the original contents of the register, i.e., prescaling by a factor of 2*n* takes only a register of *n* bits rather than 2*n* bits as is the case with the straight ring counter. However, contrary to the straight ring counter, where just one of the flip–flops is hot, the number of hot flip–flops is from 0 to *n*. Thus, the load of the power supply is not independent of the stored number, as is the case with the straight ring counter.

Ring counters and, in particular, switch-tail ring counters are typically used as decade (and binary) counters, as decimal (and binary) display decoders, and for frequency division (as *prescalers*, i.e., *divide-by-n counters*, e.g., in timers). They are also available as integrated circuits. An integrated 5-bit switch-tail ring counter will have 10 decoded outputs for decimal use.

Problem

?

5.32. What is the advantage of a switch-tail ring counter over the straight ring counter?

5.4.3 Prescaling and binary counting circuits

As mentioned in Section 5.1.2, counting is a *sequential* process based on incrementing a stored number until all elements to be counted are counted. Thus, counting contains two elements:

- incrementing, and
- storage.

T or JK flip–flops (in a register) will do just that with binary numbers. Although counting can be done unstrobed, it is often performed in relationship to a clock signal (using strobed flip–flops). Let us take a 4-bit register. After a reset signal, its contents will be 0000_2 which is 0_{10}. A signal to the first stage will toggle its content so that 0001_2 (i.e., 1_{10}) is stored in the register. The second signal will toggle the state of the first flip–flop to 0. If a transition from 1 to 0 at the output of a flip–flop is used to toggle the next flip–flop, there will be a 4-bit binary counter. Table 5.13 shows the binary pattern obtained this way.

Thus, the contents of the counting register are a 4-bit word given in the basic weighted binary code (Section 5.1.1). There are counters that deliver the result in a different code (e.g., ring counters, Sections 5.4.2.1 and 5.4.2.2, or BCD counters, Section 5.4.3.3). The counter described here is an *asynchronous counter* because only the first stage is incremented by the input signal the others are incremented by the output signal of each previous stage.

Table 5.13. Binary pattern in a 4-bit binary counter when counting up to 15.

Number of signals	Binary pattern	Inverted bit pattern
0	0000	1111
1	0001	1110
2	0010	1101
3	0011	1100
4	0100	1011
5	0101	1010
6	0110	1001
7	0111	1000
8	1000	0111
9	1001	0110
10	1010	0101
11	1011	0100
12	1100	0011
13	1101	0010
14	1110	0001
15	1111	0000

Due to its repetitive pattern, counters may be used as prescalers. A 4-bit binary counter downscales the input signal rate by $2^4 = 16$. Prescaling modulo k (with k any natural number) can be accomplished with an n-bit counter if $2^n \geq k$. To shorten the repetition cycle to a desired length there are two easy ways. The obvious one is to reset the register to 0 with the next count after the contents have become $k - 1$. Alternatively, the next count after the contents have become $2^n - 1$ sets the counter to $2^n - k$, accomplishing the same prescaling by k.

If the prescaling factor k is *fixed and even*, it is good practice to split the prescaling circuit up into two stages. The first stage should be purely binary, and the other stages should provide the prescaling by the remaining factor $k/2$. This way the higher signal frequencies are handled by straight binary prescaling without the need of any logical decisions. Such provisions reduce the dead time of a counter which is essential when statistically arriving signals are to be counted.

If a frequency is to be divided by means of a prescaler, a synchronous counter should be used.

? **Problem**

5.33. What dead time would the first stage of a scaler need so that it does not loose a single statistically arriving signal?

5.4.3.1 Asynchronous (ripple) counters

In Section 5.4.3, this type of counter was used as our model for binary counting. The external signal is applied to the first flip–flop, and the output of each flip–flop (the

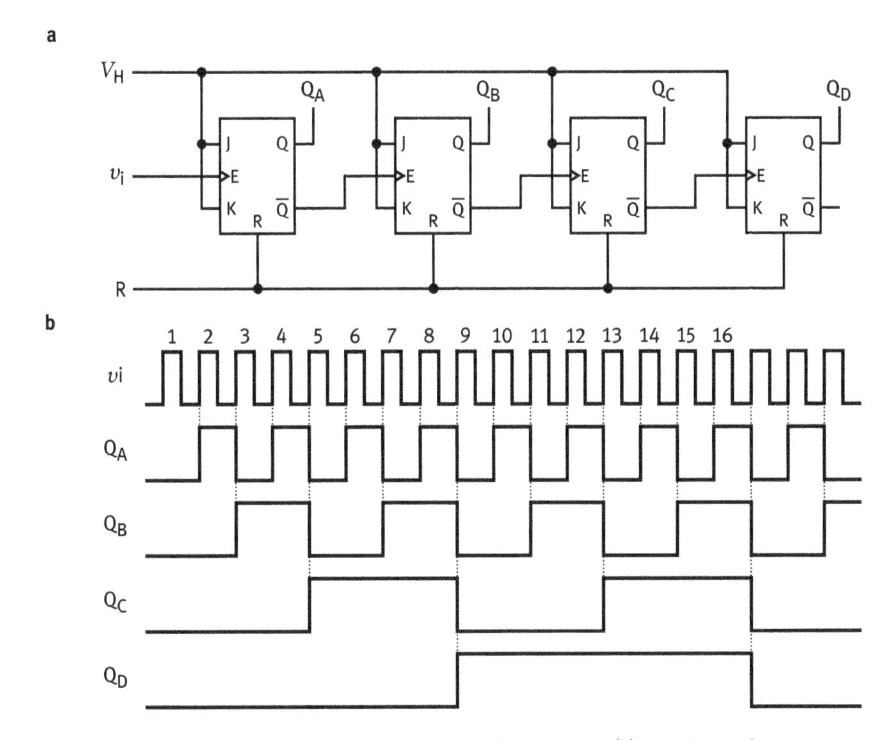

Fig. 5.23. (a) Schematic of an unstrobed asynchronous 4-bit counter (b) signal waveforms.

overflow) is connected to the input of the next one. These overflows "ripple" from stage to stage, which leads to delays at the higher data bits making this technique unfit for synchronous applications. However, it is fit for prescaling applications. Figure 5.23 shows a schematic of an unstrobed (asynchronous) 4-bit counter together with the signal waveforms.

In Figure 5.23b, the numbering of the counts starts with 1_{10} and not with 0_{10} as done in Table 5.13. Be aware of this inconsistency. One should be familiar with this problem as the first 10 digits are from 0 to 9, but one counts from 1 to 10! Besides, observe that the frequency of the output signal is highest at Q_1 (the first stage).

Therefore, the highest repetition rate that can be handled depends on the speed of the first stage.

As the output of normal counters will never have a negative number, there is no need to provide storage for negative numbers reducing the needed numbers of memory cells by a factor of 2.

Problem

5.34. What is the reason for the name asynchronous counter?

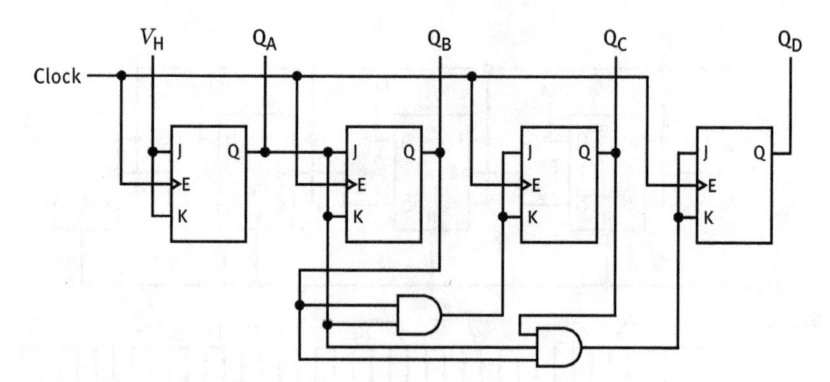

Fig. 5.24. Schematic of a strobed 4-bit synchronous binary counter.

5.4.3.2 Synchronous counters

In synchronous counters, the external input signal is applied to all the flip–flops simultaneously. By appropriate gating (Figure 5.24), it is assured that it will act only on those flip–flops that must be incremented. By AND gating the signal to be counted with the output state of all previous flip–flops, it is assured that the following stages increment only when it is due (see Table 5.13).

The schematic in Figure 5.24 shows a strobed counter, i.e., a counter that is synchronized to a clock signal that is counted. For the *first flip–flop*, the clock signal acts as an input signal whereas the parallel JK inputs are permanently high allowing the toggle operation. At any following stage the clock signal acts only then as input signal when its toggle operation is allowed. This happens when the appropriate AND gate provides a 1 for the JK inputs.

Problem

5.35. Is the timing in the output signal with regard to the input signal of a synchronous counter independent of the counted number?

5.4.3.3 Binary-coded decimal (BCD) counters

Binary signals can be decoded into decimal signals by a binary-to-decimal decoder. As we are used to decimal numbers, it is essential for easy use that the output of numbers (in calculators, in counters or digital measuring devices) is done in a decimal fashion. If the registers of counters are grouped in a way that each group represents a decade, we speak of decimal counters. A 10-bit ring counter (Section 5.4.2.1) or a 5-bit switch-tail ring counter (Section 5.4.2.2) forms such a group. The first needs a one-out-of-ten (or one-hot) output code, and the second needs encoding of all 5 bits.

As discussed in Section 5.4.3, at least 4 bits are needed for a binary coded decimal counter. Therefore, in this case, 6 out of 16 binary patterns are not used, allowing a great variety of codes representing the 10 decimal digits. For example, two of the weighted codes are the 8-4-2-1 and the 2-4-2-1 code, the number giving the weight of a bit at the place of the number, e.g., 9_{10} would be 1001_2 using the first code and 1111_2 with the second code. To operate a display from the output of a BCD counter one needs a decoder that converts the output signal at the four flip–flops to the appropriate input signals of the display. If a popular *seven-segment display* is used, such a decoder would be called a *four-to-seven decoder.*

The combination of a divide-by-two counter and a divide-by-five counter to accomplish a prescaling by a factor of ten (i.e., to design a decimal counter) is insofar superior as the highest frequency is handled without the need of any logic by a single flip–flop that can be viewed at as a fast prestage. In that case, another weighted code, the 5-4-2-1 code, is appropriate.

Problems

?

5.36. A straight binary counter uses all 16 different patterns provided by the 4 bits. A decimal counter only needs 10 to represent the decimal numbers from 0 to 9. How many variations of unique codes exist for the representation of these 10 numbers?

5.37. Give the binary equivalent of 9_{10} using the 5-4-2-1 code.

5.4.3.4 Up-down counters

From Table 5.13, it should be clear that whereas the contents of the register is incremented with each signal (from 0000 to 1111), its complementary output is decremented (from 1111 to 0000). The first action is called counting-up, and the second counting-down. In Figure 5.25, a schematic of a counter is shown where both the regular and the inverted outputs of the flip–flops are simultaneously accessible. Often an up or down selector input gives the control in which direction the counting is done.

The result obtained by any counter is basically an integer. Thus, the counted number is without uncertainty.

Problem

?

5.38. Design a divide-by-two prescaler using a down counter.

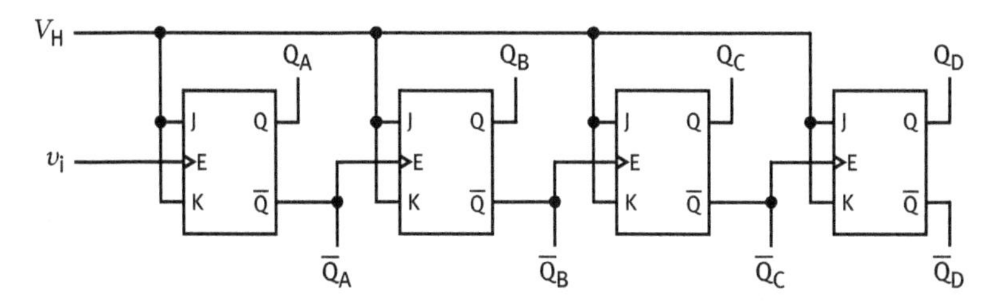

Fig. 5.25. Schematic of an asynchronous simultaneous 4-bit up/down counter based on strobed JK flip–flops.

5.5 Time response (to binary pulses)

Time is a very important parameter in binary and other circuits. It appears in several parameters
absolute time – in relation to an absolute clock,
synchronization – in relation to an internal clock,
rise and fall time – occurring in individual signals, and
pulse length – which is by itself an analog quantity.

In each of these cases, the so-called time jitter must be considered (Section 4.4.1.1).

Disregarding the amplitude there is no difference between analog and digital (rectangular) signals. However, there exist two amplitude levels for binary signals within a rather narrow band in amplitude whereas the amplitude of analog signals may be anywhere in the full dynamic range. With regard to time, only one property can be claimed to be truly digital. A truly digital signal is synchronized to a clock allowing serial operation. In this case, the timing is stringent but arbitrary, i.e., the exact frequency of the clock does not really matter. Thus, only signals synchronized by a clock are digital in time.

Problem

5.39. An analog output signal will have an intrinsic rise time (showing up in the input signal) and a circuit dependent rise time (from the transmission through the circuit). How does the intrinsic rise time of the input signal affect the output signal of a digital circuit?

5.5.1 Signal synchronization and restoration

Serial logic processes require a clock signal for synchronous operation. It must be uniquely known in any timing period which step of the protocol is executed. This sounds trivial for single tasks in isolated systems. However, if several systems have to work together or when multitasking is performed or when distributed (computer) systems must cooperate, it can become very difficult to ensure that the actions of the various systems fit together. This requires process synchronization which establishes the necessary sequence of action. In a centralized system, the clock of the central server dictates the time. In a distributed system, a global time is not easily established. The *network time protocol* is commonly used for distributed clock synchronization.

Synchronization at the lowest level, i.e., of individual circuits, is done by means of a clock signal at the enable input. One can distinguish two variants of how the timing is derived from the clock signal:

(a) Coincidence with the clock signal: enable is enacted during the (usually positive) clock signal
(b) (Rising or falling) leading edge triggering: When shortening the active clock signal (within an element) to a very short time it will be active only at the rising edge of

the clock signal. Thus, the active time is much more narrowly defined than using all of the clock pulse.

To ensure undisturbed interchip communication the propagation delay, the chip-to-chip clock skew, and the speed of the drivers must be considered. As clock signals get skewed (by capacitances) during their transmission it will be necessary the refresh them, i.e., to *restore* their original shape. Fast comparators (Schmitt triggers, Section 4.1.2.1) are commercially available to do this job.

Problem

?

5.40. What problems can occur in circuits that are expected to cooperate but have a phase difference in their clock signals?

5.5.2 Application of timers

Integrated monostable and astable relaxation oscillators are distributed as so-called *timers*. These timers are very versatile circuits. Three applications are demonstrated here.

5.5.2.1 Pulse lengthener

In Section 4.1.2.2, we came across monostable relaxation oscillators. The time constant of this circuit determines the output pulse width, i.e., the length of the output signal when the circuit is triggered with the edge of a (digital) signal depends on said time constant. Commercial circuits (*timers*) allow pulse lengths up to several hours (see also Section 4.4.3.2).

5.5.2.2 Pulse shortener

If a digital signal is to be shortened, it would be safe first to differentiate it (Section 3.4.2) which retains the signal edge and to use a monostable relaxation oscillator (a *timer*) to provide the desired (shorter) pulse length (see also Section 4.4.5).

5.5.2.3 Digital delays

Basically, any digital signal can be delayed the same way as an analog signal (Section 4.4.4). However, as only the state (L or H) must be preserved and not the amplitude, there are simple additional methods to delay a digital signal. A straightforward way is to trigger a monostable relaxation oscillator (timer) with the leading edge. If the trailing edge of the timer is used to generate a (standard) digital signal (e.g., by means of another timer), then the regenerated signal is delayed by the pulse width of the first timer.

6 Manipulations of signals (digitizing)

Digital signals excel analog signals with regard to the following properties:
- ready input for a myriad of digital instruments,
- better immunity against noise,
- easier and cheaper (long-time) storage, and
- faster signal transmission.

For this reason, converting analog information into a digital one is particularly important. The three important analog variables are as follows:
- amplitude,
- time (instant of a signal, or time difference; rise time of a pulse is appendant to time difference), and
- frequency.

As discussed in Chapter 3, signals are usually characterized by amplitude and their instant (time). As both are analog variables, there are four stages of digitization of an output signal:
analog–analog: both variables are analog,
digital–analog: the amplitude is digitized,
analog–digital: the time is digitized, and
digital–digital: full digitization.

If the instant is digitized, it means that the output signal is synchronized to a (reference) clock as done, e.g., in sampling (Section 4.4.3.1) and in the synchronous counter (Section 5.4.3.2). Digitizing a time difference (Section 6.2) is quite something else.

6.1 General properties of digitizing

Measuring without digitizing is not possible. The result of any measurement is a number, a digitized result, in most cases a floating point decimal number. However, *counting* does not need any digitization as the elements to be counted are quantized. Consequently, a counted result is an integer number. Digitization in electronics means (in almost all cases) to transform an analog signal into numerous L and H states (Section 4.1.2.1) which then are interpreted as binary digits (e.g., for positive logic as 0 and 1) to build a binary (based) number by a diversity of codes.

An *analog-to-digital converter* (in short ADC, or A/D) converts a continuous physical quantity at some moment to a number representing the quantity's amplitude. This number is a multiple of the *quantization* unit, in binary conversion the *least-significant-binary unit* (in short: *bit*, respectively LSB). The conversion introduces quantization uncertainty which is a kind of interpolation uncertainty. ADCs often perform the conversion periodically which is called sampling (Section 4.4.3.1). The result is a succession of binary codes representing the converted continuous-amplitude and continuous-time analog signal by a discrete-amplitude and a discrete-time digital signal.

An ADC may also convert an input voltage or current to a binary number proportional to the amplitude of the voltage or current. Some nonelectronic or only partially electronic devices, such as *mechanical* encoders digitizing nonelectric quantities can also be called ADCs.

Thus, one must discriminate between ADC *circuits* and ADC *devices.*

Obviously, the output of an ADC is digital. Normally the binary output is a weighted binary number proportional to the input amplitude.

The quality of a digitizing process hinges on two properties:
- resolution and quality of conversion, and
- duration of the digitizing process.

Resolution gives the number of discrete values that are produced over the full amplitude range of the analog value (its full scale FS or dynamic range). The values are usually stored by way of a binary code so that the resolution is expressed in bits (binary digits). Therefore, the number of discrete values is a power of 2. An ADC with a 10-bit output has a range of 1024 ($= 2^{10} \approx 10^3$) unique output codes defining an equal number of input signal amplitude levels. Each voltage interval lies between two consecutive code levels. Normally, the number N of voltage intervals is given by $N = 2^n - 1$, where n is the ADC's resolution in bits. Over the full range of an ADC with 10 bits, there will be exactly 1024 unique binary output numbers (from 0000000000_2 to 1111111111_2, corresponding to the decimal numbers 0_{10} to 1023_{10}). Thus, such an ADC provides a resolution of about 0.1% (10^{-3}).

The duration of the digitizing process of an ADC determines its *conversion rate or sampling frequency*. The latter term applies for ADCs that are continuously sampling analog inputs. Like resolution, the required conversion rate depends on the specific application for which the ADC is needed, i.e., on the highest frequency present in the analog input signal. When the system is too slow, fast changes of the analog signal will be missed. As discussed (Section 4.4.3.1), the highest frequency waveform an ADC can convert correctly is the *Nyquist frequency* that equals one-half of the ADC's conversion rate.

If the input signal contains a frequency that exceeds the Nyquist frequency of that ADC, the digitized signal will have a false low frequency. This phenomenon is called *aliasing*. Figure 6.1 demonstrates how aliasing comes about.

Figures 6.1a through e show cases of aliasing where there is no way whatsoever to reconstruct the original analog signal from the digital output. It should not surprise that the practical sampling rate required for a good representation of the analog signal by the digital output must be considerable higher than twice the maximum input

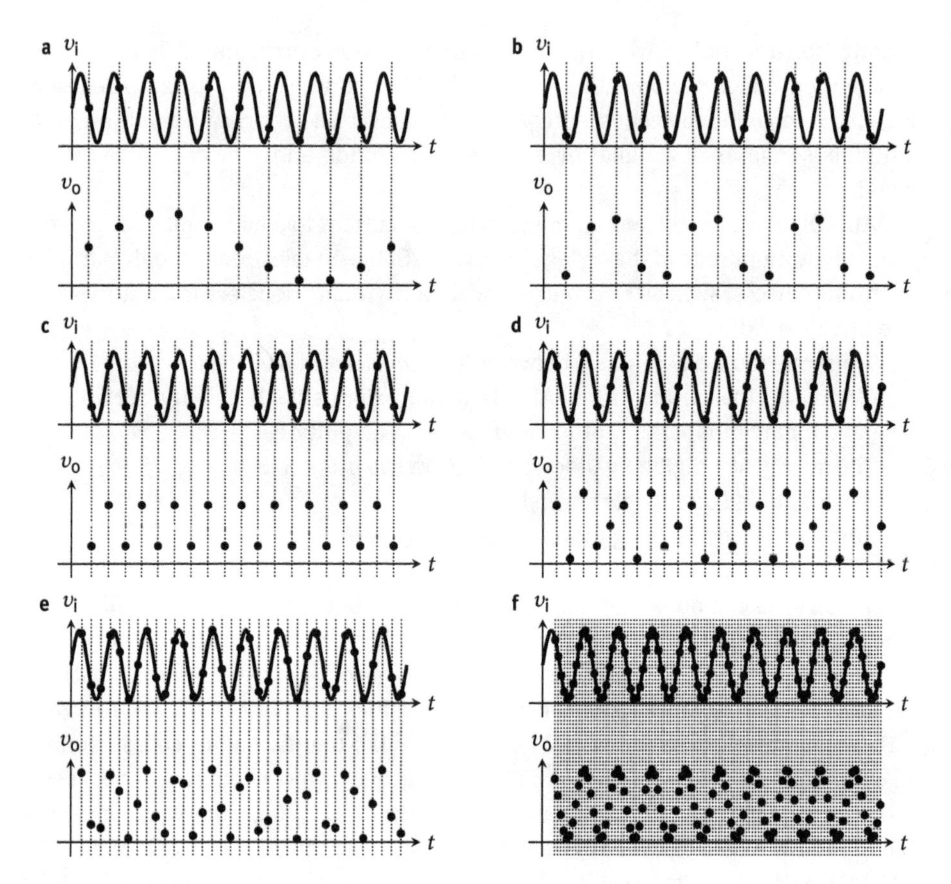

Fig. 6.1. Demonstration of cases of aliasing. The sampling rate is not high enough so that the digital output signal cannot catch the frequency of the analog input signal. Observe the difference between the analog and the digital period.

frequency (which is just the theoretical low limit). For efficiency reasons, signals are sampled at the minimum rate. Sampling at a much higher rate than demanded by the *Nyquist–Shannon sampling theorem* is called *oversampling*. Oversampling allows digital operations improving the result of the conversion.

If a sufficiently high sampling rate is not available, the too high frequencies must be removed from the input signal by means of a low-pass filter. This filter distorts the input signal by removing frequencies above half the sampling rate but the digitized output will still resemble the original analog signal to some degree. This would not be the case at all if aliasing occurs due to the admixture of higher frequencies. It is obviously better that such frequencies get lost than to have them causing aliasing producing false output signals. Such a filter is called *anti-aliasing filter* and is essential for practical ADC systems if the input signal contains higher frequency components.

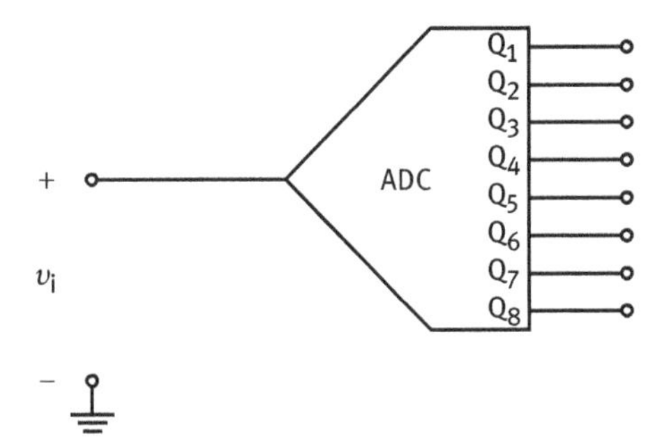

Fig. 6.2. Symbol of a device performing analog-to-digital conversion. The answer to an analog input signal is a binary output signal.

One must discriminate between a device performing analog-to-digital conversion (e.g., Section 6.5) using all kinds of electronic (or nonelectronic) processes to perform the task, and directly converting circuits (Sections 6.2 to 6.4). The symbol for an ADC is shown in Figure 6.2.

Problem

6.1. Digitizing is equivalent to an interpolation process between zero and the full amplitude value. As a consequence, there will be a digitizing (=interpolation) uncertainty. Explain why the character of results obtained by digitization requires that they are given in floating point numbers rather than in integers.

6.1.1 Duration of the digitizing process

The process of digitizing is, as any process, a sequential one, i.e., it takes time to complete it. During the conversion time, the input value must be held constant, e.g., by a circuit called sample-and-hold (Section 4.4.3.2). Monolithic ADCs usually provide the sample and hold function internally. Because of the finite conversion time, it is impossible to completely digitize a *continuous* variable. Digitization can be done either by sampling at instants given by a clock (Sampling, Section 4.4.3.1) which digitizes the instant of the sample, too, or at natural instants like the occurrence of a pulse (or of its maximum). In both cases, there will be a time span afterwards, called dead-time, in which no other conversion process is possible. This period is either called *busy time* or *dead time*. Depending on the digitizing method the dead time is either independent of the size of the signal or not. The sampling process by which amplitude values are acquired at discrete instants in time is irreversible as the timing information cannot be recovered.

Higher sampling rates require faster digitizing. Fast digitization is accompanied by inferior performance with regard to resolution. As a consequence, there does not exist a single method that is best for all applications.

The conversion time is often decisive for the choice of a particular digitizing method. However, there is a trade-off between resolution and conversion time. This difference is particularly obvious among the voltage comparing amplitude-to-digital converters (Section 6.4.2). Conversion rates vary widely, from much less than 1 megasample per second to more than 1 gigasample per second.

Problem

6.2. What is the usual trade-off for fast digitizing?

6.1.2 Performing the digitizing process

The basic digitizing process is quantization: Continuous (analog) values are converted into quantized values being multiples of the smallest quantized value, the *least-significant bit* (LSB). Quantization is irreversible as it is associated with the loss of information.

There are two limiting factors for the resolution of a digitizing process:
- the capability of the digitizing circuit, and
- the ratio of the noise level of the signal to the full amplitude (dynamic) range of the device.

Noise results in a *fuzziness* of the digitization limits. Obviously, it is counterproductive to make the quantization so small that the amount of fuzziness equals the size of the least significant bit. Then the resolution is not given by the fineness of the digitizing process but by the amount of noise. Thus,

> the useful resolution of a converter is restrained by the signal-to-noise ratio (SNR) of the signal

For some applications (in particular in the field of audio systems), admixed noise can be beneficial. *Dither*, a small amount of random (white) noise, which is added to the input before conversion, prevents that signals with a height of the order of the least significant bit are digitized either as 0 or 1. The effect of dither is to cause the state of the LSB to switch randomly between 0 and 1 rather than sticking to one of these values. Figurative this means that the least significant bit appears to be gray rather than white or black. This is beneficial when the binary signal is converted back to an audio output.

In low-bandwidth applications (e.g., audio), white noise is usually of the order of 1 μV rms. If the dynamic input range equals that of the most significant bit of a standard 2 V output signal, noise limits the useful number of bits to 20 or 21. Most commercial converters sample with 6 to 24 bits of resolution. However, as discussed, noise from the circuit components does not allow the full use of the highest resolutions.

Usually, conversion is done linearly. If a different response (e.g., logarithmic) is desired, this can either be achieved
- by nonlinear distortion of the analog signal (Section 2.5.1.2) or

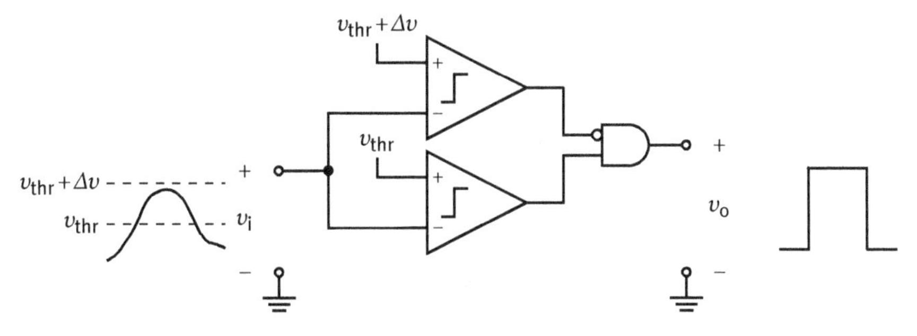

Fig. 6.3. Principle of a single-channel analyzer.

– by using a high-resolution linear ADC and modifying its output in such a manner that the desired response is obtained (with much fewer bits), at least approximately.

The usual primary output of ADCs is comparator responses. Normally, the output of each comparator is assigned a binary weight. In serial digitization, the weight would depend (in addition) on the timing code derived from the synchronizing clock pulse. Thus, the states L or H can be transferred into 0- or 1-bit with a weight that allows generating a binary number. Conversion to any binary code, as needed, is of course possible.

Problem

6.3. Does quantization introduce uncertainty?

?

6.1.2.1 Selecting analog data within a single channel

Some features of analog-to-digital conversion are easier understood by going back to the beginning, the digitizing of a very small portion of an amplitude spectrum by means of a single-channel analyzer. A single-channel analyzer consists of two comparators (Section 2.3.4.2) the output of which is fed into an anticoincidence circuit (Section 4.4.2.3). All signals recorded by the comparator with the lower threshold v_{thr} but not by the comparator with the higher threshold $v_{thr} + \delta v$ produce an output signal of the anticoincidence unit.

If 2^n comparators are stacked with the threshold voltage of each equally spaced (with each difference equalling the full scale voltage range divided by 2^n), the full scale voltage range is covered with 2^n outputs. The input signals are allotted to the outputs according to their voltages. By numbering the outputs from 1 to 2^n the signals get effectively digitized. One has a multichannel analyzer of the parallel type (Section 6.4.2.1).

It is easy to comprehend that the digitized output value is collated to the input voltage $v_{thr} + \Delta v/2$ with a channel width of $\pm\Delta v/2$. Thus, the content of the first channel (of the least significant bit) is collated to $0.5(\times FS/2^n)$ and not to 1 as one would expect from the numbering of the channels. Sometimes this slight difference must be accounted for.

Problem

6.4. How many comparators are needed to build a pulse height single-channel analyzer?

6.1.2.2 Nonlinearity

The ideal transfer function for a linear ADC is linear for all values between zero and maximum value. *Integral nonlinearity* is the worst-case deviation in the mapping of all digital output codes from the straight line. Thus, the *integral linearity* provides a measure of the closeness of the conversion to an ideal linear behavior. It should be less than 1 LSB because otherwise the monotonicity of the conversion is jeopardized. In ADCs based on principles that do not provide excellent linearity, the usual value for the integral nonlinearity is given as less than ±0.5 LSB.

There are two choices of how to construct the *nominal conversion line* to which the actual transfer curve must be compared:

1. use the zero point and the full-scale-point for the definition (in accord with a straightforward interpolation), or
2. find a straight line by the least-squares method best approximating the transfer curve.

The largest deviation of the actual *conversion* (or *transfer*) curve from the ideal line divided by the full-scale value of the conversion is called the *(fractional) integral nonlinearity* of the converter.

The integral nonlinearity is cause of an overall distortion of the shape of the converted distribution (spectrum).

The linear interpolation between the full scale value and zero is accompanied by the interpolation uncertainty. Interpolation is allowed to collate an analog value wrongly *by up to* ± one-half of an LSB. This is caused by the finite resolution of the binary representation of the signal and is *unavoidable*.

In the case of an ideal conversion, each channel having a width of 1 LSB would correspond to identically sized intervals of the analog data. Because of the nonlinearity of the conversion, a change by 1 bit in the digital value corresponds to a different size interval of the analog data, depending on the data value. This irregularity is called *differential nonlinearity*.

Whereas the integral nonlinearity is of importance for individual (independent) data, the differential nonlinearity plays a role mainly for spectra (distributions) because differential nonlinearity affects the apparent frequency of the nominal channel numbers. This results in a channel-wise distortion of the spectrum. To keep this distortion small, the differential nonlinearity must be small. The appropriate specification of the differential nonlinearity is, therefore, given in percent of the mean channel width. For some type of converters, a ±50% differential nonlinearity (expressed flatteringly as ±0.5 LSB) is common. Just consider that such a "narrow" channel occurs in the center of a peak in a pulse-height spectrum. The digital distribution would show two peaks instead of one! Therefore, the differential nonlinearity of *spectral* devices should be less than 1% to avoid crass distortions.

When grouping n data channels into one wider channel, the effect of the differential nonlinearity decreases with $\sqrt{1/n}$, *under the condition that the differential nonlinearity of one channel is independent of the next one* which will not be the case for all types of converters.

Example 6.1 (Distortion of a flat time spectrum due to differential nonlinearity). Let us assume that 10^4 random signals/s are recorded in a detector and that the time distance of each of these signals from a preceding reference signal (with a frequency of 100 kHz) is measured with a device having a time resolution (channel width) of 10 ns and a range of 1000 channels. After 1000 s, the mean frequency in each channel would be 10 000 with a statistical uncertainty of ±100. The frequency spectrum would be flat with a superposition of a statistical ripple of 1% (r.m.s.).

If the width of one specific channel would be smaller by 5% (and for compensation some other channel wider by 5%), it would result in a dip of 500 counts below the flat spectrum at this channel (and a peak of 500 counts above the flat spectrum at the other position).

Differential nonlinearity in the data conversion affects the shape of the converted *distribution*.

Figure 6.4 shows a particularly bad example of a differential nonlinearity measured as described in the above example. Obviously, the ADC is malfunctioning which can be recognized by the multiple correspondence of spurious peaks and dips.

Problems

?

6.5. Is it more likely for a fast or a slow ADC to provide poor differential linearity?

6.6. Does differential nonlinearity badly affect a single measurement, e.g., a voltage measurement performed with a digital multimeter (DMM)?

6.1.2.3 The sliding-scale method

The differential nonlinearity can be greatly reduced by employing the sliding scale or randomizing method. For some applications (e.g., the measurement of pulse-height spectra) flash and successive approximation type ADCs (Sections 6.4.2.1 and 6.4.2.5) could not be used otherwise.

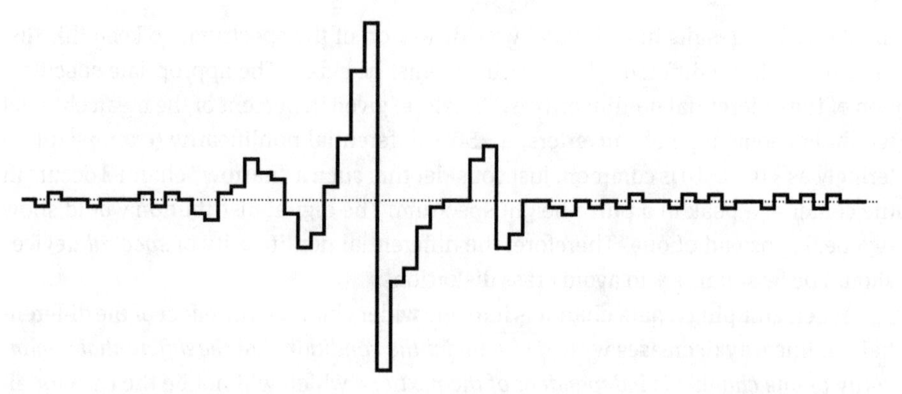

Fig. 6.4. Measured distortion in a portion of a flat amplitude distribution because of excessive differential nonlinearity.

The following example shall visualize how the sliding-scale method improves differential linearity.

Example 6.2 (Correcting a measurement done with a faulty measure). During the first term at the university, the vital statistics of the freshmen is taken. For measuring the height, a standard measure is used. Although the marking on standard measures is typically done within ±0.01 cm, let us assume that the mark for 172 cm is at 171.8 cm. Thus, when sorting the height of students into class widths of 1 cm, there will be a differential nonlinearity of ±20% affecting heights between 171 and 173 cm. There will be a surplus of 20% in the class from 172 to 173 cm, and 20% are too few in the class from 171 to 172 cm. Under the assumption that the minimum height of students is 152 cm and the maximum height is 192 cm, this measure is now moved up and down by as much as 20 cm. The amount of displacement is chosen statistically and is recorded so that the reading can be corrected for the displacement. This way it is assured that, on average, the same percentage of readings is faulty in each class, so that the effect of differential nonlinearity is (statistically) nullified.

The same averaging principle is applied in ADCs with insufficient differential linearity. The input voltage is increased with a random but known analog voltage. After conversion, the digital equivalent of the added voltage is subtracted, thus, cancelling the added signal in the digitized result. The advantage is that each conversion takes place at a random place of the conversion curve.

Problem

6.7. Is it wortwhile to apply the sliding-scale principle, when just a few voltage values are measured, e.g., with a DMM?

6.2 Direct digitizing of time difference

A serial time-to-digital converter (TDC) uses the period length of a known frequency as quantization unit (LSB). The number of period lengths within the unknown time interval is counted to get the digital equivalent of that time interval. As time is sequential, no shorter conversion is possible even when using one of the nonserial approaches. This is different for direct voltage-to-digital conversion (Section 6.4). Figure 6.5 shows such straightforward time digitizing.

The timing diagram of Figure 6.5 is self-explaining. The leading edge of the time interval to be converted generates a start signal, the trailing edge of a stop signal. The start signal opens the gate of the counter, whereas the stop signal terminates the gating signal. During the gate open signal pulses from a highly stable clock reach the counter and get counted. The output voltage of the counter is a number of clock signal that, after being counted, supply a binary number of counts, i.e., the digitized time length.

Good linearity requires a clock with a highly constant frequency so that all bits converted from individual clock periods are equally mapped from their analog counter-part. The best frequency constancy in electronic oscillators is obtained by quartz oscillators (Section 4.2.1.4). The high Q value of such a circuit does not allow synchronizing its frequency. Consequently, neither the start nor the stop signal will coincide with a clock signal which is the cause of digitizing errors, i.e., both at the beginning, and at the end there will be a time distance of less than a quantization unit that is not accounted for. Thus, the result is too small. On the other hand, the result is too large as the quantization unit is a full period, but the counted pulses are shorter. This effect makes the result too large by up to one unit, an effect that easily can be corrected.

The smaller the absolute digitization error, the smaller the quantization step is, i.e., the higher the clock frequency. But it is present, independent of the clock frequency. Ways to overcome the limitation of the resolution by the clock frequency are based on interpolation (Sections 6.2.1 and 6.2.2).

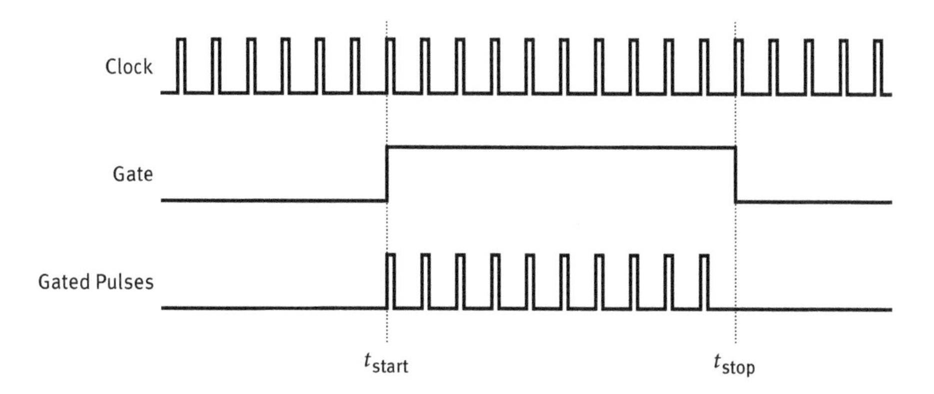

Fig. 6.5. Timing diagram of a directly converting time-to-digital converter.

If the time interval is repetitive with a repetition frequency that is not synchronous to the clock, frequency *averaging* leads to an improved resolution. As, under this condition, the start and the stop signals are random with regard to the clock signals averaging n digitized results improves the resolution by $\sqrt{n}$. Aside from resolution, dead time, minimum height, and minimum pulse width of the input signals are important specifications of a TDC.

The dead time determines an upper limit of the repetition rate of the input signals. Pulse height and width must provide enough charge to charge the input capacitance beyond the threshold of the input voltage.

Problem

6.8. Name the limitation that occurs when time differences are directly digitized.

6.2.1 Composite time-to-digital converters (TDCs)

Applying one of the two methods in Sections 4.5.1 and 4.5.2.1 allows to increase the resolution of TDCs by digitizing the missed portions at the beginning and the end of the time interval. In Section 4.5.1, a time interval is converted into a pulse height by a time-to-amplitude converter (TAC) which can then be digitized by an ADC (Section 6.4). As no significant bits are involved, the reduced precision in the conversion is of little importance.

In Section 4.5.2.3, expanding of time lengths was dealt with. Expanding the missed portions by some factor and using the same clock frequency improves the resolution by just this factor.

Problem

6.9. Name the two methods by which the resolution of TDCs can be increased.

6.2.2 The vernier method

The vernier method as used with a *caliper* allows a higher resolution (usually by a factor of 10) in a length measurement by comparing the reading on a sliding secondary scale with that on the indicating scale. The indicating scale is so constructed that when its zero point coincides with that of the sliding scale (which has slightly larger spacing of the divisions than those on the data scale) just the last division of the sliding scale coincides with a division on the data scale. Normally 10 divisions of the indicating scale would cover nine divisions of the sliding scale increasing the resolution by a factor of 10.

The same principle is applied (twice) to improve the resolution by digitizing the portions at the beginning and the end of the time interval that are shorter than one period of the clock signal. The role of the indicating scale is played by the frequency

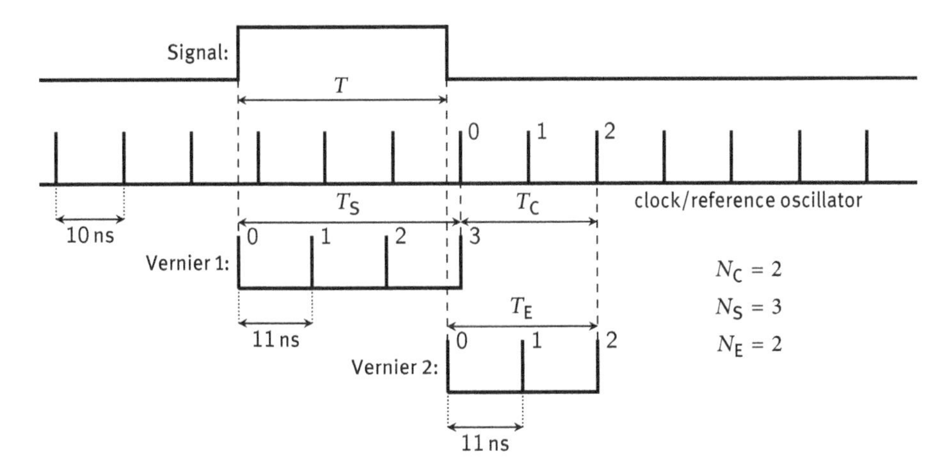

Fig. 6.6. Vernier principle applied to time lengths. (Observe that the geometric time lengths are not to scale.)

of a highly stable clock (reference oscillator). The role of the sliding scale is played by the frequency of an oscillator that has a slightly lower frequency and is started at the moment of the start signal. Its Q value must be rather low so that a prompt start is possible. However, the conversion quality of the resulting low binary bits need not be high. The simple numerical example using Figure 6.6 should be helpful in understanding the principle.

The resolution of a 100 MHz digitizer shall be improved by a factor of 10. Thus, a 90.9 MHz startable oscillator must be added to give a pulse period of 11 ns available for the vernier action. The start signal starts the vernier oscillator. The highly constant clock oscillator is permanently oscillating providing a pulse period of 10 ns. In the figure, the pulses of both oscillators coincide after three periods of the vernier oscillator making the number N_S of start oscillations $N_S = 3$. From now on, the pulses N_C of the clock oscillator are counted. From the beginning at the instant of the stop signal the number N_E of the vernier oscillations is counted until a coincidence with pulses of the clock is detected ending both the counting of the vernier counter (with $N_E = 2$) and the clock counter (with $N_C = 2$). As the stop counter counts towards the end of a period, its counts must be subtracted. The numbers counted by the vernier must be multiplied by 1.1 to make them compatible with the 10 ns period length of the clock. Thus, the length of the time interval T with improved resolution is $T = (N_C + 1.1N_S - 1.1N_E) \times 10$ ns (31 ns in this example). The resolution was effectively increased from 10 to 1 ns.

Problem

6.10. Get hold of a caliper to study the vernier principle.

6.3 Direct digitizing of frequency

Frequency is measured the same way as a time difference except for the reversal of the variables. Here, the unknown frequency is measured by counting the periods over a fixed time span provided by a timer by means of a fixed frequency, e.g., by digital rate meters. The frequency signals are shaped, e.g., by means of a Schmitt trigger (Section 4.1.2.1) and then fed coincident (Section 4.4.2) with the timer signal into a counter (scaler). The coincidence unit allows the passage of signals during a selected (digital) time interval so that the digital value of the frequency is obtained as a ratio of the contents of the scaler over the value of the selected time interval.

At low count rates, it takes too long to determine the frequency by counting the periods per time. In this case, the frequency is digitized indirectly by measuring the time length of one period (Section 6.2). In the case of a constant frequency, the uncertainty of the digitized answer can by reduced by averaging (measuring the same frequency a number of times, or by applying a longer time base) effectively increasing the resolution (at the cost of conversion time, as usually is the case). Although this digitization method is straightforward, and, therefore, generally used, it does not constitute a direct digitization. A direct frequency-comparing method would have to tune a reference voltage-controlled oscillator (VCO) (Section 4.5.3) digitally until agreement in the frequencies is observed. This can only be done serially because of the sequential nature of a frequency.

Example 6.3 (Any frequency is analog). As frequencies are measured by counters and the output of counters is an integer number, it does not surprise that a frequency is often not recognized as analog quantity.

Let us take a crass example. Why is the frequency with which the display of the seconds switches in a digital watch analog? One has to go back to the definition of a frequency (of a repetition rate). It is *events per second*. The result of counting is digital. However, time is an analog quantity making the ratio, the rate, analog. Even if the rate on the watch is exactly 1/s, the second on this watch and on any watch or clock is not exactly 1 s long. This can be expressed by some uncertainty of the time length that also enters in the uncertainty of the frequency making the result, not 1/s (which would be digital), but, e.g., the analog quantity (with d some decimal digit between 1 and 9) $(1.0000 \pm 0.000\,d)$/s. A digital (binary) number as a multiple of the LSB has NO uncertainty. Therefore, frequency cannot be digital.

Example 6.4 (Internal uncertainties in a frequency measurement). The frequency of regular rectangular signals with a width of about 1 μs and a duty cycle d of 1/20 (i.e., with an approximate repetition rate of 50 kHz) shall be measured by counting the signals. The timer is based on a quartz-controlled oscillator of 5.000 MHz. It gates the counter by way of an overlap coincidence (Section 4.4.2). What is the uncertainty in the frequency if 50 436 signals are counted during a nominal time length of 1 s ?

Note: To measure a frequency by counting, one has to count the number of *complete* periods (of length T) per time (i.e., in a time window of the known length).

Uncertainty of the length of the time window The calibration uncertainty is implicitly given as $\pm 1.0 \times 10^{-4}$, thus, the *scale* uncertainty is $\pm 1.0 \times 10^{-4}$ s. The time jitter (Section 4.4.1) can be expected to be $< 10^{-8}$ s and is, therefore, negligible.

Counting uncertainty The beginning of the time window will not coincide with the beginning of a period, nor will the two ends coincide. Consequently, there will be *digitizing* uncertainties. If the beginning of the period is given by the rising slope of the rectangular signal, then the end will be given by that of the consecutive signal. Under such conditions, the time window is effectively longer by half a length of a period T (because each incomplete period at the end of the time window has a chance to be counted) with a digitizing uncertainty of $\pm 0.50T$. At the beginning of the time window, the situation is not symmetric. There, only a fraction of the incomplete periods given by the duty factor d can be counted. Therefore, at the beginning, the time window is effectively extended by $0.50 \times d \times T$ with a digitizing uncertainty of $\pm 0.50T$.

So the scale uncertainty is $\pm 1.0 \times 10^{-4}$, i.e., ± 5.04 periods, the digitizing uncertainty is $\pm\sqrt{2} \times 0.50/\sqrt{3}$ periods, i.e., ± 0.41 periods and the correction factor $f_{\text{corr}} = (1 + d) \times 0.50 \times T$ for the leakage of incomplete signals becomes 1.04×10^{-5}. Therefore, 0.52 ± 0.41 periods were counted too many, reducing the counted result to $(50\,435.5 \pm 5.1)$ Hz. Increasing the length of the time window does not markedly reduce the percentage uncertainty because the scale uncertainty dominates.

Any digitizing process introduces *uncertainty*, at least a rounding uncertainty due to the limited number of digits used.

Problem

6.11. Is frequency an analog or digital quantity?

6.3.1 Multichannel count-rate-to-digital conversion

When an analog quantity has been converted into a (binary-coded) number, it will often be necessary to store this number for later use. The later use of sampled information will often be the reconstruction of the time dependence of the analog information (mostly much later, e.g., in audio and video application which require long-term nonelectronic storage, but this is not the subject of this book).

When concentrating on nonsampled binary-coded information, it is often of interest to know how often a certain number (binary code) was obtained during some time interval, i.e., the count-rate spectrum (or frequency spectrum, as called in statistics) of each binary code is to be known. For this end, a two-dimensional binary memory is needed to correlate the occurrence number with the specific digitized numbers, the *address numbers* in the memory. The address number is usually defined by the position of the storage cell(s) in space (parallel logic), or in time (serial logic) as defined by the corresponding code (protocol).

A memory is a physical (electronic) device that stores binary numbers (codes) (temporarily) for later use. A *random-access memory* (RAM) is needed to allow access to any address at high speed (usually provided by semiconductor electronics, e.g., bistable storage cells). The term *memory* is usually associated with the semiconductor RAM, i.e., integrated semiconductor circuits. The other (slower) not-fully-electronic storage is done with the so-called *storage* devices.

A semiconductor memory is organized into memory cells (e.g., flip–flops) with their states (L or H) used as binary information (0 or 1 when the logic is positive). Memory cells are combined into *words* of fixed length that determine the full scale, the MSB. Each word is assigned a binary coded *address* of n bits (also called *channels*), making it possible to store and access randomly 2^n words in the memory. Often memories are *volatile*, i.e., electric power is required to sustain the stored information. Memory technology is still progressing fast so that describing a status is not meaningful. Besides it is not the subject of this book.

To store digitized information in single-(or multiple-) task digital systems, not as much flexibility as in general-purpose computers is required. Thus, the location of the digitized information (the channel number) is correlated with the physical location on the memory hardware. Thus, no complex memory management is necessary speeding up the storage process.

Problem

6.12. What is the difference in the units: signals per second and Hz?

6.3.1.1 Multiscalers

In the multiscaler mode, consecutive counts (signals representing the same kind of digitized information) are used to increment the contents of a counter assigned to the appropriate channel one by one. This allows the processing of higher data rates, as to each signal only the dead time of a counting process applies and not memory dead time connected with the incrementation process of a memory cell. Thus, a multiscaler performs the task of many scalers counting digital signals each of which represents some specific digitized property coded by way of the address of each scaler. In particular, the term multiscaler is used for a device that determines a count-rate dependence by assigning the address numbers one by one to consecutive counter dwell times of constant length.

Problem

6.13. A computer-based multiscaler assigns an address to each scaler. Will the dead time of such a device be determined by the dead time of each scaler or by the dead time stemming from the assignment of the scaler in use?

6.3.1.2 Multichannel pulse-height digitizer

The digital output of an amplitude-to-digital converter (Sections 6.4 and 6.6) is sorted according to its binary value (address, channel number) and the value stored at this address is incremented by this output signal. Thus, a digitized pulse-height spectrum of the recorded signals is obtained.

6.3.1.3 Multichannel time-interval digitizer

The digital output of a time-to-digital converter (Section 6.2) is sorted according to its binary value (address, channel number) and the value stored at this address is incremented. Thus, a digitized time (difference) spectrum of the recorded signals is obtained.

6.4 Direct digitizing of amplitude

As discussed in Section 6.1, one must distinguish between ADC circuits that perform a direct conversion of an analog signal into a digital one, and ADC devices that first make an analog-to-analog conversion and then convert the converted analog signal to a digital output variable. The latter are called *composite* ADCs (Section 6.6). For most directly comparing ADCs the reference voltage is supplied by a DAC, a digital-to-analog converter.

6.4.1 Digital-to-amplitude conversion (DAC)

A digital-to-amplitude converter converts a digital (e.g., a binary coded) input signal into a voltage (or current) amplitude. It is a convenient means of supplying the reference voltages of the comparators in sequential voltage comparing ADCs (Sections 6.4.2.3 to 6.4.2.5).

Digital-to-amplitude conversion is prone to four fundamental errors:

- zero off-set error,
- scaling error,
- integral nonlinearity, and
- differential nonlinearity.

Figure 6.7 gives the ideal response of a 3-bit DAC. When the binary input is changed from 0 to 7 in equal time intervals, the voltage output should be in the form of an ideal staircase ramp with steps of equal width and equal height.

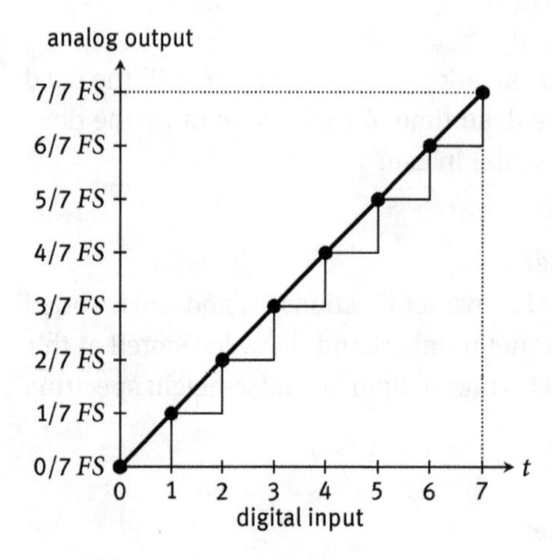

Fig. 6.7. Response of an ideal 3-bit DAC to digital inputs from 0 to 7 that arrive in consecutive constant time intervals. The straight line is the ideal conversion curve.

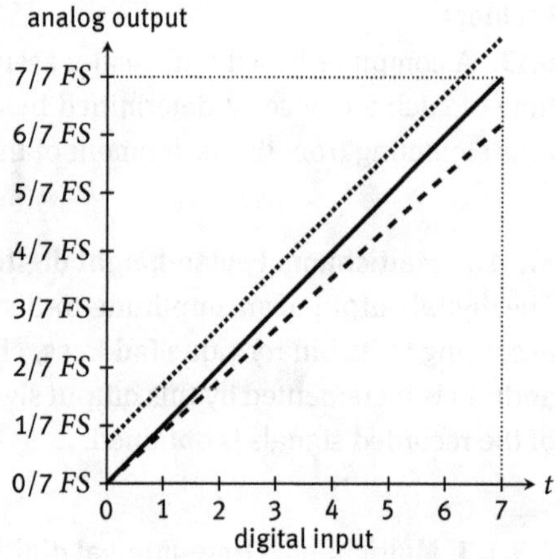

Fig. 6.8. Effect of the zero-offset error (dotted line) and scaling error (dashed line) on the conversion curve.

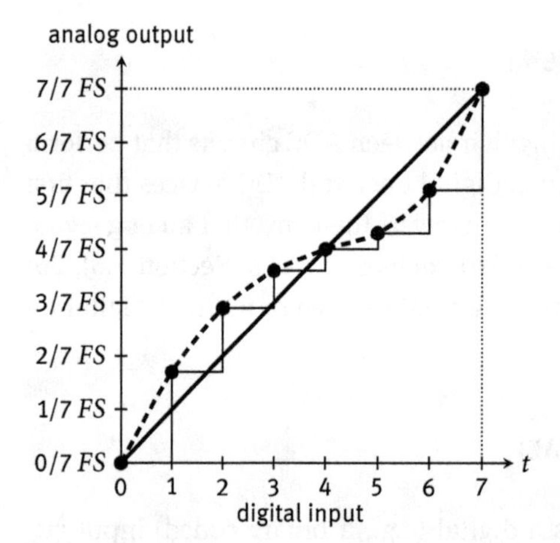

Fig. 6.9. Example of an error due to the integral nonlinearity of the actual conversion curve.

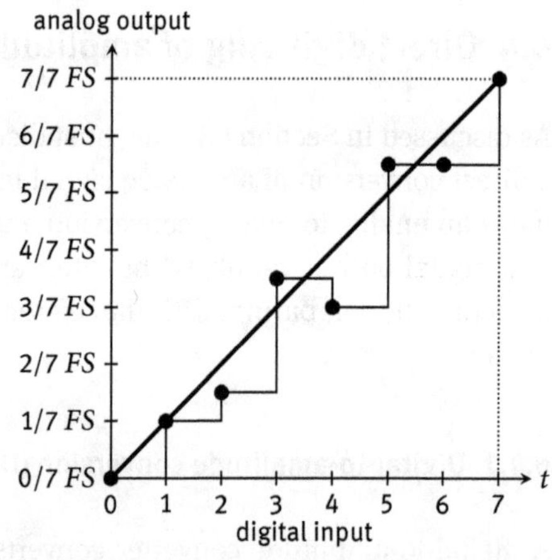

Fig. 6.10. Example of an error due to the differential nonlinearity of the actual conversion curve.

Figure 6.8 shows how a zero-offset and a scaling error affect the conversion curve. Figure 6.9 illustrates the difference between the actual and the ideal conversion curves due to integral nonlinearity (Section 6.1.2.2). Figure 6.10 illustrates the difference between the actual and the ideal conversion curve due to differential nonlinearity (Section 6.1.2.2).

The example of Figure 6.10 shows that monotonicity is not given any more if a differential nonlinearity error of 1 LSB is present. The output voltage for the number 4 is only 3 voltage step units so that it is smaller than the voltage equivalent of 3 which is 3.5 voltage step units (off by −0.5 LSB). This is avoided by the manufacturer of ADCs and DACs by allowing a maximum differential nonlinearity of not more than ±0.5 LSB (or ±50% which is gruesome but more honest). Such a specification is better than a warranty of monotonicity.

The measurement of spectra (frequency distributions) requires an excellent differential linearity.

Problem

6.14. A straight use of DACs is in displaying digital data. Name two examples of such displays.

6.4.1.1 Parallel DAC

For circuit simplicity, it is necessary that the digital input signal is available as weighted binary code (Section 5.1.1). In Figure 6.11, the symbol of a 3-bit digital-to-amplitude converter is shown.

The central part of the DAC shown in Figure 6.12 is the summing amplifier (Section 2.5.1.1), a transimpedance amplifier that converts its input current i_F to the output voltage v_o via the feedback resistor $R_F = R$. Each of the three input voltages v_k (either L or H) is supplied by a low ohmic source (e.g., a logic gate or a voltage follower, Section 2.5.2.1) so that the current i_k through the weighted source resistors $k \times R$ is summed in the virtual ground node of the amplifier to give i_F. By using only resistors with values $2^n \times R$, the maximum output voltage becomes $v_{omax} = (1 + \frac{1}{2} + \frac{1}{4}) \times v_H$ with v_H the voltage of the high state. This maximum occurs for $7 = 2^3 - 1 = 111_2$. For the other seven input combinations, the output voltage will be zero or a multiple of $0.25v_H$. The weighting factor k is proportional to the inverse of the place weight of each bit in the basic binary numeration system. Sizing the feedback resistor allows one to adjust the output voltage at will. For $v_H = 5$ V, the choice of $R_F = 0.8R$ would

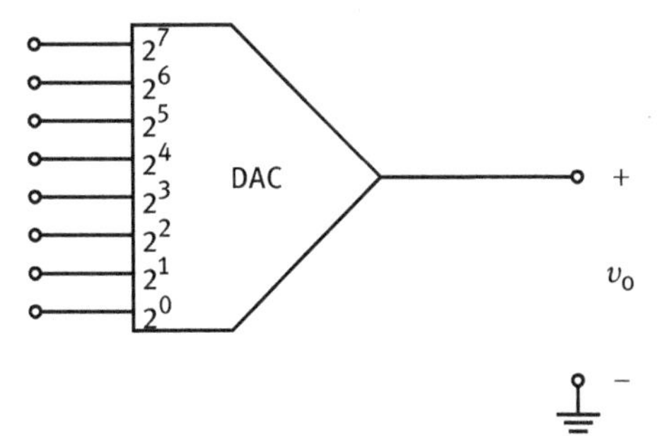

Fig. 6.11. Symbol of a parallel digital-to-amplitude converter. A binary coded input signal is converted into signal amplitude.

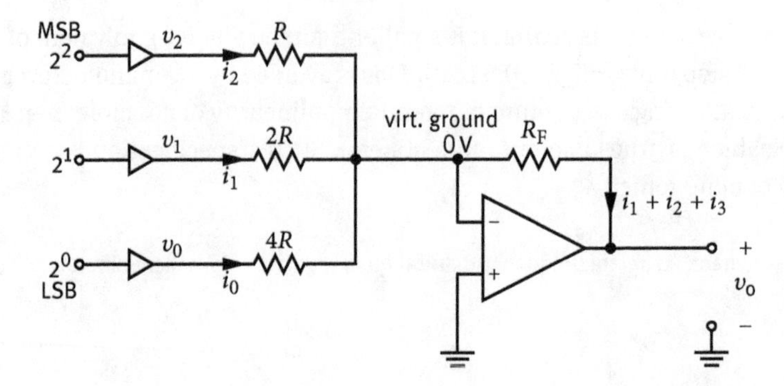

Fig. 6.12. Principle of a three bit parallel digital-to-amplitude conversion using graded resistors for achieving a binary-weighted current input.

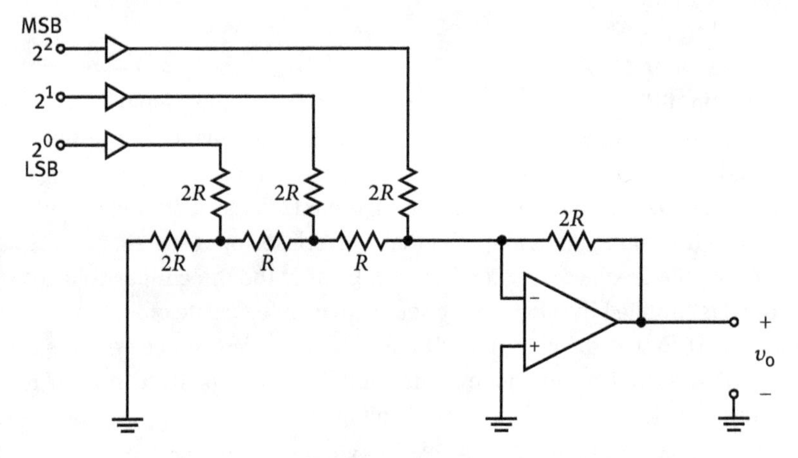

Fig. 6.13. Principle of a 3-bit parallel digital-to-amplitude conversion using an R/2R network for generating input currents with the correct binary weight.

make in this example $v_{omax} = 7$ V, i.e., the output voltage for 1 LSB would be 1 V. Thus, with the help of R_F the span of the output voltage can be adjusted.

For this circuit, it is necessary that all logical high states are stable and equal in amplitude and that the logical low states are zero. Another difficulty is the availability (and cost) of these graded resistors. Just think of building a 21-bit DAC according to this principle. The resistor for the LSB would be 2^{20}-times (i.e., $\approx 10^6$-times) larger than that of the MSB, and in addition, the uncertainty in the resistance value of the latter must be less than 10^{-6}. Obviously, a different input current adding network is necessary.

Binary weighting with only two kinds of resistor values is achieved by the so-called *R/2R-ladder* network as shown in Figure 6.13.

By applying the superposition principle and Thevenin's theorem for each bit input, the equivalence of the R/2R network with the network of graded resistors can be shown with some calculational effort. The differential nonlinearity based on the uncertainty in the resistor values is much reduced because in the case of the same value resistors not the absolute tolerance of the resistors counts but their relative deviations. As resistors of the same production batch differ very little in their values, the relative deviations will be small. Very similar considerations apply for the production as integrated circuits. The disadvantage of the R/2R-ladder network is the increased parasitic capacitance which increases the response time.

With high resolution DACs the current connected with the conversion of very small numbers is so low that charging the unavoidable stray capacities takes too long, increasing the conversion time. Boosting the currents connected with low bits by a constant current before conversion and subtracting its equivalent binary value from the converted voltage makes the circuit faster without jeopardizing the total precision of the conversion.

Problems

6.15. The full scale output voltage of an 8-bit DAC is 5.0 V. What is the step voltage for 1 bit?

6.16. The output range of a DAC is from −10 V to +10 V. How many bits are necessary to resolve 10 mV?

6.17. The output range of an 8-bit DAC is 10 V. To which voltage will the following two weighted binary codes be converted?
(a) 00100101_2
(b) 00010101_2

6.4.1.2 Serial DAC

Figure 6.14 shows a block diagram of a (compound) serial digital-to-amplitude converter. A parallel DAC receives the output of a deserializer (Section 5.4.1.1) which converts the serial logic signals into parallel logic signals. The properties of such a system are determined by the parallel DAC in use, i.e., it is actually a parallel DAC.

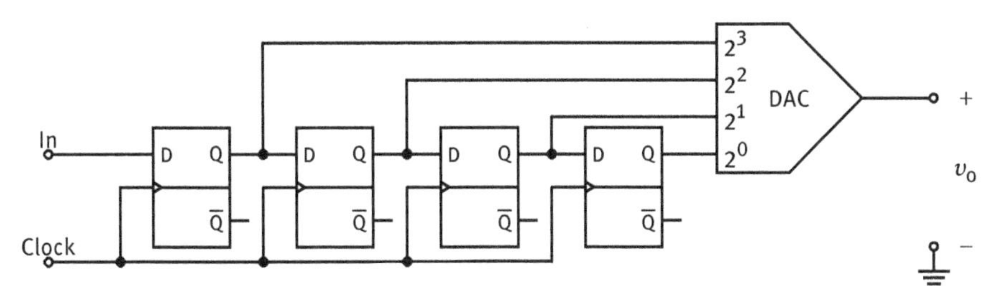

Fig. 6.14. Block diagram of a serial input digital-to-amplitude converter. Serial binary logic signals are converted into an analog signal.

An alternative is a DAC with an output by way of a staircase ramp where the voltage of the uppermost step corresponds to the number of individual input signals recorded. The *active diode pump* circuit (Section 4.5.4) is such a staircase ramp generator, i.e., it performs a serial digital-to-voltage conversion. The digital input is a string of clock pulses. The output amplitude is increased stepwise for each clock signal at the input, i.e., the instantaneous output voltage depends on the number of pulses that have been fed into the input.

Obviously, such DACs are slow because they must produce a step for each input signal. Thus, any combination with a moderately fast ADC is counterproductive. However, it could be used in a serial ADC (Section 6.4.2.3).

A genuine serial DAC has much better differential linearity (Section 6.1.2.2) than a parallel DAC because all the steps are produced the same way. The step size depends mainly on the amplitude (and rise time) of the clock signal and on the stability of the voltage division by the two capacitors. Using capacitors of equal properties, the latter contribution should not matter.

Problems

6.18. Which property makes a true serial DAC superior to other types of DACs?

6.19. Why might the conversion by serial DACs be impractical when the bit number is high, e.g., 12 bit?

6.4.2 Voltage-comparing amplitude-to-digital conversion (ADC)

Voltage-comparing *amplitude-to-digital conversion* can be done in the frame of serial logic, or parallel logic, or using an intermixture of both. The main feature of the first is the stepwise process and that of the second is the multitude of comparator responses. (For the single-channel analyzer in Section 6.1.2.1 there were two such comparator responses.)

Table 6.1 presents a simple scheme in which all five types of voltage-comparing ADCs fit. There are two classes of ADCs in correspondence with the general duality of parallel logic and serial logic: the class of parallel ADCs with t process steps and $t \times (2^{n/t} - 1)$ comparator responses and the class of serial ADCs with s responses (and sections) and $s \times (2^{n/s} - 1)$ steps. The reciprocity of the number of steps t and the number of responses s is obvious not only for the pure cases but also for the mixed cases. The number of comparator responses does not necessarily correspond to the number of comparators in the special arrangement, because (as in the case of the successive approximation), comparators can be spared by changing the size of the step and using the same comparator again and again.

Unfortunately, this nomenclature is not adhered to. So the *parallel ADC* is generally called *flash-converter*, for purely commercial reasons. In the following sections describing these ADCs the “commercial” names are given, too.

Table 6.1. Comparison of the characteristics of the five "pure" types of voltage comparing ADCs when converting a full-scale (FS) signal. The resolution range is 2^n, the number of time steps is t, and the number of comparator decisions is s.

	Parallel class			Serial class	
	Parallel ADC	Serial–parallel ADC *t* steps	ADC with successive approximation	Serial–serial ADC *s* sections	Serial ADC
Steps for a FS conversion	$t = 1$	$1 < t < n$	$t = n$	$t = s \times (2^{\frac{n}{s}} - 1)$	$t = 2^n - 1$
Responses for a FS conversion	$s = 2^n - 1$	$s = t \times (2^{\frac{n}{t}} - 1)$	$s = n$	$n > s > 1$	$s = 1$

Rather than increasing the voltage step size of the reference signal the input signal may be reduced by the equivalent amount in each step.

Problem

6.20. Name the advantage of voltage comparing ADCs.

?

6.4.2.1 Parallel ADC (flash converter)

The parallel ADC is a straight extension of the single-channel analyzer (Section 6.1.2.1). The input voltage is compared (in one step) against $2^n - 1$ fixed reference voltage levels by means of 2^n comparators. The comparator responses are interpreted by a 2^n- to n-line *priority encoder* circuit which produces a binary coded output. Figure 6.15 shows a 3-bit parallel ADC circuit.

V_{ref} is a precision reference voltage providing the individual voltage reference levels for the comparators by means of a multiple voltage divider. (By tailoring the voltage divider a variety of nonlinear amplitude-to-digital conversions by this circuit is feasible.) When the input voltage at any comparator exceeds its reference voltage, these comparator outputs will switch to the H state. The *priority encoder* selects the H output of that comparator with the highest reference voltage and generates accordingly a binary code. Conversion is very fast, capable of gigahertz sampling rates (as needed for, e.g., sampling oscilloscopes), but the resolution usually does not exceed 8 bit since the number of comparators needed, $2^n - 1$, practically doubles with each additional bit suggesting the use of two or more step converters (Section 6.4.2.2) for achieving higher resolutions. This type of ADC has high input capacitance (several 10^{-11} F) and high power dissipation. The speed of these circuits is only limited by comparator and gate propagation delays. This one-step conversion is contrasted by the need of one comparator (response) for each binary place of the output signal (see Table 6.1).

Both the integral and differential nonlinearity is usually given as ±0.5 LSB which disqualifies this kind of circuit for some applications. The stability depends on the drift of the comparator levels. It contributes to the nonlinearity of the conversion.

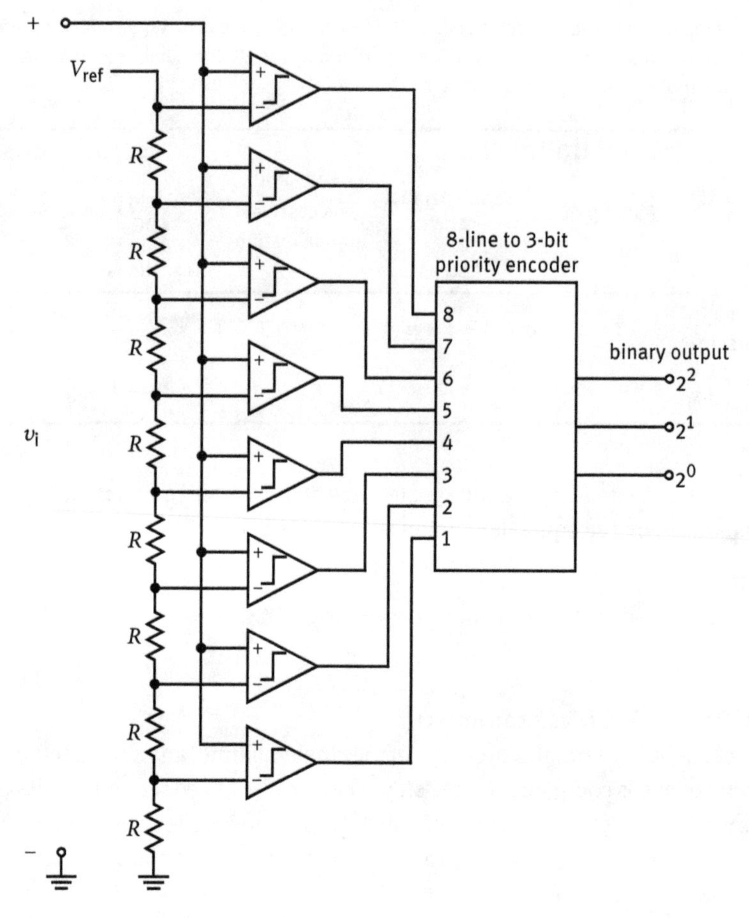

Fig. 6.15. Principle of a 3-bit parallel ADC circuit.

Problem

6.21. What is the usual method to counterbalance the rather bad differential linearity of parallel ADCs?

6.4.2.2 Serial–parallel (parallel-pipeline) ADC

As it is impractical to double the number of comparators to 512, parallel ADCs are more or less limited to a resolution of $n = 8$ bit. Using the same ADC in t steps with refined resolution of the (coarse) bit which corresponds to the amplitude of the input signal the resolution can be increased to $t \times n$ at the cost of longer conversion time. This procedure is also called *subranging*. According to Table 6.1 the input voltage is compared in t steps against $2^{n/t} - 1$ fixed reference voltage levels. In each step, the voltage intervals are reduced by a factor of $2^{n/t}$ and they are superimposed on a voltage level, produced by a DAC (digital–analog converter) from the value of the (more) significant bit(s) determined in the previous step(s). Thus, a 16-bit ADC can be obtained by using an 8-bit parallel ADCs in two steps having only twice the conversion time of a single parallel ADC. The highly advanced pocket multichannel analyzer MCA8000D of Amptec Inc. uses a 100 MHz pipelined flash ADC. It acquires 16 bits every 10 ns, i.e., the ADC conversion time is 10 ns. The system dead time is governed by the minimum rise time of the input signal which is > 500 ns. The peak detection is done digitally, requiring only a few clock signals to find the peak. The digital peak detect function is preceded by an analog shaping network. Since the processing is only peak detection, it can be done at the ADC clock rate. It is not really done in a single clock pulse but the processing is pipelined with the clock rate. As the differential linearity of parallel ADCs is mediocre at best, the sliding-scale method (Section 6.1.2.3) is applied to achieve a differential nonlinearity of < 0.6%. However, the sliding scale is not implemented after each conversion but this is done at a reduced rate. Concluding remark: there is, of course, no obligation to apply the same conversion principle in each serial step. Subranging (coarse) parallel conversion with a successive-approximation ADC for the finer conversion has been realized.

Problem

6.22. Which method must be applied to obtain a 16-bit ADC with minimum conversion time?

6.4.2.3 Serial ADC

A serial (also: *ramp-compare*, *digital-ramp*, *stairstep-ramp*, or *counter*) ADC compares the input voltage against a DAC voltage which a timer increases in incremental unit steps (with step size corresponding to the value of the least significant bit). The output of a clock is counted and the (binary) counter output is converted into amplitude by a DAC (Section 6.4.1) producing a staircase ramp. The first output voltage step that is higher than the input voltage fires a comparator stopping the counter. The binary output of the counter is then the binary equivalent of the input voltage. Just one comparator (response) is needed, but for the conversion of a full scale signal $t = 2^n - 1$ steps are necessary (see Table 6.1). The staircase ramp may be produced cheaply and precisely by a diode pump (Section 4.5.4). However, using a standard DAC (Section 6.4.1.1) gives better stability at the cost of reduced differential linearity. Obviously, the conversion time is long and the resolution and linearity depends mainly on those of the DAC. After resetting the counter, the next conversion can be done. Figure 6.16 shows the block diagram of a serial ADC. Its detailed operation is as follows.

With each counted clock pulse, the DAC output is incremented by a unit step voltage. A comparator compares this output voltage against the input voltage. As long as the input voltage is greater than the DAC output voltage the comparator's output will stay H. As soon as the DAC output voltage exceeds the input voltage the comparator's output goes L. The H-to-L transition of the comparator's output causes the shift reg-

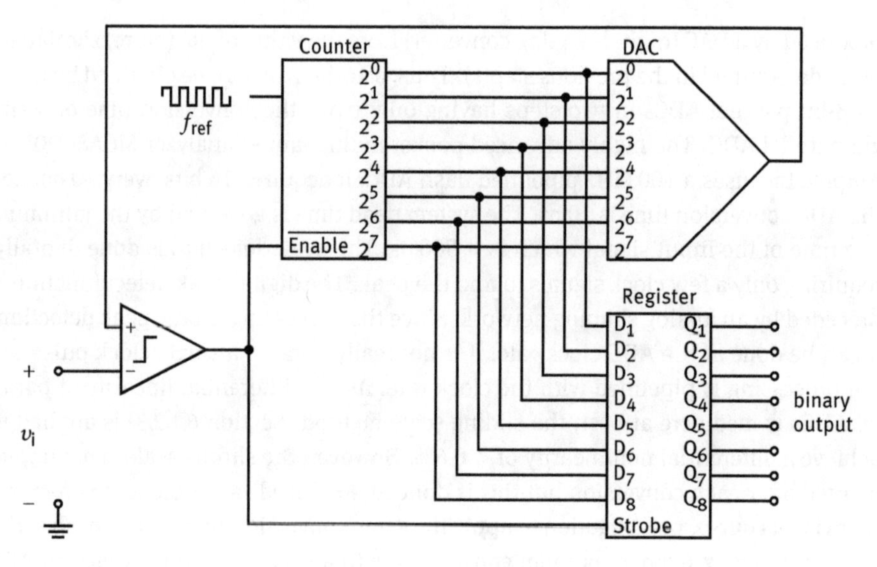

Fig. 6.16. Block diagram of a serial ADC.

ister to take over the output of the counter, thus, updating the ADC circuit's output; and the counter will receive a signal on the R (reset) input, causing the counter to reset (to resume the start value zero). Obviously, the conversion time depends on the amplitude of the input signal. This variation is not acceptable for applications requiring a constant sampling frequency. Also the long (average) conversion time places the serial ADC at a disadvantage to other ADC types. The serial ADC is in direct competition with the Wilkinson ADC (Section 6.6.1). Its advantage is that the stability of the clock frequency does not matter, its disadvantage that the absolute shape of the input signals must not change. Both the circuit complexity of these two ADCs and their specifications are comparable. However, only Wilkinson ADCs are available in monolithic form.

Problem

6.23. On which circuit is a serial ADC fully dependent?

6.4.2.4 Serial–serial (sectioned-serial, serial-pipeline) ADC

The principle of this ADC is best explained by means of an (extreme) example. A serial ADC for a 2^{18} resolution needs about 250 000 steps for the conversion of a full scale signal. By using a serial–serial ADC with three sections, with a resolution of 2^6 each, only $3 \times 64 = 192$ steps are required.

The DAC of this ADC does not produce a staircase ramp of equal step size as for the serial ADC but starts with coarse steps, then finer steps and at the end with a step size corresponding to 1 LSB. In the first section, with the coarse resolution, the comparator response gives the coarse bit of the amplitude. In the next step, this bit is scanned with higher resolution to determine a lower bit that collates to the amplitude. This is continued until in the last section the unit step size corresponding 1 LSB is reached and the conversion is finished. By using n sections for a resolution of 2^n a "ramp" of single steps of decreasing size at each step is produced. This limiting case has gained importance under the name successive-approximation ADC.

Concluding remark: there is, of course, no obligation to apply the same conversion principle for each section. Mixing serial coarse conversion with the parallel or successive-approximation principle for the finer conversion is feasible.

Problem

?

6.24. A 16-bit sectioned serial ADC is to be built in two sections. What section size is optimal?

6.4.2.5 Successive-approximation ADC

Example 6.5 (Successive approximation in the "high–low" game). A number between 1 and 1000 must be guessed. After each guess one gets the answer whether the guess was high or low. If done efficiently, it takes at most 10 guesses to arrive at the correct number as 1000 is less than 2^{10}. The scheme works by dividing the range of interest in half and choosing the new range according to the information obtained, i.e., one should start with 500.

If 500 is low, the next guess is 750, if it is high the next guess is 250 and so on. If the number to guess was 1, it would take 10 guesses to arrive at this answer. If done correctly, there is no number that requires more guesses. This efficient way to partition a quantity is found, e.g., in partitioning inches. However, overall, the decimal partitioning has its merits because the results are more easily expressed by decimal numbers.

The successive-approximation ADC is both a serial–parallel ADC with the number of steps $t = n$ and a serial–serial ADC with number of comparator responses $s = n$. The central position in the scheme of Table 6.1 takes care of this fact.

It is easier to discuss the modifications that make a serial ADC to a successive approximation ADC. Instead of counting up in binary numbers, the register that controls the DAC output uses one by one all n values of the bits starting with the most-significant bit and finishing at the least-significant bit. Throughout this process, the register supervises the comparator's output to detect whether the DAC's binary equivalent is less than or greater than the analog signal input, setting the (last) processed bit value L or H. Thus, within n steps the complete bit pattern is obtained and the conversion finished. Figure 6.17 shows the decision tree of a 3-bit successive-approximation ADC.

Figure 6.18 shows a block diagram of a successive-approximation ADC. It contains a successive-approximation processor that performs the logical decisions needed for the correct sequence of the binary input of the DAC.

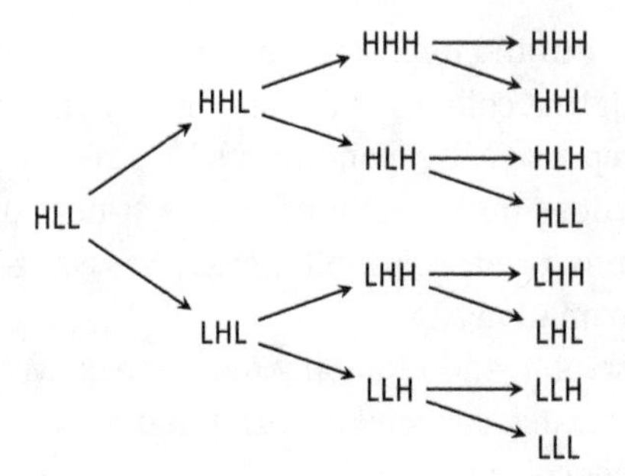

Fig. 6.17. Decision tree of a 3-bit successive-approximation ADC.

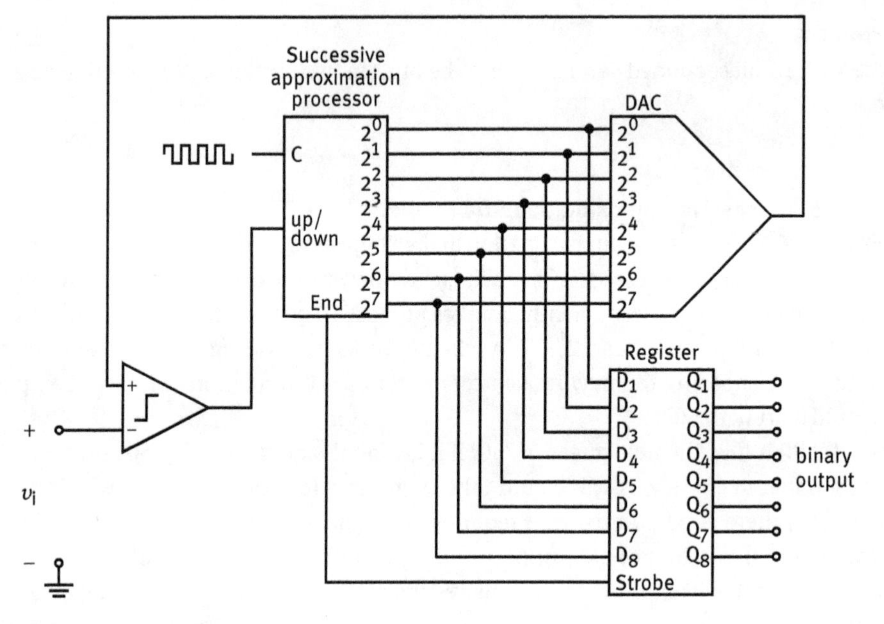

Fig. 6.18. Block diagram of a successive-approximation ADC consisting of a comparator, the successive-approximation processor, the DAC, and the output shift register.

Note that the conversion time for this ADC lasts n steps, i.e., the length of all sampling intervals is independent of the signal amplitude, by contrast to the serial ADC circuit. However, the linearity is poor. Nonlinearity arises from accumulating errors in the digital-to-amplitude conversion. For example, the amplitude difference between the converted value of 2^n and $2^n - 1$ should be the amplitude equivalent of 1 LSB. In the present case the difference in the amplitude of the nth bit and the sum of all amplitudes of the bits smaller than n should be 1 LSB. If the conversion of each bit is done with an uncorrelated (independent) uncertainty of 1%, the differential nonlinearity for the binary value just below the 8th bit would add up to 1.48 LSB. To reduce the differential nonlinearity below 0.5 LSB it would be necessary to convert each bit with an uncertainty of about 0.3%. This example is insofar typical as the largest differential nonlinearities occur for values just below each completed 2^n code. If the system is stable, these localized differential nonlinearities can be empirically corrected for.

Problem

?

6.25. A signal to the input of a 10 bit successive approximation ADC is exactly one-half of the full range. How many comparator responses are required to digitize said signal?

6.5 Differential ADCs

Digitizing just the change that has occurred since the last sample is called differential sampling. Digitizing only the difference saves time so that this differential digitizing has its merits by an increased sampling rate. The term *delta* in the name of an ADC indicates that only the change between successive samples is processed.

6.5.1 Tracking or delta-encoded ADC

For sampling applications, the serial ADC can be made more efficient for tracking continuous input amplitudes. This variation is called *tracking ADC* (Figure 6.19). As long as the comparator messages that the input signal is higher than the reference voltage provided by the output of the DAC, the up/down counter of the clock pulses is counting up. As soon as the input voltage gets below the reference voltage, the counter switches to the count down mode reducing the reference voltage (the output of the DAC). When the DAC voltage is again lower than the input voltage, the counter switches back to the count-up mode increasing the DAC voltage. Thus, each sample is not accompanied by a DAC output staircase ramp starting at zero but just a differential ramp going either

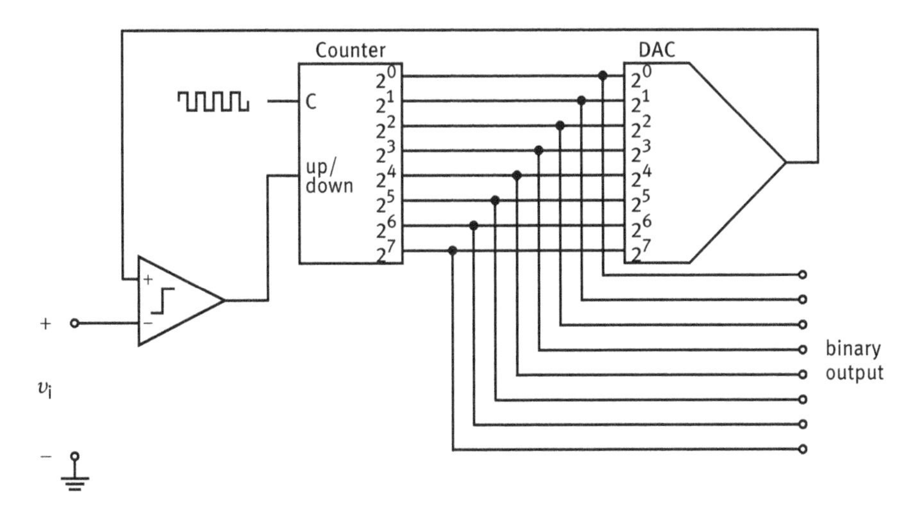

Fig. 6.19. Block diagram of a tracking ADC.

up or down allowing substantially higher sampling rates. Obviously, such a device works best for slowly changing input signals.

Tracking converters have very wide ranges and high resolution, but the conversion time is variable, and depends on the changes in the input signal. One drawback of tracking ADCs constitutes the fact that the output is not stable, it switches the states with every clock pulse, even if the input signal stays constant. This phenomenon may be problematic in some digital devices. This tracking approach may also be combined with the successive approximation method instead of the counter ramp method.

Another problem is the disproportionately slow *step recovery* response. The step recovery time is a measure of how quickly an ADC output can match a large, sudden change in the input. The tracking ADC is an example of a converter technology for which step recovery is a serious limitation.

Problem

6.26. Describe the advantage of tracking ADCs vs. standard ADCs.

6.6 Composite ADCs

If a device does function as an ADC but is *not* based on a single (electronic) ADC circuit but on the combination of at least one analog and one digital (binary) circuit, one has a composite ADC. Again, ADCs that use a combination of electronics and other technologies (e.g., optical or mechanical encoders) are disregarded. Only purely electronic devices are considered.

6.6.1 Wilkinson (single-slope or integrating) ADC

A Wilkinson ADC is nothing but a cascade of a direct voltage-to-time converter (Section 4.5.2.1) and a time digitizer (Section 6.2). This is not a circuit but a device performing (indirectly) amplitude-to-digital conversion. Although the principle has been known for several decades it still excels the competing principles by an unsurpassed differential linearity. As all has already been said in the two referenced sections the function will be summarized in a few words. The linear relation between amplitude and time of a capacitor being (dis)charged with a constant current is used to convert a voltage amplitude into a proportional pulse length. This pulse length is then digitized. The number of signals occurring during the pulse length is counted and the output of the counter delivers the binary equivalent of the signal amplitude. (The digitizing uncertainties connected with the circumstance that the beginning of the pulse length is not synchronized with the frequency of the time digitizer are discussed in Example 6.4 of Section 6.3). The conversion time for an n-bit ADC is proportional to 2^n and inversely proportional to the digitizing frequency. As it is increasingly difficult to build macro-

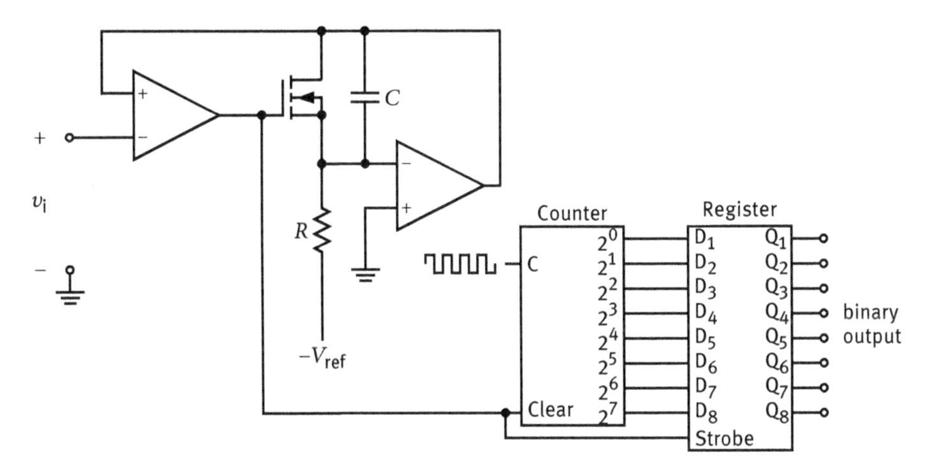

Fig. 6.20. Basic schematic diagram of a Wilkinson ADC.

scopic circuits working with frequencies in excess of 300 MHz, this frequency has remained the maximum digitizing frequency of this type of ADC over decades despite improvements in integrated circuit technology. As each channel width is correlated to one full oscillation of a highly stable harmonic oscillator (e.g., a crystal oscillator) it is not surprising that the differential linearity is inherently much better than that of voltage-comparing ADCs in particular those that depend on the rather poor differential linearity of a parallel DAC.

Figure 6.20 shows a basic schematic diagram of a Wilkinson ADC. A positive input voltage is compared by means of a comparator to the ramp voltage across the integrating capacitor C of a transimpedance amplifier. The constant charging current is supplied through a resistor R by a negative reference voltage V_{ref}. When the output voltage of the integration amplifier equals the input voltage, the comparator goes into the high state. This high state disrupts the flow of the precision clock signals to the counter and discharges the capacitor via the FET switch. The state of the counter outputs is taken over by a shift register to be available as a converted digital output. When the capacitor is fully discharged the integrator output voltage reaches zero, the comparator output switches back to the L state, clearing the counter and enabling a new conversion process. The similarity with the voltage comparing serial ADC circuit is quite artificial. Here an analog voltage ramp is used, there a binary. Here a highly constant frequency signal is needed, there not at all. Here no DAC is involved; there the DAC is essential for the operation. As the conversion is indirect (conversion of amplitude into time and digitizing of the time) a calibration process is needed to determine the analog voltage step size per LSB. This calibration is subject to aging affecting both the amplitude-to-time conversion and the time digitizing. Consequently, from time to time the calibration process must be repeated.

Problem

6.27. Which drawback does the Wilkinson ADC have when compared to a serial voltage-comparing ADC?

6.6.2 Dual-slope ADC

A dual-slope ADC is nothing but a cascade of a compound voltage-to-time converter (Section 4.5.2.2) and a time digitizer (Section 6.2). This is not a circuit but a device performing (indirectly) amplitude-to-digital conversion.

By using the compound voltage-to-time converter which uses a constant charging time t_u of the capacitor rather than the direct one (as the Wilkinson ADC) the voltage-to-digital conversion is not dependent on the capacity of the converting capacitor any more but only on the values of one resistor, the current of a constant current source and the (constant) charging time interval t_u. At the expense of conversion time the resolution can be improved by making the constant discharging current smaller which increases the run-down time t_d.

Such or similar converters are used in most digital voltmeters and other applications demanding high accuracy based on their stability, linearity, and flexibility. As discussed in Section 4.5.2.2, instabilities in the time base are not effective as they affect both the charging and the discharging time in the same proportions. This makes this principle superior to the Wilkinson ADC as the calibration drift problem is much reduced. Besides, it was shown in Section 4.5.2.2 that the effective input voltage is the integral of the instantaneous values during the charging time. This should be beneficial for various applications in particular when the input signal contains significant levels of spurious voltage dips or spikes. Other AD converters that convert instantaneous input values may by chance convert a spike or dip giving unreasonable results.

By using as reference a real constant voltage source with a resistor R_{Sd} above dependence on the input resistor R_{Su} is diminished because now one gets

$$\frac{v_i}{V_{ref}} = -\frac{N_d}{N_u} \times \frac{R_{Su}}{R_{Sd}} \tag{6.1}$$

which is the ratio of counted periods as a measure of the run-down time N_d to those of the (constant) run-up time N_u times the ratio of the two source (=input) resistances. As both counted numbers depend on the same clock, the first factor is hardly dependent on the stability of the clock and using resistors with matched properties makes the second factor stable, too. This way the above mentioned three dependences are diminished; it is a very stable arrangement. The circuit of such an enhanced dual slope converter is sketched in Figure 6.21a. Figure 6.21b sketches the time dependence of the converted voltage.

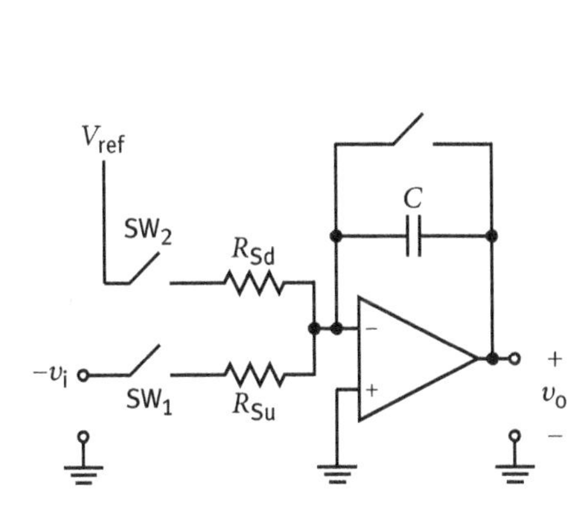

Fig. 6.21a. Principle of the amplitude-to-time conversion as used in an enhanced dual-slope ADC.

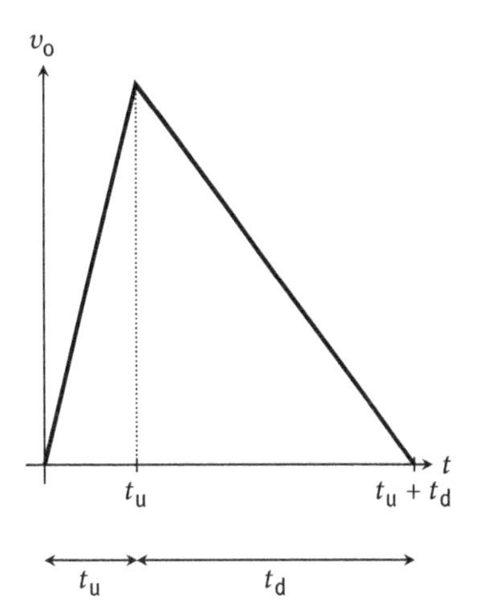

Fig. 6.21b. Time dependence of the converted voltage as used in a dual-slope ADC.

The wideband noise in the circuit and the maximum output voltage of the integrator (which is on the order of 10 V) limits the resolution of the dual-slope ADC. The noise defines how precisely the zero crossing can be determined (about 1 mV). Thus, the practical resolution of dual-slope ADCs is limited to about 14 bits.

Problem

6.28. Name instruments that take advantage of the high resolution and intrinsic stability of dual-slope ADCs.

6.6.3 Multislope ADC

Instead of using a single run-down resistor (i.e., a single slope) to completely discharge the capacitor, the multislope converter uses several resistors (i.e., multiple slopes) in sequential order to discharge the capacitor, each time more precisely. The ratio of the values of the discharging resistors increases by the same factor (e.g., 10) in each step.

Figure 6.22a shows a multislope conversion circuit based on powers of 10. Four resistors are used in this circuit, with weights of 10^3, 10^2, 10^1, and 10^0.

The time dependence of the converted signal as shown in Figure 6.22b helps with the understanding of this circuit. The first (the steepest) run-down slope (using $+V_{ref}$ and $R_b/1000$) is terminated at the end of the clock period at which zero-crossing was registered. As the time of the zero-crossing will be earlier than the end of the period

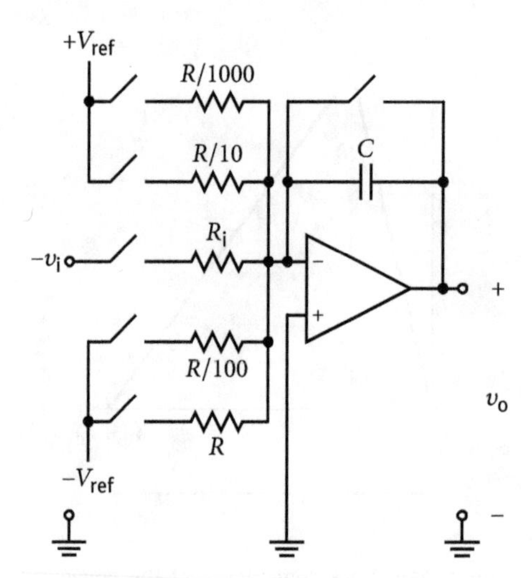

Fig. 6.22a. Principle of a four-slope amplitude-to-time converter.

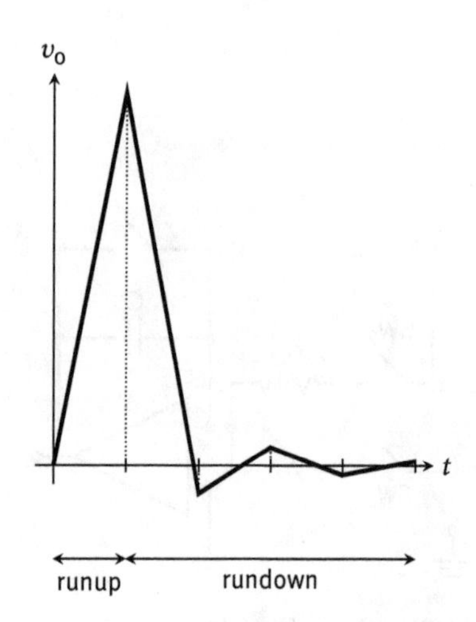

Fig. 6.22b. Time dependence of the converted voltage as found in a quadruple-slope ADC.

of the last counted signal, the capacitor will be oppositely charged at the end of the counting period, i.e., the ramp goes negative. The amplitude depends on the time difference between the zero-crossing and the end of the counting period. Now a ramp in the opposite direction is generated with a reduced slope (using $-V_{ref}$ and $R_b/100$) due to the increased resistor value (by a factor of 10). Again, the zero-crossing will not coincide with the end of the counting period; the next run-down slope with a further reduced slope is started. After the next run-up slope, again with the reduced slope, the conversion is finished. Each following slope determines the zero crossing of the output voltage of the integrator ten times more precisely than the previous slope. The attainable resolution is inversely proportional to the steepness of the slope. To get the same resolution with just one run-down slope the run-down time would have to be about 1000 times longer than that of the steep slope. Thus, the conversion time is much reduced. Obviously, the number of counted periods must be weighted with the weight assigned to each of the slopes. As at most one complete period contributes to the voltage overshoot at the capacitor and the resolution is increased by a factor of ten the duration of the last three slopes must be less than 10 periods *each*.

The present example of a multiple-slope ADC used a multiplicative factor of 10 per slope to conform to the decimal system. An optimization with regard to the shortest conversion time reveals that regardless of resolution the optimum multiplicative factor would be Euler's constant e.

Problem

6.29. Name the main reason for using multiple slope ADCs.

6.6.4 ADCs using frequency modulation

An ADC with an intermediate FM stage is a combination of a voltage-to-frequency converter (Section 4.5.3) and a frequency digitizer (Section 6.3). Longer time constants allow for higher resolutions at the cost of converter speed. The two parts of the ADC may be far apart, with the frequency signal transmitted wirelessly or nonelectronically (e.g., an optoisolator). Sine-wave, square-wave, or pulse-frequency modulation may be applied. These ADCs used to be very popular to digitize the status of a remote sensor (telemetry).

6.6.4.1 Delta–sigma (Δ–Σ) ADC

The delta–sigma ADC described here is based on a synchronous voltage-to-frequency converter with pulse-frequency modulation (Section 4.5.3) combined with a frequency digitizer (Section 6.3). Its principle is shown in Figure 6.23. The input signal v_i is fed into a summing integration amplifier (a transimpedance amplifier) producing an output voltage v_o that is the integral of the input current i_i. The input current i_A is the sum of the signal current $i_i = v_i/R$ and the (pulsed) feedback current $i_F = v_F/R$. The ramping output voltage is compared by a comparator against zero volts. If the integrator output is higher than 0 V, then the comparator's output is in the H state, else in the L state, i.e., H means that the output of the integrator is positive, *L* means negative. The comparator's output is connected to a D-type flip–flop that is clocked at a high frequency. An amplifier, acting as a level-adjusting interface between the flip–flop output and input of the summing integrator, converts the H output signal of the flip–flop into a positive rectangular voltage signal, and an *L* output signal into a negative signal with the same absolute amplitude. The current of this output signal

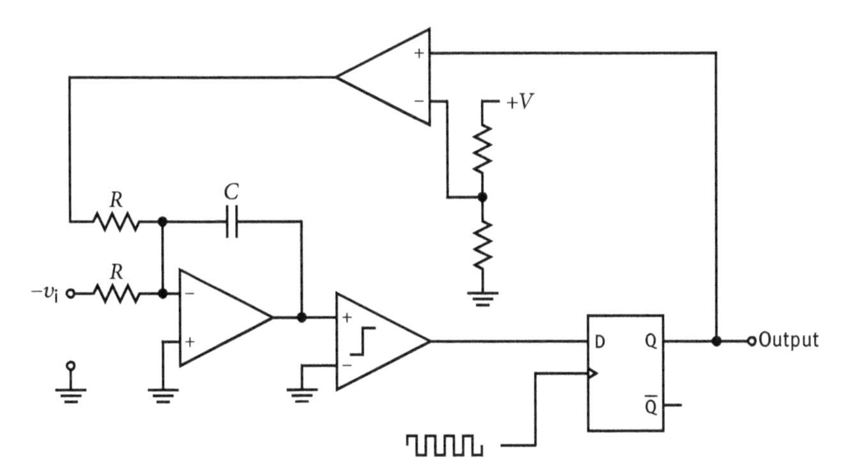

Fig. 6.23. Schematic of a delta–sigma ADC.

is added to the input current reducing the absolute output voltage of the integrator (negative feedback). Actually, the pulsed feedback signal adds to or subtracts from the capacitor with each pulse having a definite amount of charge (*charge balancing method*, Section 4.5.3). Thus, the number of pulses needed to remove the charge from the capacitor (making the integrator's output 0 V) is a measure for the size of the input voltage.

The output of the flip–flop is a serial stream of L and H states, synchronous with the frequency of the clock. For a zero input voltage the flip–flop output will oscillate between H and L, as the feedback system tries to keep the integrator output at 0 V. This circuit is very fast so that a high clock frequency may be applied allowing oversampling, i.e., using a sample rate much higher than the minimum requested by the Nyquist–Shannon sampling theorem. Thus, there is a time margin for digitizing the frequency of the H outputs, e.g., by counting the number of H signals during some given number of clock pulses (in a given time interval). The binary output of this counter is then the digital output of this ADC.

Problem

6.30. Is the delta–sigma ADC used for analyzing single pulses or for sampling a continuous input voltage?

6.7 Ranking of ADCs

There are two quite different applications of ADCs asking for quite different solutions
- sampling (of continuous signals), and
- digitizing of isolated signals (pulses).

Consequently, the requirements for an adequate conversion are widely different. The most important properties to be considered are
- resolution (precision),
- speed (sampling rate),
- accuracy (linearity),
- step recovery, and
- complexity/cost.

An ideal ADC has as many bits as needed, has no dead time (samples at ultra-fast speeds), has no nonlinearity, recovers from steps instantly, and uses only few electronic components. Such a device would satisfy all needs.

Each type of ADC has its strength and weakness. Table 6.2 gives some idea about the ranking of various ADC types with regard to some of their properties. The upper limit of the resolution of 20 bit (or more) is not so much given by the ADC itself as by the practical limit enforced by the signal-to-noise ratio.

Table 6.2. Ranking of comparable ADC types.

ADC type	Sampling rate	Resolution, Linearity	Step recovery
Parallel	1	6	1
Tracking	2	4	5
Successive approximation	3	5	2
Serial	4	3	3
Wilkinson	4	2	3
Dual slope	5	1	4

Most of the properties may be improved through better circuitry, either by increased number of components and/or by using especially fast or precise circuits. One reason for the good ranking of the compound ADCs lies in the fact that the competing voltage comparing ADCs depend on the quality of the DACs which they must use. However, precision DACs are not easily manufactured.

Fast A-D conversion can only be performed at the cost of resolution.

Problems

?

6.31. Which ADC type has the shortest dead time?

6.32. What is the best resolution obtainable with parallel ADCs?

6.33. Voltage comparing ADCs use comparators which need reference (threshold) voltages.
(a) How do parallel ADCs obtain these voltages?
(b) How are they obtained by the other ADCs?

6.34. To make the conversion time short, integrating ADCs use a high digitizing frequency.
(a) For a long time now there has been no progress with regard to the highest external frequency used in these ADCs. What is the approximate frequency limit?
(b) Name a reason why it is difficult to go beyond this limit.

6.8 ADCs in measurement equipment

Passive measurement devices (Section 2.1.4) do not need auxiliary power for proper operation (they utilize part of the signal power). Active devices get their power either from batteries or from mains. The main advantage of active meters is the increased signal range by the use of a signal amplifier. Besides, the instrument is designed for minimum loading of the circuit under investigation. Thus, only in rare cases a correction of the measured values (Section 2.1.4) is needed. In special applications, the input impedance of the instrument can be made independent of the input range in use. As the result of any measurement is a number, it comes naturally that an ADC is included in active measuring devices. Thus, the result is available in a numerical (digital) form to be used in a display or for further electronic manipulations.

6.8.1 Digital multimeter (DMM)

The name multimeter indicates that a multiple number of tasks can be performed with such an instrument. The function of a voltmeter, an ammeter, and an ohmmeter are usually combined in such an instrument. As such, a multimeter may be passive or active.

In active instruments, additional functions (measurement of capacitance, inductance, temperature) can easily be included. A measurement of the capacitance is easily performed by measuring the time it takes to charge the capacitance linearly (with a constant current i_S) to a certain voltage level V_C (Section 4.5.2). $C = i_S/V_C \times t = cst. \times t$. Thus, most active multimeters will need the following components:

- power supply;
- measurement amplifier;
- ADC;
- stable voltage reference for the ADC;
- set of calibrated resistors R to convert input currents into voltages: $v_x = i_x \times R$;
- set of calibrated current sources is to convert resistances into voltages: $v_x = R_x \times i_S$, and for the capacitance measurement;
- a stable oscillator, a comparator and a counter; and
- a display or a computer interface for the data output.

For the measurement of AC voltages or currents it is necessary to convert the AC values into DC values, e.g., by means of active precision rectifiers (Section 2.5.2.3).

Next to the amplifier which makes the instrument active, the ADC plays a central role because it is needed for the straight voltage measurement (usually with an input impedance of 10 MΩ shunted by a capacitance of less than 100 pF), the measurement of the voltage across the shunt resistors of the ammeter and the input voltage when a current is fed into an external resistance R_x. A measurement with a multimeter need not be fast. Therefore, any slow ADC may be used. When a wide dynamic range is required, high resolution is imperative. A popular choice is the multiple-slope ADC which provides high resolution with little effort. High-quality instruments are protected against excessive voltages or currents and are self-calibrating and self-ranging, i.e., they find the appropriate measuring range by themselves.

The usual display is numerical, sometimes aided by an analog display (using a DAC, Section 6.4.1) that allows a faster recognition of changing input values.

6.8.2 Digital (sampling) oscilloscope (DSO)

Traditionally, the main purpose of an oscilloscope is the display of the time dependence of the voltage applied to its input. In analog oscilloscopes this is achieved by using a cathode ray tube which displays the time dependence of the voltage.

Digital oscilloscopes are based on sampling (Sections 4.4.3.1 and 6.1) of the input signal. The sampling rate depends on the desired time resolution. The amplitude of the signal is digitized by means of an ADC and the binary result stored in a memory for later use (display, all kinds of operations). The usual size of the display is of the order of 10 cm × 10 cm. Thus, a visual resolution of somewhat less than 1% suffices. An ADC of at least 8 bit provides such a resolution. Parallel ADCs allowing sampling rates in access of 100 MHz are the choice for high-speed digital oscilloscopes. As all sampling data are stored, numerical averaging and other computational techniques can be used for improving the effective resolution.

The time resolution of repetitive signals is not tied to the speed of the ADC. Using the moment of the trigger response as fiducial time mark the moment of sampling is delayed by one timing unit for each consecutive signal. For each sampling event a (consecutive) signal is used. The (binary) sampling results are stored in a memory which is read after the scan has been completed.

The bandwidth of a digital oscilloscope is limited by the bandwidth of the input amplifier and the Nyquist frequency (Section 6.1) of the sampling process. To observe the latter is very important to avoid aliasing.

This feature is the only disadvantage of digital vs. analog oscilloscopes. The minimum rise time (of any electronic instrument) is determined by this bandwidth (Section 3.3.1). If the signal rise time is comparable to the rise time of the instrument, a correction for the instrument rise time is feasible in view of the quadratic addition of rise times (Section 3.3.1).

The vertical sensitivity of an oscilloscope is changed by using attenuators or by changing the gain of the input amplifier. At very high sensitivities, there might be a reduced bandwidth due to the constant gain-bandwidth product (Section 3.6). The horizontal resolution depends primarily on the sampling rate (given in samples per second, S/s). It may be prescaled for a coarser resolution. When looking at slowly changing signals over long periods of time a low sampling rate is mandatory.

An essential part of each digital oscilloscope is the data memory. The number of memory cells determines the length of the record that may be stored, i.e., the number of waveform points that may be taken for one waveform record. Obviously, there is a trade-off between record length and record detail. There is either a detailed picture of a signal over a short time interval (high sampling rate, the memory is filled quickly) or a less detailed picture over a longer time interval.

6.8.3 Spectrum analyzers

A spectrum analyzer performs Fourier analysis (see Section 3.1.1) of the time-dependent (voltage) signal and displays the frequency spectrum of the said signal. For the mathematical operations which are involved it is necessary that the time dependence is available in the numerical form. This feature is basic in digital oscilloscopes (Section 6.8.2). Therefore, advanced digital oscilloscopes offer as an option spectrum analysis.

After repeatedly sampling the signal wave-form and filling a memory (at least 15 bit deep) with the recorded time samples in a FIFO (first-in, first-out) memory (e.g., Section 5.4.1) this digital information is used to convert the time spectrum into a (discrete) frequency spectrum. Obviously, the sampling frequency must exceed the Nyquist frequency (the capture bandwidth) so that slow ADCs disqualify. To cover a high dynamic range, the resolution must be adequate. A good linearity is needed to achieve sufficient fidelity in the conversion process.

In digital signal processing, the discrete Fourier transform (DFT) converts any signal that varies over time and sampled over a finite time interval (the *window function*) from the time domain to the frequency domain. The samples taken from the input are real numbers converted to complex numbers, and the coefficients obtained as output are complex, too (see Section 3.1.1.2). The frequencies of the output sinusoidals are integer multiples of a fundamental frequency with a period corresponding to the length of the sampling interval. As the input and output sequences are both finite the Fourier analysis is that of finite-domain discrete-time functions. This discrete Fourier transform is usually applied to perform Fourier analysis in many instruments. Since it deals with a finite number of data, it can easily be implemented by numerical algorithms in digital processors or by (hardware) programming of appropriate circuits. Usually, these implementations are based on efficient fast Fourier transform algorithms which are available for data samples with block sizes of 2^n.

Solutions

Problems of Chapter 1

1.1 Passive voltmeter, passive ammeter, and crystal radio receiver.

1.2 (a) It is heated up. (b) Look up in a chemistry book.

1.3 It does not make sense to assign the attribute *small* or *large* to a resistance in absence of other resistances. A resistance of 100 Ω can only be regarded as small or large relative to another resistance.

Problems of Chapter 2

2.1 The flow of current through an ideal current source in series with any other element is fixed by the ideal current source. Since the voltage between the terminals of an ideal current source is not defined (it is floating) the voltage across a series connection of an ideal current source with any other element can take any value. Any series configuration with an ideal current source has the identical property as the current source by itself.

2.2 The voltage across an ideal voltage source in parallel with any other element is fixed by the ideal voltage source. The current flow through the ideal voltage source can have any value. Any parallel configuration that contains one ideal voltage source has the same electrical properties as the ideal voltage source by itself.

2.3 There is no difference in electronic behavior between a real linear voltage source and a real linear current source.

2.4

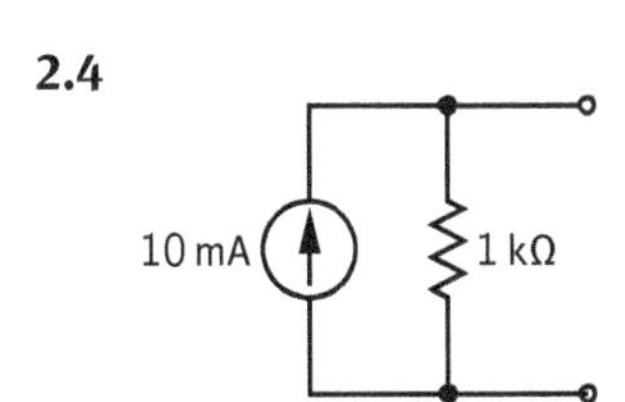

2.5 (c) Of course, in both cases $v_o = 16$ V.

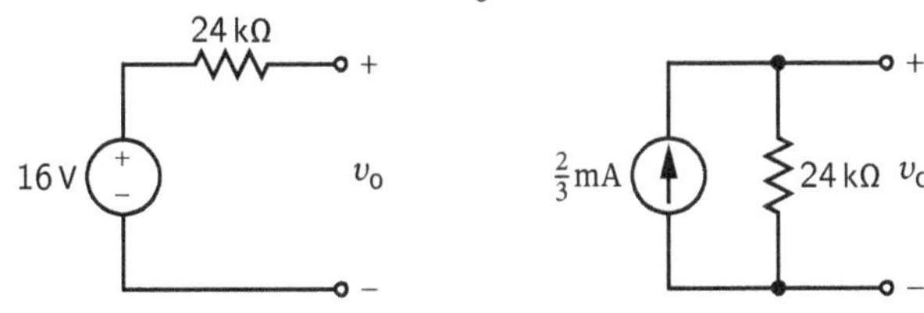

2.6 (a) 10 mA (b) 10 V

2.7 $v_R = 100$ V

2.8
(a) Original and simplified circuit:

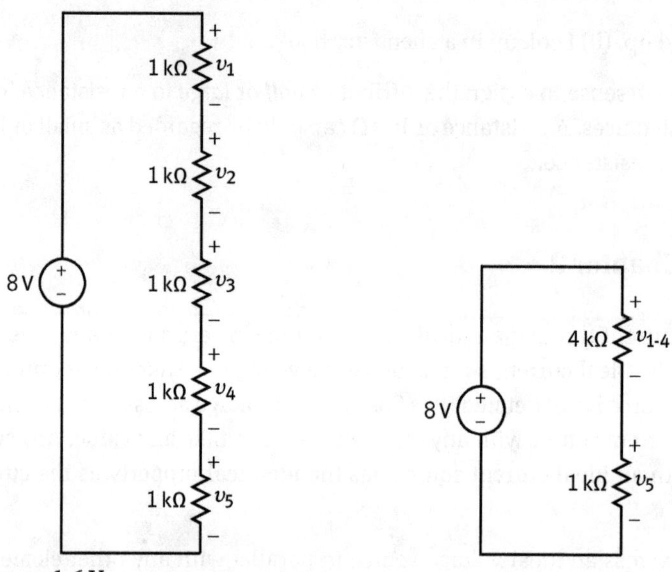

(b) $v_5 = 1.6$ V
(c) Equivalent circuit according to the theorem of Thevenin:

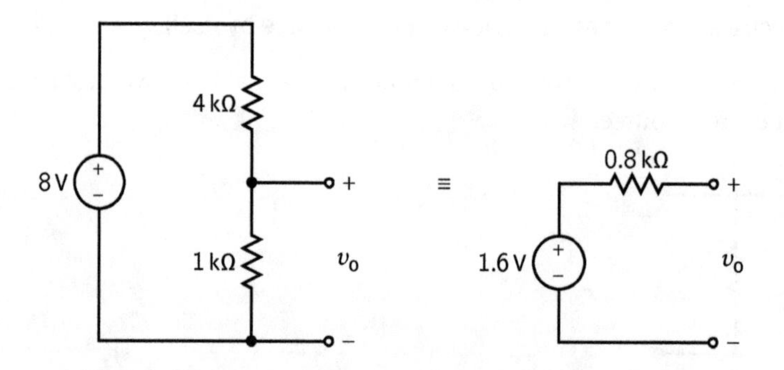

(d) Equivalent circuit following Norton's theorem:

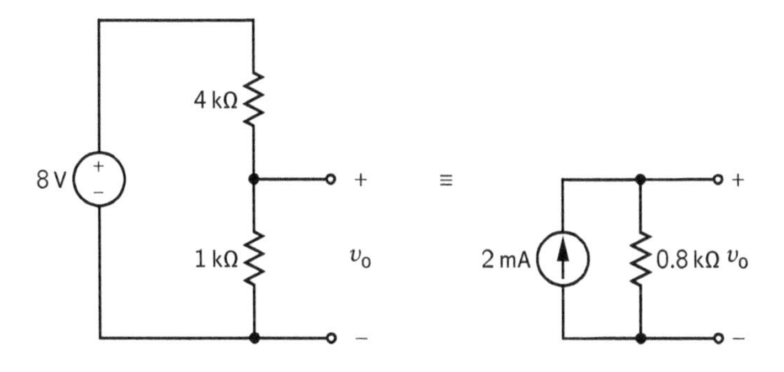

2.9 A real current source with load conductance G_L:

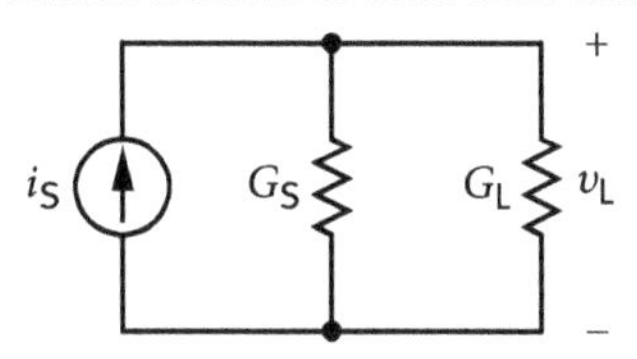

2.10 $\dfrac{G_1}{G_1 + G_2} = \dfrac{R_2}{R_1 + R_2}$ with $R_1 = \dfrac{1}{G_1}$ and $R_2 = \dfrac{1}{G_2}$

2.11

(a) $i_{sc} = i_S$

(b) The equation for the open-circuit voltage v_{oc} is dual to that of the short-circuit current: $v_{oc} = v_S$

2.12 (a) v_S (b) $v_S \times 11.11\ \text{mS} = v_S/90\ \Omega$

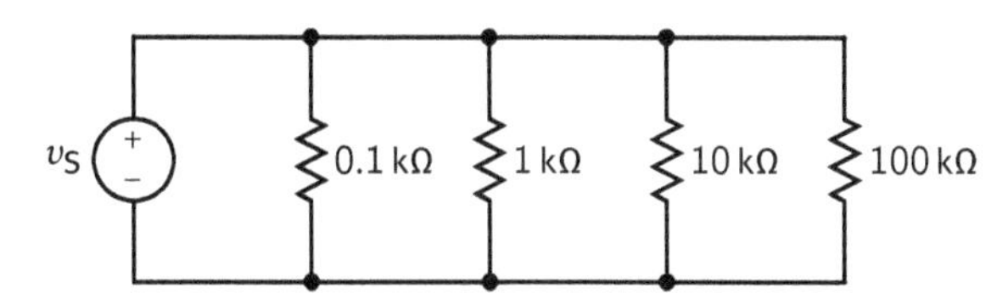

2.13 (a) 1 mA (b) 10 V

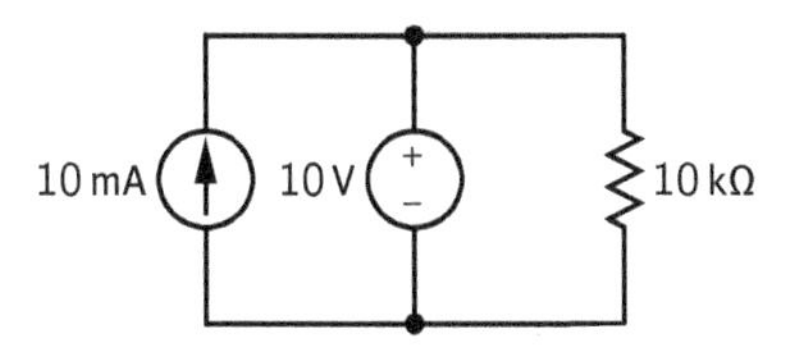

2.14 No, they do not add up linearly.

2.15 No, both theorems are important. In practice real voltage sources appear more frequently than real current sources and thus one might be tempted to do all analysis by the "voltage view." The "current view," however, is appropriate just as often as the "voltage view."

2.16 $i_1 = 2.1$ mA, $i_2 = 0.9$ mA, $i_L = 3$ mA

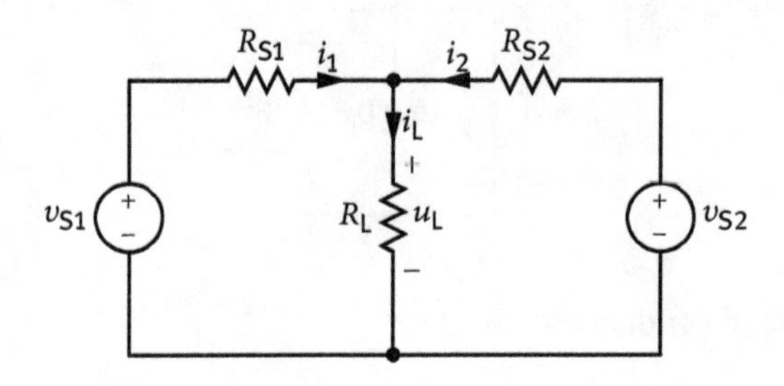

2.17 Switch open: $v_x = 2.5$ V, switch closed: $v_x = 4$ V

2.18 Switch open: $i_x = 4$ mA, switch closed: $i_x = 2.5$ mA

2.19 $V_{op} = 5$ V with $v_S = 2$ V and $V_{op} = 7$ V with $v_S = -2$ V

2.20 A series resistance with 2.4975 MΩ that tolerates 40 µA current (4 mW power) makes the meter suitable for measuring voltages up to 100 V.

2.21 A conductor of 9.9996 S, i.e. a resistor of 0.100 04 Ω, in parallel, i.e. as a shunt.

2.22 No – the correction of < 1 pV is much smaller than the resolution of the instrument of 10 nV.

2.23 Yes – the conductance of the instrument of 1.82 µS is even larger than that of the circuit, therefore, the correction will be large and the uncertainty of the result as well.

2.24 (a) $R_x = v_m/(i_m - v_m G_V)$ (b) 0.99% vs. 3.85%

2.25 (a) 0.5 (b) It is a ratio that is independent of the unit with which the two values were measured.

2.26 (a) 0.5 mA, 3/80 mS (b) 297/80 mS = 269 Ω (c) 0.495 mA (d) 0.5 mA

2.27 (0 V, 0 mA)

2.28 Since the element is nonlinear the (small-signal) impedance cannot be derived from the knowledge of two operating points.

2.29 $V_{Th} = 6.75$ V, $R_{Th} = 195$ Ω, $r_Z = 2.5$ Ω, $V_Z = 5.96$ V; operating point (5.97 V, 4 mA)

2.30 (a) 25 mW (b) between (10 V,0 mA) and (0 V,10 mA) (c) 0 W at 0 V

2.31 (a) $V_{S\,max} = 20$ V (b) 0.4 W

2.32 No, the transformer as a two-port is symmetric with regard to input and output.

2.33 (a) $g_i = -1$ (b) $Z_o = 100\,\text{k}\Omega$

2.34 (a) $g_v = 50$ (b) $Z_o = 100\,\text{k}\Omega$

2.35 (a) $i_x = 5\,\text{V}/(10\,\text{k}\Omega + R_x)$ (b) at $R_x = 10\,\text{k}\Omega$, $p_{\text{max}} = 0.625\,\text{mW}$

2.36 (a) $R_x = 9\,\text{k}\Omega$ (b) $p_{\text{max}} = 1\,\text{mW}$

2.37 $p_{\text{Smax}} = 4p_{\text{Lmax}}$

2.38 $1/\sqrt{2}$

2.39 (a) $289.54\,\Omega$ (b) $g_v = 1$, under any condition

2.40 (a) $Z_{\text{ch}} = 80\,\Omega$ (b) $-9.5\,\text{dB}$

2.41 (a) $Z_{\text{ch}} = 420\,\Omega$ (b) $-7.96\,\text{dB}$

2.45 (a) Yes. (b) V_{STh}

2.46 (a) No. (b) $v_{\text{thr}} = v_{\text{S}} \times \frac{R_{\text{L}}}{R_{\text{L}}+R_{\text{S}}} + 0.7\,\text{V}$ (depending on the diode type)

2.48
(a) At the position with maximum current, i.e. at the top-most position.
(b) The potentiometer.

2.49 Because it deals with currents just opposite to voltages.

2.50 MOS FETs

2.51 It produces multiple responses.

2.52 g_v

2.53
(a) $A_{\text{F}} = \dfrac{A_1A_2}{1 - B_1B_2A_1A_2}$

(b) $A_{\text{F}} = \dfrac{A_1A_2}{1 - BA_1A_2 - B_1A_1}$

(c) $A_{\text{F}} = \dfrac{A_1A_2A_3A_4}{(1 - B_1A_1A_2) \times (1 - B_2A_3A_4) - B_3A_1A_2A_3A_4}$

2.54
(a) With $B = 2 - 1/A$.
(b) B must be active to get $B > 1$.

2.55 Not at all.

2.56 The sign of the closed-loop gain signals which kind of feedback is present.

2.58 (a) $g_{iF} = g_{iA}$ (b) $g_{vF} = \frac{R_L}{R_F} \times \frac{g_{iA}}{g_{iA}+1}$

2.59
(a) No.
(b) No.

2.60 The current does not change.

2.61 No, it doesn't.

2.62 (a) $1/Z_i = 1/R + (1 - g_v)/R_S$ (b) $g_v = 1 + R_S/R$

2.63 (a) 2 μA (b) −2 mV

2.64 (a) No, it doesn't. (b) No. (c) No. (d) Transadmittance.

2.65 Series-parallel feedback with $AB = -39$.

2.66 Negative series–parallel feedback with $AB = -11$.

2.67 $g_{vF} = g_v = -1000$, disregarding loading

2.70 (a) $R_X = 0$ (b) (0.699 V, 0.05 mA; 6 V, 5 mA)

2.71
(a) v_{o1}: parallel–series, v_{o2}: parallel–parallel
(b) v_{o1}: current gain, v_{o2}: transimpedance
(c) input impedance: low in both cases, output impedance:, Z_{o2} low, Z_{o1} medium ($\approx R_L$)

2.72 Negative.

2.73 (a) parallel–parallel (b) not at all

2.74 A negative input voltage will drive the differential amplifier into the cut-off region, i.e. the positive input becomes (much) more negative than the negative input.

2.75 The closed-loop gain must be less than one, under all circumstances.

2.78 Because the effective configuration depends on the position of the input.

2.79 A change in biasing acts as a "DC-noise," i.e. the same feedback configuration as in the case of noise is effective.

2.80 Today's semiconductor technologies are optimized for voltage electronics.

2.81 The output is floating, i.e. it is not grounded.

2.82 The closed-loop gain is a product of A and B; it does not matter where A is located

2.83 g_i is closer to one and the output impedance is higher.

2.84 It provides voltage with low output impedance.

Problems of Chapter 3

3.1 $a(t) = -1 + 2\frac{t_p}{T}$ with $t_p > 0$ and $t_p = t - nT$

3.2 $a(t) = \begin{cases} +1 & \text{for} t_p < T/2, \text{and} \\ -1 & \text{for} t_p \geq T/2 \end{cases}$ with $t_p > 0$ and $t_p = t - nT$

3.3 $a(t) = \begin{cases} -1 + 4\frac{t_p}{T} & \text{for} t_p < T/2, \text{and} \\ +1 - 4\frac{t_p}{T} & \text{for} t_p \geq T/2 \end{cases}$ with $t_p > 0$ and $t_p = t - nT$

3.4 It is charged in the switching process.

3.5 analog, unchangeable in time

3.6 Energy = frequency

3.7 steady-state sinusoidals

3.8 in those cases in which the integral of the Fourier transform does not converge

3.9 Imaginary unit ($\mathrm{j} = \sqrt{-1}$)

3.10 Maximum amplitude, frequency

3.11 Yes.

3.12 $g_v(\omega) = \frac{\omega\tau}{\sqrt{1+4\omega^2\tau^2}}$, $\varphi(\omega) = \arctan\frac{1}{2\omega\tau}$ with $\tau = RC$

3.13 For DC there are no losses due to the capacity of the power lines.

3.14 $g_i(\mathrm{j}\omega) = \dfrac{1}{2 + \dfrac{R}{\mathrm{j}\omega L}}$

3.15 (a) no, no, no (b) $i_o = 9/\sqrt{26}$ mA

3.16
(a) 10 mW
(b) When the period length is shorter than or on the order of the thermal time constant.

3.17 (a) $d = 0.1$ (b) 10 mW

3.18 (a) 10 mW (b) There is no difference.

3.19 When connecting the battery there will be a transient current that charges the capacitors.

3.20 (a) $v_1 = \frac{9}{3-4\mathrm{j}}$ V (b) $v_2 = (1.64 - 0.48\mathrm{j})$ V (c) $i = \frac{0.3}{3-4\mathrm{j}}$ A

3.21 (a) $\tau_- = 10\,\mu\text{s}$, $\tau_+ = 20\,\mu\text{s}$ (b) $t_{rs} = 8\,\mu\text{s}$, $t_f = 22\,\mu\text{s}$ (c) $v_{omax} = -0.393$ V

3.22 No.

3.23 It has zero conductance.

3.24 When the power supply is switched on the transient charges the capacitor.

3.25 At a given supply voltage the impedance can be higher; less power is consumed.

3.26 This transition is made of very high frequency components for which the impedance of a capacitor is close to zero.

3.27 Current.

3.28 Because diodes behave like current check valves.

3.29 $i_1 = 18.08$ mA, $i_2 = 11.5$ mA, $i_L = 8.1$ mA

3.30 (a) $Z_i = \infty$ (b) $Z_i = R$ (c) $Z_i = \infty$ (d) $Z_i = R$ (e) $Z_i = \infty$ (f) $Z_i = \infty$

3.31 (a) $Z_i = R$ (b) $Z_i = \infty$ (c) $Z_i = R$ (d) $Z_i = \infty$ (e) $Z_i = \infty$ (f) $Z_i = \infty$

3.32 (a) $Z_o = \infty$ (b) $Z_o = R$ (c) $Z_o = R$ (d) $Z_o = 0$ (e) $Z_o = 0$ (f) $Z_o = \infty$

3.33 (a) $Z_o = 0$ (b) $Z_o = R$ (c) $Z_o = R$ (d) $Z_o = \infty$ (e) $Z_o = \infty$ (f) $Z_o = 0$

3.34 (a) $Z_i = 0$ (b) $Z_i = R$ (c) $Z_i = 0$ (d) $Z_i = R$ (e) $Z_i = 0$ (f) $Z_i = 0$

3.35 (a) $Z_i = R$ (b) $Z_i = 0$ (c) $Z_i = R$ (d) $Z_i = 0$ (e) $Z_i = 0$ (f) $Z_i = 0$

3.36 (a) $Z_o = R$ (b) $Z_o = \infty$ (c) $Z_o = R$ (d) $Z_o = \infty$ (e) $Z_o = \infty$ (f) $Z_o = \infty$

3.37 (a) $Z_o = \infty$ (b) $Z_o = R$ (c) $Z_o = \infty$ (d) $Z_o = R$ (e) $Z_o = \infty$ (f) $Z_o = \infty$

3.38 (a) $f_u = 1/(2\pi\tau)$ (b) $g_v(10f_u) = 0.1$ (c) $\varphi(10f_u) = -84.29°$, $\varphi(10f_u) \approx -90°$

3.39 (a) $g_v(0.1\omega_x) = -3$ dB ≈ 0 dB (b) $BW = \frac{0.1\omega_x}{2\pi}$

3.40 (a) 1 μs (b) $\sqrt{2}$ μs

3.41 (a) $v_{omax} = v_o(\tau/2) = 3.935$ V (b) $t_{rs} = 7.96$ μs (c) $t_f = 44$ μs

3.42 $t = \tau = 1$ ms

3.43 $\omega = 2000\ \mathrm{s}^{-1}$

3.44 $g_v(f=0) \times BW = g_m(f=0) \times \dfrac{R_L}{2\pi R_L C_L} = \dfrac{g_m(f=0)}{2\pi C_L}$

3.45 (a) See Figure 3.30a. (b) See Figure 3.30b. (c) $BW = \frac{R}{2\pi L}$

3.46
(a) one-half
(b) $\tau = \dfrac{R \times C_1 \times C_2}{C_1 + C_2}$

(c) $g_v(\omega) = \dfrac{\tau}{R \times C_2 \times \sqrt{1+\omega^2\tau^2}}$ and $\varphi(\omega) = \arctan(-\omega\tau)$

3.47 L in parallel to $G = 1/R$, a current high-pass filter.

3.48 (a) (Z_1: C = 1 nF & Z_2: R = 10 kΩ), (Z_1: R = 10 kΩ & Z_2: L = 100 mH) (b) 0°, 90° (c) 45°, 45°

3.49 (a) one-half (b) $\tau = G \times (L_1 + L_2)$ (c) $1/g_v(\omega) = 0.5 \times \sqrt{(1 + \frac{1}{4\omega^2\tau^2}}$, $\varphi(\omega) = \arctan(1/2\omega\tau)$

3.50 $v_\mathrm{i} = \frac{3\sqrt{5}}{13}$ V, $v_\mathrm{o} = \frac{9}{\sqrt{26}}$ V

3.51 Avoiding undershoot at the output of a high-pass filter for unipolar signals at the input.

3.52 They must be equal.

3.53 $g_v = \frac{L_2}{L_1+L_2}$

3.54 $g_v = \frac{C_1}{C_1+C_2}$

3.55 $g_i = \frac{L_1}{L_1+L_2}$

3.56 $g_i = \frac{C_2}{C_1+C_2}$

3.57 R = 40 kΩ, L = 113.2 mH, C = 89.5 pF

3.58 (a) f_r = 1.027 MHz (b) 0.1667 µS (c) $g_v = -107.2$ (d) $Q = 68.9$

3.59 (a) $Q = 50$ (b) G_p = 40 µS

3.60 R = 25 Ω, C = 0.4 µF, L = 1.0 mH

3.61
(a) $\dfrac{1}{g_v(\mathrm{j}\omega)} = 1 + \mathrm{j}RC \times (\omega - \frac{1}{\omega LC})$
(b) $\dfrac{1}{g_v(\omega)} = \sqrt{1 + R^2C^2 \times (\omega - \frac{1}{\omega LC})}$
(c) $\varphi(\omega) = \arctan(RC(\omega - \dfrac{1}{\omega LC}))$
(d) $\omega = \dfrac{1}{\sqrt{LC}}$
(e) $Y_\mathrm{i} = 0$
(f) $Y_\mathrm{o} = 0$

3.62 Yes.

3.63 $1 - \frac{\pi}{4} = 12.3°$

3.64 (a) $BW_a = 15.9$ kHz (b) $BW_{ph} = 23.5$ kHz

3.65 (a) $L = 100$ mH, $R = 10$ kΩ (b) $v_{omax} = 6.321$ V (c) (d) 10 V – 6.321 V

3.66 (a) 1.7 V (b) 5 V, 45°

3.67 No.

3.68 No.

3.69 Common emitter + common emitter

3.70 Wiring.

3.71 (a) 1 V (b) 1 V (c) 100 ns
(d) Input: one rectangular pulse, 1 V high, 100 ns long; output: 0 V all the time
(e) Input: at first 20 mA, after 100 ns 40 mA; output: after 50 ns 40 mA

3.72 75%, −12 dB attenuation

3.73 (a) Three resistors of 50/3 Ω each. (b) 6 dB

3.74 Put a resistor of 108 Ω in parallel to the input.

3.75 1000 V A

3.76 (a) 50 mW (b) 100 mW (c) 50 mW (d) yes – no

3.77 (a) $R = \frac{1}{5\pi}$ Ω (b) $G = 20\pi$ μS

3.78 It depends on the feedback configuration.

3.79 $G_v(j\omega) = \dfrac{1 + j\omega R_2 C}{1 + j\omega R_2 C \times (1 + \frac{R_1}{R_2})}$

3.80 $G_v(j\omega) = \dfrac{1 + j\omega R_2 C}{1 + j\omega R_2 C \times (1 + \frac{R_2}{R_1})}$

3.81
(a) Apply series–parallel positive feedback with a closed-loop gain of 0.9.
(b) $t_{rs} = 1$ μs
(c) No.

3.82 There are three identical stages with 60° phase shift each; 20%

3.83 $B < (1/25)^3$

3.84 $t_{rs} \approx t_{rsF}$

3.85 (a) 4 MΩ, 32.5 pF (b) 5 MΩ ∥ 26 pF (c) If otherwise the probe causes a frequency instability,

3.86 (a) $f_l = 0$, $1/f_u = 0$ (b) 19.61 pF

3.87 (a) $v_{o\,max} = 5\,V$, $v_{o\,min} = 0\,V$ (b) $V_{op} = 2.5\,V$

3.88 $A \times BW = A_F \times BW_F$

3.89

- dynamic change of Z_F
- smaller BW results in less thermal noise
- tailored BW gives better feedback

3.90 Parallel conductance and series inductance.

Problems of Chapter 4

4.1 Analog.

4.2 Amplitude, phase, frequency, and signal length.

4.3 No.

4.4 The transfer function of a Schmitt trigger has a hysteresis.

4.5 Parallel–parallel.

4.6 It depends on the position of the quiescent operating point.

4.7 The amount of positive closed-loop gain.

4.8 No.

4.9 $\frac{1}{(1+\frac{\tau_2}{\tau_1})}$ with τ_i the two time-constants.

4.10 Equal time constants.

4.11 $Q_1 - R_{B2} - Q_2 - R_{B1}$.

4.12 Current.

4.13 Such transformers must have high bandwidth.

4.14 Because transformers do not transfer DC.

4.15 Because transformers do not transfer DC.

4.16 Because they use a transformer that can supply high voltages.

4.17 Constant pulse distance.

4.18 No.

4.19 If the closed-loop gain at $\omega = 0$ is greater than 1, the quiescent operating point will move to a position where the closed-loop gain is less than 1.

4.20 Parallel–parallel feedback; it results in a node that is virtually grounded.

4.21
(a) A differential amplifier.
(b) Negative feedback to conform to the stability criterion, and capacitive positive feedback for the frequency selection.

4.22 By the amplitude.

4.23 $Z(\omega_o) = j \times (\sqrt{L/C} - \frac{1}{\sqrt{C/L}}) = 0$

4.24 The input port of a standard LC oscillator is not uniquely defined.

4.25 A quartz has a high quality factor, i.e. its oscillating frequency is well defined. To start an oscillator takes low frequencies which are difficult to supply with a quartz circuit.

4.26 By both.

4.27 Because the closed-loop gain is less than 1 at this frequency.

4.28 The negative feedback branch.

4.29 By the amplitude.

4.30 Current-based electronics has not gained much importance.

4.31 Transformers need an active element with output and input.

4.32 Inductive.

4.33 Capacitive.

4.34 An Armstrong oscillator requires an active element with separated input and output; one-ports do not qualify in this regard.

4.35 Yes.

4.36 No; a transformer requires an active element with separated input and output.

4.37 Both of them are important.

4.38 An infinite number.

4.39 Because it takes a frequency component of infinite frequency to form the step.

4.40 A discriminator deals with the analog information pulse-height, a trigger with the analog information time.

4.41 It is noise in the position of a timing mark.

4.42 Leading edge triggers.

4.43 Bipolar.

4.44 Very probably it will retrigger, i.e., it will deliver multiple output signals for one input signal.

4.45 Time as analog information is preserved.

4.46 None.

4.47 Bothe circuit: the active elements are in series, the output current is identical for all elements; Rossi circuit: the active elements are in parallel, the output voltage is identical for all elements

4.48 Yes, predominantly.

4.49 Yes.

4.50 A linear gate retains the information on the amplitude of the input signal.

4.51 Yes.

4.52 Cascode circuit.

4.53 (a) At least 16 kHz. (b) Yes.

4.54 Droop.

4.55 Transmission lines, sample-and-hold circuits, charge-coupled devices.

4.56 Higher count-rate capabilities by avoiding overlap.

4.57 A short positive rectangular pulse at the beginning and a negative counterpart after the end of the input signal.

4.58 Analog electronics.

4.59 None, both signals are equivalent.

4.60 No, it is an analog process.

4.61 Yes.

4.62 No.

4.63 Yes.

4.64 Because it is used for both the signal generation and the measurement.

4.65 (a) Negative feedback increasing the capacitor dynamically. (b) Positive feedback in a boot-strap circuit.

Problems of Chapter 5

5.1 No.

5.2 Resistor, switch, gyrator.

5.3 No, they are equivalent.

5.4 $10_{10} = 11_9 = 12_8 = 13_7 = 14_6 = 20_5 = 22_4 = 101_3 = 1010_2 = 1111111111_1$

5.5 Parallel logic, used serially.

5.7 Power supply voltages, amplitudes of the H-state and the L-state, power consumption, speed – propagation delay.

5.8 Schottky type.

5.9 They use common base and common collector transistor stages which have higher bandwidth due to negative feedback.

5.10 Negligible quiescent power consumption – increased propagation delay.

5.11 None.

5.14 Invert inputs and output by NOR gates used as NOT circuits.

5.15

A	B	Q
L	L	H
H	H	L

5.16 See Figure 5.4.

5.17 See Figure 5.4.

5.19 See Figure 5.4.

5.20 The use of an independent clock signal.

5.21 Use Figure 5.13

5.22 To have, at the same time, the present state and the previous state of the flip–flop available.

5.23 For incrementing, i.e., counting.

5.25
(a) Length, bandwidth.
(b) No.

5.26

(a) If the shift occurs toward higher order bits it is multiplied.

(b) The highest order bit must be 0 before the shift, and 0 must be entered to the LSB during the shifting process.

5.27 Serial data transmission takes less hardware but takes longer.

5.28 5

5.29 It takes more than twice the number of flip–flops.

5.30 Only one bit is active at any time, the current load of the power supply stays the same for each number.

5.31 01_{10}

5.32 Only half as many flip–flops are needed.

5.33 It is not possible to count statistically arriving pulses without loss.

5.34 The higher a bit is, the later it will be switched.

5.35 Yes.

5.36 Many (29 059 430 400).

5.37 1100_{5421}

5.38 A T-flip–flop with an additional Set input and the output at $\overline{Q}$.

5.39 Not at all. It may have an effect on the propagation delay.

5.40 Misalignment of the protocols.

Problems of Chapter 6

6.1 Integers have no uncertainty.

6.2 Resolution.

6.3 Yes, it is called quantization or rounding uncertainty.

6.4 Two.

6.5 Fast.

6.6 No.

6.7 No.

6.8 The time resolution depends on the frequency of the reference oscillator.

6.9 Expanding the unconverted time portion by means of time interval expansion, or applying time-to-amplitude conversion of these portions.

6.11 Analog.

6.12 Hz is the unit of a regular (sinusoidal) frequency.

6.13 Depending on the count rate the importance of either contribution to the dead time will vary.

6.14 Monitor, display of a digital oscilloscope.

6.15 20 mV.

6.16 11 bit

6.17 (a) 1.48 V (b) 0.84 V

6.18 Differential linearity.

6.19 Relatively long conversion time.

6.20 Their *ratiometric* behavior does not require a calibration of the pulse height response by the user.

6.21 Sliding scale method.

6.22 Serial parallel ADCs, i.e., pipelined flash converters.

6.23 A serial DAC.

6.24 8 bit

6.25 10

6.26 Faster response because only changes will be converted.

6.27 It needs calibration of the pulse height response.

6.28 Electronic multimeters.

6.29 Improved resolution without increased reference frequency.

6.30 For sampling.

6.31 Parallel.

6.32 8 bit

6.33 (a) Voltage divider. (b) From DACs.

6.34 (a) 300 MHz. (b) Parasitic capacitances.

List of examples

2.1 Analyzing dual circuits based on voltages and currents — 20
2.2 Replacing a voltage divider by a real voltage source using Thevenin's theorem — 23
2.3 Combined voltage signal — 25
2.4 Measurement of the resistance of a linear resistor by measuring simultaneously voltage and current — 27
2.5 Linear model of a junction diode — 33
2.6 Maximum power of resistors in a network — 61
2.7 Measuring the voltage across a dynamic impedance — 88
2.8 Series–series feedback as source of the common-mode rejection in a long-tailed pair — 92
2.9 All four types of feedback configurations in one circuit — 94
2.10 Maximum series–parallel feedback of a common-emitter circuit yields the common-collector circuit — 97
2.11 Bootstrapping with a common collector circuit — 99
2.12 Ideal vs. real operational amplifier — 102
2.13 Importance of reverse loop gain — 108
2.14 Output impedance of an operational amplifier aided current source — 119
2.15 Current mirror aided by an operational amplifier — 121

3.1 Charging a capacitor with constant current — 143
3.2 Identification of (exactly) two linear elements which are not accessible — 145
3.3 Charging of a capacitor involving a diode — 151
3.4 Charging of a capacitor involving an emitter follower — 152
3.5 Comparison of the two types of resonant circuits made of real components — 173
3.6 Signal splitting — 195
3.7 Analytical proof that (negative) feedback has no impact on very short signals — 200
3.8 Slewing due to an excessive signal amplitude — 202
3.9 Attenuating probes — 210

4.1 Elementary analysis of the transition in a flip–flop from L to H — 231
4.2 Wien's bridge oscillator is an undamped resonant circuit — 241
4.3 Output impedance of a Wien's bridge oscillator at ω_0 — 242
4.4 Degeneration of the three basic circuits of three-terminal components to a single circuit when used in an LC feedback oscillator circuit — 255
4.5 Relation between quality factor (width of the resonance curve) and the rise time of the oscillation amplitude — 257
4.6 Phase shift oscillator using three equal unloaded filter sections — 262
4.7 Unreflected use of a constant fraction discriminator — 273
4.8 Adjusting the full-scale (FS) range of a TAC to the need at hand — 285

5.1 Consequence of the sequential nature of information — 296
5.2 Performance of a 4-bit switch-tail ring counter — 318

6.1 Distortion of a flat time spectrum due to differential nonlinearity — 333
6.2 Correcting a measurement done with a faulty measure — 334
6.3 Any frequency is analog — 338
6.4 Internal uncertainties in a frequency measurement — 338
6.5 Successive approximation in the "high–low" game — 351

Index

3D-IC 300

A

absolute time *see* time, absolute
AC 362
accumulator 5, 316
accuracy 360
acquisition time 278, 280
active element 5, 7, 43, 47, 81, 121
active high-pass *see* filter, high-pass, active
active low-pass *see* filter, low-pass, active
active probe 211
actuator 2, 65
ADC 282, 326–336, 338–364
– composite 290, 341, 354–360
– delta–sigma 359
– differential 353, 354
– dual-slope 356, 357
– high-resolution 331
– multiple-slope 362
– multi-slope 357, 358
– parallel 299, 346–349, 363
– ranking of 360
– serial 346, 349–353, 355
– serial–parallel 347, 348, 351
– serial–serial 347, 350, 351
– successive-approximation 347, 349, 351
– tracking 353, 354
– voltage-comparing 346–352
– Wilkinson 350, 354–356
– with frequency modulation 359
adder *see* half-adder
address 340, 341
address number 340
admittance 9, 23, 30, 37, 73, 84, 86–88, 95, 108, 115, 140–142
– complex 243
– dynamic 88, 115
– imaginary 142
– input 38, 39, 103
– negative 46, 87, 115
– output 39, 41, 50, 103
– source 46
aliasing 279, 327, 328, 363
alignment 9
alternating current *see* AC
ammeter 3, 26, 362
amplification 32, 42–46, 213, 236
amplifier 60–74
– biased 67–70
– buffer 262, 285
– charge sensitive 215, 292, 293
– classification of 66
– current 62, 121, 182
– difference *see* amplifier, operation, difference
– differential 70
– fully differential 68, 70, 71–73
– linear 78
– logarithmic *see* amplifier, operation, logarithmic
– non-inverting *see* amplifier, operation, non-inverting
– operation 74, 100–117
  – difference 111, 112, 283
  – inverting 101
  – logarithmic 105
  – noninverting 73, 107, 112, 118
  – summing 104, 343
– operational 68–70, 73, 74, 86, 88, 90, 92, 98, 100–103, 107, 112, 113, 115, 117–119, 121, 122, 184, 197, 208, 212, 214–217, 222, 228, 230, 236, 241, 265, 304
– operative *see* amplifier, operational
– push-pull 70
– standard operational 70
– summing *see* amplifier, operation, summing
– threshold 67, 69
– transimpedance 292, 343, 355, 359
– triangular symbol 255
– voltage 90, 116
  – noninverting 223
– with A operating point 66, 109, 110, 279
– with B operating point 67, 69, 110, 154
– with C operating point 67, 224, 225, 231, 232, 278, 305
amplitude stability 241
amplitude transfer function 206
analog delay *see* delay, analog
analog-analog 123, 326
analog-digital 123, 326

AND gate *see* gate, AND
angular frequency 123, 124, 126, 132, 133, 139, 141, 146, 163, 167, 173, 195, 217, 221, 224
– basic 132
anticoincidence 278, 297, 331
aperture effect 280, *see also* window function
ARC
– constant fraction trigger 273
attenuation 45, 53, 170, 193, 194, 211, 263
– power 45
– specific 193
– voltage 45
attenuation factor 211
attenuation per section 54
attenuator 53–55, 363
averaging 334, 336, 338, 363

B
bandwidth 150, 164, 170, 175, 177, 179, 184, 190, 193, 194, 197, 201, 214, 217, 249, 256, 283, 303, 330, 363
– capture 364
– phase 177
baseline 155, 169, 271
baseline restoring 155, 156
battery 15
BCD *see* binary-coded-decimal
Bel 45
bias point *see* operating point
biasing 56, 57, 60, 66, 152, 181, 184, 225, 230, 266, 268, 276
BiCMOS 301, 304
bidirectional 52, 96
binary circuit 64, 294–325, 296, 304, 305
binary decimal counter *see* counter, BCD
binary digit 123, 295, 297, 298, 315, 326, 327
binary logic 295, 296
binary number
– weighted 327
binary pattern 320
binary-coded-decimal 298
bipolar junction transistor *see* transistor, bipolar junction
bistable circuits 309–323
bit pattern 318
BJT *see* transistor, bipolar junction
black-box 145
Bode plot 204–207, 215
bootstrap circuit 87, 99
Bothe circuit 275–277
bouncing 65, 229, 272
break frequency 164
bridge 157, 236
building block 301
burden 10, 53
bus transceiver 304
busy time 296, 297, 329
Butterworth filter 164
– first-order 165, 167

C
calculator 322
caliper 336
capacitance
– dependent 288
– input 208
– internal 300
– negative 115
– output 208
– parasitic 345
– specific 188
– stray 144, 175, 179, 202
capacitive coupling 183
capacitor 9, 141–144
– bypass 66
– ceramic 66, 250
– charged 142, 151
– decoupling 234
– electrolytic 66
– mica 250, 285
– polystyrene 250, 281
– reservoir 158
– smoothing 157, 158
– speed-up 231, 232
– storage *see* capacitor, smoothing
– tantalum 66
– variable 210, 211, 240, 261, 288
capacity *see* capacitance
carry bit 308
cascade 36, 69, 78, 102, 111, 112, 158, 184, 213, 275, 277, 303, 315, 354, 356
cascode 184
CCD *see* charge-coupled device
ceramic core 250
channel 331, 332, 333, 340, 341
channel number 341
channel width 355

characteristic 9
- forward 56
- nonlinear 34, 105
- transfer 57, 67, 80
characteristics 62–64, 66, 115–117, 121, 147, 151, 188, 267, 269
charge balance VFC 289
charge balanced method 288
charge-coupled device 282
chatter 65, *see also* bouncing
check valve 31, 33, 157
circuit diagram 5, 6, 29, 57
- simplifying of 29
clamping 154, 155
class A amplifier *see* amplifier, with A operating point
class B amplifier *see* amplifier, with B operating point
class C amplifier *see* amplifier, with C operating point
clipping 55–59, 57, 274
clock 282, 300
- internal rate 299
- master 294
- two-phase 313
- ytterbium 221
clock generators 230
clocked VFC 290
CMOS 73, 277, 300, 301, 303, 304
CMOS technology 277
CMRR *see* common-mode-rejection ratio
coaxial cable 189–194, 282, 283
- input impedance 193
Cockcroft–Walton 159
code 297–299, 322, 326, 327, 332, 340
- 2-4-2-1 322
- 5-4-2-1 323
- 8-4-2-1 322
- one-hot *see* code, one-out-of-ten
- one-out-of-ten 317, 322
- weighted 297, 298, 322
code converter 294
coincidence 221, 274, 275, 278, 290, 294, 296, 299, 337
- delayed 274
- overlap 338
coincidence circuit 274, 275
coincidence unit 338
collector
- series resistor 207
collector resistor 93
common mode 71–73, 108, 112
common mode rejection ratio 108
common-base circuit 47, 50, 69, 96, 182
common-collector circuit 47, 50, 69, 96, 97, 99, 182
common-drain circuit 98
common-emitter circuit 47, 48, 51, 62, 95–97, 99, 121, 182
common-mode rejection 241
common-mode-rejection ratio 72, 73, 85, 92, 93
common-source circuit 95
comparator 69, 70, 223, 225, 226, 235, 269, 286, 292, 304, 331, 341, 346–351, 353, 355, 359, 362
- fast 325
comparator level 347
comparator response 346, 351
complement output 318, 323
compliance range 120
compliance voltage 120, 121
computer 2
- distributed 324
- general-purpose 340
computer chips 301
computer interface 362
computer language 297
conductance 9
- damping 268
- specific 188
conducting loops 189
conjugate variable 146
continuous spectrum *see* spectrum
conversion
- analog 283–293
conversion factor 219, 286, 287
conversion rate 327
converter
- amplitude-to-digital 330, 341, *see also* ADC
- amplitude-to-frequency 287–290
- amplitude-to-time 285, 287
- compound
  - voltage-to-time 286
- current-to-voltage
  - logarithmic 121
- digital-analog 349
- digital-to-amplitude *see* DAC

- direct
  - voltage-to-time 286
- frequency 293
- frequency-to-amplitude 291–293
- parallel-to-serial 316
- serial-to-parallel 316
- time-to-amplitude 287, 336
- time-to-digital 335
  - composite 336
- voltage-to-current
  - exponential 120, 121

corner frequency 164, 167, 177, 191, 198, 199, 208, 214, 215, 217, 218
Coulomb 3
counter 317–323
- asynchronous 319, 320
- BCD 317, 319, 322, 323
- binary 313, 319–323, 320
- clock 337
- divide-by-n 319, *see also* prescaling
- ring *see* ring counter
- ripple *see* counter, asynchronous
- synchronous 320, 322
- up/down 353
- up-down 323
- vernier 337
- weighted binary 317

counter ADC *see* ADC, serial
counter dwell time 340
counting 298, 326
counting rate 284
counting-down 323
counting-up 323
count-rate spectrum *see* spectrum, count-rate
CRC *see* cyclic redundancy check
cross talk 271
current
- reverse 55
- short-circuit 21

current boost 345
current divider 24, 161, 165, 168
current electronics 117–122
current filter 161, 167, 263
current flow 4, 5, 13
current mirror 120–122, 121
current source *see* source, current
current step 159, 160
cut-off region 63, 64, 153, 235, 275, 277
cycle 123, 139, 147, 221, 270
cycle time 221
cyclic redundancy check 297

D

DAC 304, 341–346
- parallel 343–345, 346, 355
- serial 345, 346

damping 115, 179, 267
- critical 180

Darlington *see* transistor, Darlington
data storage 309
data transparency 313
data word *see* word, binary
dB *see* deciBel
DC *see* direct current
dead time 296, 316, 320, 329, 336, 340, 360
debouncing 229, 272
deciBel 45
decision tree 352
decoder 323, *see also* code
- binary-to-decimal 322
- four-to-seven 323

decomposition of signals 127
decrement 323
delay 176, 191, 200, 227, 235, 272, 273, 283
- analog 282
- digital 325

delay line 235, 283
- lumped 283

delay time 176, 191, 235, 270
delta-encoded ADC *see* ADC, tracking
dependent source *see* source, dependent
deserializer 294, 316, 345
detector
- peak 280

DFT *see* Fourier transform, discrete
differentiator 141, 167
digital – *see also* binary –
digital circuit 304, 308
digital delay *see* delay, digital
digital multimeter *see* multimeter, digital
digital oscilloscope *see* oscilloscope, digital
digital-analog 123, 326
digital-digital 123, 290, 326
digital-ramp ADC *see* ADC, serial

digitizing 123, 279, 283, 326–364, 327, 329–331, 335, 336, 353, 360
– process 330, 339
 – duration of 327, 329
 – quality of 327
– time 355
digitizing error 335
digitizing of amplitude
– direct 341–352
digitizing of frequency
– direct 338, 339
digitizing of time
– direct 335–337
diode 55, 110, 158, 159
– clipping 57, *see also* clipping
– Esaki *see* diode, tunnel
– impedance of 156
– junction 29, 33, 34
– Schottky *see* Schottky technology
– tuning *see* capacitor, variable
– tunnel 47
– varactor 288, *see also* capacitor, variable
– Zener 32, 58, 110
diode discriminator *see* discriminator
diode limiter 105, *see also* clipping
diode logic 275
diode pump 291, 292, 349
– active 346
direct current 4, 12, 123, 142, 157, 232, 362
– pulsating 157
discharging 158
discrete spectrum *see* spectrum
discriminator 226, 269–271, 273, 274
– parallel-diode 59
– serial-diode 59
display 26, 28, 323, 362, 363
display decoder 319
dissipative element 4, 7, 12, 19
distribution 271, 332, 333
dither 330
divide by 2 316
DMM *see* multimeter, digital
double differentiation 169
drain current 65
drain voltage 65
drain–source resistance 241
drift 72, 74, 347, 356
– thermal 72
driver circuit 294
droop rate 280, 281
DSO *see* oscilloscope, sampling
duality 7–9, 12, 14, 15, 17, 19, 22, 28, 30, 37, 52, 84, 85, 87, 89, 90, 96, 140, 162, 173, 181, 219, 247, 249, 263, 264, 269, 299, 346
duty cycle 147, 229, 230, 338
duty factor 339, *see also* duty cycle
dwell time *see* counter dwell time
dynamic range 35, 60, 67, 71, 72, 88, 110, 202, 203, 272, 274, 324, 327, 364
– small-signal 31

E

earth 29, 190
earth potential 29, 190
ECL 301, 303
electronic switch *see* switch
emitter 153
emitter follower 98, 99, 122, 152–154, 185, 187, 192, 207
– complementary 153
emitter resistance 50
– external 234
– internal 184, 234
emitter resistor 93
emitter-coupling 303
enable input 296, 309, 310
enable signal 304
encoder *see also* code
– mechanical 327, 354
– optical 354
EOR gate *see* gate, exclusive-OR
error
– scaling 341
– zero off-set 341
Esaki diode *see* diode, tunnel
Euler's identity 131
events per second 338
EXOR gate *see* gate, exclusive-OR
exponential voltage-to-current converter *see* converter, voltage-to-current, exponential

F

fall time 147, 149, 151–153, 221, 233, 302, 324, *see also* rise time
False 295
fan-in 302
fan-out 302
Farad 7

faulty measure 334
feedback
- current 81
  - current–current 81
  - current–voltage 81, 93
- external 83–86
- frequency independent negative 199
- internal 37, 83
- local 78
- negative 51, 73, 74, 76–80, 79, 92, 98, 101, 113, 114, 116, 120, 198, 199, 201, 202, 204, 217, 240, 243, 250, 263, 264, 360
- parallel–parallel 81
- parallel–series 81
- positive 69, 76, 81, 87, 113, 114, 203, 217, 222–224, 226–228, 231, 232, 234, 236, 240, 241, 243, 244, 255, 257, 259, 263–265, 310
  - local 113
  - stable 80, 81
- series–parallel 81, 97, 208
- series–series 81, 92
- static 74–100, 198, 200
- three-terminal components 95–100
- voltage 81
  - voltage–current 81
  - voltage-voltage 81
feedback element 81
feedback loop 74, 76, 79, 89, 102, 105, 113, 144, 156, 224, 250, 259, 310
feedback oscillator 236, 251, 257, 264, 265
- high-frequency 265
ferrite core 250, 283
FET *see* transistor, field-effect
FFT *see* Fourier transform, fast
field-effect transistor 277
field-programmable gate arrays 301
FIFO *see* first-in, first-out
filter 161
- active 212–218, 213, 214
- anti-aliasing 328
- band-elimination *see* filter, notch
- band-pass 170, 173, 217, 218
- band-reject *see* filter, notch
- band-stop *see* filter, notch
- current 161, 162, 165, 167, 263
- high-pass 161, 162, 166, 167, 169, 170, 177, 183, 217, 218, 261, 262, 271, 283
  - active 216–218
- low-pass 161–164, 170, 176, 177, 191, 198–203, 210, 214, 215, 218, 328
  - active 214, 215, 284
  - current 179
- notch 170
- passive 160–179, 213
- resonant 172, 174, 175
- voltage 161, 165, 167
filter section 176
firmware 301
first-in, first-out 364
fixed-wired circuit 301, *see also* hardware programmed
flash converter *see* ADC, parallel
flip-flop 231, 288, 301, 309, 310, 313, 315, 317, 318, 319, 320, 322, 323, 340, 359, 360
- D 290, 310, 312, 314, 315
- edge-triggered 309
- general type 314
- JK 310, 314
- master–slave 312
- SR 229, 286, 310–312, 314
- synchronous 309
- T 310, 313, 314
floating 85, 119
floating voltage 72
forbidden band 302
forward quantities 161
forward voltage 33
forward voltage gain *see* gain, voltage, forward
forward-linear region 64
four-emitter input 302
Fourier analysis 127, 135, 221, 364
Fourier series 127, 131–133
- exponential 132
Fourier transform 123, 126, 127, 133–135, 137, 138, 221
- continuous 127
- discrete 364
- fast 364
frequency 198, 269–283
- averaging 336
- cut-off 164
- unity-gain 236
frequency compensation 198, 208–211, 210, 226, 231, 232
frequency devision 319
frequency domain 126, 146–179, 364
frequency equation 260

frequency instability 189, 203
frequency instabilty 202
frequency modulation *see* oscillator, voltage-controlled
frequency spectrum *see* spectrum, frequency
frequency stability 250, 256, 259
frequency-to-voltage conversion *see* converter, frequency-to-amplitude
full scale 327, 331, 340
full-scale range 285
full-scale signal 347
full-wave rectification 157, 158
fuzziness 330

G
gain 115–117
– closed-loop 51, 108, 204, 207, 210, 231, 232, 244, 245, 264
– common mode 72
– current 37, 43, 50, 81, 82, 89, 108, 117, 120–122, 162, 182
  – closed-loop 264
  – forward 41
  – reverse 38
– forward-loop 96
– open-loop 116, 204, 231, 232
– power 42–46, 48, 50, 52, 95, 108, 182, 183
– reverse-loop 96, 108
– source-voltage 116, 215
– voltage 37, 50, 81, 82, 92, 98, 115–117, 122, 146, 182, 197, 215
  – cascode 184
  – closed-loop 264
  – forward 38
  – reverse 41
gain-bandwidth product 197
– constant 199, 363
Galvanic isolation 190
gate
– AND 274–276, 284, 306
– EOR *see* gate, exclusive-OR
– exclusive-OR 308
– EXOR *see* gate, exclusive-OR
– linear 278, 279, 283, 285
– logic 274–279, 300, 301, 304, 343
– NAND 276, 305, 306
– NOR 305, 306
– NOT 305
– OR 306
– XOR *see* gate, exclusive-OR
gate circuits 305–308
General Mean Time 221
glow lamp 47
GMT *see* General Mean Time
g-parameter 38, 39
Graetz bridge 157
graphical methods 34–36
ground 29
ground loop 190
ground terminal 48
grounded-emitter circuit *see* common-emitter circuit
grounding 70, 189, 190
– single-point 190
group delay 176
– constant 176
group velocity 191
grouping 333
gyrator 219, 242, 265

H
half-adder 308
half-wave rectification *see* rectifier, half-wave
hardware programmed 364, *see also* fixed-wired circuit
Henry 8
hertz 221
hierarchy
– voltage transfer 116
high-pass *see* filter, high-pass
high-pass filter sections 265
h-parameter 40, 41, 48
hysteresis 69, 222, 223, 225, 226

I
IC *see* integrated circuit
ideal source *see* source, ideal
imaginary part 140, 146, 173, 174, 188, 203, 237, 260
impedance 9, 30
– characteristic 54, 188, 189, 191, 192, 194, 195, 283
– complex 138–140, 146, 181, 202, 208, 209, 255
– dynamic 42, 50, 86–89, 99, 103, 114, 115, 212, 241
– dynamic negative 32, 113

– generalized complex 138, 139
– input 40, 41
– load 10, 42, 44, 50, 55, 91, 97, 99, 161
– negative 32, 46, 47, 87, 88, 113, 115, 142, 219, 242, 266–268
– output 38, 40, 119, 243
impedance bridging 180, 181, 187
impedance matching 44, 181
impedance mismatch 180, 183, 192
increment 323
incrementing 319, *see also* counter
indicator 26
inductance 8, 142, 144, 188, 189, 220, 234–236, 252, 254
– high quality 258
– ideal 219
– parasitic 250
– specific 188
inductivity *see* inductance
inductor 9, 144–146, 159, 160
inhibit circuit 278
input offset 73, 215
input offset voltage 74
input stage 73
input variables 219
instant of an event 221, 269
integrated circuit 299–304
integrating ADC *see* ADC, Wilkinson
integrator 141, 355, 357–360
interface circuit 294
interfacing 180–195, 301
interference
– electromagnetic 271
inverter 102
– logic 305, 307
irregular state 311

J

JFET *see* transistor, JFET
jitter 256, 270, 271, 272, 273, 280, 324, 339
Johnson ring counter *see* ring counter, switch-tail
junction 16
junction transistor 255, 299, *see* transistor, bipolar junction

K

Kirchhoff's laws 9, 16, 17, 124

L

Laplace transform 138, 139
latch 309, 310
– JK 311
– NAND-gate 310
– R 311
– S 311
– slave 313
– SR 311
– transparent 312
LC oscillator 264
least-significant bit 326, 330, 344, 351
level shifter 294
limiter 105, *see also* clipping
linear mode 65
linear model 33, 34
linear phase 176
linear region 32, 64
linearity 77, 79, 213, 360
linearization 6, 7, 34, 62
live time 297
load 9
– capacitive 153, 196, 210, 211
– complex 9
load impedance 161
load line 34, 35, 66, 267
load resistor 56, 57, 158
logarithmic current-to-voltage converter *see* converter, current-to-voltage, logarithmic
logic
– binary 64, 274, 308
– negative 295
– parallel 299, 340, 345, 346
– positive 295, 326
– serial 297, 299, 340, 345, 346
logic code *see* code
logic families 301–304
logic gate 300
logical circuit 294
logical level 295, 301, 302
logical symbol 306, 307, 311, 313, 314
long-tailed pair 67–70, 92, 117, 184, 185, 225, 228, 230, 232, 279, 281, 303
loop 17
loss *see* attenuation
loss tangent 142, 174
loudspeaker 115
lower corner frequency 217
low-pass 260, *see* filter, low-pass

low-pass characteristic 197
low-pass filter sections 265
LSB *see* least-significant bit
L-section 53
LSI 300

M

magnetic coupling factor 252
magnitude 140, 161, 164, 237
magnitude criterion 236, 264, 265
mass production 300
Maxwell's equations 4, 7, 188, 202
measurement 3, 26, 27, 145, 210, 211, 326, 333, 336, 362
– current 14
– multichannel 339–341
– of capacitance 362
– of inductance 362
– of temperature 362
– voltage 14
measurement equipment 361–364
memory 294, 340, 363
– binary
 – two-dimensional 340
– random-access 340
– serial 299
– volatile 309
memory cell 309, 312, 321, 340, 363
memory chips 300
memory dead time 340
memory effect 9, 10
memory hardware 340
memrestivity 9, 10
memristor 9, 10
mesh 9
meta-stability 302, 310
meter 14, 26–28
– active 26
– digital 322
– passive 26
microphonics 115
microprocessor 297, 304
Miller effect 87, 144, 184, 212, 214, 217, 280
miniaturization 299–304
mismatch 73, 180, 183, 192
mobile phone 1
modulation
– pulse-frequency 288, 359
– sine-wave 359
– square-wave 359
modulo 320
MOSFET *see* transistor, MOSFET
most-significant bit 316, 340, 344
MSB *see* most-significant bit
MSI 300
multichannel analyzer 331, 349
multimeter 362
– digital 362
multiple output signals 272
multiply by 2 316
multiscaler 340
multitasking 324
multivibrator 224, 227, 232, 235, 268
– astable 229, 230, 231, 235
 – symmetric 230
– bistable 231, 232, 268, 309
– current-steering 288
– free running 230
– monostable 224, 227–229, 268, 274, 289
– one-shot *see* multivibrator, monostable

N

NAND gate *see* gate, NAND
n-channel MOS 300
negation bar 306
negative impedance converter 113, 114, 219
negative logic *see* logic, negative
negative number 321
neon bulb 266
network model 36–59
network time protocol 324
NIC *see* negative impedance converter
NMOS 301
node 6, 9, 16, 17, 84, 85, 104, 124
noise 73, 77, 79, 80, 110, 116, 117, 189, 190, 193, 194, 198, 214, 223, 235, 269, 271, 280, 288, 326, 330, 357
– admixed *see* dither
nonlinear phase 176
nonlinearity 47, 56, 249, 250, 332, 333, 360
– differential 332, 333, 334, 341, 342, 345, 347, 349, 352
– integral 332, 333, 341, 342, 347
NOR gate *see* gate, NOR
Norton's theorem 9, 15, 23, 24, 208, 295
notch frequency 170, 263
N-shape 9, 87, 88, 266, 267, 269

N-shaped characteristic 115
NTP *see* network time protocol
Nyquist criteria 203, 204, 210
Nyquist frequency 279, 327, 364
Nyquist plot 203, 204
Nyquist–Shannon sampling theorem 279, 328, 360

O
Ohm 8
ohmic mode 65
ohmmeter 362
Ohm's law 8, 21, 27, 30, 60, 87
– inverse 21
one-port 1, 4, 7, 12–34, 42, 46, 47, 135, 140, 146, 266–269, 268
– active 32, 266
– frequency dependent 139–146
– fusing of 29
– i-v characteristic 29–32
– i-v non linear 29–32
– linearization 32–34
– passive 135
– small-signal behavior 32–34
one-ports
– in parallel 22
– in series 21
opaque latch 309
open-circuit 9
open-circuited end 191
operating point 30, 62, 66, 67
– instantaneous 60
– quiescent 30, 60, 66, 67, 154, 215
operating range
– small-signal 31
operation amplifier 73, 109, 116, 158, 213, 222, *see* amplifier, operation
– inverting 88, 212
OR gate *see* gate, OR
orthogonality 128
oscillation frequency 261
oscillator 113, 221–269
– amplitude stability 241
– Armstrong 254
– blocking 225, 232–234
  – astable 234
  – free-running 232, 234
  – monostable 233, 234
  – triggered 233, 234
– Clapp 251
– Colpitts 251, 288
– delay-line 235
– feedback 254
– free 224
– frequency stability 256, 259
– gated 224
– harmonic 267, 268
– harmonic feedback 236–265, 256, 264
– Hartley 252
– high-frequency 48, 235, 236
– LC 244–256
– Meissner 254
– one-port 266–269
– phase-shift 259–263, 265
– positive feedback 266
– quartz 256–259, 335, 338
– RC-feedback 265
– reference 337
– relaxation 221, 223, 224, 227, 232, 249, 268, 269, 287, 288, 325
  – monostable 325
  – one-port 268, 269
– start 235
– startable 337
– stop 235
– twin-T 263–265
– using three-terminal devices 236–263
– voltage-controlled 287
– Wien 237–243
oscilloscope
– digital 347, 363, 364
  – high-speed 279, 363
– sampling *see* oscilloscope, digital
output pulse width 325
output variables 219
oven
– temperature-controlled 259
Overbeck ring counter *see* ring counter, straight
overflow 321
overlap coincidence 274, 278
overlap method 284
oversampling 328, 360
overshoot 147, 180, 201, 202, 358

P
parallel ADC *see* ADC, parallel
parallel DAC *see* DAC, parallel

parallel in–serial out *see* converter, parallel-to-serial
parallel logic *see* logic, parallel
parallel resonant circuit *see* resonant circuit, parallel
parallel-pipeline ADC *see* ADC, serial–parallel
parity check 297
pass-band 164, 167
passive element 43
p-channel MOS 300
period 123, 126, 127, 132, 133, 139, 147
period length 235, 335
permability 250
permittivity 190
phase 221
phase criterion 236, 264, 265
phase deviation 176
phase distortion 176, 177
phase lag circuit 204
phase lead circuit 204
phase response 145, 164, 167, 168, 170
phase shift 127, 131, 139–141, 164, 165, 167, 168, 170, 221
phase-shift spectrum *see* spectrum, phase-shift
piezoelectricity 257
pile-up 155
PISO *see* converter, parallel-to-serial
PLD *see* programmable logic device
polarized devices 5
pole-zero cancellation 169
positive logic *see* logic, positive
potentiometer 106
power 195–197
- apparent 196
- consumption 300
- dissipated 4, 60, 300, 301, 303
- reactive 196, 197
- real 196
power amplifier 109, 110
power factor 196
power matching 44, 181
power supply 23, 32, 43, 66, 109–111, 157–159, 222, 303, 319
- regulated 110
power transfer
- optimal 181
precision 360
prefix 5
prescaling 313, 317
priority encoder 347
probe 210, 211
- active 211
- attenuating 210, 211
- current 14
process
- serial logic 324
processor
- arithmetic 294
- digital 364
programmable logic device 301
propagation 4, 190, 191, 200, 202
propagation delay 146, 191, 200, 221, 235, 270, 282, 300, 307, 310, 325, 347
protocol 297, 298, 316, 324
pull-up resistor 187, 301
pulse length 234, 324, 325, 354
pulse lengthener 325
pulse shortener 283, 325
pulse-height spectrum *see* spectrum, pulse-height
pulse-lengthener 281

Q

Q-point *see* operating point
quality factor 143, 174, 175, 249, 256–258
quantization 330, 335
quantization unit 326, 335
quiescent operating point *see* operating point, quiescent
quiescent point *see* operating point

R

R/2R-ladder network 344
rail-to-rail input range 226
RAM *see* memory, random-access
ramp-compare ADC *see* ADC, serial
random access 315
randomizing method 333, 334
rate meter 291
- digital 338
reactance 140, 244, 247, 255, 260, 265
- stray 227
reactances
- distributed 188
reactive element 7, 161, 172, 197
real part 140, 146, 203, 237
real source *see* source, real
receiver circuit 294

recovery
– of clock signals 227
recovery time 110
rectification *see* rectifier
rectifier
– half-wave 157–159
– precision half-wave 110, 111
redundancies 297
redundancy 297
reference voltage 362
reflected current 192
reflection 191, 192, 194, 195, 283
register 294, 309, 315–323, 316, 317, 319, 320, 322, 351
– cyclic 317, 318
regulator
– active voltage 158
reliability 300
remote sensing 109
repetition frequency 147, 230, 336
repetition rate 336
resistance 9
– saturation 234
– specific 188
resistor
– damping 267
– variable 65
– voltage-controlled 65
resolution 269, 336, 358, 360
– high 345, 362
– increase 337
– visual 363
resolution time 227, 228
resolving time 296
resonant circuit 115, 162, 172–175, 179, 242, 245, 248–252, 254, 257, 258, 265, 267, 268
– ideal 267
– parallel 162, 173, 179, 248–252, 258, 266, 268
 – ideal 245
– series 162, 172–175, 225, 248–250, 257, 266, 267
 – ideal 245
 – impedance of 174
– undamped 241, 242
resonant frequency 173–175, 179, 242, 248–252, 254, 256–259, 267
response time 345
restoration 324
return difference 76, 78–80, 92, 98, 113, 199
reverse current *see* current, reverse
reverse current gain *see* gain, current, reverse
reverse impedance 33
reverse quantities 161
reverse-linear region 64
ring counter 317, 319, 322
– Johnson *see* ring counter, switch-tail
– Overbeck *see* ring counter, straight
– straight 317
– switch-tail 318, 322
– twisted *see* ring counter, switch-tail
ripple counter *see* counter, asynchronous
rise time 146, 147, 149–151, 179, 193, 197, 198, 200–202, 271, 272, 324, *see also* fall time
– intrinsic 149
rise-time correction 363
root-mean-square 196
Rossi circuit 276, 277

S

sample and hold 280–282, 329
sample-and-hold 329
samples per second 363
sampling 279–282, 326–331, 327, 329, 353, 360, 363, 364
– continuous 327
sampling frequency 279, 327, 364
– constant 350
sampling interval 352
sampling oscilloscope *see* oscilloscope, sampling
sampling rate 279, 327–329, 347, 353, 354, 360, 363
saturation 62, 64, 73
saturation region 64, 235
saw-tooth signal 124
scaler *see* counter
Schottky technology 301
second 221
sectioned-serial ADC *see* ADC, serial–serial
selfdual 219
self-duality 8
sense line 109
sensing 9
sensing device 14
sequential 296
serial ADC *see* ADC, serial
serial DAC *see* DAC, serial

serial in–parallel out *see* converter, serial-to-parallel
serial logic *see* logic, serial
serializer 294, 316
serial-pipeline ADC *see* ADC, serial–serial
series regulator 110
series resonant circuit *see* resonant circuit, series
settling time 147
seven-segment display 323
shielding 189, 190, 193
– double 193
shift register 315–317, 349, 355
– analog 282
– serial 299
short-circuit 9, 191, 283
short-circuit current gain 48, 62
short-circuited end 191
shunt compensation 179, 180
shunt regulator 110
shunt resistor 362
Siemens 7
signal
– output
 – multiple 274
signal conditioning 187
signal delay 187
signal inversion 187, 232, 236
signal length 221
signal power 195–197
signal rate 270, 320
signal restoration 187
signal splitting 195
signal synchronization 187
signal-to-noise ratio 80, 194, 330, 360
silicon steel core 250
simultaneity 221, 269, 274
single-channel analyzer 331, 346, 347
single-slope ADC *see* ADC, Wilkinson
SIPO *see* converter, serial-to-parallel
skin effect 189
slew rate 143, 197, 198, 223, 302
slewing 143, 198, 202
sliding-scale method 333, 334, 349
small-signal 6, 10, 126, 182, 203, 213
small-signal parameters 203
small-signal response 6
smoothing 157, 158
SNR *see* signal-to-noise ratio
source
– current 13, 68, 118–120, 119, 362
 – dependent 179, 180
 – ideal 13, 15, 19, 20, 118
 – invariable 216
 – voltage controlled 288
– dependent 36–42
– ideal 13, 14
– real linear 15, 16
– voltage 13
 – dependent 180, 197, 259
 – ideal 13, 15, 19, 20, 25, 109, 110
 – real 19
source follower 98, 187, 288
spectrum 332
– count-rate 339
– frequency 131, 133, 333, 339, 364
 – continuous 133
 – discrete 131
– phase-shift 131
– pulse-height 341
– time interval 274
spectrum analyzer 364
speed of light 188, 190
square wave 124, 147, 154, 155, 287
S-shape 9, 88, 266–268
S-shaped characteristic 115
SSI 300
stability 77, 213
stability equation 260
stairstep-ramp ADC *see* ADC, serial
start-stop method 284, 285
steady-state signal 123, 124, 126, 195, 221
step recovery 360
step recovery response 354
step response 147, 152
– current 179
– voltage 177, 178
step signal 146, 147, 149, 150, 160, 169, 192, 200, 210, 269, 271, 282, 283
stop-band 164, 167
storage 157, 319
– nonelectronic 339
storage circuit 294
storage device 280
storage element
– binary 315
storage time 302
stray properties 174

strobe input 296
strobe signal 304
subranging 348
sum bit 308
superposition theorem 22–26, 88, 97, 98, 109, 111, 117, 118
supply voltage 73, 144, 301
– additional 186
switch 64–66
– change-over 288
– electronic 65, 228
switching 187
switching rate 303
synchronization 221, 296, 304, 324
synchronous operation 324

T
TAC *see* converter, time-to-amplitude
tank circuit *see* resonant circuit
TDC *see* converter, time-to-digital
telemetry 288–290, 359
temperature coefficient 250
temperature dependence 174, 250, 259, 285
termination 189, 191, 193, 194
termination resistor 189, 191
Thevenin's theorem 9, 15, 19, 23, 34, 101, 150, 245, 344
three-terminal component 65, 95, 117, 181, 236, 255, 256, 265, 275
three-terminal devices
– active 70
three-terminal element 47–52
threshold 67, 222, 223, 225, 269, 271, 273, 289, 331, 336
threshold voltage 222, 226, 271, 331
time 269–283
– absolute 221, 324, *see also* General Mean Time
time constant 149–157, 160, 163, 166, 169, 170, 177, 197, 199, 200, 202, 208, 210, 217
– discharging 156, 158
– dominating 198, 203
– effective 214
time difference 221
time dispersion 270
time domain 126, 146–179, 296, 364
time interval 221
time jitter 270
time resolution 269, 280, 333, 363
time response 324
time spectrum 341, 364
time-interval magnification 287
time-interval spectrum *see* spectrum, time-interval
time-length expander *see* time-interval magnification
time-length modulator *see* converter, amplitude-to-time
timer 297, 319, 325, 338, 349
time-to-amplitude converter 284, 285
timing 269
T-model 48, 89
T-parameter 49
tracking ADC *see* ADC, tracking
transadmittance 37, 82, 95, 102, 103, 117, 141, 185, 186
– forward 39
– reverse 39
transducer 1
transfer
– reverse 74
transfer characteristic 56, 116, 138, 222, 224
transfer function 69, 77, 79, 95, 121, 141, 160, 161, 165, 167–170, 195, 198–200, 203, 204, 212–214
– closed-loop 204
– complex 199, 245
– complex closed-loop 264
transformer 60, 157, 158, 224, 232–236, 254
– pulse 233
transimpedance 37, 82, 90, 92, 95, 101–103, 116, 117, 141, 212, 215
– forward 40
– reverse 40
transistor
– bipolar junction 47, 181
– Darlington 184
– FET *see* transistor, field-effect
– field-effect 47, 65, 69, 73, 277, 281, 304, 355
– JFET 47, 181
  – n-channel 241
– MOSFET 47, 65, 181, 299, 301, 304
– Schottky 299, 302
transmission line 188–195, 189, 282
transmitter circuit 294
triangular wave 124
trigger 270–274
– constant-fraction 272–274

– leading-edge 270, 271
– Schmitt 70, 223–227, 232, 266, 268–270, 288–290, 302, 325, 338
– zero-crossing 271, 272
trigger response 363
True 295
truth table 64, 274, 275, 278, 295, 296, 305–308, 310–314
T-section 53
TTL 301–303
tuned circuit *see* resonant circuit
tuning diode *see* capacitor, variable
tunnel diode *see* diode, tunnel
turnover frequency 163, 166
twisted ring counter *see* ring counter, switch-tail
two-port 2, 36–74, 46, 83, 84, 86, 89–91, 141, 146, 147, 160, 161, 180, 195, 267
– active 86, 90, 91, 195–220
 – real 60–74
– *i-v* characteristic 62–64
– matrix 36
– nonlinear 56
– passive 52–59
 – nonlinear 55–59
two-port convention 36, 120

U
ULSI 300
uncertainty 26, 221, 323, 338
– calibration 339
– digitizing 339, 354
– dominating 339
– interpolation 327, 332
– of resistance value 344
– percentage 339
– quantization 327, *see also* uncertainty, digitizing
– rounding 339
– scale 339
– statistical 333
– uncorrelated 352
undamping 115
undershoot 155, 156, 169, 271
unidirectional 52, 157
universal gate 305
up/down selector 323
upper corner frequency 198, 199, 208, 214, 217

V
vacuum triode 47
varactor *see* capacitor, variable
VCO 338, *see* oscillator, voltage controlled
vernier method 336
veto signal 278
VFC *see* converter, amplitude-to-frequency
virtual ground 84, 85, 101, 102, 112, 215, 292, 343
virtual open-circuit 71, 85
vital statistics 334
VLSI 300, 301
voltage
– addition of 14
voltage divider 21, 23, 53, 61, 112, 151, 161, 165, 167, 180, 210
– complex 213, 237
– frequency selective 244
– multiple 347
– reactive 247
voltage divider equation 21, 24
voltage follower 98, 99, 107, 176, 184, 192, 207, 262, 263, 288, 292, 343
voltage limiting 57
voltage multiplication 158, 159
voltage regulator 304
voltage signal
– combined 25
voltage source *see* source, voltage
voltage-to-freqency conversion *see* converter, amplitude-to-frequency
voltmeter 3, 26, 88, 89, 362
– digital 356

W
walk 271–273
Watt 196
weighted code *see* code, weighted
weighting factor 343
Wilkinson ADC *see* ADC, Wilkinson
winding sense 234
window function 364, *see also* aperture effect
wire level 302
word
– binary 315, 316, 340
– sequential 316
word length 315

X

XOR gate *see* gate, exclusive-OR

Y

y-parameter 39, 40, 86
ytterbium clock *see* clock, ytterbium

Z

zero-crossing discriminator 357
z-parameter 40, 54, 89